T0342542

Urban Pollution

Urban Pollution

Science and Management

Edited by

Susanne M. Charlesworth
Coventry University
Coventry, United Kingdom

Colin A. Booth
University of the West of England
Bristol, United Kingdom

This edition first published 2019

Registered Office(s)
John Wiley & Sons, Inc., 111 River Street, Hoboken, NJ 07030, USA
John Wiley & Sons Ltd, The Atrium, Southern Gate, Chichester, West Sussex, PO19 8SQ, UK

Editorial Office
9600 Garsington Road, Oxford, OX4 2DQ, UK

For details of our global editorial offices, customer services, and more information about Wiley products visit us at www.wiley.com.

Wiley also publishes its books in a variety of electronic formats and by print-on-demand. Some content that appears in standard print versions of this book may not be available in other formats.

Library of Congress Cataloging-in-Publication Data

Names: Charlesworth, Susanne M., editor. | Booth, Colin (Colin A.), editor.
Title: Urban pollution : science and management / edited by Susanne Charlesworth, Colin Booth.
Description: Hoboken, NJ : John Wiley & Sons, 2018. | Includes bibliographical references and index. |
Identifiers: LCCN 2018014793 (print) | LCCN 2018028469 (ebook) | ISBN 9781119260462 (pdf) | ISBN 9781119260509 (epub) | ISBN 9781119260486 (cloth)
Subjects: LCSH: Urban pollution.
Classification: LCC TD177 (ebook) | LCC TD177 .U73 2018 (print) | DDC 628.509173/2–dc23
LC record available at https://lccn.loc.gov/2018014793

Cover Design: Wiley
Cover Image: © Melinda Nagy/Shutterstock

Set in 10/12pt WarnockPro by SPi Global, Chennai, India
Printed in Singapore by C.O.S. Printers Pte Ltd

10 9 8 7 6 5 4 3 2 1

For the future: our children (Ruth, Hugh, and Thomas; Esmée, Edryd, and Efren), and their children (Rónán, Ella-Rose, Aidan, and Douglas), and their children ...

Contents

List of Contributors

Isaac I. Akinwumi
Department of Civil Engineering
Covenant University
Nigeria

Rodrigo Álvarez
Mining and Exploration Department
School of Mines
University of Oviedo
Spain

Kwasi Gyau Baffour Awuah
School of the Built Environment
The University of Salford
United Kingdom

Fernando Barrio-Parra
Laboratorio de Investigación e Ingeniería Geoquímica Ambiental (LI2GA)
Environmental Geochemistry Research & Engineering Laboratory
Universidad Politécnica de Madrid
Spain

Lynn Besenyei
Faculty of Science and Engineering
School of Sciences
University of Wolverhampton
United Kingdom

Ranjan Bhattacharyya
Centre for Environment Science and Climate Resilient Agriculture (CESCRA)
Indian Agricultural Research Institute
India

Colin A. Booth
Architecture and the Built Environment
University of the West of England
United Kingdom

Susanne M. Charlesworth
Centre for Agroecology
Water and Resilience
Coventry University
United Kingdom

Chung Choi
Department of Earth Sciences
University of Bristol
United Kingdom

Akinwale O. Coker
Department of Civil Engineering
Faculty of Technology
University of Ibadan
Nigeria

Natasha Constant
Sustainable Places Research Institute
Cardiff University
United Kingdom

Stephen J. Coupe
Centre for Agroecology
Water and Resilience
Coventry University
United Kingdom

Paulo Roberto Bairros Da Silva
Federal University of Santa Maria
Brazil

Eduardo De Miguel
Laboratorio de Investigación e Ingeniería Geoquímica Ambiental (LI2GA)
Environmental Geochemistry Research & Engineering Laboratory
Universidad Politécnica de Madrid
Spain

A. A. ElKordi
Faculty of Engineering
Beirut Arab University
Lebanon

Fabio Falchi
ISTIL - Light Pollution Science and Technology Institute
Italy

Martin Fenn
Environment Agency
United Kingdom

Felippe Fernandes
University of Sao Paulo
Brazil

Michael A. Fullen
Faculty of Science and Engineering
School of Architecture and Built Environment
University of Wolverhampton
United Kingdom

Avijit Ghosh
Centre for Environment Science and Climate Resilient Agriculture (CESCRA)
Indian Agricultural Research Institute
India

Ileen Gladstone
GEI Consultants, Inc.
Massachusetts
United States

Andrew Gooding
Wessex Water
Bath
United Kingdom

Enda Hayes
Geography and Environmental Management
University of the West of England
United Kingdom

Ryan S. Hoffman
GEI Consultants, Inc.
Massachusetts
United States

Rosemary Horry
College of Life & Natural Sciences
University of Derby
United Kingdom

Katherine Hyde
School of Construction Management and Engineering
University of Reading
United Kingdom

Miguel Izquierdo-Díaz
Laboratorio de Investigación e Ingeniería Geoquímica Ambiental (LI2GA)
Environmental Geochemistry Research & Engineering Laboratory
Universidad Politécnica de Madrid
Spain

Jamal M. Khatib
Faculty of Science and Engineering
School of Architecture and Built Environment
University of Wolverhampton
United Kingdom

and

Faculty of Engineering
Beirut Arab University
Lebanon

Viktória Kovács Kis
Institute of Technical Physics and Materials Science
Centre for Energy Research
Hungary

Jianmin Ma
Key Laboratory for Environmental Pollution Prediction and Control Gansu Province
College of Earth and Environmental Sciences
Lanzhou University
China

Zoltán May
Institute of Materials and Environmental Chemistry
Research Centre for Natural Sciences
Hungarian Academy of Sciences
Hungary

Anne-Marie McLaughlin
Centre for Agroecology, Water and Resilience
Coventry University
United Kingdom

Catherine M. Malagrida
GEI Consultants, Inc.
Massachusetts
United States

Juan Mingot
Laboratorio de Investigación e Ingeniería Geoquímica Ambiental (LI2GA)
Environmental Geochemistry Research & Engineering Laboratory
Universidad Politécnica de Madrid
Spain

Tibor Németh
Institute for Geological and Geochemical Research
Research Centre for Astronomy and Earth Sciences
Hungarian Academy of Sciences
Hungary

Alan P. Newman
Centre for the Built and Natural Environment
Coventry University
United Kingdom

Ernest O. Nnadi
GITECO-UC
University of Cantabria
Santander
Spain

Adebanji S. Ogbiye
Department of Civil Engineering
Covenant University
Nigeria

Oluwapelumi O. Ojuri
Department of Civil & Environmental Engineering
Federal University of Technology
Akure
Nigeria

David Oloke
Faculty of Science and Engineering
University of Wolverhampton
United Kingdom

Almudena Ordóñez
Mining and Exploration Department
School of Mines
University of Oviedo
Spain

Cristiano Poleto
Universidade Federal do Rio Grande do Sul
Brazil

Ann Power
The BioEconomy Centre
University of Exeter
United Kingdom

Kenneth Pye
Kenneth Pye Associates Ltd
Blythe Valley Innovation Centre
United Kingdom

Jose L. Rubio
Centro de Investigaciones sobre Desertificación
Universitat de Valencia
Spain

Safaa Baydoun
Research Centre for Environment and Development
Beirut Arab University
Lebanon

Z. Abou Saleh
Faculty of Engineering
Beirut Arab University
Lebanon

Manoj Shrivastava
Centre for Environment Science and Climate Resilient Agriculture (CESCRA)
Indian Agricultural Research Institute
India

Andrew B. Shuttleworth
SEL Environmental Limited
Lancashire
United Kingdom

S.D. Singh
Centre for Environment Science and Climate Resilient Agriculture (CESCRA)
Indian Agricultural Research Institute
India

Péter Sipos
Institute for Geological and Geochemical Research
Research Centre for Astronomy and Earth Sciences
Hungarian Academy of Sciences
Hungary

Matthew Smith
School of Construction Management and Engineering
University of Reading
United Kingdom

Fraser Torpy
School of Life Sciences
University of Technology Sydney
Australia

Rebecca Wade
Urban Water Technology Centre
School of Science Engineering and Technology
Abertay University
Scotland
United Kingdom

J. Webb
Faculty of Science and Engineering
School of Architecture and Built Environment
University of Wolverhampton
United Kingdom

Sara Wilkinson
Faculty of Design Architecture and Building
University of Technology Sydney
Australia

Craig D. Williams
Faculty of Science and Engineering
University of Wolverhampton
United Kingdom

Annie Worsley
Strata Environmental
Wester Ross
Scotland
United Kingdom

Jianzhong Xu
State Key Laboratory of Cryospheric Sciences
Cold and Arid Regions Environmental and Engineering Research Institute
China

Norbert Zajzon
Institute of Mineralogy and Geology
Faculty of Earth Science and Engineering
University of Miskolc
Hungary

1

Insights and Issues into the Impacts of Urban Pollution

Colin A. Booth[1] *and Susanne M. Charlesworth*[2]

[1] *Architecture and the Built Environment, University of the West of England, United Kingdom*
[2] *Centre for Agroecology, Water and Resilience, Coventry University, United Kingdom*

1.1 Introduction

Urban pollution can be defined as the presence or introduction of contaminant material (solid, liquid, gas) or energy (heat, noise, light, radiation) into the built environment, either directly or indirectly, by natural sources and/or anthropogenic activities, which are likely to have harmful or poisonous effects on people, property, and/or the environment. This encompasses pollution of the air we breathe, pollution of the water we drink, pollution of the soil that grows the food we eat, pollution of plants we are reliant upon to perform photosynthesis, pollution of the buildings we live and work in and, ultimately, pollution that changes our weather/climatic systems.

Pollution events occur every day as a result of spills, accidents, negligence, or vandalism (Environment Agency, 2013). However, the effects can be devastating and long-lasting for both humans and the environment (e.g. radiation exposure from Chernobyl, Ukraine). Typically, some of the worst places to suffer from pollution are towns and cities (Table 1.1). Poor air quality is prevalent many days of the year in many cities around the world. For instance, Marylebone Road is a major arterial route (A501) for traffic and pedestrians in the City of Westminster, Central London, where the roadside buildings create an asymmetric street canyon with a height-to-width ratio of ~0.8 (Charron *et al.*, 2007) and, as a consequence the area has consistently high daily mean PM_{10} level that regularly exceeded the EU (1993/30/EC) Air Quality Directive (47 incidents in 2007; 29 in 2008; 36 in 2009; 15 in 2010; 34 in 2011; and 27 in 2012). These exceedances are attributed to high traffic flows, congestion, and vehicle combustion particulates (AQEG, 2005; Crosby *et al.*, 2014).

As much as 54% of the total global population was estimated to live in urban areas in 2014, which represents an increase of 20% since 1960, and this percentage is expected to grow to 66% by 2050, adding a further 2.5 billion people to urban populations (United Nations, 2014). Many cities are expanding at rates that exceed their capacity to accommodate their growing populations. This means some cities are experiencing a growth of informal settlements on their periphery where they lack services and infrastructure. The density of cities increases the chance that one source of pollution will affect a great many people (WHO, 2016a).

Intensive urbanisation brings with it increased pollution from a variety of sources including industry, traffic, domestic heating, coal and oil combustion, incineration, construction activities, road weathering, and

Urban Pollution: Science and Management, First Edition. Edited by Susanne M. Charlesworth and Colin A. Booth.

Table 1.1 The world's ten most polluted cities.

	City/country	Type of pollution
1.	Linfen, China	Coal
2.	Tianying, China	Heavy metals
3.	Sukinda, India	Hexavalent chromium
4.	Vapi, India	Chemicals and metals
5.	La Oroya, Peru	Sulphur dioxide, lead, copper, and zinc
6.	Dzerzhinsk, Russia	Chemicals and toxic by-products, such as sarin and VX gas
7.	Norilsk, Russia	Air pollution, such as particulates and sulphur dioxide
8.	Chernobyl, Ukraine	Radiation
9.	Sumgayit, Azerbaijan	Organic chemicals, heavy metals, and oil
10.	Kabwe, Zambia	Cadmium and lead

Derived from http://www.blacksmithinstitute.org/

maintenance activities such as street sweeping and gully emptying. Inevitably, this leads to increased release of polluted particulates, dissolved contaminants, nutrients, new and emerging pollutants (such as hormones and personal care products), as well as inhalable and respirable particles, among others.

Air pollution is a major risk for many people, as it can cause cardiovascular diseases, strokes, chronic obstructive pulmonary disease, lung cancer, and acute respiratory infections. Moreover, an estimated 3.0 million deaths in 2012 were caused by exposure to outdoor pollution, specifically ambient air pollution, and an estimated 4.3 million deaths were caused by household air pollution. These mortality rates vary regionally – with Georgia, North Korea, Bosnia and Herzegovina, Bulgaria, Albania, China, and Sierra Leone among the highest per 100,000 of their population (WHO, 2016b).

Most global cities lack adequate wastewater management, such that unsafe water, sanitation, and hygiene were responsible for an estimated 871,000 deaths in 2012. Most of these deaths were linked to diarrhoeal diseases, together with malnutrition, intestinal nematode infections, and schistosomiasis, caused mainly by contamination of drinking water, waterbodies, and soil. Sanitation for urban populations in the world's least developed regions is limited. As a consequence, mortality rates are greatest in Africa – with Angola, Congo, Somalia, Chad, Sierra Leone, Niger, and Burundi among the highest per 100,000 of their population (WHO, 2016b).

With urban pollution being responsible for so many deaths annually, and with an estimated one in every three people expected to be living in cities with at least half a million inhabitants by 2030 (United Nations Human Settlements Programme, 2016), there is a pressing need to explore, understand, expose, and address pollution sources, pathways, and receptors in the urban setting and identify how best to manage and police the issues and impacts associated with them. The next section provides some notable examples of urban pollution that have taught some harsh lessons enabling changes to be made in management and policy.

1.2 Examples of Urban Pollution

There has been a plethora of pollution disasters around the world that have diseased populations, infected landscapes, and contaminated resources above and below ground. As a means of introducing the severity of urban pollution around the world, detailed in the following is an array of examples where air pollution, water pollution, and soil pollution has been devastating and long-lasting for those affected.

1.2.1 Air Pollution in London, United Kingdom

Air quality in towns and cities has been a global problem for many centuries (Brimblecombe,

1998). One of the most widely reported examples of urban air pollution is the 'Great Smog' of London. On 4 December 1952, an anticyclone descended over a windless London, causing a temperature inversion with cold, stagnant air trapped under a layer of warm air. The period of cold weather meant the residents of London burned more coal than usual so they could stay warm. The resultant chimney smoke, mixed with fog, culminated in a blanket of smog forming over the city. However, as concentrations of air pollutants built up in the air, this aided the condensation of water and thus decreased temperatures causing concomitant increases in demands for further heating. As a consequence, air quality quickly deteriorated alongside the health of vulnerable groups (the very young and elderly) or those with pre-existing respiratory problems. By the time the weather changed on 9 December 1952, it was estimated the event had caused (or advanced) the death of 4,000 people, mainly due to respiratory tract infections from hypoxia (low oxygen levels in blood) and due to mechanical obstruction of the air passage from lung infections caused by the smog (Logan, 1953). Similar international examples include the 1930 Meuse Valley fog in Belgium, which killed 60 people (Nemery *et al.*, 2001); the 1948 Donora smog in Pennsylvania, United States, which killed 20 people (Snyder, 1994); the 1966 New York City smog in the United States, which killed 168 people; the 2013 Harbin smog in China (Nunez, 2013); and the 2013 Shanghai smog in China. Unfortunately, in many cases, it takes fatalities attributed to extreme pollution disasters before policies and regulations are even initiated or updated (e.g. the Great Smog (1952) led to the creation of the Clean Air Act (1956) in the United Kingdom).

1.2.2 Air Pollution in Bhopal, India

On 3 December 1984, the world's worst industrial disaster occurred when a toxic gas plume leaked from the Union Carbide Corporation (now Dow Chemical) pesticide plant in Bhopal, in the state of Madhya Pradesh, India, and descended on the residents of the surrounding urban area, as they slept in their beds (Agarwal and Narain, 1985; Labib and Champaneri, 2012). More than 40 tons of methyl isocyanate gas fumes drifted into the city. People woke with burning eyes and lungs. The incident directly caused at least 3,800 deaths with significant morbidity and premature death for many thousands more (Broughton, 2005). After almost 20 years of legal haggling, compensation was awarded to 554,895 people for injuries received and 15,310 surviving relatives of those killed. The average amount paid to victims was a paltry 25,000 rupees (£300; US$390; 360 euros), with a shameful 100,000 rupees (£1,200; US$1,560; 1,440 euros) to families of the dead (Kumar, 2004). It was only following the event that the Indian Government created the Environment Protection Act (1986), which gave powers to the Ministry of Environment and Forests for administering and enforcing environmental laws and policies (Broughton, 2005).

1.2.3 Water Pollution in London, United Kingdom

Claimed by many as the first epidemiological study, Dr John Snow, a public health worker, investigated a severe cholera outbreak in Soho, London, in 1854, that was responsible for killing 616 people (Newsom, 2006; Tulodziecki, 2011). Around this time, London had already suffered incapacitating outbreaks of cholera (1832 and 1849), which had killed thousands of people. To unearth the evidence he needed, he set about mapping the location of those people who contracted the deadly disease and was able to link their movements and activities to those who drank from a water well on Broad Street (now Broadwick Street). Dr Snow was convinced that contaminated water (and not foul air) was responsible for spreading the infectious disease. Conclusive by modern-day standards, it took the removal of the pump handle and subsequent abatement of the outbreak to convince some officials that the disease was waterborne. In fact, further investigation revealed the well was sited close to a cesspit that was discharging into groundwater and, by doing so, highlighted that faecal waste was somehow responsible for the contamination of the drinking water and for the original cholera outbreak. Other international modern-day examples of cholera outbreaks include those in Iraq (2007), in Congo and Zimbabwe (both 2008), in Haiti (2010), and 2012 in Sierra Leone (Mason, 2009; Nguyen *et al.*, 2014; Piarroux *et al.*, 2011).

1.2.4 Water Pollution in Minamata, Japan

Toxic discharges of industrial wastewater effluent (since 1932) from a petrochemical factory (owned by the Chisso Corporation) in the city of Minamata, Japan, contained methylmercury, which bioaccumulated in shellfish and fish that were then eaten by local residents (D'itra, 1991; Harada, 1995). By 1956, it was observed that many of these people had developed Minamata disease (or Itai-Itai disease, Japanese for 'ouch-ouch') – a neurological syndrome that causes dysfunctional muscle movement together with hearing and speech loss, instigated by severe mercury poisoning. Sadly, the disease was responsible for 21 fatalities over the next two years and, despite accumulating evidence indicating the point source of the pollution, no controls were ever imposed on production and processes at the factory until it ceased operating in 1968. By 1975, there were 800 verified victims of this long–term pollution event, of which 107 were fatalities, and a further 2800 possible additional victims (Mance, 1987).

1.2.5 Soil Pollution in Missouri, United States

A dioxin disaster forced the resident of Times Beach, in Missouri, United States, to abandon their homes and town forever (since 1982) (Belli *et al.*, 1989; Lower *et al.*, 1990). The United States Environmental Protection Agency (EPA) found high levels of a toxic chemical called dioxin (unwanted by-products of industrial and combustion processes) had contaminated many parts of the town. The roads in this suburban town were unpaved and dusty so, in 1972, in an effort of control the dust, the town employed the services of a waste oil haulier to spray its dirt roads with oil. Unfortunately, the haulier, who normally applied used motor vehicle oil for spraying, had been subcontracted to remove oily residues from the processing activities of a pharmaceutical and chemical company (NEPACCO) in Verona. The chemical waste (heavily contaminated with dioxin) was then mixed with waste motor vehicle oil, stored in tanks before being sold or, in the case of Times Beach, being applied as a dust control (Hites, 2011). It is estimated that ~160,000 gallons of waste oil was sprayed over a four-year period. It was almost a decade later that the EPA investigated and found contaminated soils along the network of roads surrounding the homes of the entire community. In the interests of safety, officials were forced to evacuate the town and opted to buy out the 800 residential properties and 30 businesses at an estimated cost of US$36.7 million (£28.5 million; 33.8 million euros).

1.3 Structure of This Book

This book comprises five sections, which are collated into 31 chapters. The first part of the book provides *An Introduction* that offers some initial insights into the impacts and issues associated with urban pollution (Chapter 1). Since many nations now have legislation and policies in place for their designated environmental protection bodies to control and manage pollution to regulated and permitted guidelines, Section 2 exposes *Policy and Pollution* through chapters that take a historical view of pollution and offer insight into relevant air policies, water polices, and soil policies (Chapters 2–5), and demonstrate a range of types of pollution (Chapters 6–12). Section 3 assembles options for *Monitoring, Remediation, and Management* through a collection of chapters concentrating on river ecology, urban hay meadows, ecosystems, waste disposal sites, building materials, zeolites, bioremediation, environmental impact assessment, and citizen science (Chapters 13–22). Section 4 contextualises *International Case Studies* with examples from the United States, China, India, Brazil, Hungary, Ghana, Nigeria, and Lebanon (Chapters 23–30). Finally, Section 5 converges with a *Summary of the Book* that distils lessons that can be learned to protect people and property in the built environment.

1.4 Conclusions

This edited book aims to develop an appreciation of the diverse, complex, and current themes of the urban pollution debate across the built environment, urban development, and management continuum. While it cannot cover every type of pollutant and revisit every notable pollution incident in every country, it does offer an integration of physical and environmental sciences, combined with social, economic, and political sciences and the use of case studies to provide a unique resource, useful to policy experts, scientists, engineers, and subject enthusiasts. Many of these chapters have been written by academics with expertise in the field, but this book also contains chapters authored or co-authored by practitioners with a wealth of practical experience. It is, therefore, of interest to those involved in the impacts and management of urban pollution issues worldwide, such as environmental toxicologists, chemists, and health experts working in government agencies, academia, and commercial consultancies.

References

Agarwal, A. and Narain, S. (1985) *The State of India's Environment 1984–85: A Second Citizen's Report*. Centre for Science and Environment, New Delhi.

AQEG (2005) *Particulate Matter in the United Kingdom*. DEFRA, London.

Belli, G., Cerlesi, S., Kapila, S., Ratti, S.P., and Yanders, A. (1989) Geometrical description of the TCDD contamination of Times Beach. *Chemosphere*, 18, 1251–1255.

Brimblecombe, P. (1998) History of urban air pollution. In: Fenger, J., Hertel, O., and Palmgren, F. (eds.), *Urban Air Pollution, European Aspects*. Kluwer Academic Publishers, Dordrecht, pp. 7–20.

Broughton, E. (2005) The Bhopal disaster and its aftermath: A review. *Environmental Health*, 4 (1), 6 pages.

Charron, A., Harrison, R.M., and Quincey, P. (2007) What are the sources and conditions responsible for exceedences of the 24 h PM10 limit value at a heavily trafficked London site? *Atmospheric Environment*, 41, 1960–1975.

Crosby, C.J., Booth, C.A., Fullen, M.A., and Searle, D.E. (2014) A dynamic approach to urban road deposited sediment pollution monitoring (Marylebone Road, London, UK). *Journal of Applied Geophysics*, 105, 10–20.

D'itra, F.M. (1991) Mercury contamination – What have we leaned since Minamata? In: Lee, H.K. (ed.), *Proceedings of the Fourth Symposium on Our Environment*, Singapore, pp.165–182.

Environment Agency (2013) *Pollution Prevention Pays in England and Wales*. EA.

Harada, M. (1995) Minamata disease: Methyl mercury poisoning in Japan caused by environmental pollution. *Critical Reviews in Toxicology*, 25, 1–24.

Hites, R.A. (2011) Dioxins: An overview and history. *Environmental Science and Technology*, 45, 16–20.

Kumar, S. (2004) Victims of gas leak in Bhopal seek redress on compensation. *British Medical Journal*, 329 (7462), 366.

Labib, A.W. and Champaneri, R. (2012) The Bhopal disaster – learning from failures and evaluating risk. *Maintenance and Asset Management*, 27, 41–47.

Logan, W.P.D. (1953) Mortality in the London fog incident, 1952. *The Lancet*, 261 (6755), 336–338.

Lower, W.R., Thomas, M.W., Puri, R.K., Judy, B.M., Zacher, J.A., Orazio, C.E., Kapila, S., and Yanders, A.F. (1990) Movement and fate of 2,3,7,8–tetrachlorodibenzo–p–dioxin in fauna at Times Beach, Missouri. *Chemosphere*, 20, 1021–1025.

Mance, G. (1987) *Pollution Threat of Heavy Metals in Aquatic Environments*. Elsevier Applied Science Publishers Ltd., US.

Mason, P.R. (2009) Zimbabwe experiences the worst epidemic of cholera in Africa. *The Journal of Infection in Developing Countries*, 3, 148–151.

Nemery, B., Hoet, P.H.M., and Nemmar, A. (2001) The Meuse Valley fog of 1930: An air pollution disaster. *The Lancet*, 357 (9257), 704–708.

Newsom, S.W.B. (2006) Pioneers in infection control: John Snow, Henry Whitehead, the Broad Street pump, and the beginnings of geographical epidemiology. *Journal of Hospital Infection*, 64, 210–216.

Nguyen, V.D., Sreenivasan, N., Lam, E., Ayers, T., Kargbo, D., Dafae, F., Jambai, A., Alemu, W., Kamara, A., Sirajul Islam, M., Stroika, S., Bopp, C., Quick, R., Mintz, E.D., and Brunkard, J.M. (2014) Cholera epidemic associated with consumption of unsafe drinking water and street–vended water – Eastern Freetown, Sierra Leone, 2012. *American Journal of Tropical Medicine and Hygiene*, 90, 518–523.

Nunez, C. (2013) Harbin Smog Crisis Highlights China's Coal Problem. *National Geographic*, October Issue.

Piarroux, R., Barrais, R., Faucher, B., Haus, R., Piarroux, M., Gaudart, J., Magloire, R., and Raoult, D. (2011) Understanding the cholera epidemic, Haiti. *Emerging Infectious Diseases*, 17, 1161–1168.

Snyder, L.P. (1994) The Death–Dealing Smog over Donora, Pennsylvania: Industrial Air Pollution, Public Health Policy, and the Politics of Expertise, 1948–1949. *Environmental History Review*, 18, 117–139.

Tulodziecki, D. (2011) A case study in explanatory power: John Snow's conclusions about the pathology and transmission of cholera. *Studies in History and Philosophy of Science Part C: Studies in History and Philosophy of Biological and Biomedical Sciences*, 42, 306–316.

United Nations (2014) *World Urbanisation Prospects: The 2014 Revision, Highlights*. Department of Economic and Social Affairs, Population Division of the United Nations (ST/ESA/SER.A/352).

United Nations Human Settlements Programme (2016) *Urbanization and Development: Emerging Futures – The World's Cities in 2016*. UN-Habitat, Nairobi, Kenya.

World Health Organisation (2016a) *Global Report on Urban Health: Equitable Healthier Cities for Sustainable Development*. WHO Press, Geneva, Switzerland.

World Health Organisation (2016b) *World Health Statistics 2016: Monitoring Health for the Sustainable Development Goals*. WHO Press, Geneva, Switzerland.

2

Historical Urban Pollution

Ann Power[1] *and Annie Worsley*[2]

[1] *The BioEconomy Centre, University of Exeter, United Kingdom*
[2] *Strata Environmental, Wester Ross, Scotland, United Kingdom*

2.1 Introduction

Pollution released from anthropogenic activities is a major environmental concern at global, regional and local scales. The natural environment has been contaminated by human activity for thousands of years; however, modern industrial development, advancements in transport technologies, population increases and the subsequent urban sprawl have led to unprecedented levels of pollution experienced since the twentieth century. A changing complexity of contaminants has been released to land, waterways, and air, from industry, transport and residential sources within built environments over time, especially since the Industrial Revolution post 1800 in Europe and the United States (Brimblecombe, 2005). Urban populations residing in close proximity to industrial complexes and transport infrastructure are most at risk to the long- and short-term health effects of exposure to toxic pollutants. Air pollution is ubiquitous within the urban environment and comprises principally noxious gases and particulate matter (PM). Exposure is unavoidable and dangerous even at low levels (Shi *et al.*, 2016). Air pollution is recognized as the single largest global environmental health risk, responsible for 7 million annual global deaths (Silva *et al.*, 2013; Austen, 2015), making it the focus of this chapter.

Much of what we understand about past urban air pollution is derived from documented historical evidence of known industrial and urban activities and only relatively recent (post 1960) pollution monitoring programmes. Environmental archives such as lakes, peat bogs and ice sheets therefore provide invaluable evidence of pollution emissions. Natural sinks for atmospheric contaminants, they allow the reconstruction of past pollution trends beyond what is possible with conventional monitored data. Records from sediment archives reveal intricate relationships between technological advancements and air pollution since ancient times (Brännvall *et al.*, 1999). As new technologies and industries have emerged and expanded, so too have new types of pollutants, especially post WWII.

Urban pollution varies from place to place, depending on specific past urban and industrial activities experienced in countries and cities. Sedimentary archives record local, regional and global pollution signals. In this chapter, case studies of urban pollution are presented that focus primarily on UK cities since Britain is renowned for its extensive industrial and urban past. A general history of urban air pollution is presented, supported by evidence

Urban Pollution: Science and Management, First Edition. Edited by Susanne M. Charlesworth and Colin A. Booth.

from sedimentary archives and key events that span ancient civilizations, the Industrial Revolution and the twentieth century.

2.2 Historical Pollution Monitoring using Environmental Archives

Once emitted into the environment, pollutants are transported via air or water and eventually can settle in natural archives, such as ice sheets, estuaries, lakes, and peat bogs (MARC, 1985) (Figure 2.1). A robust record of pollution deposition can be preserved within naturally accumulating sediments, inferred from a range of environmental proxies and atmospheric contaminants stored within extracted sediment/ice cores (Figure 2.1, Table 2.1). These records allow the reconstruction of spatially specific, long-term pollution histories spanning decades to millennia. Lake and peat bog records from around the world have revealed local, regional and global impacts of industry and urbanization on pollutant emissions. Contamination detected in archives from even the most remote, 'pristine' regions (e.g. the Arctic) demonstrates the long-range transport and far-reaching impacts of air pollution (Rose *et al.*, 2004).

2.3 Ancient Air Pollution

Urban pollution is not a recent phenomenon. Palaeolithic cave dwellings would have been polluted with smoke from wood burning, particularly from 300,000 to 400,000 years ago when fire became a common hominin technology (Boros *et al.*, 2003; Roebroeks and Villa, 2011). As early settlements were established, domestic activities would have caused deterioration of air quality at a very localized scale. The earliest significant regional air pollution caused by pre-historic anthropogenic activities is detected in Swiss lake sediments. Enhancements in Pb flux occur during the Neolithic (3800 BC) due to the release of Pb from soil due to early agriculture and Bronze Age (2500 BC) metallurgy (Thevenon *et al.*, 2011).

A 7000-year metal history reconstructed from Chinese lake sediments detects emissions from the start of the Bronze Age at ~3000 BC (+/− 328) and identifies a period of rapid metal enhancement in the late Bronze Age (475 BC–220 AD) from the production of weapons, vessels, and tools (Lee *et al.*, 2008). Temporal shifts in the metallic characteristics of lake stratigraphies also distinguish between early Bronze Age technology (Cu) and later (1100–1300 AD) silver mining activity (Cd, Ag, Pb, and Zn) in China (Hillman *et al.*, 2015).

Air pollution increased as a consequence of the rise of ancient civilizations. Peaks in airborne Pb recorded in European (Swiss and Swedish) sediment records are attributed to the Greek (800–146 BC) and Roman (300 BC–100 AD) empires from Pb and Sn smelting (Brännvall *et al.*, 1999; Thevenon *et al.*, 2011). Mining regions in central Spain, exploited for gold and silver during the past 2000 years, are also highlighted as a potential ancient source of atmospheric Pb and Hg releases (Thevenon *et al.*, 2011).

Permanent increases in Pb deposition are observed during the medieval period (900–600 AD) in Europe, reflecting population increases and the expansion of mining activities (Brännval *et al.*, 1997). Periods of relatively low anthropogenic emissions reflect episodes of reduced economic and industrial activity due to population declines during the Great Famine (1315–1321 AD) and Black Plague (1346–1383 AD) (Brännvall *et al.*, 1999; Thevenon *et al.*, 2011). Further Pb reductions ~1600 AD are attributed to the relocation of metal exploitation from Europe to America. This is reflected in Peruvian environmental archives, which reveal localized, low-scale metallurgical activity during the Inca Empire (1400–1533 AD), with regional metal enhancement due to intensified extensive silver exploitation during the onset of the colonial period (post 1540 AD) (Uglietti *et al.*, 2015).

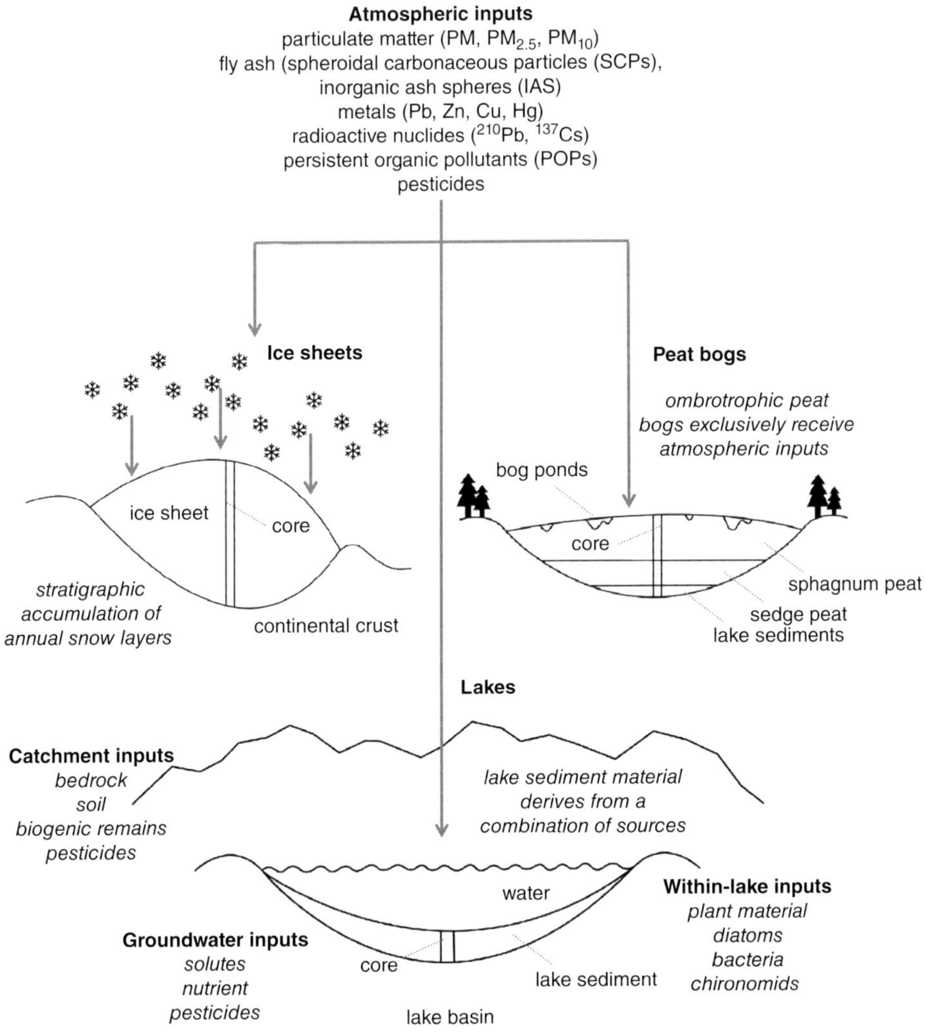

Atmospheric inputs
particulate matter (PM, $PM_{2.5}$, PM_{10})
fly ash (spheroidal carbonaceous particles (SCPs),
inorganic ash spheres (IAS)
metals (Pb, Zn, Cu, Hg)
radioactive nuclides (^{210}Pb, ^{137}Cs)
persistent organic pollutants (POPs)
pesticides

Ice sheets
ice sheet
core
stratigraphic accumulation of annual snow layers
continental crust

Peat bogs
ombrotrophic peat bogs exclusively receive atmospheric inputs
bog ponds
core
sphagnum peat
sedge peat
lake sediments

Lakes
Catchment inputs
bedrock
soil
biogenic remains
pesticides
lake sediment material derives from a combination of sources
water
Within-lake inputs
plant material
diatoms
bacteria
chironomids
Groundwater inputs
solutes
nutrient
pesticides
core
lake sediment
lake basin

Figure 2.1 Atmospheric contaminants and environmental proxies (Table 2.1) used for reconstructing air pollution histories from ice sheets, peat bogs and lakes.

Permanent ice sheets: Falling snow scavenges aerosols (including metal and radioactive isotopes) from the atmosphere. Snow deposits on ice sheets and freezes to produce a stratified annual record of atmospheric deposition.

Ombrotrophic peat bogs: The surface layers of 'rain-fed' (ombrotrophic) bogs exclusively receive atmospheric contaminants, with no catchment interference or groundwater flow. Peat accumulates over time from the decomposition of plant material, capturing a history of atmospheric pollution deposition. Peat bog stratigraphies include the upper sphagnum peat layers, older sedge peat and underlying lake sediments from which the peat bog originally formed.

Lake sediments: Lakes receive inputs from the atmosphere, groundwater, and catchment (surrounding terrestrial environment), as well as contributions within the lake. An overriding atmospheric signal can be recorded if the catchment disturbance and input are minimal. Particulates deposit over lakes and settle through the water column, onto the lake basin. Particles and associated pollution proxies are stored within the naturally accumulating sedimentary matrix.

Table 2.1 Atmospheric pollutants found in environmental archives.

Pollutants	Description
Persistent organic pollutants (POPs) PAHs, PCBs, OCPs	Synthesized organic chemicals include PAH (polycyclic aromatic hydrocarbons), PCBs (polychlorinated biphenyls), and OCPs (organochlorine pesticides) that persist and bioaccumulate in the environment. Released as products or by-products from industrial processes and road traffic, they adsorb to particulate matter and, if deposited in lakes, mainly adsorb within the organic, humic fraction of sedimentary deposits (Huang *et al.*, 2012). Environmental archives have played a crucial role in determining POP sources and the global extent of POP contamination since their manufacture during the twentieth century.
Metals Cd, Cr, Cu, Pb, Hg, Ni, Zn,	Metals are released to the atmosphere from a range of industrial activities including mining, metallurgy, coal combustion metal recycling, industrial processes, and traffic (Nriagu, 1996). Metal enrichment factors (normalized for natural, background metal concentration) are used to distinguish anthropogenic inputs (Norton *et al.*, 1992). Pb is released from mining Pb-rich ores and smelting of silver and bronze and, as such, is a suitable tracer for mining activities dating back to ancient times. Cd, Cr, Cu, Pb, Hg and Zn are typically associated with a range of industrial processes and fossil fuel combustion. Metal isotope ratios can also distinguish between different pollution sources (McConnell *et al.*, 2014).
Fly ash SCPs, IASs	The high-temperature combustion of fossil fuels produces fly ash particles. Spheroidal carbonaceous particles (SCPs) and inorganic ash spheres (IASs) are components of fly ash that can be used as indicators of coal and oil combustion, making them useful tracers of pollution since the Industrial Revolution. SCPs are carbon particles, usually porous, with a near-spherical shape, formed from the incomplete combustion of fossil fuels. They are unambiguous tracers of anthropogenic emissions (Rose, 2001). IASs are formed by the fusing of non-combustible inorganic minerals within fuel to form droplets (spheres) comprising aluminosilicate with varying amounts of Fe and other trace elements (Rose, 2001; Goodarzi and Sanei, 2009).
Radioactive nuclides ^{210}Pb, ^{137}Cs, ^{14}C, ^{241}Am	The decay of naturally occurring radioisotopes such as ^{210}Pb (in the uranium ^{238}U decay series) is commonly used for accurate dating of sediments back to 150 years (Appleby and Oldfield, 1978; Appleby, 1993). ^{137}Cs and ^{241}Am are markers for artificial nuclear isotopes and are also used to corroborate ^{210}Pb dating profiles. Peaks signify the height of nuclear weapons testing (in 1963) (Appleby, 2008). For longer timescales, radiocarbon (^{14}C) dating is used.

The long-range extent of ancient contamination is reflected in ice cores from Greenland (Rosman *et al.*, 1997). The global impact of early urban activities are, however, relatively minor with, for example, Pb levels several hundred times lower (Shotyk *et al.*, 1998) when compared to the unprecedented pollution enhancement experienced throughout the nineteenth and twentieth centuries due to the Industrial Revolution in Europe and the United States (McConnell *et al.*, 2014; Uglietti *et al.*, 2015).

2.4 Industrial Revolution

Industrialization has spread across the world at different times since the eighteenth century (Rose, 2015), marking a period of enhanced human population, technological advancement, urbanization and, subsequently, urban pollution. For example, regions of the Southern Hemisphere, such as Australia, have much more recent industrial pollution histories. Australian peat bog records show enhanced metal pollution since the 1840s following the settlement of Europeans and subsequent rapid industrialization (initially Pb mining) (Marx *et al.*, 2010). Prior to this, indigenous people were mainly hunter-gatherers with little environmental impact.

Britain was the world's first industrialized nation and typifies the stages in industrial progression and urban pollution experienced throughout the world. The rapid pace of industrialization was made possible by the consumption of coal, which replaced wood as the dominant fuel used in the United Kingdom from the thirteenth century. 'Sea coal' produced large amounts of smoke upon combustion (Department of the Environment, 1974) and was soon acknowledged as a serious problem, prompting a Royal Proclamation in 1306 restricting its use. However, by the sixteenth century, coal was extensively used by a growing population for domestic heating and fuelled small-scale industries (Brimblecombe, 1987).

The visible dark smoke emitted during coal combustion comprises PM containing fly ash (including spheroidal carbonaceous particles (SCP) and inorganic ash spheres (IAS), Table 2.1). Sulphur dioxide and nitrogen dioxide are also produced, which react with hydrogen to form sulphuric and nitric acids. The first major scientific publication highlighting the environmental consequences of urban coal smoke was John Evelyn's study on the air pollution of London: 'Fumifugium; or, The Inconvenience of the Aer, and Smoake of London Dissipated' in 1661. It was extremely perceptive for the time, associating coal combustion with building damage, killing bees and flowers, preventing fruit growth on trees and, prior to substantial medical research, suggested links with respiratory health effects, attributing air pollution as the cause of half of the deaths of London's inhabitants (Evelyn, 1661).

Air pollution was, however, viewed as a necessary consequence of industrial development, especially during the eighteenth and nineteenth centuries as coal continued to fuel the Industrial Revolution. Smoke and noxious gases were indiscriminately released into the air via steam-powered locomotives, industrial stacks and domestic chimneys. Notable pollution enhancement from coal burning at this time is demonstrated by initial increases in SCP concentrations from the mid-nineteenth century recorded in lakes from the United Kingdom (Rose and Appleby, 2005), Europe, and North America (Rose, 2015).

2.4.1 Case Study: Chemicals in Merseyside, NW England

NW England is the birthplace of the Industrial Revolution in the United Kingdom. Improvements in transport links such as canals and railways across Britain during the nineteenth century resulted in the expansion of barge and shipbuilding and exports of stone and coal to other towns and cities. In NW England, industrial activity included quarrying, shipbuilding and low-scale industries such as tanneries, tool making, breweries and file cutting. The city of Liverpool was established as an international trading port. During the early 1800s, chemical works were constructed along the River Mersey, and the region subsequently developed as the epicentre of the early chemical industry (Warren, 1980). Chemicals soon dominated industry here, particularly in the 'Chemical Towns' of Runcorn and Widnes.

Early chemicals were produced by the Leblanc process. Powered by localized coal burning at factories, salt was heated with sulphuric acid in lead chambers to produce salt cake, which was then heated with limestone to form black ash, from which alkali was recovered (Dingle, 1982). The establishment and subsequent expansion of the chemical industry due to increased demand for chemicals throughout the nineteenth century brought immediate consequences to the urban landscape. Smoke and hydrochloric acid gas were released from low chimneys with ineffective pollution dispersal and a solid waste by-product containing high amounts of sulphur and arsenic (known locally in NW England as 'galligu') was piled high into 'mountains'. At the time, there were no strict pollution legislations or controls founded in the United Kingdom, and waste was indiscriminately released to waterways, land and the air. Substantial population increases, due to demands in workforce, compounded the problem of urban smoke since chemical factory workers resided in close proximity to the factories, releasing residential smoke from chimneys at low heights (Figure 2.2):

> "Houses and streets spread themselves over the open spaces around the works, and in a very few years Widnes was transformed from a pretty, sunny riverside hamlet... into a settlement of thousands of labouring men ... with dingy, unfinished streets of hastily constructed houses, with works that were belching forth volumes of the most deleterious gases, and clouds of black smoke from chimneys of inadequate height with trees that stood leafless in June, and hedgerows that were shrivelled in May. The air reeked with gases offensive to the sight and smell, and large heaps of stinking refuse began to accumulate." (Allen, 1906)

By 1888 Widnes was described in a national newspaper as 'the dirtiest, ugliest and most depressing town in England' (Diggle, 1961), reflecting the consequences of Leblanc chemical production at its peak in the region.

The adverse health effects of air pollution were soon felt by factory workers and landowners, who noticed their crops being damaged by the noxious fumes. This prompted the introduction of the Alkali Acts in 1863 to control pollution from chemical factories. The act was updated in 1906 to include the majority of industrial emissions and was the main legislative control of industrial pollution in the United Kingdom (Department of the Environment, 1974).

Recessions in chemical trade during the late nineteenth century resulted in Leblanc closures. Leblanc production was obsolete by the twentieth century, replaced by electrolytic alkali production (e.g. chlor alkali). Major chemical companies were established, manufacturing a diverse range of inorganic and organic modern chemicals (Carter, 1964; Jones, 1969), marking a new industrial era.

Figure 2.2 The urban landscape of a chemical town during the late nineteenth century. Industrial and domestic smoke emissions in Widnes, NW England, that experienced significant expansion in the chemical industry with 22 Leblanc works constructed between 1847 and 1884 showing the combined visible impact of industrial and domestic coal emissions on the urban environment (Hardie, 1950).

2.5 Twentieth-Century Urban Pollution

Extensive urbanization and industrialization has released a complex mixture of pollutants to the air during the twentieth century. The diversification of industries and manufacturing processes, increased power generation, fuel consumption trends and rising road and air travel have influenced the composition and concentration of urban air pollution. Pollutant releases have inevitably varied from place to place and over time depending on the history of urban development, types of industry, transport infrastructure and the implementation of pollution controls at given localities.

2.5.1 Coal Consumption and the Rise of Urban Smog

Despite efforts to reduce urban smoke, for example, the 1926 Public Health (Smoke Abatement) Act in the United Kingdom, the burden of coal combustion on public health was not unequivocally realized until a series of smog events occurred during the first half of the twentieth century. The 1930 Meuse Valley fog (Belgium), 1939 St Louis smog (United States), 1948 Dohora smog (United States), and a succession of smogs in London between 1948 and 1962 confirmed links between air pollution and mortality within local populations (Nemery *et al.*, 2001). The infamous London smog of 1952, however, is one of the most important air pollution episodes in history since it demonstrated an irrefutable link between air pollution and mortality and influenced public perception, science and government legislation.

2.5.2 Case Study: London Smog 1952

London experienced a slow-moving, high-pressure weather system from 5 to 9 December in 1952. Smoke from industry and domestic coal fires, during an unusually cold winter, was compounded by diesel emissions from buses, which had recently replaced electric trams in London. These conditions created a smog that persisted for several days, trapping high concentrations of PM and sulphur dioxide at ground level, reducing visibility to almost zero (Brimblecombe, 1987; Bell *et al.*, 2004).

The smog was initially thought to have caused between 3,000 and 4,000 deaths during the following winter months. However, recent research suggests that air pollution and mortality rates remained elevated for several months with 12,000 deaths a more likely estimate (Bell *et al.*, 2004). Analysis of archived lung material from people known to have died from the smog show ultrafine carbon agglomerates forming PM <1 μm derived from coal and diesel combustion as well as metallic (e.g. Pb-containing) particles (Hunt *et al.*, 2003).

The high death rate directly linked to the smog instigated the establishment of the Clean Air Act in 1956. Domestic chimneys at low heights were considered more detrimental to public health than taller industrial stacks. Therefore, smoke control areas were introduced prohibiting the release of smoke from homes. Coal was subsequently replaced by natural gas, and power was generated from centralized nuclear, oil, and coal-fired power stations due to a rapid demand for electricity following WWII (Department of the Environment, 1974). Coal use peaked around 1960 in the United Kingdom and has since been mainly used by power stations, with consumption declining gradually to the present day (Figure 2.3). Declines in black smoke and sulphur dioxide (Figure 2.3), as recorded by the world's first nationwide air quality monitoring network, 'The National Survey', reflect declining coal consumption trends, combustion efficiency and the implementation of pollution controls and legislations.

The National Survey measured black smoke and sulphur dioxide levels at over 1200 sites in the United Kingdom post

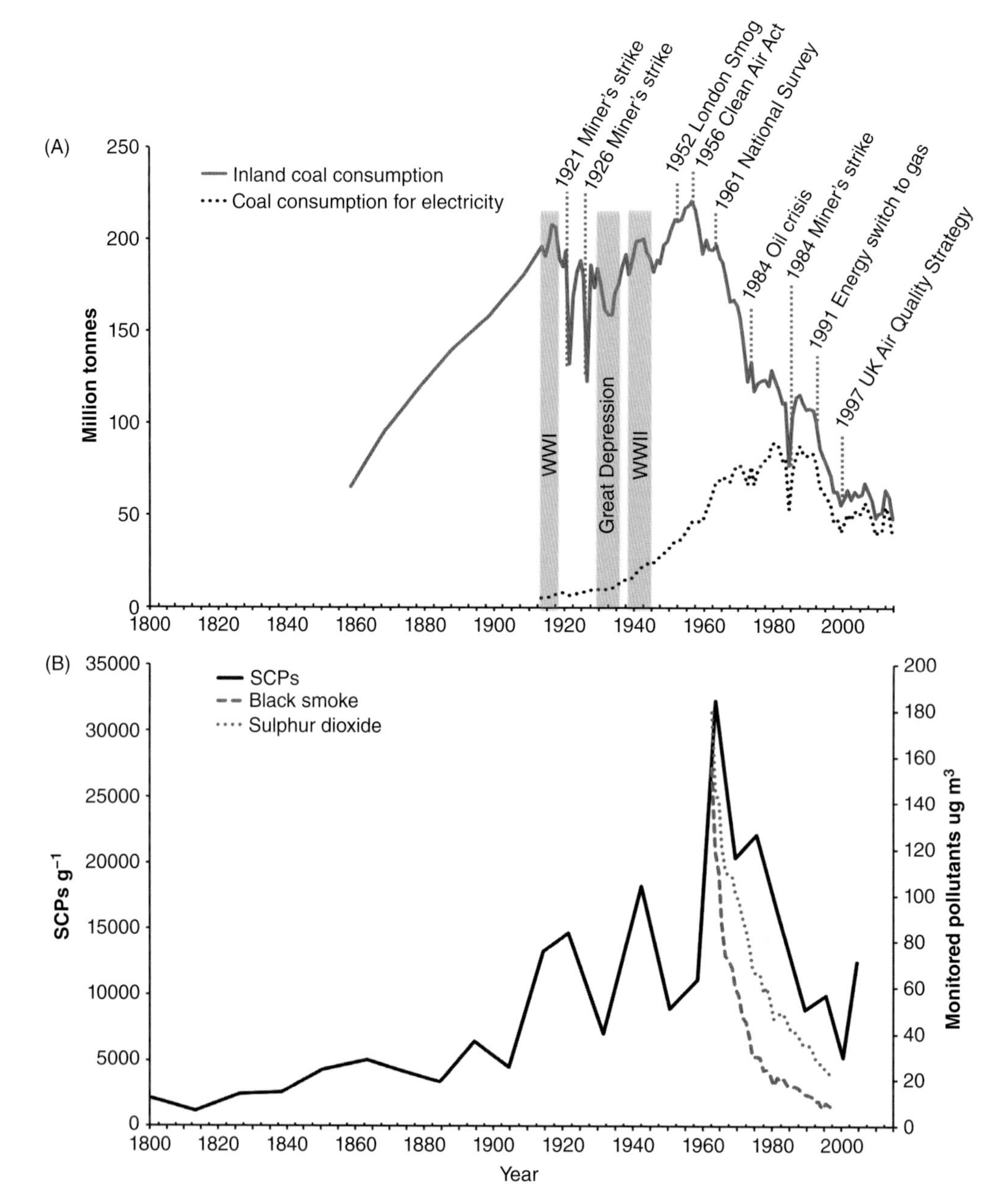

Figure 2.3 A history of coal consumption and air pollution in the United Kingdom.
A: Total coal consumption and coal used for electricity production (http://uk–air.defra.gov.uk (1) and http://www.carbonbrief.org (2)) with key industrial and legislative events superimposed.
B: Historical national averages of sulphur dioxide and black smoke measurements for the United Kingdom recorded post 1961 as part of the National Survey (3) are presented with spheroidal carbonaceous particle (SCP) concentrations recorded in an urban pond from NW England (authors' data). The start of the SCP record occurs during the nineteenth century, followed by rapid increases during the twentieth century, mid-twentieth century peaks and subsequent decline to present day.
(1) http://uk–air.defra.gov.uk/assets/documents/repo rts/cat05/0408161000_Defra_AQ_Brochure_2004_s.pdf
(2) http://www.carbonbrief.org/uk–coal–use–to–fall–to–lowest–level–since–industrial–revolution)
(3) http://www.gov.uk/government/statistical–data–sets/historical–coal–data–coal–prodcution–availability–and–coal–consumption–1853–to–2014

1961 using light reflectance of filter papers and acid titration techniques, respectively (http://uk–air.defra.gov.uk/networks/brief–history). The British government considered the decline in coal smoke a triumph in pollution control, and that the urban pollution problem had been solved, claiming 'London air is as clean as that in many rural areas' (Department of the Environment, 1974).

SCP profiles from British lakes provide further evidence of the historical release of particles from high-temperature combustion of fossil fuels. Rapid increases in SCP concentrations during the mid-twentieth century and 1970–1980 peaks are observed across the United Kingdom, reflecting peak coal consumption and electricity generation by coal and oil-fired power stations. Subsequent declines highlight the effects of air quality legislations and particle removal from flue gases (Rose *et al.*, 1995; Rose *et al.*, 1999; Rose, 2001) (Figure 2.3). Corresponding increases in hematite and magnetite in peat bogs and lakes from Europe and the United States are indicative of IAS deposition from the combustion of coal. Initial post-1800 increases mark the Industrial Revolution (as seen in SCP records), with further enhancement post 1950 (Oldfield *et al.*, 2014; Oldfield, 2015). Coal-fired power stations also release trace metals including Hg, As, Se, Pb, and Cu, and polycyclic aromatic hydrocarbons (PAHs) into the environment. Increases in these pollutants are identified in recent lake sediments corresponding with the operation of regional coal burning power plants (Goldberg *et al.*, 1981; Griffin and Goldberg, 1983; Donahue *et al.*, 2006).

2.5.3 Post-1950 Urban Pollution – A Complex Signal

The post-1950 period marks the 'Great Acceleration', a time of unprecedented global increases in urban population, energy and water use, coupled with corresponding shifts in earth system indicators including CO_2, NOx and methane levels, surface temperature, and ocean acidification (Steffen *et al.*, 2015). Rapid increases in fossil fuel combustion, industrial activity, vehicle use, and urbanization during the mid-to-late twentieth century is reflected globally in sediment archives by pronounced post-1950 increases in soot and PAH (Han *et al.*, 2016), persistent organic pollutants (POPs) (Bigus *et al.*, 2014), elevated increases in metals (Tylmann, 2004), and prominent SCP concentrations within environmental archives from Asia, North Africa, Southern Hemisphere, North America, and Greenland (Rose, 2015).

Urban air pollution episodes were no longer necessarily directly linked to a single, visible source such as factory chimneys. Instead, modern urban air pollution is invisible, derived from a complexity of stationary and mobile sources with local and long-range pollutants (Brimblecombe, 2005). As well as sulphurous smogs such as those experienced in London, photochemical smogs were also becoming a common occurrence in the urban environment, formed as a secondary pollutant from chemical transformations in the atmosphere as a complexity of vehicle and industrial pollution sources react with sunlight. Los Angeles (LA) is famed for its photochemical smogs, owing to its unique basin topography, heavy traffic intensities, and industrial emissions (Brimblecombe, 2014). Increasingly stringent emission standards and vehicle technology have resulted in an improvement in LA's air quality over the past 20 years (Pollack *et al.*, 2010); however, modern smogs still occur globally, and the importance of low-level long-term exposure to fine 'invisible' $PM_{2.5}$ has emerged as a major environmental health risk, even at low levels (Kelly and Fussell, 2015).

Ambient $PM_{2.5}$ is responsible for 3.7 million premature deaths worldwide (Lelieveld *et al.*, 2015). Anthropogenic sources of $PM_{2.5}$ include the combustion of coal, oil and biomass for power generation and industrial processes, diesel and petrol for road and air travel, and domestic fuel, as well as fugitive dusts from industry, construction, agriculture, quarrying, and non-exhaust particles from road traffic (AQEG, 2012). The composition, shape, size and consequent toxicity of ambient $PM_{2.5}$ may vary in different urban areas over time, due to changing industrial activities (chemical, petrochemical, metal, mineral, waste, nuclear and radioactive industries) and the introduction of new pollutants, patterns of fuel consumption and composition, transport infrastructure, and the implementation of pollution control legislations (Charlesworth *et al.*, 2011; Heim and Schwarzbauer, 2013; Lewis and Maslin, 2015; Rose, 2015).

Although industrialization in Europe and the United States started during the mid-eighteenth century, the Industrial Revolution was not globally uniform. For example, in China, rapid industrialization did not occur until the establishment of the People's Republic of China in 1949–1950 and is continuing today. China's reliance on coal burning for energy (47% global coal consumption in 2012), combined with heavy industry, soil dusts, densely populated cities, road transport and agricultural burning practices, has resulted in severe air quality degradation from $PM_{2.5}$ pollution in mainland China (Zhang *et al.*, 2014). Resultant smogs are a common occurrence in China's cities (Kulmala, 2015). The most polluted megacity in the world is currently Delhi, India, with the highest recorded $PM_{2.5}$ levels derived from a complex mix of coal, gas, diesel and biomass combustion, and industrial and construction dust (Subramanian, 2016; WHO, 2016).

2.6 Industrial Emissions

2.6.1 Metals

Metals are predominantly released into the atmosphere from various industrial processes, mining, and combustion of fossil fuels and fuel additives (Nriagu and Pacyna, 1988; Morwaska and Zhang, 2002), making them reliable tracers of human activities, especially Cd, Cu, Hg, Ni, Pb and Zn (Nriagu, 1996). Some metals have known toxicological effects (e.g. Pb, Cd and Hg), and high metal loadings are observed within $PM_{2.5}$ (Al-Rajhi *et al.*, 1996; Allen *et al.*, 2001), highlighting inhalation as an important pathway for human exposure to metals in the urban environment.

As technology has advanced, different types of metals have been utilized, resulting in the releases of a range of metals from mining, metallurgical, and manufacturing processes (Artiola, 1996; Gao *et al.*, 1996; Alloway and Ayres, 1997). The establishment of specific industries can immediately impact the local environment with, for example, Hg enrichment detected in lake sediments coinciding with the operation of nearby chlor alkali chemical plants (Kemp *et al.*, 1976), Pb enhancement associated with local waste incineration activities (Chillrud *et al.*, 1999), distinct phases of metal enrichment related to mining operations (Couillard *et al.*, 2004; 2008), and Zn and Cu detected in recent peat sediments from local steelworks (Gilbertson *et al.*, 1997). Furthermore, relatively recent industrial processes such as the recycling of scrap metal and electronic waste is potentially releasing metallic dust reflecting the composition of alloys (Fe, Cr, Cu, Co, Mn, and Ni), surface coatings (As, Cd, Cr, Pb, Hg, and Se) and electronic devices (containing rare earth elements such as Au and Y) into the urban environment (Raun *et al.*, 2013). Stationary fossil fuel combustion sources are the main source of Cr, Hg, Mn, Sb, Se, Sn, and Tl (coal) and Ni and V (oil) in the atmosphere with non-ferrous metal production being the largest source of As, Cd, Cu, and Zn (Pacyna and Pacyna, 2001).

Global atmospheric metal emissions generally demonstrate increases in the twentieth century, with overall declines observed post 1970 due to pollution controls (Harland *et al.*, 2000; Mahler *et al.*, 2006). Despite these declines, values are still elevated above pre-industrial levels (Norton *et al.*, 1992; Connor and Thomas, 2003). Metal emissions vary regionally, with declines observed in Europe and the United States, where legislative controls are relatively strict; yet increases are observed into the late twentieth century in, for example, Asia, which is experiencing extensive industrialization (Pirrone *et al.*, 1996), confirmed by continued metal enhancement in sedimentary archives from Japan (Kuwae *et al.*, 2013) and China (Yuan *et al.*, 2011).

2.6.2 Persistent Organic Pollutants

Technological advancements post WWII shaped the chemical industry, with the synthesis of organic chemical compounds and petrochemicals. Diverse organic products including fertilizers, plastic, polymers, pesticides and pharmaceuticals, dyestuffs, chemicals, printing materials, paints, polymers, chlorine, chlorinated solvents, fluorinated refrigerants, aerosols, fire retardants, insecticides and pesticides expanded throughout the twentieth century. The industrial synthesis of organic compounds resulted in the production and release of POPs (Table 2.2), well known for their persistence and bioaccumulation in the environment, biomagnification in the food chain, and toxicity to humans (Nadal *et al.*, 2011; Qiu, 2013; Odabasi *et al.*, 2015).

Rachel Carson's publication *Silent Spring* in 1962 raised initial concerns regarding POPs, detailing the detrimental environmental consequences, death of wildlife, and cancer-causing potential of synthesized pesticides, including DDT (dichlorodiphenyltrichloroethane) (Carson, 1962). DDT had been indiscriminately applied in the United States post WWII; however, it was banned in most industrialized countries by the 1970s. Despite the subsequent restricted use of

Table 2.2 Description of persistent organic pollutants.

Type of POP	Description
Organochlorine pesticides (OCPs)	A broad group of chemicals extensively used between the 1940s and 1960s in US and EU agriculture and mosquito (malaria) control. Include aldrin, dieldrin, endrin, lindane, methoxychlor, endosulfans and DDT. The majority of OCPs have been restricted or banned in most countries (Botella *et al.*, 2004).
Polycyclic aromatic hydrocarbons (PAHs)	A group of >100 chemicals containing two or more benzene rings, with both natural – biomass combustion, forest fires and volcanoes – and anthropogenic – sewage discharge, fossil fuel combustion, coal and coke production, and high-temperature combustion processes including oil refineries, incineration of waste and road traffic sources (Slezakova *et al.*, 2010; Huang *et al.*, 2012; Bigus *et al.*, 2014).
Polychlorinated biphenyls (PCBs)	PCBs have a molecular formula $C_{12}\,H_{10-x}\,Cl_x$ containing between one to five chlorine atoms. Commercially produced between the 1930s and 1980s (Bigus *et al.*, 2014), the majority of PCB uses were banned internationally in 1986 (EU directive 96/59/EC). PCBs were released by chemical factories, smelters, incinerators, pesticides and wood treatments and used for plasticizers in paints, plastics and rubber, pigments, dyes, electrical transfer equipment (such as cable insulation), heat insulation (fire retardants, fibre glass and foam), oil used in motors and fluorescent lights. PCBs are classified as human carcinogens (IARC, 2016). Although banned, PCBs still persist in the environment, animals and humans.
Dioxins	Dioxins, polychlorinated dibenzodioxins (PCDDs) and polychlorinated dibenzofurans (PCDFs) are highly stable and toxic POPs. Produced as unwanted by-products of industrial processes including the production of chorine and organochlorine chemicals, paper bleaching, waste incineration and smelting and also generated naturally by forest fires and volcanoes.

POPs, they are still prevalent in the environment and in human tissue and fluids (Botella *et al.*, 2004; Ruzzin, 2012; Grindler *et al.*, 2015) with exposure mainly occurring via the food chain (Nakata *et al.*, 2002). The most dangerous organic compounds are restricted in terms of production and use under the Stockholm Convention on POPs (2001).

The synthesis of organic chemicals began in the 1920s and rapidly increased post WWII, making them characteristic pollutants of the twentieth century (El-Shahawi *et al.*, 2010). This is reflected by the detection of POPs post the 1930s in a range of river, lake, and estuarine sediments (Andersson *et al.*, 2014; Bigus *et al.*, 2014; Arinaitwe *et al.*, 2016; Lorgeoux *et al.*, 2016). Their trans-boundary, long-range transport is confirmed by POP contamination detected in the Arctic (Evenset *et al.*, 2007; Ma *et al.*, 2015; Yuxin *et al.*, 2015).

The production and use of POPs varies in different regions. Generally, peak concentrations are mainly influenced by the industrial manufacture and use of PAHs post 1930, with expansion to include PCBs and OCPs post 1950, for example, the manufacture of brominated fire retardants after 1953 (Lebeuf and Nunes, 2005; Evenset *et al.*, 2007). POP releases were enhanced by the effects of urbanization (Feo *et al.*, 2011; Lorgeoux *et al.*, 2016), increased fossil fuel combustion, and traffic emissions (Boonyatumanond *et al.*, 2007).

Wartime (WWII) gunfire and bombing (Frignani *et al.*, 2005; Zhang *et al.*, 2009), the use of the herbicide Agent Orange in Vietnam (1961–1971), and nuclear weapons testing (post 1945) (Quiroz *et al.*, 2005) are also associated with increased PAHs and PCBs in the environment. The restricted use of POPs post 1970 resulted in a decline in releases (Hartmann *et al.*, 2005; Kim *et al.*, 2008); however, due to their environmental persistence, levels have not returned to pre-production values. Furthermore, recent increases in PCBs are derived from the recycling of electrical waste contaminated with PCBs (Yang *et al.*, 2012).

2.7 Transport

Improvements in transport infrastructure (construction of motorways and airport expansions), technological advancements in motor car and aviation technology, and increased numbers of road vehicles have occurred since the 1950s. Road transport is now a major mobile source of global urban pollution, and city airports are recognized as also impacting urban air quality. Railway stations and rail traffic are emerging as important sources of urban $PM_{2.5}$ due to diesel exhaust emissions (Chong *et al.*, 2015), and shipping activities in city harbours are a growing concern (Mueller *et al.*, 2011; Contini *et al.*, 2015); however, they have received little attention compared to road and air travel.

2.7.1 Road Transport

Changes in dominant Pb sources have occurred over time. Initial Pb enhancement from early metallurgy, to coal consumption and smelting of Pb metal ores, to the intensified use in motorcars post WWII and the addition of tetraethyl lead to petrol as an anti-knock agent (Heim and Schwarzbauer, 2013) are key phases identified by Pb isotope ratios and PAH enhancements in lake sediments (Vesleý *et al.*, 1993; Renberg *et al.*, 2002; Leorri *et al.*, 2014). Traffic-derived Pb during the late twentieth century exceeds previous Pb releases from metallurgy by a factor of three as indicated by South American ice cores (Eichler *et al.*, 2015). In the United Kingdom alone, there was an increase in motor vehicles from 2.3 million in 1930 to 15.0 million in 1970, rising to 34.2 million in 2008 (Office for National Statistics, 2010), with the distance travelled by cars and vans increasing by 613 billion passenger km from 1950 to 2007 (Table 2.3).

During the 1970s, the UK government believed that despite the unpleasant smell from vehicle emissions there was 'no evidence that present levels of emissions constitute a threat to health' (Department of the

Table 2.3 Distances (billion passenger km) travelled by people in the United Kingdom in 1952 and 2007 by mode of transport, showing percentage of total (Office for National Statistics, 2010).

	1952		2007	
Mode of transport	**Billion passenger km**	**%**	**Billion passenger km**	**%**
Bus and coach	92	42	50	6
Car	58	27	689	84
Rail	38	17	59	7
Air (domestic)	0.2	0.9	10	1
Total	218	100	817	100

Environment, 1974). Today, however, Pb is classified as a probable carcinogen (IARC, 2006) with a range of health effects associated with exposure including neurological development in children (Lidsky and Schneider, 2003). Pb has been reduced in ambient air in the United Kingdom via the updated Clean Air Act of 1974, a series of legislations (EC Directive 85/884/EEC, Air Quality Standards Regulations 1989, The Environment Act 1995, and EC Directive 78/611/EEC), limitations on motor fuel composition (The Motor Fuel Regulation 1981), and the introduction of unleaded petrol in 1985 (EC Directive 85/210/EEC). Leaded petrol is gradually being phased out of use globally; however, not until the early 2000s were there nationwide bans in China, Russia and India.

As well as CO, NOX, benzene, PAHs and hydrocarbons, PM emitted from road transport is a major component of urban pollution. The composition of road-derived PM varies from city to city and is dependent upon traffic volumes and vehicle fleet composition as well as local climate and geology (Pant and Harrison, 2013). Combustion particles are generally composed of fine carbonaceous aerosols ($PM_{2.5}$) including ultrafine particles (UFPs, <100 nm) containing organic carbon, elemental carbon, and smaller amounts of trace elements (Kumar *et al.*, 2014). Emissions from diesel and petrol/gasoline engines vary in composition, with diesel engines emitting a greater mass and number of UFPs compared to petrol/gasoline vehicles (Pant and Harrison, 2013).

Non-exhaust PM from road traffic is also an important source of metallic dust in the post-1950 urban environment due to increases in road transport. Tyre, break, clutch, and road wear contribute as much to ambient PM in cities as exhaust emissions (Amato *et al.*, 2014; Adachi and Tainosho, 2004). PM from beak wear is coarser than exhaust emissions (<6 μm) with a metallic signature including Fe, Cu, Zn, and Pb as well as minor contributions (<0.1 wt%) of Ba, Mg, Mn, Ni, Sn, Cd, Cr, Ti, K, and Sb (Grigoratos and Martini, 2014). Vehicle abrasion (e.g. of tyres and galvanized car body coatings) is reported to be a major contributor of Zn in the urban environment (Thorpe and Harrison, 2008). The physical movement of vehicles on roadways also re-suspends road dust, which includes asphalt and crustal material, into the atmosphere (Thorpe and Harrison, 2008). Platinum group elements are distinct metallic tracers of vehicle emissions since the 1980s, as modern cars have been equipped with catalytic converters to reduce CO emissions releasing Pt (platinum), Pd (palladium), and Au (gold) into the urban environment (Varrica *et al.*, 2003). This is reflected by post 1990 Pt, Pd, and Rh (rhodium) enhancements in urban lake sediments (Boston, Massachusetts, United States) (Rauch *et al.*, 2004).

2.7.2 Air Transport

Airports are important hubs of air pollution, especially those located in major cities that are visited by millions of passengers each day. Air transport has rapidly increased since the 1950s with the rise of international travel. In the United Kingdom, there was a 100-fold increase in passengers travelling though UK airports from <200,000 passengers in 1950 to 211,000,000 in 2010 (Rutherford, 2011). Airports generate heavy road traffic from visitors and require maintenance equipment, facilities, and fuel depots on the ground (Whitelegg and Williams, 2000), and have been shown to impact particle deposition up to 16 km downwind (Hudda *et al.*, 2014).

Aircraft emit combustion particles, mainly comprising carbonaceous UFPs with S, Cl, and K (Mazaheri *et al.*, 2013). Jet engine exhaust particles are also reported to contain a range of heavy metals including Be, Co, Cu, Pb, V, and Zn derived from metal impurities in aviation turbine fuel (Mazaheri *et al.*, 2013), confirmed by metal enhancement in coastal wetland samples in close proximity to airports (Boyle, 1996). Pb additives (tetraethyl lead) are still applied to aviation fuel (also known as Avgas or aviation gasoline) to achieve high-octane fuel for general aircraft with piston engines (Ebbinghaus and Wiesen, 2001) and are therefore an important modern atmospheric source of Pb since industrial and road traffic sources have been steadily phased out. Combusted aviation residues recorded in lake sediments (Graney and Eriksen, 2004) highlight airports as potential sources of metals, especially Pb, to the urban environment post 1950. Abrasion particles from tyre and break wear from takeoff and landings also contribute to larger (>1 μm) particles containing Ba, Cu, Sb, Mo, and Zn, and Al flakes from aircraft bodies deposited adjacent to aircraft runways (Amato *et al.*, 2010; Mazaheri *et al.*, 2013).

2.8 Conclusions

Air pollution has existed since the rise of early civilizations, and changes in the composition and concentrations of air pollution across the globe have been reconstructed using environmental archives. Histories of urban air pollution vary from place to place, over time. Generally, industrialization in Europe and the United States fuelled by coal combustion increased urban pollution during the nineteenth century. Unprecedented increases in pollutants have been experienced globally since the mid-twentieth century with advances in industrial technologies, extensive urbanization and the rise of car travel and global transport. Although air pollution has declined overall due to the introduction of pollution controls (post the 1960s), levels are still elevated above pre-industrial values. However, pollution deposition is increasing in parts of the world currently experiencing extensive industrial development. Urban air pollution remains a major environmental concern. The nature of air pollution has altered considerably over time: visible air quality has generally improved; however, the types of pollutants have changed so that invisible atmospheric contamination derived from a complexity of sources now pose new problems, both to human health and global environmental quality.

References

Adachi, K. and Tainosho, Y. (2004) Characterization of heavy metal particles embedded in tire dust. *Environment International*, 30, 1009–1017.

Air Quality Expert Group (2012) Fine Particulate Matter ($PM_{2.5}$) in the United Kingdom. DEFRA, available from http://uk–air.defra.gov.uk/assets/documents/

reports/cat11/1212141150_AQEG_Fine_Particulate_Matter_in_the_UK.pdf (June 2016).

Al-Rajhi, M.A., Al-Shayer, S.M., Seaward, M.R.D., and Edwards, H.G.M. (1996) Particle size effect for heavy metal pollution analysis of atmospherically deposited dust. *Atmospheric Environment*, 30,145–153.

Allen, (1906) Some Founders of the Chemical Industry. Sherrat and Hughes.

Allen, A.G., Nemitz, E., Shi, J.P., Harrison, R.M., and Greenwood, J.C. (2001) Size distributions of trace metals in atmospheric aerosols in the United Kingdom. *Atmospheric Environment*, 35, 4581–4591

Alloway, B.J. and Ayers, D.C. (1997) *Chemical Principles of Environmental Pollution.* Blackie Academic and Professiona, London.

Amato, F., Moreno, T., Pandolfi, M., Querol, X., Alastuey, A., Delgado, A., Pedrero, M., and Cots, N. (2010) Concentrations, sources and geochemistry of airborne particulate matter at a major European airport. *Journal of Environmental Monitoring*, 12, 854–862.

Amato, F., Cassee, F.R., Denier van der Gon, H.A.C., Gehrig, R., Gustafsson, M., Hafner, W., Harrison, R.M., Jozwicka, M., Kelly, F.J., Moreno, T., Prevo, A.S.H., Schaap, M., Sunyer, J., and Querol, X. (2014) Urban air quality: The challenge of traffic non-exhaust emissions. *Journal of Hazardous Materials*, 275, 31–36.

Andersson, M., Klug, M., Eggen, O.A., and Ottesen, R.T. (2014) Polycyclic aromatic hydrocarbons (PAHs) in sediments from lake Lille Lungegårdsvannet in Bergen, western Norway; appraising pollution sources from the urban history. *Science of the Total Environment*, 470, 1160–1172.

Appelby, P.G. and Oldfield, F. (1978) The calculation of lead-210 dates assuming a constant rate of supply of unsupported ^{210}Pb to the sediment. *Catena*, 5, 1–8.

Appleby, P.G. (1993) Forward to the lead-210 dating anniversary series. *Journal of Paleolimnology*, 9, 155–160.

Appleby, P.G. (2008) Three decades of dating recent sediments by fallout radionuclides: A review. *The Holocene*, 18, 83–93.

Arinaitwe, K., Rose, N.L., Muir, D.C.G., Kiremire, B.T., Balirwa, J.S., and Teixeira, C. (2016) Historical deposition of persistent organic pollutants in Lake Victoria and two alpine equatorial lakes from East Africa: Insights into atmospheric deposition from sedimentation profiles. *Chemosphere*, 144, 1815–1822.

Artiola, J.F. (1996) Industrial sources of pollution in Pepper, I.L., Gerba, C.P., and Brusseau, M.L. (eds.), *Pollution Science.* Academic Press, London, pp. 267–277.

Austen, K. (2015) Pollution Patrol. *Nature*, 517, 136–138.

Bell, M.L., Davis, D.L., and Fletcher, T. (2004) A retrospective assessment of mortality from the London smog episode of 1952: The role of influenza and pollution. *Environmental Health Perspectives*, 112, 6–8.

Bigus, P., Tobiszewski, M., and N Namieśnik, J. (2014) Historical records of organic pollutants in sediment cores. *Marine Pollution Bulletin*, 78, 26–42.

Boonyatumanond, R., Wattayakorn, G., Amano, A., Inouchi, Y., and Takada, H. (2007) Reconstruction of pollution history of organic contaminants in the upper Gulf of Thailand by using sediment cores: First report from Tropical Asia (TACO) project. *Marine Pollution Bulletin*, 54, 554–565.

Borsos, E., Makra, L., Béczi, R., Vitányi, B., and Szentpéteri, M. (2003) Anthropogenic air pollution in the ancient times. *Acta Climatologica et Chorologica*, 36, 5–15.

Botella, B., Crespo, J., Rivas, A., Cerrillo, I., Olea-Serrano, M.F., and Olea, N. (2004) Exposure of women to organochlorine pesticides in southern Spain. *Environmental Research*, 96, 34–40.

Brännvall, M.-L., Bindlwe, R., and Renberg, I. (1999) The medieval metal industry was the cradle of modern large-scale atmospheric lead pollution in northern Europe. *Environmental Science and Technology*, 33, 4391–4395.

Brimblecombe, P. (1987) *The Big Smoke.* Methuen, London.

Brimblecombe, P. (2005) The globalization of local air pollution. *Globalization*, 2, 429–411.

Brimblecombe, P. (2014) Deciphering the chemistry of Los Angeles smog 1945–1995. In Flemmng, J.R. and Johnson, A. (eds.), *Toxic Airs: Body, Place, Planet in Historic Perspective.* University of Pittsburgh Press, Pittsburgh, pp. 95–108.

Carson, R. (1962) *Silent Spring.* Houghton Mifflin, Boston.

Carter, D.H. (1964) *Mond Division.* ICI Magazine, pp. 2–7.

Charlesworth, S., De Miguel, E., and Ordóñez, A. (2011) A review of the distribution of particulate trace elements in urban terrestrial environments and its application to considerations of risk. *Environmental Geochemistry and Health*, 33, 103–123.

Chillrud, S.N., Bopp, R.F., Simpson, H.J., Ross, J.M., Shuster, E.L., Chaky, D.A., Walsh, D.C., Choy, C.C., Tolley, L.-R., and Yarme, A. (1999) Twentieth century atmospheric metal fluxes into Central Park Lake, New York City. *Environmental Science and Technology*, 33, 657–662.

Chong, U., Swanson, J.J., and Boies, A. (2015) Air quality evaluation of London Paddington train station. *Environmental Research Letters*, 10, 1–11.

Connor, S.E. and Thomas, I. (2003) Sediments as archives of industrialisation: Evidence of atmospheric pollution in coastal wetlands of southern Sydney, Australia. *Water, Air and Soil Pollution*, 149, 189–210.

Contini, D., Donateo, A., Gambaro, A., Argiriou, A., Melas, D., Cesari, D., Poupkou, A., Karagiannidis, A., Tsakis, A., Merico, E., Cesari, R., and Dinoi, A. (2015) Impact of ship traffic to PM2.5 and particle number concentrations in three port-cities of the Adriatic/Ionian area. International *Journal of Chemical, Nuclear, Materials and Metallurgical Engineering*, 9, 529–534.

Couillard, Y., Cattaneo, A., Gallon, C., and Courcelles, M. (2008) Sources and chronology of fifteen elements in the sediments of lakes affected by metal deposition in a mining area. *Journal of Paleolimnology*, 40, 97–114.

Couillard, Y., Courcelles, M., Cattaneo, A., and Wunsam, S. (2004) A test of the integrity of metal records in sediment cores based on the documented history of metal contamination in Lac Dufault (Québec, Canada). *Journal of Paleolimnology*, 32, 149–162.

Department of the Environment (1974) *Clean Air Today*. Her Majesty's Stationary Office, London.

Diggle, Rev. G.E. (1961) A History of Widnes. Corporation of Widnes.

Dingle, A.E. (1982) 'The monster nuisance of all': Landowners, alkali manufacturers and air pollution 1828–64. *Economic History Review*, 35, 529–548.

Donahue, W.F., Allen, E.W., and Schindler, D.W. (2006) Impacts of coal-fired power plants on trace metals and polycyclic aromatic hydrocarbons (PAHs) in lake sediments in central Alberta, Canada. *Journal of Paleolimnology*, 35, 111–128.

Ebbinghaus, A. and Wiesen, P (2001) Aircraft fuels and effect upon engine emissions. *Air and Space Europe* 3, 101–103.

Eichler, A., Gramlich, G., Kellerhals, T., Tobler, L., and Schwikowski, M. (2015) Pb pollution from leaded gasoline in South America in the context of a 2000-year metallurgical history. *Science Advances*, 1, doi:10.1126/sciadv.1400196

El-Shahawi, M.S., Hamza, A., Bashammakh, A.S., and Al-Saggaf, W.T. (2010) An overview on the accumulation, distribution, transformations, toxicity and analytical methods for the monitoring of persistent organic pollutants. *Talanta*, 80, 1587–1597.

Evelyn, J. (1661) Fumifugium; or *The inconveniency of the aer and smoake of London dissipated.* G. Bedd and T. Collins, London.

Evenset, A., Christensen, G.N., Carroll, J., Zaborska, A., Berger, U., Herzke, D., and Gegor, D. (2007) Historical trends in persistent organic pollutants and metals recorded in sediment from Lake Ellasjøen, Bjørnøya, Norwegian Arctic. *Environmental Pollution*, 146, 196–205.

Feo, M.L., Sprovieri, M., Gherardi, S., Sammartino, S., and Marsella, E. (2011) Polycyclic aromatic hydrocarbons and polychlorinated biphenyls in the harbour of Naples (Southern Italy): time and spatial distribution patterns. *Environmental Monitoring and Assessment*, 174, 445–459.

Frignani, M., Bellucci, L.G., Favotto, M., and Albertazzi, S. (2005) Pollution historical trends as recorded by sediments at selected sites of the Venice Lagoon. *Environment International*, 31, 1011–1022.

Gao, N., Hopke, P.K., and Reid, N.W. (1996) Possible sources for some trace elements found in airborne particles and precipitation in Dorset, Ontario. *Air and Waste Management Association*, 46, 1035–1047.

Gilbertson, D.D., Grattan, J.P., Cressey, M., and Pyatt, F.B. (1997) An air pollution history of metallurgical innovation in iron and steel making: a geochemical archive of Sheffield. *Water, Air and Soil Pollution*, 100, 327–334.

Goldberg, E.D., Hodge, V.F., Griffin, J.J., and Minoru, K. (1981) Impact of fossil fuel combustion on the sediments of Lake Michigan. *Environmental Science and Technology*, 15, 466–471.

Goodarzi, F. and Sanei, H. (2009) Plerosphere and its role in reduction of emitted fine fly ash particles from pulverized coal-fired power plants. *Fuel*, 88, 382–386.

Graney, J.R. and Eriksen, T.M. (2004) Metals in pond sediments as archives of anthropogenic activities: a study in response to health concerns. *Applied Geochemistry*, 19, 1177–1188.

Griffin, J.J. and Goldberg, E.D. (1983) Impact of fossil fuel combustion on the lake sediments of Lake Michigan: A reprise. *Environmental Science and Technology*, 17, 244–245.

Grigoratos, T. and Martini, G. (2014) *Non-Exhaust Traffic Related Emissions. Brake and Tyre Wear PM.* Joint Research Centre, the European Commission, Publications Office of the European Union, Luxembourg.

Grindler, N.M., Allsworth, J.E., Macones, G.A., Kannan, K., Roehl, K.A., and Cooper, A.R. (2015) Persistent organic pollutants and early menopause in US women. *PLOS ONE* DOI:10.1371/journal.pone.0116057

Han, Y.M., Wei, C., Huang, R.-J., Bandowe, B.A.M., Ho, S.S.H., Cao, J.J., Jin, Z.D., Xu, B.Q., Gao, S.P., Tie, X.X., An, Z.S., and Wilcke, W. (2016) Reconstruction of atmospheric soot history in inland regions from lake sediments over the past 150 years. *Scientific Reports*, 6, 19151.

Hardie, D.W.F. (1950) *A History of the Chemical Industry in Widnes*. Imperial Chemicals Industries Limited, General Chemicals Division.

Harland, B.J., Taylor, D., and Wither, A. (2000) The distribution of mercury and other trace metals in the sediments of the Mersey Estuary over 25 years 1974–1998. *Science of the Total Environment*, 253, 45–62.

Hartmann, P.C., Quinn, J.G., Cairns, R.W., and King, J.W. (2005) Depositional history of organic contaminants in Narragansett Bay, Rhode Island, USA. *Marine Pollution Bulletin*, 50, 388–395.

Heim, S. and Schwarzbauer, J. (2013) Pollution history revealed by sedimentary records: A review. *Environmental Chemistry Letters*, 11, 255–270.

Hillman, A. L., Abbott, M.B., Yu, J., Bain, D.J., and Chiou-Peng, T. (2015) Environmental legacy of copper metallurgy and Mongol silver smelting recorded in Yunnan lake sediments. *Environmental Science and Technology*, 49, 3349–3357.

Huang, W., Wang, Z., and Yan, W. (2012) Distribution and sources of polycyclic aromatic hydrocarbons (PAHs) in sediments from Zhanjiand Bay and Leizhou Bay, South China. *Marine Pollution Bulletin*, 64, 1962–1969.

Hudda, N., Gould, T., Hartin, K., Larson, T.V., and Ruin, S.A. (2014) Emissions form n international airport increase particle number concentrations 4-fold at 10 km downwind. *Environmental Science and Technology*, 48, 6628–6635.

Hunt, A., Abraham, J.L., Judson, B., and Berry, C.L. (2003) Toxicologic and epidemiologic clues from the characterization of the 1952 London smog fine particulate matter in archival autopsy lung tissues. *Environmental Health Perspectives*, 111, 1209–1214.

IARC (2006) Monographs on the evaluation of carcinogenic risks to humans. Volume 87: Inorganic and organic lead compounds. World Health Organization, Lyon, France.

IARC (2016) Polychlorinated biphenyls and polybrominated biphenyls. Volume 107. IARC Monographs on the evaluation of carcinogenic risks to humans, International Agency for Research on Cancer, Lyon France.

Jones, A.D. (1969) *Industry & Runcorn 1750 to 1960.* Publicity & Information services Department. Halton Borough Council (January 1969).

Kelly, F.J. and Fussell, J.C. (2015) Air pollution and public health: Emerging hazards and improved understanding of risk. *Environmental Geochemistry and Health,* 37, 631–649.

Kemp, A.L.W., Thomas, R.L., Dell, C.I., and Jaquet, J.-M. (1976) Cultural Impact on the Geochemistry of Sediments in Lake Erie. *Journal of the Fisheries Research Board of Canada*, 33, 440–462.

Kim, Y.-S., Eun, H., and Katase, T. (2008) Historical distribution of PCDDs, PCDFs, and coplanar PCBs in sediment core of Ariake Bay, Japan. *Archives of Environmental Contamination and Toxicology*, 54, 395–405.

Kulmala, M. (2015) China's chocking cocktail. *Nature*, 526, 497–499.

Kumar, P., Morawska, L., Birmili, W., Paasonen, P., Hu, M., Kulmala, M., Harrison, R.M., Norford, L., and Britter, R. (2014) Ultrafine particles in cities. *Environment International*, 66, 1–10.

Kuwae, M., Tsugeki, N.K., Agusa, T., Toyoda, K., Tani, Y., Ueda, S., Tanabe, S., and Urabe, J. (2013) Sedimentary records of metal deposition in Japanese alpine lakes for the last 250 years: Recent enrichment of airborne Sb and In in East Asia. *Science of the Total Environment*, 442, 189–197.

Lebeuf, M. and Nunes, T. (2005) PCBs and OCPs in sediment cores from the lower St. Lawrence Estuary, Canada: evidence of fluvial inputs and time lag in delivery to coring sites. *Environmental Science and Technology*, 39, 1470–1478.

Lee, C.S.L., Qi, S.-H., Zhang , G., Luo , C.-L., Zhao, L.Y.L., and Li, X.-D. (2008) Seven thousand years of records on the mining and utilization of metals from lake sediments in Central China. *Environmental Science and Technology*, 42, 4732–4738.

Lelieveld, j., Evans, J.S., Fnais, M., Giannadaki, D., and Pozzer, A. (2015) The contribution of outdoor air pollution sources to premature mortality on a global scale. *Nature*, 525, 367–371.

Leorri, E., Mitra, S., Irabien, M.J., Zimmerman, A.R., Blake, W.H., and Cearreta, A. (2014) A 700 year record of combustion-derived pollution in northern Spain: Tools to identify the Holocene/Anthropocene transition in coastal environments. *Science of the Total Environment*, 470, 240–247.

Lewis, S.L. and Maslin, M.A. (2015) Defining the Anthropocene. *Nature* 519, 171–180.

Lidsky, T.I. and Schneider, J.S. (2003) Lead neurotoxicity in children: Basic mechanisms and clinical correlates. *Brain* 126, 5–19.

Lorgeoux, C., Moilleron, R., Gasperi, J., Ayrault, S., Bonté, P., Lefèvre, I., and Tassin, B. (2016) Temporal trends of persistent organic pollutants in dated sediment cores: Chemical fingerprinting of the anthropogenic impacts in the Seine River basin, Paris. *Science of the Total Environment*, 541, 1355–1363.

Ma, Y., Halsall, C.J., Crosse, J.D., Graf, C., Cai, M., He, J., Gao, G., and Jones, K. (2015) Persistent organic pollutants in ocean sediments from the North Pacific to the Arctic Ocean. *Journal of Geophysical Research: Oceans*, 2723–2735.

Mahler, B., Van Metre, P.C., and Callender, E. (2006) Trends in metals in urban and reference lake sediments across the United States, 1970 to 2001. *Environmental Toxicology and Chemistry*, 25, 1698–1709.

Marx, S.K., Kamber, B.S., McGowan, H.A., and Zawadzki, A. (2010) Atmospheric pollutants in alpine peat bogs record a detailed chronology of industrial and agricultural development on the Australian continent. *Environmental Pollution*, 158, 1615–1628.

Mazaheri, M., Bostrom, T.E., Johnson, G.R., and Morawska, L. (2013) Composition and morphology of particle emissions from in-use aircraft during takeoff and landing. *Environmental Science and Technology*, 21, 5235–5242.

McConnell, R.J., Maselli, O.J., Sigl, M., Vallelonga, P., Neumann, T., Anschütz, H., Bales, R.C., Curran, M.A.J., Das, S.B., Edwards, R., Kipfstuhl, S., Layman, S., and Thomas, E.R. (2014) Antarctic-wide array of high-resolution ice core records reveals pervasive lead pollution began in 1889 and persists today. *Scientific Reports*, 4, 5848.

MARC (1985) *Historical Monitoring.* Monitoring and Assessment Research Centre, University of London.

Morwaska, L. and Zhang, J. (2002) Combustion sources of particles.1: Health relevance and source signatures. *Chemosphere*, 49, 1045–1058.

Mueller, D., Uibel, S., Takemura, M., Klingelhoefer, D., and Groneberg, D.A. (2011) Ships, ports and particulate air pollution – An analysis of recent studies. *Journal of Occupational Medicine and Toxicology*, 6, 31–36.

Nadal, M., Schuhmacher, M., and Domingo, J.L. (2011) Long-term environmental monitoring of persistent organic pollutants and metals in a chemical/petrochemical area: Human health risks. *Environmental Pollution*, 159, 1769–1777.

Nemery, B., Hoet, P.H.M., and Nemmar, A. (2001) The Meuse Valley fog of 1930: An air pollution disaster. *The Lancet*, 357, 704.

Norton, S.A., Beinert, R.W., Binford, M.W., and Kahl, J.S. (1992) Stratigraphy of total metals in PIRLA sediment cores. *Journal of Paleolimnology*, 7, 191–214.

Nriagu, J.O. (1996) A history of global metal pollution. *Science*, 272, 223–224.

Nriagu, J.O. and Pacyna, J.M. (1988) Quantitative assessment of worldwide contamination of air, water and soils by trace metals. *Nature*, 333, 134–139.

Odabasi, M., Cetin, B., and Bayram, A. (2015) Persistent organic pollutants (POPs) on fine and coarse atmospheric particles measured at two (urban and industrial) sites. *Aerosol and Air Quality Research*, 15, 1894–1905.

Office for National Statistics (2010) Social Trends 40: Chapter 12: Transport available from http://webarchive.nationalarchives.gov.uk/20160108204507/http://www.ons.gov.uk/ons/rel/social-trends-rd/social-trends/social-trends-40/index.html.

Oldfield, F. (2014) Can the magnetic signatures from inorganic fly ash be used to mark the onset of the Anthropocene? *The Anthropocene Review*, 1–11.

Oldfield, F., Gedye, S.A., Hunt, A., Jones, J.M., and Richardson, N. (2015) The magnetic record of inorganic flay ash deposition in lake sediments and ombrotrophic peats. *The Holocene*, 25, 215–225.

Pacyna, J.M. and Pacyna, E.G. (2001) An assessment of global and regional emissions of trace metals to the atmosphere from anthropogenic sources worldwide. *Environmental Reviews*, 9, 269–298.

Pant, P.H. and Harrison, R.M. (2013) Estimation of the contribution of road traffic emissions to particulate matter concentrations from field measurements: A review. *Atmospheric Environment*, 77, 78–97.

Pirrone, N., Keeler, G.L., and Nriagu, J.O. (1996) Regional differences in worldwide emissions of mercury to the atmosphere. *Atmospheric Environment*, 30, 2981–2987.

Qiu, J. (2013) Organic pollutants poison the roof of the world. *Nature* 11 April 2013 doi:10.1038/nature.2013.12776

Quiroz, R., Popp, P., Urruita, R., Bauer, C., Araneda, A., Treutler, H.-C., and Barra, R. (2005) PAH fluxes in the Laja Lake of south central Chile Andes over the last 50 years: Evidence from a dated sediment core. *Science of the Total Environment*, 349, 150–160.

Rauch, S., Hemond, H.F., and Peucker-Ehrenbrink, B. (2004) Recent changes in platinum group element concentrations and osmium isotopic composition in sediments from an urban lake. *Environmental Science and Technology*, 38, 396–402.

Raun, L., Pepple, K., Hoyt, D., Richner, D., Blanco, A., and Li, J. (2013) Unanticipated potential cancer risk near metal recycling facilities. *Environmental Impact Assessment Review*, 41, 70–77.

Renberg, I., Brännvall, M.L., Bindler, R., and Emteryd, O. (2002) Stable lead isotopes and lake sediments – A useful combination for the study of atmospheric lead pollution history. *Science of the Total Environment*, 292, 45–54.

Roebroeks, W. and Villa, P. (2011) On the earliest evidence for habitual use of fire in Europe. *PNAS*, 106, 5209–5214.

Rose, N.L. (2001) Fly-ash particles. In Last, W.M. and Smol, J.P. (eds.), *Tracking Environmental Change Using Lake Sediments. Volume 2: Physical and Geochemical Methods.* Dordrecht, The Netherlands, Kluwer Academic Publishers.

Rose, N.L. (2015) Spheroidal carbonaceous fly ash particles provide a globally synchronous stratigraphic marker for the Anthropocene. *Environmental Science and Technology*, 4155–4162.

Rose, N.L. and Appleby, P.G. (2005) Regional applications of lake sediment dating by spheroidal carbonaceous particle analysis I: United Kingdom. *Journal of Paleolimnology*, 34, 349–361.

Rose, N.L., Harlock, S., Appleby, P.G., and Battarbee, R.W. (1995) Dating of recent lake sediments in the United Kingdom and Ireland using spheroidal carbonaceous particle (SCP) concentration profiles. *The Holocene*, 5, 328–335.

Rose, N.L., Juggins. S., and Watt. J. (1999) The characterisation of carbonaceous fly-ash particles from major European fossil-fuel types and applications to environmental samples. *Atmospheric Environment*, 33, 2699–2713.

Rose, N.L., Rose, C.L., Boyle, J.F., and Appleby, P.G. (2004) Lake-sediment evidence for local and remote sources of atmospherically deposited pollutants on Svalbard. *Journal of Paleolimnology*, 31, 499–513.

Rosman, K.J.R., Chisholm, W., Hong, S., Candelone, J.-P., and Boutron, C.F. (1997) Lead from Carthaginian and Roman Spanish mines isotopically identified in Greenland ice dated from 600 B.C. to 300 A.D. *Environmental Science and Technology*, 31, 3413–3416.

Rutherford, T. (2011) Air Transport Statistics. House of Commons Library SN/SG/3760.

Ruzzin, J. (2012) Public health concern behind the exposure to persistent organic pollutants and the risk of metabolic diseases. *BMC Public Health*, 12, 298–306.

Shi, L., Zanobetti, A., Kloog, I., Coull, B.A., Koutrakis, P., Melly, S.J., and Schwartz, J.D. (2016) Low concentration $PM_{2.5}$ and mortality: estimating acute and chronic effects in a population-based study. *Environmental Health Perspectives*, 124, 46–52.

Shotyk, W., Weiss, D., Appleby, P.G., Cheburkin, A.K., Frei, R., Gloor, M., Kramers, J.D., Reese, S. and Van Der Knaap, W.O. (1998) History of atmospheric lead deposition since 12,370 (14)C yr BP from a peat bog, Jura Mountains, Switzerland. *Science*, 281, 1635–1640.

Silva, R.A., West, J.J., Zhang, Y., Anenberg, S.C., Lamarque, J.-F., Shindell, D.T., Collins, W.J., Dalsoren, S., Faluvegi, G., Folberth, G., Horowitz, L.W., Nagashima, T., Naik, V., Rumbold, S., Skeie, R., Sudo, K., Takemura, T., Bergmann, D., Cameron-Smith, P., Cionni, I., Doherty, R.M., Eyring, V., Josse, B., MacKenzie, I.A., Plummer, D., Righi, M., Stevenson, D.S., Strode, S., Szopa, S., and Zeng, G. (2013) Global premature mortality due to anthropogenic outdoor air pollution and the contribution of past climate change. *Environmental Research Letters*, 8, 1–11.

Slezakova, K., Castro, D., Pereira, M.C., Morais, S., Delerue-Matos, C., and Alvim_Ferraz, C. (2010) Influence of traffic emissions on the carcinogenic polycyclic aromatic hydrocarbons in outdoor breathable particles. *Journal of the Air and Waste Management Association*, 60, 393–401.

Steffen, W., Broadgate, W., Deutsch, L., Gaffeny, O., and Ludwig, C. (2015) The trajectory of the Anthropocene: The Great Acceleration. *The Anthropocene Review*, 1–18.

Subramanian, M. (2016) Can Delhi save itself from its toxic air? *Nature* 534, 166–169.

Thevenon, F., Guédron, S., Chiaradia, M., Loizeau, J.-L., and Poté, J. (2011) (Pre-) historic changes in natural and anthropogenic heavy metals deposition inferred from two contrasting Swiss Alpine lakes. *Quaternary Science Reviews*, 30, 224–233.

Thorpe, A. and Harrison, R.M. (2008) Sources and properties of non-exhaust particulate matter from road traffic: A review. *Science of the Total Environment*, 400, 270–282.

Tylmann, W. (2004) Heavy metals in recent lake sediments as an indicator of 20th century pollution: Case study on lake Jasień. *Limnological Review*, 4, 261–268.

Uglietti, C., Gabrielli, P., Cooke, C.A., Vallelonga, P., and Thompsona, L.G. (2015) Widespread pollution of the South American atmosphere predates the industrial revolution by 240 y. *PNAS*, 112, 2349–2354.

Varrica, D., Dongarrà, G., Sabatino, G., and Monna, F. (2003) Inorganic geochemistry of roadway dust from the metropolitan area of Palermo, Italy. *Environmental Geology*, 44, 222–230.

Vesleý, J., Almquist-Jacobson, H., Miller, L.M., Norton, S.A., Appleby, P., Dixit, A.S., and Smol, J.P. (1993) The history and impact of air pollution at Certovo Lake, southwestern Czech Republic. *Journal of Plaeolimnology*, 8, 211–231.

Warren, K. (1980). *Chemical Foundations: The Alkali Industry in Britain to 1926*. Oxford, Clarendon Press.

Whitelegg, J. and Williams, N. (2000) *The Plane Truth; Aviation and the Environment.* London, The Ashden Trust & Transport 2000 Trust.

World Health Organization (2016) WHO's urban ambient air pollution database – Update 2016 available from http://www.who.int/phe/health_topics/outdoorair/databases/AAP_database_summary_results_2016_v02.pdf

Yang, H., Zhuo, S., Xue, B., Zhang, C., and Liu, W. (2012) Distribution, historical trends and inventories of polychlorinated biphenyls in sediments from Yangtze River Estuary and adjacent East China Sea. *Environmental Pollution*, 169, 20–26.

Yuan, G.-L., Chen, L., and Yang, Z. (2011) Inputting history of heavy metals into the inland lake recorded in sediment profiles: Poyang Lake in China. *Journal of Hazardous Materials*, 185, 336–345.

Zhang, P., Song, J., Fang, J., Liu, Z., Li, Z., and Yuan, H. (2009) One century record of contamination by polycyclic aromatic hydrocarbons and polychlorinated biphenyls in core sediments from the southern Yellow Sea. *Journal of Environmental Science*, 21, 1080–1088.

Zhang, D., Lui, J., and Li, B. (2014) Tackling air pollution in China – What do we learn from the Great Smog of 1950s in London? *Sustainability*, 6, 5322–5338.

3

Evolution of Air Quality Policy and Management in Urban Areas

Enda Hayes

Geography and Environmental Management, University of the West of England, United Kingdom

3.1 Introduction

Air pollution is intrinsically linked to the day-to-day practices, activities, and behaviours of citizens. History has shown that the public willingness to accept a particular amount of air pollution changes over time, and the air pollution experienced in urban areas is a consequence of a dynamic and evolving inter-relationship between the political, economic, and societal choices made. Subsequently, these choices made regarding air pollution have often lagged behind awareness and understanding of the science and impacts of air pollution.

This chapter explores the evolution of air quality policy and management approaches in urban areas focusing on 'traditional' air pollution such as nitrogen dioxide, particulate matter, sulphur dioxide, and ozone rather than the greenhouse gases more commonly associated with climate change. The sources of urban air pollution, the effects of air pollution particularly on public health, the historical and current legislative environment across Europe and in the United Kingdom, and the future challenges for the urban environment and subsequently the need for further evolution of air quality policies and management to meet these challenges will be discussed.

3.2 Sources of Urban Air Pollution

Air pollution can be considered as any substance in the atmosphere that is not naturally present, out of natural proportions, or can cause problems (i.e. impacts to health or sensitive ecosystems). The UK Air Quality Strategy, 2000, recognised that 'Clean air is an essential ingredient of a good quality of life. People have the right to expect that the air they breathe will not harm them' (DEFRA, 2000), but the short- and long-term outlook suggests that air pollution and its related ecosystems impacts and environmental health risks are partially on track to achieving these key policy targets. However, transport demand and related environmental impacts such as air pollution are largely not on track to achieve key policy targets (EEA, 2015a).

There are three 'types' of air pollution urban environments:

1) Primary pollution – a result of the combustion of fossil fuels including materials which pass from the fuel to the air (e.g. sulphur, metals), incomplete combustion (e.g. carbon monoxide, hydrocarbons), and the product of combustion (e.g. NO_x, CO_2)

Urban Pollution: Science and Management, First Edition. Edited by Susanne M. Charlesworth and Colin A. Booth.

2) Secondary pollution – this is not generally emitted directly but is predominantly formed through chemical reactions in the atmosphere (e.g. NO_2, O_3)
3) Natural pollution – this includes pollution that may originate outside of urban areas but impacts them through trans-boundary air mass movements by storms (e.g. Saharan dust storms), volcanoes (e.g. Eyjafjallajökull ash cloud), and fires (e.g. South East Asia biomass burning).

Natural sources include volcanoes, dust storms, soil erosion, forest fires, and vegetation. Anthropogenic sources such as fugitive releases, mechanically raised dust, and agriculture are relevant for urban areas as they can have an impact, but this chapter will primarily focus on sources in urban areas such as industry/power generation, transport, and domestic (heating and cooking) emissions.

3.3 Health Implications of Urban Air Pollution

The World Health Organization estimates that over 7 million premature deaths are caused by poor air quality globally, accounting for one in every eight deaths (WHO, 2014). In Europe, this equates to approximately 400,000 premature deaths per annum and an estimated cost of €1.7 trillion to the economy (EEA, 2015b). In 2013, the EU limit and target values for PM were exceeded in large parts of Europe while the annual limit value for NO_2 was widely exceeded across Europe particularly in locations close to roads. The EEA (2015b) reported that in 2013, 17% of the EU population was exposed to PM_{10} concentrations above the daily limit value while 61% were exposed to concentrations exceeding the WHO daily guideline for PM_{10}. In the case of NO_2, 90% of the EU population live in areas that exceed the annual EU limit value and the WHO annual guidelines (EEA, 2015b).

3.4 Historical Context of Air Quality Policy and Management

The evolution of air pollution policy and management approaches has in one respect changed substantially in recent years, but they have largely retained the general principles upon which air quality management is based. John Evelyn's pivotal *Fumifugium* (Evelyn, 1661) sets out the three key principles of air quality management that still apply today – smokeless fuel, fuel substitution, and the separation of polluting sources from receptors. During the Industrial Revolution, the application of these principles primarily focused on the management of air pollution at source and specifically considered industrial emissions from coal burning and cotton industries in the United Kingdom. These three key principles can still be found in modern air quality policies and legislations such as the UK Clean Air Act 1956 (HM Government, 1956), but the middle of the twentieth century saw the need for a new way of thinking as the most significant challenge for modern air quality management in urban areas emerged – the motor vehicle and personal mobility.

3.4.1 Towards Modern Air Quality Management in Europe

The emergence of the motor vehicle, the geographical location of sources, and the composition of air pollution changed in urban areas. It was recognised that a legislative change was required from a source control process to a system that placed greater emphasis on ambient air quality concentrations while also recognising the need for an effects-based approach to air quality management expressed in terms of air quality standards and management regimes.

International agreements such as the International Convention on Long Range Transboundary Pollution (1979) introduced controls to mitigate the transboundary effects of acid rain and limited emissions of acidifying pollutants; subsequent protocols have been well received and largely successful. In 1987, the World Health Organization set health-based guidelines for a number of pollutants including particulate matter, nitrogen dioxide, ozone and sulphur dioxide, among others, providing absolute thresholds for health exposure (WHO, 1987). On the basis of these WHO guidelines, the EU Air Quality Framework Directive (96/62/EC) (European Commission, 1996) and subsequent daughter directives were developed, which member states were required to translate into national legislation by 1998. Today, modern air quality legislation in Europe is governed by a substantial number of directives that influence air pollution through industrial, transport, fuel regulation, and so on, but primarily there are five historical, current, and future key directives, strategies, and policy packages which provide the framework for air quality management in urban areas:

1) The Thematic Strategy on Air Pollution (European Commission, 2005). This set objectives for reducing certain pollutants and reinforced the legislative framework for combating air pollution via two main routes: improving community environmental legislation and integrating air quality concerns into related policies. The strategy was one of the seven thematic strategies provided for in the Sixth Environmental Action Programme adopted in 2002 and was based on research carried out under the Clean Air For Europe (CAFE) Programme.
2) The Ambient Air Quality Directive (2008/50/EC) consolidated the 1996 Framework Directive and its first three daughter directives. The aim of the directive was to protect human health and the environment. This meant taking active measures to monitor the purity of ambient (or outside) air and removing any pollutants. To achieve this, the Ambient Air Quality Directive set out legally binding limit and target values for ground-level concentrations of a number of pollutants (see Table 3.1). The directive set provisions which oblige European member states to prepare and implement plans and programmes in the case of non-compliance. When a zone or agglomeration exceeds a limit value, the directive requires member states to take all necessary measures not entailing disproportionate costs (European Commission, 2008).
3) The Environment Impact Assessment Directive (85/337/EEC) prescribed a list of development projects for which an environmental assessment, including air quality, would be required, and that the outcomes of the assessment would be reported in the public domain. This directive was amended in 1997, 2003, and 2009; these amendments were codified in Directive 2014/52/EU (European Commission, 2014).
4) The National Emissions Ceiling Directive (2001/81/EC) aimed to reduce national emissions through the application of ceiling limits for certain atmospheric pollutants such as SO_2, NO_x, NMVOC, and NH_3. This directive was replaced by Directive 2016/2284/EU. These ceilings have to be met by member states and are supported by a range of sectoral measures such as fuel quality or product standards to control emissions from road transport, industry, agriculture, and so on (European Commission, 2016).
5) The Clean Air Policy Package 2030 includes measures to ensure that existing targets are met in the short term (2015, 2020); new air quality objectives for the period up to 2030 imply no exceedance of the WHO guideline levels for human health. Support measures helped cut air pollution, with a focus on improving air quality in cities, maintaining research and innovation, and promoting international cooperation. A revised NECD directive was brought in with stricter national emission ceilings for six main pollutants, provisions for black carbon, and a proposal for a new directive to reduce

Table 3.1 UK air quality objectives and EU limit and target values.

		United Kingdom	**Europe**
	Measured as	**Objective/to be achieved by**	**Objective/to be achieved by**
Nitrogen dioxide (NO_2)	1 hour mean	200 μg/m^3 not to be exceeded more than 18 times per annum to be achieved by 31/12/2005	200 μg/m^3 not to be exceeded more than 18 times per annum to be achieved by 01/01/2010
	Annual mean	40 μg/m^3 to be achieved by 31/12/2005	40 μg/m^3 to be achieved by 01/01/2010
Particulates (PM_{10})	24 hour mean	50 μg/m^3 not to be exceeded more than 35 times per annum to be achieved by 31/12/2004	50 μg/m^3 not to be exceeded more than 35 times per annum to be achieved by 01/01/2005
	Annual mean	40 μg/m^3 to be achieved by 31/12/2004	40 μg/m^3 to be achieved by 01/01/2005
Particulates ($PM_{2.5}$) exposure reduction	Annual mean	25 μg/m^3 to be achieved by 2020	Target value of 25 μg/m^3 to be achieved by 2010
	Annual mean	Target of 15% reduction in concentrations at urban background to be achieved between 2010 and 2020	Target of 20% reduction in concentrations at urban background to be achieved between 2010 and 2020
Ozone (O_3)	8 hour mean	100 μg/m^3 not to be exceeded more than 10 times per annum to be achieved by 31/12/20045	Target of 120 μg/m^3 not to be exceeded more than 25 times per annum averaged over 3 years to be achieved by 31/12/2010
Sulphur dioxide (SO_2)	15 minute mean	266 μg/m^3 not to be exceeded more than 35 times per annum to be achieved by 31/12/20045	–
Sulphur dioxide	1 hour mean	350 μg/m^3 not to be exceeded more than 24 times per annum to be achieved by 31/12/2004	350 μg/m^3 not to be exceeded more than 24 times per annum to be achieved by 01/01/2005
Sulphur dioxide	24 hour mean	125 μg/m^3 not to be exceeded more than 3 times per annum to be achieved by 31/12/2004	125 μg/m^3 not to be exceeded more than 3 times per annum to be achieved by 01/01/2005

pollution from medium-sized combustion installations. It is expected that the benefits to society will be more than 20 times the cost of implementing the legislation including avoiding 58,000 premature deaths each year by 2030 with an approximate saving of €550 million in healthcare costs and €2 billion due to fewer working days lost. Beyond the health benefits, the environmental benefits include the protection of almost 20,000 km^2 of forest from acid rain and substantial reductions in nitrogen pollution (European Commission, 2013).

From a policy and a management perspective, European legislation has been the driving force behind improvements in air quality in urban areas for the last two decades. These directives have led to air quality assessments being required for a wide range of new developments, including transport, industry, and energy generation emissions, and have encouraged the introduction of measures to reduce emissions. Effective mitigation measures to tackle air pollution in urban areas require action to be taken at many different scales – international, national, local, and individual. The legislative responses to air pollution also operate at different levels, but fitting these scales of governance together can be complex. The transboundary nature of air pollution means that effective and efficient management can only be attained through shared responsibility and ownership not only of the problem but also of the solutions. As such, it is the responsibility of each member state to transpose EU directives into their national legislation.

3.4.2 Towards Modern Air Quality Management in the United Kingdom

There have been some pivotal moments in urban air quality management and the evolution of modern air quality policies – most notable was the London Smog event of 1952. On 5 December, a slow-moving anticyclone resulted in a severe fog over London. The combination of the anticyclone, the fog, and pollution generated from industrial and domestic stacks and transport resulted in smog that lasted five days (Brimblecombe, 1987; NSCA, 1998). The number of deaths has subsequently been the subject of some debate, but the smog of 1952 produced increased respiratory mortality and morbidity not just during the five-day event but in subsequent weeks. Hansell *et al.* (2016) suggest that exposure to high air pollution concentrations carries long-term risks to health. Bell and Davis (2001) found that the mortality rates in London during and after the 1952 Smog increased by 50%–300% from the previous year with a death rate more than three times normal for this period, resulting in an estimated 3,000 deaths during the event and 12,000 in the months following.

The London Smog was the key driver for a number of Clean Air Acts (1956, 1968, 1993) following the establishment of the Beaver Committee in 1953. While the 1956 Clean Air Act focused on reducing smoke pollution, it actually reduced sulphur dioxide levels at the same time through the introduction of a number of mitigation strategies such as introducing smoke control areas; increasing electric and gas usage; decreasing the use of solid fuels; burning cleaner, low-sulphur coal; relocation of power stations to rural areas with tall chimney stacks to enhance dispersion; and the continued decline in heavy industry. These interventions also echo the air pollution control principles identified by Evelyn in Fumifugium (Evelyn, 1661).

In the United Kingdom, the primary legislation setting the comprehensive programme designed to meet air quality standards is the Environment Act, 1995 (part iV) (HM Government, 1995). This act was devised to respond to a multi-source, multi-pollutant, dynamic, and evolving air quality challenges and provided a radical shift from a source to an effects-based control system that included health-focused air quality objectives (Longhurst *et al.*, 2009). The key elements were as follows:

1) **Section 80:** This obliges the Secretary of State (SoS) to publish a National Air Quality Strategy.
2) **Section 82:** This requires local authorities to review air quality in their administrative area and assess whether the air quality standards and objectives are being achieved. Areas where standards fall short must be identified.
3) **Section 83:** This requires a local authority, for any area where air quality standards are not being met, to issue an order designating it an Air Quality Management Area (AQMA).
4) **Section 84:** This imposes duties on a local authority with respect to AQMAs. The local authority must draw up an action plan specifying the measures to be carried out and the time scale to bring air quality in the area back within limits.
5) **Section 87:** This provides the SoS with wide ranging powers to make regulations concerning air quality. These include standards and objectives, the conferring of powers and duties, the prohibition and restriction of certain activities or vehicles, the obtaining of information, the levying of fines and penalties, the hearing of appeals, and other criteria.

The result of the Environment Act has been the introduction and implementation of an efficient yet only partially effective Local Air Quality Management process. The process has been efficient in that clear guidance, templates, reporting timelines, and support mechanisms have been provided to local authorities. It has been effective to the point of identification of problem areas with >700 geographically distinct AQMAs being identified, predominantly for exceedances of the nitrogen dioxide annual mean objective and related to road transport (Longhurst *et al.*, 2016). However, the LAQM process is not effective in solving identified air quality hot spots due to a perceived lack of political will, a lack of funding, and of clarity regarding responsibilities and integration into other agendas such as transport and health (Chatterton *et al.*, 2007; Olowoporoku *et al.*, 2010; Brunt *et al.*, 2016). Despite some streamlining of process reporting and modifications of air-quality-objective timescales, values and/or exceedance limits, LAQMs have remained largely unchanged since their inception (Barnes *et al.*, 2014).

The UK government uses the Automatic Urban and Rural Network (a monitoring network) and the Pollution Climate Mapping model to assess air quality against EU limit values set out in the Ambient Air Quality Directive (2008/50/EC). In 2013, the United Kingdom was exceeding EU limit values for the NO_2 annual mean at 31 of its 43 zones and agglomerations (38 excluding the Margin of Tolerance). In compliance with the directive, the United Kingdom is required to produce an Air Quality Plan to tackle this problem. The 2015 plan, which was heavily criticised, acknowledged that full compliance would not be achieved until 2025 with exceedances after 2020 predicted in six zones and agglomerations (Birmingham, Leeds, Southampton, Nottingham, Derby, and London). To address this, the UK government has suggested a new approach to air quality management through the introduction of Clean Air Zones which will target the most polluting vehicles through a charging mechanism should they enter city centres (Longhurst *et al.*, 2016).

3.5 Future Urban Challenges

The total global population is increasing and is expected to reach 9.6 billion by 2050, with an estimated 70% living in towns and cities (United Nations, 2015). These global trends of population growth coupled with rapid urbanisation and rising living standards are placing increasingly competitive demands upon finite natural resources for agriculture, energy, and industrial production. If these trends continue, by 2050 water demand is projected to increase by 55%, energy demand is projected to increase by 80%, and food demand is projected to increase by 60% with the growing global middle classes being the main consumers of electricity, oil, food, beverages, household appliances, cars, and other goods and services, suggesting an increasing disproportionately large demand on natural resources (Rockström *et al.*, 2009). This rapidly urbanising and industrialising world, with its rapidly growing world population, demand for more energy, a global population travelling more and requiring personal mobility, globally interconnected aspiring to Western levels of goods, services, and comfort is a recipe for increased emissions and poor air quality. Recent scientific discoveries regarding the application of vehicle emission factors, vehicle fleet dynamics, and the need to consider human behaviour illustrates that current legislation, or more accurately, the interpretation and implementation of current legislation, may not fit the purpose.

3.5.1 Current Vehicle Emission Factors

Emission factors are rates of pollutant emissions for a specified year, road type, vehicle speed, and vehicle fleet composition and are used to assess the contribution of vehicles to air pollution. Across Europe, the COPERT software tool, coordinated by the EEA (http://emisia.com/products/copert-4), is used to calculate air pollutant and greenhouse gas emissions from road transport. Some member states use this tool in their national-level modelling to determine compliance against the Ambient Air Quality Directive (2008/50/EC) with the broadly acknowledged assumption that emissions from Euro 6 cars are 2.8 times higher than the EU standard, based on testing a small selection (ca. 6 cars) of early models in 2013. These regulations will become more stringent over time and define limits for exhaust emissions of new light-duty vehicles sold in the EU and EEA (European Economic Area). Recent analysis by the Dutch research organization, TNO, suggested that COPERT (a software tool used worldwide to calculate air pollutant and greenhouse gas emissions from road transport) vehicle emission factors for Euro 6 vehicles (which are unlikely to dominate the national fleets until at least 2025) are too low and likely to be between 30% and 50% higher for urban driving conditions. Williams *et al.* (2016) reviewed the literature on real-world emission testing studies and concluded that vehicle emissions from Euro 6 vehicles were on average 5–7 times greater than type approval results suggested. Considering this evidence, and in light of the 2016 revelation of vehicle manufacturers (e.g. Volkswagen) using defect devices to achieve compliance with type approval tests, it is evident that more robustness and accountability is required in the current transport sector and related vehicle emission legislation.

3.5.2 Vehicle Fleet Dynamics

In developing air quality management plans, national governments always consider the national vehicle fleet turnover. This refers to the introduction of new 'cleaner' vehicles into the national fleet and the removal of older 'dirty' vehicles. The issue with this assumption is that the air quality impact assessment is often rather simplistic in its application. Chatterton (2016), identified a number of underlying assumptions that should be considered:

- The distances travelled by different vehicle classes
- The volume of pollution each vehicle class is responsible for
- The robustness of the assumptions regarding vehicle scrappage rates and vehicle upgrading
- The vehicle owners' willingness to pay to upgrade their vehicles to more modern and cleaner technology (i.e. the public's willingness to pay principle)

3.5.3 Human Behaviour

Traditional approaches to air quality management have always taken a very technocratic approach, and subsequently the legislation, national policies, and mitigation strategies are biased towards addressing air pollution challenges through technology-oriented solutions without considering human behaviour. This is particularly evident in the way air pollution is source apportioned. For example, transport pollution may be apportioned by fuel type (petrol, diesel, LNG, etc.), vehicle type (cars, buses, HGVs, etc.), or Euro standard (Euro 4, 5, 6). The failure to factor in people and their daily activities, behaviours, and practices in air quality legislation has led to the over-reliance on technological innovation to the detriment of social innovation, with insufficient consideration given to how societies and cities can change the way they operate and function. Future cities, its citizens, and their demands will be unlike what they are today; therefore, to truly understand how to meet the air quality challenges, it is necessary to consider what, when, and why citizens use energy and understand how end-uses are changing. The answer to these questions can lead to a better understanding of the air pollution at both point of generation and point of exposure while also providing a clearer delineation of city citizen ownership and responsibility to both air quality challenges and more importantly, solutions.

3.5.4 Environmental Justice

Air pollution also needs to be contextualised in the debate around environment justice and the emerging understanding of current and generational social inequity. Concepts such as energy decadence (the profligate use of energy where wealth and circumstance allow high energy consumption through choice) indicates the presence of a strong relationship between levels of poverty and levels of resource consumption: resource consumption and resource availability decrease as poverty indices increase. These inequalities increase the pressure on – and importance of – decision-makers making informed and balanced choices not only for natural resources but also in working towards achieving Millennium Development Goals and the more recent Sustainable Development Goals (2015). The argument for a parallel air quality management approach that considers both traditional hotspots and wider exposure reduction approaches needs to be considered especially as it draws comparison with the universal proportionalism concept in the public health debate.

3.6 Conclusions

There is no quick-fix solution to the current air quality management challenge in urban areas; rather what is required is to start thinking outside of the 'silo' for a clean, low-carbon, healthy future, a future with shared ownership not only of the air pollution challenges but, more importantly, the solutions and the impact that individuals' choices can have on stressed global systems. Addressing these growing challenges requires a proactive legislative approach that is not only proportionate to the scale of the public health challenge but is also scientifically robust.

References

Barnes, J.H., Hayes, E.T., Chatterton, T.J., and Longhurst, J.W.S. (2014) Air quality action planning: Why do barriers to remediation in local air quality management remain? *Journal of Environmental Planning and Management*, 57 (5), 660–681.

Bell, M.L. and Davis, D.L. (2001) Reassessment of the lethal London fog of 1952: Novel indicators of acute and chronic consequences of acute exposure to air pollution. *Environmental Health Perspectives*, 109 (Suppl 3), 389.

Brimblecombe, P. (1987) The Big Smoke. Routledge, 185 pp.

Brunt, H., Barnes, J., Longhurst, J., Scally, G., and Hayes, E.T. (2016) Local Air Quality Management policy and practice in the UK: The case for greater public health integration and engagement. *Environmental Science & Policy*, 58, pp. 52–60. ISSN 1462–9011.

Chatterton, T.J., Longhurst, J.W.S., Leksmono, N.S., Hayes, E.T., and Symons, J.K. (2007) Ten years of Local Air Quality Management experience in the UK: An analysis of the process. *Clean Air and Environmental Quality*, 41 (2), 26–31.

Chatterton, T. (2016) A Review of the Incorporation of Vehicle Fleet Dynamics within Defra's Final Air Quality Plan. Consultancy Report for Gatwick Airport Ltd.

DEFRA (2000) The Air Quality Strategy for England, Scotland, Wales and Northern Ireland. Department of the Environment, Transport and Regions, Scottish Executive, National Assembly for Wales, Department of Environment for Northern Ireland. CM 4548. London, TSO.

EEA (2015a) The European Environment – State and Outlook 2015: Synthesis Report. European Environment Agency, Copenhagen.

EEA (2015b) Air Quality in Europe – 2015 Report. European Environment Agency, Copenhagen, Publications Office of the European Union. Luxembourg.

European Commission (1996) Council Directive on Ambient Air Quality Assessment and Management (96/62/EC).

European Commission (2005) Thematic Strategy on Air Pollution, COM(2005)446.

European Commission (2008) Council Directive on Ambient Air Quality and Cleaner Air for Europe (2008/50/EC).

European Commission (2013) Clean Air Policy Package Press Release, Brussels, 18 December 2013. http://europa.eu/rapid/press–release_IP–13–1274_en.htm

European Commission (2014) Council Directive on the Assessment of the Effects of Certain Public and Private Projects on the Environment (2014/52/EC).

European Commission (2016) Council Directive on the Reduction of National Emissions of Certain Atmospheric Pollutants (2016/2284/EU).

Evelyn, J. (1661) Fumifugium. http://www.gyford.com/archive/2009/04/28/www.geocities.com/Paris/LeftBank/1914/fumifug.html

Hansell, A., Ghosh, R.E., Blangiardo, M., Perkins, C., Vienneau, D., Goffe, K., Briggs, D., and Gulliver, J. (2016). Historic air pollution exposure and long-term mortality risks in England and Wales: Prospective longitudinal cohort study. *Thorax*, doi:10.1136/thoraxjnl–2015–207111, 2016

HM Government (1956) Clean Air Act, 1956, Chapter 52. London, HMSO.

HM Government (1995) Environment Act 1995, Chapter 25 (part IV). London, HMSO.

Longhurst, J.W.S., Irwin, J.G., Chatterton, T.J., Hayes, E.T., Leksmono, N.S., and Symons, J.K. (2009) The development of effects based air quality management regime. *Atmospheric Environment*, 43 (1), 64–78.

Longhurst, J., Barnes, J., Chatterton, T., Hayes, E.T., and Williams, B. (2016) Progress with air quality management in the 60 years since the UK Clean Air Act, 1956: Lessons, failures, challenges and opportunities. *International Journal of Sustainable Development & Planning*, 11 (4), 491–499. ISSN 1743–7601, 2016

NSCA (1998) Clearing the Air. 100 Years of the National Society for Clean Air and Environmental Protection. NSCA, Brighton. 22 pp.

Olowoporoku, A.O., Hayes, E.T., Leksmono, N. S., Longhurst, J.W.S., and Parkhurst, G. (2010) A longitudinal study of the links between Local Air Quality Management and Local Transport Planning policy processes in England. *Journal of Environmental Planning and Management*, 53 (3), 385–403.

Rockström, J., Steffen, W., Noone, K., Persson, Å., Chapin III, F.S., Lambin, E., Lenton, T.M., Scheffer, M., Folke, C., Schellnhuber, H., Nykvist, B., De Wit, C.A., Hughes, T., van der Leeuw, S., Rodhe, H., Sörlin, S., Snyder, P.K., Costanza, R., Svedin, U., Falkenmark, M., Karlberg, L., Corell, R.W., Fabry, V.J., Hansen, J., Walker, B., Liverman, D., Richardson, K., Crutzen, P., and Foley, J. (2009) Planetary boundaries: Exploring the safe operating space for humanity. *Ecology and Society*, 14 (2), 32.

United Nations (2015) World Population Prospects: Key Findings and Advance Tables, Department of Economic and Social Affairs, New York. https://esa.un.org/unpd/wpp/publications/

Williams, B., Barnes, J., Chatterton, T., Hayes, E.T., and Longhurst, J. (2016) A critical review of the robustness of the UK governments air quality plan and expected compliance dates. *WIT Transactions on Ecology and the Environment*, 207, 1–9. ISSN 1743–3541

World Health Organization (1987) Air Quality Guidelines for Europe. Copenhagen, WHO Regional Office for Europe, 1987 (WHO Regional Publications, European Series, No. 23).

World Health Organization (2014) *Health and the Environment: Addressing the Health Impact of Air Pollution*. http://apps.who.int/gb/ebwha/pdf_files/EB136/B136_15–en.pdf.

4

UK and EU Water Policy as an Instrument of Urban Pollution

Anne-Marie McLaughlin, Susanne M. Charlesworth, and Stephen J. Coupe

Centre for Agroecology, Water and Resilience, Coventry University, United Kingdom

Acronyms used in This Chapter

AA	annual average
BOD	biochemical oxygen demand
CLEA	Contaminated Land Exposure Assessment Model
CSO	combined sewer overflow
DEFRA	Department for Environment, Food and Rural Affairs
DFT	Department for Transport
DWI	Drinking Water Inspectorate
EA	Environment Agency
EQS	environmental quality standard
MAC	maximum allowable concentration
NEP	new and emerging pollutant
PAP	particulate associated pollutant
PCP	personal care product
PPS	pervious paving system
PS	priority substance
RBMP	River Basin Management Plan
RST	run-off specific threshold
SGV	soil guideline value
SPP	Scottish Planning Policy
SuDs	sustainable drainage system
TEC	total electrical conductivity
TSS	total suspended solids
UKTAG	UK Technical Advisory Group
UWWTD	Urban Waste Water Treatment Directive
WFD	Water Framework Directive

4.1 Introduction

There are three major water challenges that water policy aims to control: water scarcity, abundance of water, and pollution. In terms of controlling pollution to receiving waterbodies, policy regulates using the different frameworks set out in Table 4.1. Water in the environment naturally contains particulate matter, dissolved compounds, and organisms, but its quality is affected by both human and natural processes, and the resultant impacts have become a global issue. There are therefore no catch-all quality standards for 'the environment' as a whole; this is reflected in separate requirements for individual freshwater environments, such as surface water and groundwater, and specific environments such as trout and salmon rivers. There are also separate standards for so-called 'transitional' waters (i.e. estuaries, brackish water, etc.) as well as the coast, and many other sub-categories covering the term 'water quality'. When defining the water quality of discharge into the environment, it is therefore the *receiving environment* which is the driver. Hence, it is difficult to separate out 'urban' specifically from the plethora of guidelines covering pollution of the aquatic environment; this chapter will therefore cover those considered to be the most relevant.

Urban Pollution: Science and Management, First Edition. Edited by Susanne M. Charlesworth and Colin A. Booth.

Table 4.1 Definitions for different frameworks to regulate water policies.

Legislation	Legislative frameworks, such as the EU Water Framework Directive (WFD), are essential to establish requirements for successful water quality management (European Commission, 2012). In the United Kingdom, the Environment Agency (UK EA) is responsible for the enforcement of regulations derived from these legislative frameworks.
Regulation	EU Directives have been implemented in the United Kingdom through regulation by statutory instruments. This is to ensure that institutions in charge (such as the UK EA) meet these requirements through planning and appropriate systems.
Policy	The intention to implement a procedure or protocol.
Guidelines	Guidelines, similar to recommendations developed by the UK Technical Advisory Group (UKTAG), assist in the implementation of the WFD (Defra, 2014). Guidelines can be used if there are no other statutory values to comply with.
Standards	Standards have been set by the United Kingdom to meet the objectives of the WFD. UKTAG have defined standards for water quality as numerical limits on concentrations of chemicals or measurements of biological communities (WFD, 2014). Standards are needed to ensure institutions can calculate limits for pollutants in waterbodies without causing harm to the environment or public health (Defra, 2014).

4.2 Determining Water Quality

Water quality is described in terms of sets of determinands which include both physical and chemical properties (collectively: physicochemical), such as pH, temperature, turbidity, total electrical conductivity (TEC), chemical pollutants, and microbiological parameters such as the numbers of bacteria in a specified volume of water. In terms of potential pollutants, it is important to consider the *form* in which they are carried. Some are readily dissolved, such as nutrients (nitrate, nitrite), but many of the more toxic elements are preferentially transported in association with particulates (particulate associated pollutants or PAPs), for example, the heavy metals (e.g. Zn, Ni, Pb, Cd, Cu). Currently, the risk of PAPs associated with sediment in waterbodies, for example, whether suspended or deposited, is assessed using a total acid digestion, which breaks down the whole of the sediment liberating all the absorbed and bound elements into solution, referred to as 'total concentrations'. These concentrations are then compared with Contaminated Land Exposure Assessment Soil Guideline Values, or CLEA SGVs (ICRCL, 1987; Cole and Jeffries, 2009); trigger concentrations; or other relevant guideline concentrations to assess their potential to be a hazard to the environment. However, it is unlikely that all of these elements are released under normally changing environmental conditions, and not all of them are likely to be available to impact human, or environmental health. Thus, methods have been refined to focus on *bioavailable* concentrations, unlike other methods that were considered too stringent as they focused on total concentrations. Bioavailability gives an estimate of the amount of potentially toxic elements capable of being incorporated into the body via ingestion, inhalation, or dermal absorption and is usually divided into those which are known carcinogens (e.g. Hg, As) and those which are not (De Miguel *et al.*, 2012).

Measurement of turbidity or total suspended solids (TSS) can give an assessment of contamination by PAPs, particularly in surface run-off from urban areas, and their potential to be transported to the receiving environment. For example, traffic-related sources could include the main heavy metals as listed above, but also platinum group elements (PGEs) associated with catalytic convertors, for which there are currently no standards. The standards for traffic-related deposition of oil and grease as hydrocarbons on the road surface generally require there to be no visible film on the surface of the water, and no odour of petrol or diesel. There are therefore a great deal of inter-related determinands to take into account, and also many environments defined by the standards to which, individually and collectively, various determinands apply. Water policies should be consistent across all governing bodies on both a national and international scale. It is therefore important to engage key stakeholders, alongside regulatory parties, such as policy makers, to ensure that unauthorised and undesirable pollution does not occur.

WHO (2011) states that if national standards for water quality regulations are not stringent enough, then the drinking water guidelines should be referred to, particularly in catchments used for the abstraction of potable water. EQS Directive 2008/105/EC and WFD 2000/60/EC (Section 4.3.2) also argue that catchments used for the abstraction of water should be as close to standards suitable for drinking water as possible to save on treatment costs.

The following sections address water policy at the UK level, complicated by different implementation in England, Scotland, Wales, and Northern Ireland, and across the EU. An example of the application of UK policy using sustainable drainage systems (SuDS) is then given to illustrate this complexity.

4.3 UK Water Policy

The discharge of polluting substances into controlled waterbodies is an offence according to The Water Resources Act 1991 and the Control of Pollution Act 1974. Regulation 3(1) in Schedule 21 of the Environmental Permitting Regulations 2010 (Defra, 2010) defines water discharge activity as that entering inland freshwaters, coastal waters, or territorial waters (not groundwater) of any pollutant matter, waste matter, or trade effluent. Environmental quality standards (EQSs) from the applicable directives should be applied to fulfil the Environmental Permitting Regulations 2010 (Defra, 2010). Guidelines to reduce pollution entering watercourses from non-domestic buildings follow Pollution Prevention Guidelines to ensure the implementation of legislation preventing the discharge of pollutants into waterbodies. Additionally, key legislation and policies through the Department for Transport (DFT) ensures that road drainage is not detrimental to the water environment include the WFD (2000/60/EC), Groundwater Daughter Directive (2006/118/EC), PPS25 (England), SPP7 (Scotland), PPS15 (Northern Ireland), and TAN15 (Wales) (DFT, 2009). Although the Shellfish Directive, PPS25 (England), Freshwater Fish Directive, and Dangerous Substances Directive have been revoked, and replaced with the National Planning Policy Framework (DCLG, 2012), their objectives will continue to be incorporated into River Basin Management Plans (RBMPs) through the WFD.

4.3.1 The EU Water Framework Directive

The EU WFD (2000/60/EC) drives the protection of freshwater in the United Kingdom; through it, all waterbodies have a requirement for no deterioration in status and to reach 'good ecological status' or 'good ecological potential' by 2015 through active management. The WFD does not define 'Good'; rather, this is achieved through associated directives, such as Drinking Water, Bathing Waters and Groundwater, implemented through national legislation. The WFD was initially implemented into UK law through The Water Environment (Water Framework Directive) (England and Wales) Regulations 2003, The Water Environment (Water Framework Directive) (Northern Ireland) Regulations 2003, The Water Environment and Water Services (Scotland) Act 2003, and The Water Environment (Controlled Activities) (Scotland) Regulations 2005. The regulations, standards, and guidelines needed to take into account site-specific factors. For example, threshold values for groundwater needed to allow for drainage of treated sewage if the site was near a sewage treatment works. Although specific values are nationally implemented, local authorities need to be consulted for any strictly local requirements.

The Natural Environment White Paper (Defra, 2014) suggested that 32% of waterbodies by 2015 and the majority by 2027 would be of good ecological status. However, Table 4.2 gives an indication of the status that waterbodies across the United Kingdom have achieved up to 2015, showing that only groundwater in Northern Ireland was close to 100% of 'good' status. There are specific circumstances under Articles 4 and 5 of the WFD which will allow for the 2015 time limit to be extended or allows the application of alternative environmental objectives. As a result of concerns related to the WFD, the United Kingdom has implemented RBMPs to include statutory objectives to improve water quality for the 2015–2021 period. This approach is also being used due to growing concern for increased urban pollution as climate change and the expectation of three million new homes built in England and Wales by 2020 (DCLG, 2007) could exacerbate water pollution issues.

Table 4.2 WFD status of UK waterbodies (Priestley, 2015).

Country in the United Kingdom	Surface waterbodies achieving 'good' or 'high' status	Year	Reference
England	20%	2015	Defra (2016)
Scotland	52%	2013	SEPA (n.d.)
Northern Ireland	22% of rivers and streams 14% of lakes 97% of groundwater	2012	Northern Ireland EA (2014)
Wales	42%	2014	National Assembly for Wales (2015)

An issue in the United Kingdom is the management structure of the water sector as there are multiple national companies responsible for water resources, dependant on location, and a range of other water management bodies (e.g. Environment Agency; SEPA), which can complicate the implementation of water policies. Table 4.3 establishes the different regulators assigned to monitor water quality of waterbodies in the United Kingdom and their most recently reported status.

4.3.2 Drinking Water Standards

The Water Supply (Water Quality) Regulations (2010) established requirements for drinking water in England and Wales to be 'wholesome', defined as water that does not contain any micro-organism or substance that can potentially harm human health. Schedule 1 of The Water Supply (Water Quality) Regulations (2010) was adopted from the Drinking Water Directive. Examples of the 26 chemical parameters are given in Table 4.4. There are 9 other requirements such as taste, odour, and colour and 12 'indicator parameters' including conductivity, total organic carbon, and turbidity.

If the national standards are insufficient, or if there are no guidelines for certain chemicals, the WHO (2011) drinking guidelines should be referenced. Guidelines for certain substances are still under consideration as their concentrations currently remain below values that could potentially affect human health. The more stringent guidelines for groundwater (particularly for sources used for drinking water abstraction) and bathing waters can therefore be used in this case.

To achieve 'wholesome' water, according to the most recent regulations, very few to no micro-organisms or parasites should be present in drinking water. However, the Water Supply (Water Quality) Regulations 2000 (amended in 2007) omitted any requirement

Table 4.3 Regulation, regulators, and latest reported status of waterbodies in the United Kingdom.

	Regulations	Regulator	Latest reported status of waterbodies
England	Water Environment (WFD England and Wales) Regulations 2003 (SI 2003 No. 3242) (as amended)	Environment Agency	20% of surface waterbodies (25% in 2010), 27% of lakes, and 26% of estuaries and coastal waterbodies were classified as being of 'good' status in 2015.[a]
Wales	Water Environment (Water Framework Directive England and Wales) Regulations 2003 (SI 2003 No. 3242) (as amended)	Natural Resources Wales	42% of waterbodies in Wales achieved 'good' ecological status in 2014.[b]
Scotland	Water Environment and Water Services (Scotland) Act 2003 (as amended)	Scottish Environment Protection Agency	65% of waterbodies were classified as being of 'good' status in 2015[c]
Northern Ireland	Water Framework Directive (Classification, Priority Substances and Shellfish Waters) Regulations (Northern Ireland) 2015	Department of Environment Northern Ireland	Overall 37% of waterbodies meet 'good' ecological status.[d]

a) Defra (2016).
b) National Assembly for Wales (2015).
c) SEPA (2016).
d) Cave and McKibbin (2016).

Table 4.4 Examples of UK requirements for chemical parameters where the point of compliance is at consumers' taps.

Parameters	Maximum concentration μg/L
Arsenic	10
Cadmium	5.0
Copper	2.0
Lead	10
Nickel	20

to monitor *Cryptospordium* to achieve <1 oocyst per 10 L as water companies are required to have adequate treatment facilities. The WHO (2011) also criticised the use of *E. Coli* as an indicator for pathogens, as well as others, such as *Giardia*, as cysts can be resistant to disinfectants. Therefore, the most recent regulations may not be robust enough to prevent the occurrence of pathogens in drinking water supplies.

4.3.3 Regulations to Protect Groundwater

As a result of Article 7.1 of the WFD (2000/06/EC), all groundwater bodies in England and Wales are designated Drinking Water Protected Areas and must conform to the Drinking Water Inspectorate. If these groundwater bodies are used for the abstraction of drinking water, under Article 7.3 of the WFD (2000/06/EEC) deterioration of quality must be avoided to reduce purification treatment. Additionally, direct discharges of pollutants into groundwater are prohibited under the WFD without authorisation (2000/60/EC).

The Water Resources Act 1991 allowed the implementation of statutory water protection zones for both surface water and groundwater to prohibit or restrict polluting matter. Regulation 13(1) of The Groundwater (England and Wales) Regulations (2009) states it is an offence to cause or knowingly discharge any hazardous substance or non-hazardous pollutant into a groundwater source without a permit from the EA. This was a result of the Integrated Pollution Prevention and Control Directive (2008/1/EC) brought in to meet the objectives of other EU directives dealing with water policy. Therefore, source protection zones have been designated according to groundwater sources, including wells, boreholes, and springs, to protect these waterbodies from pollution (EA, 2009). They have been defined as follows:

- SPZ1 – The Inner Protection Zone is defined as a 50-day travel time of groundwater from any point below the water table to the source or a minimum 50 m radius from the source. The most stringent controls are in this zone.
- SPZ 2 – The Outer Protection Zone is defined as a 400-day travel time of the groundwater or 25% of the source catchment.
- SPZ 3 – The Source Catchment Protection Zone needs to be protected from long–term groundwater recharge.

There are two classifications of substances under the Groundwater Directive (80/68/EEC): those which must be prevented from entering (List I) and those which must be controlled to prevent pollution (List II), and these are listed in Table 4.5. According to Groundwater Regulations 1998, List I substances derived from the Groundwater Directive (80/68/EEC) can be allowed into a groundwater body if it is no longer in use.

A substance is also in List II if it is considered inappropriate for List I, in regard to toxicity, persistence, and bioaccumulation. Studies commissioned by the Drinking Water Inspectorate found some of the substances listed are not of concern to drinking water as they occur in low concentrations (DWI, 2008). Compliance for List I substances under the Groundwater Regulations 1998 is applied to the unsaturated zone. In terms of List II substances, these may apply for a short distance in the direction of groundwater flow. Under the WFD (2000/60/EC) and the Priority Substances Directive (2013/39/EU), a majority of these substances have been classified as either Priority

Table 4.5 List I and List II substances as classified under the Groundwater Directive.

	List I	List II
a)	Organohalogen compounds and substances in the aquatic environment	Metalloids, metals, and compounds (Zn, Cu, Ni, Cr, Pb, Se, As, Sb, Mo, Ti, Sn, Ba, Be, Bo, U, V, Co, Tl, Te, Ag)
b)	Organophosphrous compounds	Biocides and their derivatives not on List I
c)	Organotin compounds	Substances with a deleterious effect on the taste or odour of groundwater, and compounds which can cause such substances to form in groundwater, rendering it unfit for human consumption
d)	Substances with carcinogenic, mutagenic, or teratogenic properties in the aquatic environment (this includes substances with properties which would otherwise be in List II)	Toxic or persistent organic compounds of silicon, and substances which may cause the formation of such compounds in water, excluding those which are biologically harmless or are rapidly converted in water into harmless substances
e)	Hg and its compounds	Inorganic phosphorous compounds as well as elemental phosphorous
f)	Cd and its compounds	Fluorides
g)	Mineral oils and hydrocarbons	Ammonia and nitrites
h)	Cyanides	

Substances (PSs), priority hazardous substances, or specific pollutants. The Nitrates Directive (91/676/EEC) designates Nitrate Vulnerable Zones to reduce groundwater pollution on the basis of a value of 50 mg/L nitrate in groundwater not being exceeded. In addition to the Groundwater Directive (2006/118/EC), the Drinking Water Directive (80/778/EEC as amended by 98/83/EC) is referred to for groundwater sources used for abstraction. Additionally, EQS standards should be considered, even though there are no statutory groundwater quality standards (EA, 2013). GP3 (Principles and Practice for Groundwater Protection) sets out key policy statements to reduce the pollution of groundwater (EA, 2013).

Inputs of non-hazardous pollutants must be controlled or limited to prevent pollution of groundwater. Although these are not defined in the WFD (2000/60/EC), they are considered to be substances that can potentially pollute groundwater. Standards for these particular pollutants depend on the receptor. For example, if the receptor is surface waters or abstraction for human consumption, the groundwater standard will correspond to surface water or drinking water, respectively (UKTAG, 2013).

4.3.4 Road Run-Off and the Development of Run-Off Specific Thresholds

Research by Johnson and Crabtree (2007) assessed the effects of soluble pollutants from road drainage on receiving ecological waters. Table 4.6 presents the run-off specific thresholds (RSTs) derived from the study and are designed to be used with EQSs for soluble pollutants.

As a result of the RBMPs derived from the WFD (2000/60/EC), EQSs were developed by UKTAG (2008) to comply with legislation and were implemented by the River Basin Districts Typology, Standards and Groundwater Threshold Values (Water Framework Directive) (England and Wales) Directions 2010; in Northern Ireland through the Water Framework Directive (Priority Substances and Classification) Regulations (Northern Ireland) 2011; and in Scotland by the Scotland River Basin District (Surface Water Typology, Environmental Standards, Condition Limits and Groundwater Threshold Values) Directions 2009. To provide specific water quality standards for rivers, there are seven typologies defined by alkalinity and altitude as shown in Table 4.7 which are subsequently used to describe water quality standards for rivers.

Six tables then detail the specific standards in terms of dissolved oxygen, pH (different standards apply for England and Wales; Scotland and Northern Ireland), phosphorus, reactive phosphorus, and temperature. The application of these standards is therefore highly complex and site specific.

4.3.5 Heavily Modified Waterbodies and Artificial Waterbodies

These waterbodies are commonly found in urban areas, and their water quality classification is referred to as 'Good Ecological

Table 4.6 Run-off specific thresholds (RSTs) for short-term exposure.

		Zinc (μg/L)		
		Hardness		
Name of threshold	Copper (μg/L)	Low (<50 mg $CaCO_3$/L)	Medium (50 to 200 mg $CaCO_3$/L)	High (>200 mg $CaCO_3$/L)
RST 24 hour	21	60	92	385
RST 6 hour	42	120	184	770

Table 4.7 Criteria for identifying the types of river to which the dissolved oxygen, biochemical oxygen demand, and ammonia standards for rivers apply.

Site altitude	Alkalinity (as mg/L $CaCO_3$)				
	Less than 10	10 to 50	50 to 100	100 to 200	Over 200
Under 80 metres	Type 1	Type 2	Type 3	Type 5	Type 7
Over 80 metres			Type 4	Type 6	

Potential' or 'Moderate Ecological Potential' under the WFD (2000/60/EC), which assesses these using UKTAG 2008 by:

- Identifying the impacts affecting the waterbody
- Identifying the mitigation measure necessary that could be taken to improve the ecology
- Assessing whether mitigation measures have been taken

The EQSs for physicochemical quality elements and specific pollutants apply in the same way to these Heavily Modified Water Bodies and Artificial Water Bodies except classifying 'good ecological status' as 'good ecological potential'. The relevant standards should refer to the closest comparable surface waterbody under Annex V of the WFD (2000/60/EC). However, these standards may not be appropriate in relation to the hydromorphological characteristics of these waterbodies, and should therefore be reviewed before using these measures for classification.

4.4 Sustainable Drainage Systems (SuDS)

The complexity of identifying which standards to apply can be illustrated in relation to drainage water from SuDS devices, such as pervious paving systems (PPSs). Rather than encouraging water to leave the urban area as quickly as possible via pipes, straightened and channelised streams, and gutters and gulleys, SuDS are designed to infiltrate, detain, retain, and slowly convey water to the receiving waterbody (Charlesworth and Booth, 2017). It will thus reduce flooding, improve water quality, and provide opportunities for amenity and space for biodiversity. SuDs can be both urban or rural in application, but are generally associated with urban areas, where its impact on flooding has substantial potential. This section therefore focuses on policy and guidance applicable to SuDS, bearing in mind that knowledge of the receiving waterbody is required in order to apply the correct standard and that, under the Environmental Permitting (England and Wales) Regulations 2010, it is an offence to knowingly discharge effluent of poor water quality.

4.4.1 Run-Off Destination

The National Standards for SuDs provide guidance for the reduction of surface water run-off and their potential pollutants (Defra, 2011; Warwick, 2017). These standards are again dependent on the receiving waterbody and have been categorised by run-off destination, including discharge into the ground, surface waterbody, water sewer, and combined sewer.

4.4.1.1 Discharge into the Ground

Surface run-off must be discharged into the ground except where one or more of the following can be demonstrated:

a) The rate of surface run-off is greater than the rate at which water can infiltrate into the ground. In this case, as much of the water as reasonably practicable must be discharged by infiltration; or

b) There is an unacceptable risk of ground instability or subsidence; or
c) There is an unacceptable risk of pollution from mobilising existing contaminants on the site; or
d) Infiltration is not compliant with the water quality requirements; or
e) There is an unacceptable risk of groundwater flooding; or
f) The infiltration system would create a high risk of groundwater leakage into the combined sewer.

4.4.1.2 Discharge to a Surface Waterbody

Surface run-off not discharged into the ground must be discharged to a surface waterbody except where it can be demonstrated that:

a) It is not reasonably practicable to convey the run-off to a surface waterbody; or
b) Pumping of the surface run-off, either on site or further downstream, would be required and there is a reasonably practicable alternative; or
c) Discharge would result in an unacceptable risk of flooding from the surface waterbody.

4.4.1.3 Discharge to a Surface Water Sewer or Local Highway Drain

Surface run-off that cannot be discharged into the ground or to a surface water body must be discharged to a surface water sewer or local highway drain, except where it can be demonstrated that it is not reasonably practicable to do so.

4.4.1.4 Discharge to a Combined Sewer

Surface run-off that cannot be discharged into the ground, a surface waterbody, or a surface water sewer or local highway drain must be discharged to a public, combined sewer system.

Surface run-off must not be discharged to a separate foul sewer.

4.4.1.5 Effective Treatment

Depending on the potential hazards, a number of treatment stages are permitted, but this depends on the sensitivity of the receiving waterbody (Tables 4.8 and 4.9). However, those hazards considered to be 'high' (Table 4.8) would not be permitted to infiltrate into groundwater.

Table 4.8 Level of hazard (Defra, 2011).

Low	Roof drainage
Medium	Residential, amenity, commercial, and industrial uses including car parking and roads
High	Areas used to handle and store waste, chemicals and fuels, including scrap yards. Lorry, bus, or coach parking or turning areas

Table 4.9 Minimum number of treatment stages for infiltration of drainage into groundwater (Defra, 2011).

		Low	Medium	High
G1	Source Protection Zone I, within 50 m of a well, spring, or borehole that supplies potable water.	1	3	**Consult the EA**
G2	Into or immediately adjacent to a sensitive receptor that could be influenced by infiltrated water. Includes designated nature conservation, heritage and landscape sites – including Biodiversity Action Plan (BAP) habitats and Protected Species.	1	3	
G3	Source Protection Zone II or III or Principal Aquifer			
G4	Secondary Aquifer	1	2	
G5	Unproductive strata	1	2	

4.4.1.6 Infiltration into the Ground

The National Standards state that the surface run-off from roof drainage needs to be isolated from other sources to ensure it is not significantly contaminated when it is directed to groundwater categories G1 and G2 (Table 4.9), where infiltration may only be allowed when a risk assessment has been undertaken and the design of the SuDS effectively addresses any risk(s) identified.

4.4.1.7 Surface Waterbody

The required treatment stages prior to the discharge of surface run-off to a surface waterbody or surface water sewer must be as shown in Table 4.10.

The treatment stages given in Tables 4.9 and 4.10 are the *minimum* suggested, rather than the absolute numbers required. Unfortunately, it would appear that under some circumstances, rather than design a suite of SuDS to address issues at a site, these treatment stages have been taken as absolute and as a result the full approach may not have been taken.

Table 4.10 Minimum number of treatment stages of surface water (Defra, 2011).

Hazard	Normal surface water	Sensitive surface water
Low	0	1
Medium	2	3
High	Consult the EA	

4.4.2 Rainwater Harvesting

Currently, there is interest in the use of rainwater harvesting, not only as an alternative source of water, but also as a means of storing excess surface water, and allowing it to dissipate slowly, thus attenuating the storm peak. However, currently, there are no regulatory water quality standards for rainwater use in England and Wales, which would not be considered drinkable without referring to the potable quality standards (EA, 2010). Table 4.11 presents the guideline values for bacteriological monitoring of harvested rainwater and corresponds to the traffic light system given in Table 4.12. The guideline values for general system monitoring are given in Table 4.13; its associated traffic light system is given in Table 4.14.

The British Standards Institute (Code of Practice BS 8515:2009) states that UV or chemical treatment might be beneficial if human exposure is likely, and a backflow device must to be fitted to the rainwater pipework to prevent the rainwater from entering the public mains supply. There is always a risk of contamination in harvested rainwater due to the characteristics of the collection surface; surfaces that have the potential to cause

Table 4.11 Guideline values for bacteriological monitoring (EA, 2010).

	Guideline values by use		System type
Parameter	In garden sprinklers and pressure washers	Flushing WC and watering gardens	Single site and communal domestic systems unless otherwise stated
Escherichia coli N° 100 mL^{-1}	1	250	
Intestinal enterococci N° 100 mL^{-1}	1	100	
Legionella N° 100 L^{-1}	100		If risk assessment indicates analysis is necessary
Total coliforms N° 100 mL^{-1}	10	1000	

Table 4.12 Interpretation of results from bacteriological monitoring (EA, 2010).

Sample result [1]	Status	Interpretation
<G	Green	System under control
G to 10G	Amber	Re-sampling required in order to confirm result and enable investigation of system operation
>10G [2]	Red	Rainwater to be suspended pending resolution of problem

1) G = guideline value from Table 4.11.
2) If *E. coli*, intestinal *enterococci*, and *Legionella* are absent, and if appropriate, the system can remain in use if coliforms >100 times guideline limits.

Table 4.13 Guideline values for general system monitoring (EA, 2010).

Parameter	Guideline values	System type (all systems unless otherwise stated)
Stored rainwater dissolved oxygen	>10% saturation or >1 mg L^{-1} oxygen (whichever is the least) all uses	
Suspended solids	Clear visually; free from floating debris for all uses	
Colour	Not objectionable for all uses	
Turbidity	<10 NTU all uses (<NTU if UV disinfection is used)	
pH	5–9 for all uses	Includes single site as well as communal domestic systems
Residual chlorine	<0.5 mg L^{-1} garden watering <2mg L^{-1} all other uses	
Residual bromine	<2mg L^{-1} all uses	

Table 4.14 Interpretation of results from general system monitoring.[a]

Sample result[2]	Status	Interpretation
<G	Green	System under control
>G	Amber	Re-sample to confirm result and investigate system operation

a) When monitoring pH, the system is under control (green) when levels are within the range recommended in Table 4.13. For levels outside of this range, the system's status then becomes amber, when re-sampling is required. If colour or suspended solids are present at questionable levels, the operation of the system must be investigated to resolve any problems.
G = guideline value (from Table 4.13).

contamination include the following (EA, 2010; Charlesworth *et al.*, 2014):

- Asbestos–cement roofs.
- Metal roofs (except stainless steel), which can release small amounts of leachates, and can cause staining of water fixtures.
- Bitumen felt or coated roofs can cause discolouration and odour problems.
- Grass roofs (and other vegetation) have the potential to cause discolouration.

Therefore, rainwater has not been used as a source of potable water in the United Kingdom, owing to strict drinking water guidelines (Charlesworth *et al.*, 2014).

4.5 European Policy

4.5.1 The Water Framework Directive

The WFD (2000/60/EC) was established to improve water policy for the protection of inland surface, transitional, coastal waters, and groundwater. Environmental objectives are to be met through the implementation of RBMPs, emission limit values, and EQSs. The WFD has been implemented to reduce the deterioration of waterbodies (Article 4) and also reduce priority pollutants. Both good 'Ecological classification' and 'Chemical classification' must be accomplished to achieve 'good status' under the WFD by 2015. This includes biological, chemical, physicochemical, and hydromorphological quality elements and specific pollutants. However, the WFD has insisted the poorest result would determine the overall classification of the waterbody. Member states are expected to implement directives into their national legislation through the use of environmental standards to assess the risks to ecological quality of the water environment. These environmental standards are used to assess the ecological status of rivers and should work in collaboration with other EU directives that are applicable to specific waterbodies (e.g. the Groundwater Directive), although it was necessary to amend certain aspects of the legislation after the introduction of the WFD.

PSs under the WFD and EQS Directive (2008/105/EC) were amended by Directive 2013/39/EU to reduce pollution of receiving waters. This resulted in the addition of 12 new substances to the original 33 PSs. Member states are required to comply with 50 EQSs, examples of which are given in Table 4.15, established in Annex II of Directive (2013/39/EU) according to Article 3(1). However, EQS Directive (2008/105/EC) recognises that indirect effects, such as secondary poisoning are not included for hexachlorobenzene (HCB), hexachlorobutadiene (HCBD), and mercury. Additionally, EQS Directive (2008/105/EC) highlighted the importance of deriving sediment quality standards for PSs to identify specific areas of concern. Some updated EQSs for PSs have been implemented in the United Kingdom under the RBMP2 (Defra, 2014). As reported by Whalley (2015), from December 2015

Table 4.15 Environmental quality standards for Priority Substances and certain other pollutants (heavy metals used as examples).

Substance N° under Directive	Name of substance	CAS [a] number	AA-EQS [b] Inland surface waters [c] (μg/L)	AA-EQS [b] Other surface waters (μg/L)	MAC-EQS [d] Inland surface waters [c] (μg/L)	MAC-EQS [d] Other surface waters (μg/L)
(6)	Cd and compounds (depends on water hardness class) [e]	7440-43-9	≤0.08 (Class 1) 0.08 (Class 2) 0.09 (Class 3) 0.15 (Class 4) 0.25 (Class 5)	0.2	≤0.45 (Class 1) 0.45 (Class 2) 0.6 (Class 3) 0.9 (Class 4) 1.5 (Class 5)	≤0.45 (Class 1) 0.45 (Class 2) 0.6 (Class 3) 0.9 (Class 4) 1.5 (Class 5)
(20)	Pb and compounds	7439-92-1	1.2 [f]	1.3	14	14
(21)	Hg and compounds	7439-97-6			0.07	0.07
(23)	Ni and compounds	7440-02-0	4 [f]	8.6	34	34

a) Chemical Abstracts Service.
b) Annual Average EQS value. Unless otherwise specified, it applies to the total concentration of all chemical species.
c) Inland surface waters encompass rivers and lakes and related artificial or heavily modified waterbodies.
d) Maximum Allowable Concentration for EQS; if marked 'not applicable', the AA-EQS values are considered protective against short-term pollution peaks in continuous discharges since they are significantly lower than the values derived on the basis of acute toxicity.
e) For Cd and its compounds, EQS values vary depending on the hardness of the water as specified in five classes: Class 1: <40 mg $CaCO_3$/L; Class 2: 40 to <50 mg $CaCO_3$/L; Class 3: 50 to <100 mg $CaCO_3$/L; Class 4: 100 to <200 mg $CaCO_3$/L; Class 5: ≥200 mg $CaCO_3$/L.
f) Bioavailable concentrations.

there are revised EQSs for anthracene, BDEs, DEHP, fluoranthene, lead, naphthalene, nickel, PAHs (i.e. BaP), and trifluralin. Others require supplementary monitoring programmes, but new EQSs for dicofol, PFOS, Quinoxyfen, dioxins, aclonifen, bifenox, cybutryne, cypermethrin, dichlorvos, HBCDD, heptachlor, and terbutryn should be decided upon by 2018. However, this was before the vote for Britain to leave the EU (Section 4.6.3).

4.5.2 Drinking Water

The Drinking Directive 98/83/EC provides legislation on the quality of water intended for human consumption, using the WHO (2011) guidelines in order to establish its standards. It provides specific parameters and their values to be implemented into national legislation; examples of some of the 26 chemical parameters are given in Table 4.16.

4.5.3 Groundwater

The Groundwater Directive (2006/118/EC) produced Groundwater Quality Standards (Table 4.17) as a result of Article 17 of the WFD (2000/60/EC). Additionally, Annex II of the Groundwater Directive (2006/60/EC) requires member states to establish threshold values for pollutants that could cause groundwater sources to fail.

Table 4.16 Examples of chemical parameters and their concentrations for the Drinking Water Directive.

Parameter	Value
Arsenic	10 μg/L
Cadmium	5.0 g/L
Copper	2.0 g/L
Lead	10 μg/L
Mercury	1.0 g/L
Nickel	20 μg/L

Table 4.17 Groundwater Quality Standards according to the Groundwater Directive (2006/118/EC).

Pollutant	Quality standards
Nitrates	50 mg/L
Active substances in pesticides, including relevant metabolites, degradation, and reaction products[a]	01. μg/L 0.5 μg/L (total)[b]

a) Pesticides includes plant protection and biocidal products (defined in Article 2 of Directive 91/414/EEC and in Article 2 of Directive 98/8/EC, respectively).
b) 'Total' means *all* individual pesticides detected and quantified during the monitoring procedure, including relevant metabolites, degradation, and reaction products.

4.5.4 Treatment of Urban Wastewater

The Urban Waste Water Treatment Directive (UWWTD) (91/271/EEC) sets requirements for the treatment of wastewater discharges to reduce pollution into receiving waters. However, more stringent standards than those shown in Table 4.18 (Annex I of the UWWTD) can be set to satisfy the objectives of other directives. However, Table 4.18 is not applicable if wastewater discharges are from high mountainous regions, where treatment is difficult, or in sensitive areas, where there needs to be at least a 75% reduction in both total phosphorous and nitrogen. Less stringent treatment is needed for wastewater discharges in less sensitive areas if the discharge is not detrimental to the environment. Less sensitive and sensitive areas are defined in Annex II of this directive.

4.6 The Future

As for the future, standards are likely to change as new pollutants are identified and established ones are refined. New and emerging pollutants (NEPs) include pharmaceuticals such as ibuprofen and hormones, and personal care products (PCPs)

Table 4.18 Requirements for discharges from urban wastewater treatment plants subject to Articles 4 and 5 of the directive and sensitive areas subject to eutrophication. Concentration or % reduction.

Parameters	Concentration (p.e. – population equivalent)	Minimum percentage reduction [a]
Biochemical Oxygen Demand (BOD at 20°C without nitrification[b])	25 mg/L O_2	70–90
Chemical Oxygen Demand	125 mg/L O_2	75
Total suspended solids	35 mg/L [c] (>10,000 p.e.) 60 (2000–10,000 p.e.)	90 [c] (>1,000 p.e.) 70 (>2,000–10,000 p.e.)
Total Phosphorous	2 mg/L P (10,000–100,000 p.e.) 1 mg/L P (>100,000 p.e.)	80
Total nitrogen[d]	15 mg/L N (10,000–100,000 p.e.) 10 mg/L N (>100,000 p.e.) [e]	70–80

a) Reduction in relation to the load of the influent.
b) The parameter can be replaced by another parameter: total organic carbon or total oxygen demand if a relationship can be established between BOD and the substitute parameter.
c) This requirement is optional.
d) Total nitrogen: the sum of total Kjeldahl–nitrogen (organic N + NH_3), nitrate (NO_3)– nitrogen, and nitrite (NO_2)–nitrogen.
e) Alternatively, the daily average must not exceed 20 mg/L N. This requirement refers to a water temperature of 12°C or more during the operation of the wastewater treatment plant biological reactor. As a substitute for the temperature conditions, it is possible to apply a limited time of operation, which takes into account regional climatic conditions.

such as cosmetics and perfumes. These compounds have become a nuisance in freshwaters in the last decade, but as yet there are no standards to control them. In terms of the impacts of climate change on water quality, and on legislation in the future, this is less certain. However, if storminess increases as is predicted, pollution will increase from the increased incidence of combined sewer overflows, the erosion of soils and sediments, and the removal of polluted dusts from urban areas. Temperatures are likely to increase, and with it, algal blooms and eutrophication.

4.6.1 Water Quality and Climate Change

In terms of water quality, the IPCC (2013) report provided limited information or advice for future water policies as it primarily concentrates on water resources. Whitehead *et al.* (2009) compiled a report for the EA on the potential effects of climate change on river water quality, indicating areas to focus on for future standards. The report concluded that:

- Flow regimes will increase concentrations from point discharges as lower minimum flows will reduce dilution of substances. Evidence from the River Tame, downstream from Birmingham, West Midlands, United Kingdom, indicated an increase in phosphorous during the summer months where flow levels decreased, lessening dilution of wastewater effluents.
- Dissolved oxygen levels will be affected as growth of algal blooms will be enhanced as a result of the reduced influx of water. This will result in an increase in BOD and a decrease in dissolved oxygen.
- Increased incidences of combined sewer overflows (CSOs) discharging highly polluted waters into receiving waterbodies from more frequent storm events, changes in short duration rainfall intensity, and sea-level rise.

- The most immediate effect of climate change on waterbodies are increases in temperature and their particular impact on aquatic organisms (Hammond and Pryce, 2007). Higher temperatures make chemical reactions and bacteriological processes more rapid, particularly acidification.
- Increased sediment yields, suspended solids, and associated contaminant metal fluxes from intense rainfall and flooding.
- Increase in nutrient loads. Ammonia levels will fall as nitrate concentrations increase.
- Cyanobacteria blooms will increase as oxygen levels decline in shallow lakes.
- Although research is currently scarce, it has been suggested that toxins may become a problem in streams, lakes, and sediments. It has been argued that substances can be re-mobilised in increased storm events, even though these pollutants are prohibited or restricted in use.

Modelling future scenarios of water quality is affected by uncertainty, making requirements for water policies to include mitigation of climate change difficult.

4.6.2 Potential Impacts of Brexit: Britain's Exit from the European Union

On 29 March 2017, the United Kingdom triggered Article 50 of the Lisbon Treaty, the formal process to begin its transition out of the EU. The impacts of Brexit on the aquatic environment are highly uncertain, making prediction of the outcomes extremely difficult. However, this chapter has shown that large numbers of EU policies have been adopted into English and Scottish law, and these would still apply unchanged, although the British Parliament and devolved governments would be able to make changes to them if they wished. Also, if the United Kingdom were to remain a member of the European Economic Area (EEA), then most environmental laws would still apply. On completion of Brexit, it may no longer be a requirement to report to the EU on progress with the WFD, but this may be replaced by inspection by the European Free Trade Association (EFTA) Surveillance Authority, which monitors compliance with rules in EEA countries: Iceland, Liechtenstein, and Norway.

In a report prepared for the Institute for European Environmental Policy on the potential impacts of Brexit on the environment and policy, Baldock *et al.* (2016) acknowledge the positive influence of EU legislation on the UK aquatic environment, such as significant improvements in the water quality of bathing waters, rivers, and coasts; going so far as to say that these have been 'dramatic'. However, there may also be pressure to relax some of the more difficult WFD targets, since compliance with these would entail expense and further changes to working practices. Baldock *et al.* (2016) highlight 'significant uncertainty' and the potential for 'substantial risks' in the future for the UK environment.

4.7 Conclusions

There is no catch-all set of standards for the environment in general; consideration needs to be given to the specific receiving environment, for example, surface water, groundwater, transitional waters, and coastal waters. Within that specific environment, standards are applied to the use of that water, whether for drinking, for discharge into Bathing Waters, for industrial use (e.g. abstracted) or for specific species of fish (e.g. ciprinid or salmonid) or shellfish.

Such standards take account of water chemistry, and also properties such as turbidity (cloudiness), temperature, and TEC (total ions in solution). Different pollutants are transported around the environment in different ways, and thus some of the standards may be useful proxies; for example, for particulate associated pollutants, it is useful to measure TSS, or turbidity. To assess the likelihood of their release into the environment, it is also useful to measure pH, since, for example, acidic pHs (e.g. <5) lead to their release and dissolution. However, even if it is not necessary for the water to be of potable quality, the WHO (2011) drinking water standards may be used as a comparator; at the very least, it should be harmless if discharged directly into the receiving water environment.

The future for aquatic environmental policy in general is very uncertain, with the impacts of climate change a global phenomenon, the local effects are difficult to predict. A shorter-term impact specifically applied to the United Kingdom is that of its exit from the European Union, and the unknown bearing this may have on policy applied to the urban environment. The potential for pollution from PCPs and other NEPs requires further research to feed into the development of policies and guidelines to ensure their control to reduce any impacts.

References

Baldock, D., Buckwell, A., Colsa–Perez, A., Farmer, A., Nesbit, M., and Pantzar, M. (2016) The Potential Policy and Environmental Consequences for the UK of a Departure from the European Union. Institute for European Environmental Policy (IEEP). London, Brussels. 104pp. Available at: http://www.ieep.eu/assets/2000/IEEP_Brexit_2016.pdf

Cave, S. and McKibbin, D. (2016) River Pollution in Northern Ireland: An Overview of Causes and Monitoring Systems, With Examples of Preventative Measures. Paper 20/16. The Stationary Office: Belfast.

Charlesworth, S.M. and Booth, C.A. (eds.) (2017) *Sustainable Surface Water Management: A Handbook for SuDS.* Wiley-Blackwell.

Charlesworth, S.M., Booth, C.A., Warwick, F., Lashford, C. and Lade, O. (2014) Rainwater harvesting – reaping a free and plentiful supply of water. In: Booth, C.A and Charlesworth S.M. (eds.), *Water Resources in the Built Environment – Management Issues and Solutions.* Wiley-Blackwell.

Cole, S. and Jeffries, J. (2009) Using Soil Guideline Values. Better regulation science programme science report: SC050021/SGV Introduction. Environment Agency. Available at: https://www.gov.uk/government/uploads/system/uploads/attachment_data/file/297676/scho0309bpqm–e–e.pdf

DCLG (2012) National Planning Policy Framework. Available at: http://tinyurl.com/o5s4ydt

DCLG (2007) Homes for the Future: More Affordable, More Sustainable. London, Department for Communities and Local Government.

Defra (2010) Environmental Permitting Guidance: Water Discharge Activities. Available at: http://archive.defra.gov.uk/environment/policy/permits/documents/ep2010waterdischarge.pdf

Defra (2011) *National Standards for Sustainable Drainage Systems. Designing, Constructing, Operating and Maintaining Drainage for Surface Runoff.* Defra, London.

Defra (2014) *Water Framework Directive implementation in England and Wales: New and Updated Standards to Protect the Water Environment.* Defra, London.

Defra (2016) *England Natural Environmental Indicators.* Crown Copyright, London.

De Miguel, E., Mingot, J., Chacón, E. and Charlesworth, S.M. (2012) The relationship between soil geochemistry and the bioaccessibility of trace elements in playground soil. *Environmental Geochemistry and Health,* 34, 677–687.

DFT (2009) Road Drainage and the water Environment. In: *Design Manual for Roads and Bridges.* Available from: http://www.dft.gov.uk/ha/standards/dmrb/vol11/section3/hd4509.pdf

DWI (2008) Review of England and Wales Monitoring Data for Which a National or International Standard Has Been Set. Available from: http://dwi.defra.gov.uk/research/completed-research/reports/DWI70_2_215_Monitoring_Data.pdf
Environmental Permitting (England and Wales) Regulations 2010.
Environment Agency (2007) The Unseen Threat to Water Quality: Diffuse Water Pollution in England and Wales report. Environment Agency, Bristol; 21 pp.
Environment Agency (2009) Groundwater Source Protection Zones – Review of Methods Integrated Catchment Science Programme Science report: SC070004/SR1. Available at: https://www.gov.uk/government/uploads/system/uploads/attachment_data/file/290724/scho0309bpsf-e-e.pdf
Environment Agency (2010) Harvesting Rainwater for Domestic Uses: An Information Guide. Available at: http://webarchive.nationalarchives.gov.uk/20140328084622/http:/cdn.environment-agency.gov.uk/geho1110bten-e-e.pdf
Environment Agency (2012) Review of Urban Pollution Management Standards against WFD Requirements. Available from: https://www.gov.uk/government/uploads/system/uploads/attachment_data/file/291496/LIT_7373_b2855a.pdf
Environment Agency (2012) Catchment Based Approach for a Healthier Water Environment. Available at: https://www.gov.uk/government/uploads/system/uploads/attachment_data/file/204231/pb13934-water-environment-catchment-based-approach.pdf
Environment Agency (2013) Groundwater Protection: Principles and Practice (GP3) August 2013 Version 1.1. Available at: https://www.gov.uk/government/uploads/system/uploads/attachment_data/file/297347/LIT_7660_9a3742.pdf
European Commission (2012) Report from the Commission to the European Parliament and the Council on the Implementation of the Water Framework Directive (2000/60/EC) – River Basin Management Plans. Brussels.
Hammond, D. and Pryce, A.R. (2007). Climate Change Impacts and Water Temperature. Environment Agency Science Report No. SC060017/SR.
ICRCL (1987) *Guidance on the Assessment and Redevelopment of Contaminated Land.* Guidance Note 59–83. 2nd edition. DETR Publications.
IPCC (2013) Climate Change 2013: The Physical Science Basis. Contribution of Working Group I to the Fifth Assessment Report of the Intergovernmental Panel on Climate Change [Stocker, T.F., Qin, D., Plattner, G.-K., Tignor, M., Allen, S.K., Boschung, J., Nauels, A., Xia, Y., Bex, V. and Midgley, P.M. (eds.)]. Cambridge University Press, Cambridge, United Kingdom and New York, NY, USA, 1535 pp.
Johnson, I. and Crabtree, R.W. (2007) Effects of Soluble Pollutants on the Ecology of Receiving Waters, WRc Plc, Report No.: UC 7486/1, UK Highways Agency.
National Assembly for Wales (2015) Water Quality in Wales. [online] Available from: http://www.assembly.wales/research%20documents/qg15-004-water%20quality%20in%20wales/qg15-004.pdf
Northern Ireland Environment Agency (2014) Northern Ireland Water Management Facts & Figures. Available at: http://www.nienvironmentlink.org/cmsfiles/policy-hub/files/documentation/Marine/water-facts-and-figures-booklet-2014-final-for-web.pdf
Priestley, S. (2015) Water Framework Directive: Achieving Good Status of Water Bodies. CBP 7246. House of Commons Library: London. Available at: http://www.legco.gov.hk/general/english/library/stay_informed_overseas_policy_updates/water_framework.pdf
SEPA (2016) *State of Scotland's Water Environment 2015: WFD Classification Summary Report.* [online] Available from: https://www.sepa.org.uk/media/219474/state-of-scotlands-water-environment-wfd-classification-summary-report.pdf

UKTAG (2008) UK Environmental Standards and Conditions (Phase 2). Available at: https://www.wfduk.org/sites/default/files/Media/Environmental%20standards/Environmental%20standards%20phase%202_Final_110309.pdf

UKTAG (2008) UK Environmental Standards and Conditions. Available from: http://www.wfduk.org/sites/default/files/Media/Environmental%20standards/Environmental%20standards%20phase%201_Finalv2_010408.pdf

UKTAG (2013) River Flow for Good Ecological Potential: Final Recommendations. Available at: http://www.wfduk.org/sites/default/files/Media/Assessing%20the%20status%20of%20the%20water%20environment/UKTAG%20River%20Flow%20for%20GEP%20Final%2004122013.pdf

Warn, T., Heaney, T., Batty, J. and Davies, G. (2010) Water Framework Directive: An approach to the Revoked Directives: – the Freshwater Fish Directive, the Shellfish Directive and the Dangerous Substances Directive. Available from: http://www.wfduk.org/sites/default/files/Media/Environmental%20standards/UKTAG%20%20approach%20to%20revoked%20directives_Draft_160210.pdf

Warwick, F. (2017) Surface Water Strategy, Policy and Legislation. In: Charlesworth S.M. and Booth, C.A. (eds.), *Sustainable Surface Water Management; A Handbook for SuDS*. Wiley-Blackwell.

Water Supply (Water Quality) Regulations 2000. Available at: http://www.legislation.gov.uk/uksi/2000/3184/contents/made

WFD/ UKTAG (2014) Updated recommendations on Environmental Standards River Basin Management (2015–21). Available at: http://www.wfduk.org/sites/default/files/Media/Environmental%20standards/UKTAG%20Environmental%20Standards%20Phase%203%20Final%20Report%2004112013.pdf

Whalley, C. (2015) Update on Priority Substances – EU. Defra. Available at: http://www.eic–uk.co.uk/Documents/Files/Working%20Groups/Environmental%20Laboratories%20Group/4%20March%202015/15–02–02%20PS%20stakeholders%20update.pdf

Whitehead, P., Wade, A.J., and Butterfield, D. (2009) Potential impacts of climate change on water quality and ecology in six UK rivers. *Hydrology Research*, 40, 113–122.

WHO (2011) Guidelines for Drinking-Water Quality. 4th edition. Available from: http://whqlibdoc.who.int/publications/2011/9789241548151_eng.pdf?ua=1

5

Soil Quality and Policy

J. Webb[1], Jose L. Rubio[2], and Michael A. Fullen[1]

[1] Faculty of Science and Engineering, School of Architecture and Built Environment, University of Wolverhampton, United Kingdom
[2] Centro de Investigaciones sobre Desertificación, Universitat de Valencia, Spain

5.1 Introduction

Urban soils are very diverse and found in gardens, parks, cemeteries, allotments, grass verges, playing fields, and sometimes derelict and commercial land. Commercial land may include disposal sites, demolition and building sites, waste and derelict land, rubbish tips, spoil heaps, canal and railway land, collieries, docklands, power-station land, shipbuilding land, scrap yards, dried-out industrial lagoons, sewage works, and land associated with mining, smelting, and manufacture.

Of the main threats to soil sustainability (soil erosion; decreasing soil organic matter (SOM) content; loss of biodiversity; contamination; sealing; compaction; salinisation and floods and landslides), sealing, and contamination are the greatest threats to soils in urban areas. In this chapter, we concentrate on the sources of pollution (contamination) of urban soils, the consequences for urban dwellers of polluted soils, and how policies developed or proposed to reduce soil pollution impact urban soils.

Urban areas were once the main locations of industry. Formerly, little or no attempt was made to limit hazardous emissions from that industry, and so many urban soils became contaminated with a range of pollutants. Domestic heating systems using coal, oil, or wood; waste treatment facilities as well as road traffic have all contributed to the contamination of urban soils. The diversity, fragmentation, and complexity of urban soils may lead to contamination by numerous pollutants and also considerable variation in the degree of contamination.

5.2 Soil Pollutants and Their Sources

The major pollutants found in urban soils are:

- Heavy metals
- Hydrocarbons
- Polychlorinated biphenyls (PCBs)
- Dioxins
- Platinum group elements (PGEs)
- Rare earth elements (REEs)
- Nanoparticles

When assessing contamination of urban soils, soil sampling procedures developed for rural areas may be inappropriate due to the large spatial variability of contaminants, the fragmented distribution of soils, problems of accessibility, and rapid and unpredictable changes in land use (Ajmone-Marsan and Biasioli, 2010).

Urban Pollution: Science and Management, First Edition. Edited by Susanne M. Charlesworth and Colin A. Booth.

5.3 Consequences of Urban Soil Pollution

Due to their contamination, some urban soils can be sources of pollution following soil disturbance (e.g. Biasioli *et al.*, 2006). This effect can be apparent for distances of 15–20 km, depending on the prevailing wind direction (Blum, 1998). Cachada *et al.* (2016) considered the assessment of potential risks to the environment and human health of contaminants present in urban soils can be difficult due to the heterogeneity and complexity of the matrix, the existence of multiple point and diffuse sources, and the presence of mixtures of contaminants. Below we summarise the size, origin, and health impacts of the major soil contaminants.

5.3.1 Heavy Metals

Alekseenko and Alekseenko (2014) reviewed the metal contents of over 300 soils from cities and settlements from every continent and found that the metal contents of urban soils reflected those in the underlying Earth's crust. Webb *et al.* (2012) that in general urban soil types are broadly similar to those occurring elsewhere in the region. However, in addition to the underlying geology, the spectrum of metals will be influenced by the activities currently or previously carried out in the urban area (Galušková *et al.*, 2014). Consequently, the amounts of some potentially toxic elements (PTEs) are commonly larger than those in natural soils. For example, De Miguel *et al.* (1998; cited in Charlesworth *et al.*, 2011) report enrichment factors of 2.3, 2.6, and 4.0 for zinc (Zn), copper (Cu), and lead (Pb), respectively, in the urban soil of Madrid relative to natural background levels. The mean concentrations of Cu and Zn in urban soils in the urban and country parks of Hong Kong (24.8 and 168 mg/kg, respectively) were at least four and two times greater than those of rural soils (5.17 and 76.6 mg/kg, respectively), while the mean Pb concentration of urban soils (89.9 mg/kg) was one order of magnitude greater than that of rural soils (8.66 mg/kg) (Li *et al.*, 2016).

5.3.1.1 Sources

Metals continue to be emitted directly from traffic due to the combustion of fossil fuels, and indirectly due to the erosion of road surfaces and abrasion of vehicle components (Wiseman *et al.*, 2015). Formerly, a major source of heavy metals was the lead (Pb) added to petrol. The implications of Pb pollution of urban soils are dealt with in a separate subsection. Wiseman *et al.* (2015) concluded from work in Toronto that amounts of cadmium (Cd), antimony (Sb), and Pb were continuing to increase due to traffic.

5.3.1.2 Characterisation

Evidence suggests that the behaviour of elements predominantly of anthropogenic origin (e.g. Cu, Pb, and Zn) differs from that of elements primarily of geochemical origin (e.g. iron (Fe) and manganese (Mn)) (Madrid *et al.*, 2009).

Rodrigues *et al.* (2013) distinguished between the availability in soil for plant uptake and leaching and via oral ingestion. The pH of urban soils is usually higher than that of rural soils due to the addition of calcareous materials from building (e.g. Rodrigues *et al.*, 2013). The concentrations of reactive PTE, determined by 0.43 M HNO_3, were reported to indicate bioaccessibility by Rodrigues *et al.* (2013). Hong *et al.* (2016) found that bioaccessibility of PTEs was better associated with neurodevelopmental conditions in children than total concentrations.

Due to their observed sensitivity to heavy metals, plants may be good indicators of environmental pollution. For example, Diatta *et al.* (2003) evaluated dandelions and concluded they are suitable plants to indicate contamination.

5.3.1.3 Health Risks

The PTEs in urban soils may be a significant source to the atmosphere, particularly in the form of street dust. Metals retained by soils and subsequently dispersed may lead to ingestion by the human population long after the industrial activity that led to the original emissions has ceased. The accumulation of the PTEs Cd, Cu, Pb, and Zn increases the risk of exposure to PTEs, since PTEs from anthropogenic sources tend to be more available than those from natural sources (Rodrigues *et al.*, 2013), and PTEs are persistent as they are not biodegradable (Diatta *et al.*, 2003). The exposure of children to trace metals can increase greatly through their ingestion of metal-laden soil particles and dust via frequent hand-to-mouth activities (Wong *et al.*, 2006).

Lead Of the PTEs in urban soils Pb is considered to pose the greatest and most persistent threat to human health (Cai *et al.*, 2016). There are concerns that soils are a permanent repository for Pb (Watmough *et al.*, 2004 cited in Walraven *et al.*, 2014). Nevertheless, Walraven *et al.* (2014) reported that Pb concentrations in soils have decreased following the decline in Pb emissions from traffic. However, if soils are not always permanent sinks for Pb, the loss of Pb from urban soils may lead to groundwater pollution. Walraven *et al.* (2014) calculated that 35%–90% of the anthropogenic Pb deposited from 1962 to 2003 had entered groundwater. Flooding may mobilise PTEs in soil leading to adverse effects on aquatic biota in streams, ponds and lakes (Mukwaturi and Lin, 2015).

The total concentration of Pb in soil is not the best predictor of the impacts of soil ingestion on children (Cai *et al.*, 2016), since the availability of Pb in soils differs according to many factors including soil texture, pH, SOM content, and the reactive iron (Fe) content. Cai *et al.* (2016) found the gastric bioaccessibility (GB) of Pb to be <20% in all of the size fractions of the urban soils analysed while the gastro-intestinal bioaccessibility (GIB) was generally <15%. The percentage GB tended to decrease with increasing SOM. Walraven *et al.* (2014) concluded that only the litter layer and topsoil retain anthropogenic Pb.

5.3.2 Polycyclic Aromatic Hydrocarbons

Polycyclic aromatic hydrocarbons (PAHs) in urban soils originate from both industrial activities that burn hydrocarbon fuels and from road traffic (Nadal *et al.*, 2004). There may be ≤~10 times the concentrations of PAHs in urban soils than in rural soils (Nadal *et al.*, 2004; Stajic *et al.*, 2016). Being generally insoluble, PAHs can be adsorbed rapidly onto soil particles, especially on SOM (Means *et al.*, 1980, cited in Stajic *et al.*, 2016). Due to the dispersion by surface run-off and dust production, soils are a source of PAH in the atmosphere (Tang *et al.*, 2005, cited in Stajic *et al.*, 2016).

5.3.2.1 Health Risks

PAHs are of concern due to their carcinogenic and/or mutagenic potential (Cachada *et al.*, 2016, and references cited therein).

5.3.3 Polychlorinated Biphenyls (PCBs)

Soil is an important environmental receptor of PCBs, particularly soils rich in organic matter (Schuster *et al.*, 2011, cited in Glüge *et al.*, 2016). Although the worldwide production and usage of PCBs was prohibited in 2004, substantial amounts of PCBs are still emitted from primary sources in cities or landfills (Glüge *et al.*, 2016). As a result, some workers have concluded that PCBs may originate from urban areas (e.g. Jamshidi *et al.*, 2007 cited in Glüge *et al.*, 2016)). However, on the basis of the results of a modelling study, Glüge *et al.* (2016) concluded that for all PCB congeners, emissions from environmental reservoirs do not exceed primary emissions.

5.3.3.1 Health Risk

PCBs are fat-soluble substances to which people are exposed through ingesting animal fats, inhalation, or dermal contact. Exposure to PCBs suppresses the immune system, with carcinogenic, mutagenic, or endocrine disrupting consequences. Exposure to PCBs, especially during foetal and early life, can reduce IQ and alter behaviour. After ingestion, PCBs can alter thyroid and reproductive functions in both genders, thereby increasing the risk of developing cardiovascular and liver disease and diabetes. Women exposed to PCBs have an increased risk of giving birth to infants of low birth weight, who are at high lifetime risk for several diseases (Carpenter, 2006).

5.3.4 Dioxins

Urban *et al.* (2014) considered that dioxin-like compounds are ubiquitous in soils due to the wide variety of sources that have contributed to background levels. Data reported by Urban *et al.* (2014) for the United States indicate that the toxic equivalent (TEq) concentrations in background rural soils ranged from 0.1 to 22.9 ng/kg, while mean rural TEq concentrations ranged from 1.1 to 7.1 ng/kg across the 14 studies that reported data for rural soils. While rural mean concentrations were relatively small, 4 of the 14 studies had maximum concentrations over 20 ng/kg. The concentrations of dioxins in background urban/suburban soils were substantially larger and more variable than those in rural soils, with TEQ concentrations ranging from 0.1 to 186.2 ng/kg. It was also noted that the data for urban soils were considered less robust than for rural soils. The range of mean TEq concentrations in urban/suburban soils was also substantially higher and ranged from 2.2 to 56.6 ng/kg. Importantly, 4 of the 11 studies with data for background urban/suburban soils reported maximum concentrations that exceed 100 ng/kg.

5.3.4.1 Health Risks

Dioxins are considered to be carcinogenic and are also reported to have adverse cardiovascular- and endocrine-related effects (Bertazzi *et al.*, 2001). Most of the exposure (~90%) to dioxins in the general population is from diet (mainly animal products), whereas <10% comes from exposure to dioxins in other sources (water, inhalation of air, ingestion of soil, soil dermal contact, and vegetable fat intake) (Lorber *et al.*, 2009 cited in Urban *et al.*, 2014). Hence, the greater concentrations in urban soils do not pose a significant risk to urban populations.

5.3.5 Platinum Group Elements

Catalytic converters were introduced in the mid-1980s in Europe. With increasing use of multi-element analytical techniques such as inductively coupled plasma atomic emission spectroscopy (ICP–AES), it was realised that PGEs, which included platinum (Pt), palladium (Pd), rhodium (Rh), ruthenium (Ru), iridium (Ir), and osmium (Os), had begun to accumulate in the environment (Ravindra *et al.*, 2004, cited in Charlesworth *et al.*, 2011).

Wong *et al.* (2006) cited UK results between 1982 and 1998 which demonstrated that there had been an increase in PGEs in road dust. Further indications that traffic was the source of Pt were obtained by comparing Pt with gold (Au) in soils and dust sampled in the London Borough of Richmond in 1994 (Farago *et al.*, 1995, 1996). Concentrations of Pt, like those of Pb, which originate from traffic, were greater in road dust than in soil samples. For Au, which does not originate from traffic, concentrations were greater in soils than in road dust. Ajmone-Marsan and Biasioli (2010) considered the data available to be inconsistent due to their being based on different sampling strategies and analytical procedures combined with the extreme variability of urban soils. They concluded that a 'sampling design adapted to local urban patterns, a prescribed sampling depth, and a minimum set of elements that deserve to be measured could be the core of a common methodology'.

5.3.6 Rare Earth Elements

The most frequently detected anthropogenic REE is Gadolinium (Gd), which is issued from magnetic resonance imaging and released into the environment through hospital effluents (Brioschi *et al.*, 2013).

5.3.7 Particulate Matter

Soil particles dislodged from the soil matrix can be a component of the coarse fraction of particulate matter (PM) in urban areas (Charlesworth *et al.*, 2011). Both the particle size and chemical composition determine the potential health impacts of PM. Coarser (>10 μm) particles are usually considered to be trapped in the nose, throat, and upper respiratory tract. The <10 μm (PM_{10}) is considered 'inhalable', reaching the alveoli of the lungs and potentially causing irritation and disease. The <2.5 μm ($PM_{2.5}$) fraction is regarded as 'respirable', since they can be drawn deep into the respiratory system, generally beyond the body's natural clearance mechanisms, and are more likely to be retained and absorbed. Consequently, $PM_{2.5}$ is associated with adverse health effects, such as asthma and even death (Kappos *et al.* 2004, cited in Charlesworth *et al.*, 2011).

5.4 Soils Legislation

5.4.1 The European Strategy for Soil Protection

The European Commission (EC) has developed proposals to improve environmental quality for many years, but soil protection was not a specific objective of any EU legislation. Nevertheless, soil protection features in various legislation as a secondary objective. To close the gap, the commission launched a communication entitled 'Towards a Thematic Strategy for Soil Protection' in April 2002. This document is a comprehensive analysis of the state of Europe's soils, highlighting the importance of soil as a threatened and non-renewable natural resource. This EC approach was noticeably innovative in incorporating new and wider environmental perspectives including the identification of the following soil functions: (a) biomass production, including agriculture and forestry; (b) storing, filtering, and transforming nutrients, substances, and water; (c) biodiversity, habitats, species, and genes; (d) physical and cultural environment for humans and human activities; (e) source of raw materials; (f) carbon pool; and (g) archive of ecological and archaeological heritage. The communication also identified soil contamination as one of the eight soil threats expressed in the thematic strategy and the proposed directive.

To develop the communication, an advisory forum was created, a secretariat and five technical working groups (Monitoring, Erosion, Organic Matter, Contamination, and Research) which prepared several important documents as background information on the status of soil and soil problems including the one related to Pollution and Land Management (Van Camp *et al.*, 2004; http://www.ec.europa.eu/environment/soil).

After several years of political discussions, negotiations, and confrontations due to the different economic, political, and social interests of the EU member states (MSs) and stakeholders related to soil, in 22 September 2006, the EC proposed a Soil Thematic Strategy (STS) (COM (2006) 231) and a Soil Framework Directive (SFD) (COM (2006) 232) with the objective of formulating a pan-European common framework for soil protection.

The objective of the STS was the 'protection and sustainable use of soil by preventing further soil degradation and preserving its functions and also restoring degraded soils to its functionality and considering cost implications' (COM 2006, 231). The STS includes four pillars: (a) the development of framework legislation, (b) integration and horizontal implications, (c) identification of knowledge gaps on soils, and (d) the need to enhance public awareness in relation to soil. The STS also includes the identification of 'specific risk areas' related to erosion, organic matter decline, soil sealing, contamination, and landslides.

The STS was received favourably by the Committee of the Regions (February 2007) and the European Economic and Social Committee (April 2007), and in November 2007 it was adopted by the European Parliament (EP), which supported the commission in its general approach and underlined the importance for climate change mitigation, biodiversity loss, and desertification. The EP also adopted a favourable opinion on the directive. In the amendments adopted, the EP maintained all the key elements of the commission proposal, providing more flexibility in some provisions and strengthening others.

Despite these positive and important decisions, the soil strategy faced many problems. The development of a proposed SFD produced not only conceptual and technological disagreements among EU MSs, but also societal problems related to the absence of a shared common vision on soil protection under legislation, scope of protection, economic considerations, definitions, and governance aspects.

Another long period of intense discussions among all soil actors and MSs followed. Substantial changes were introduced, adding flexibility for the MSs. However, the EC did not reach a political agreement by 20 December 2007. Five MSs (Germany, France, the United Kingdom, The Netherlands, and Austria) formed a blocking minority against the favourable opinion of the majority of MS. The negative decision was for reasons of subsidiarity in the case of Germany, The Netherlands, and Austria and of proportionality/costs in the case of the United Kingdom and France.

The blocking status of the SFD implied that discussions would be continued. Meanwhile, the EP was in favour of an SFD. The EC was functioning with the communication on the Soil Thematic Strategy for Soil Protection (22 September 2006) as a common legislative framework for soil 'to halt and reverse the process of soil degradation ensuring that EU soils stay healthy for future generations and remain capable of supporting the ecosystems on which our economic activities and our well–being depend' (JRC, 2014)

However, repeated discussions could not secure agreement by a qualified majority for an SFD, and in October 2013 the EC adopted the communication on 'Regulatory Fitness and Performance (REFIT),' in which it noted that 'the proposal for a SFD had been pending for eight years during which time no effective action has resulted.' On 30 April 2014, the EC took the decision to withdraw the proposal of a SFD. The decision entered into force 21 May 2014. However, using diplomatic wording, the EC stated that it 'remains committed for soil protection and will examine options on how best achieve this.'

The current situation is one of internal reflection by the commission, including soil global issues (food security, climate change, …) and the implications of outcomes from Rio+20 and Post 2015 Sustainable Development Goals. At present, the EC is working along two lines, which include soil contamination as one of the main concerns:

a) preparation of a communication on 'Land as a Resource' with focus on more efficient land use planning including environmental, economic, and social aspects
b) Further developing the sustainable management of soil as an ecosystem

5.4.2 EU Legislation Related to Soil Contamination

As there was no agreement for the establishment of the SFD, legal requirements for the general protection of soil, including contamination, only exist in individual MSs. As a general principle, arising from Article 191(2) of the Treaty on the Functioning of the European Union, EU policy on the environment aims at a high level of protection taking into account the diversity of situations in the various regions of the Union, and is based on the precautionary principle and on the principles that preventive action should be taken, that environmental damage should, as a priority, be rectified at source and that the polluter should pay (Payá Pérez *et al.*, 2015). In general, the protection of urban soil contamination is indirectly addressed as part of other environmental protection policies, such as the EU Water Framework Directive

(WFD), the Waste Directive, and the Integrated Pollution and Prevention Control Directive (IPPC), not aimed directly at soil protection, but providing indirect controls on soil contamination.

There are important difficulties in the appraisal and development of soil policies for contaminated sites in urban environments. These are due to the wide range of contaminated sites in municipalities and the enormous heterogeneity of the criteria by which different countries define contaminated soils, quantify acceptable risks, and characterise tools and adopted methodologies. To improve the situation, some efforts have been made regarding the identification and management of contaminated sites. For example, Panagos *et al.* (2013) described an initiative of the European Environment Agency on the development of an indicator of soil contamination (Progress in the Management of Contaminated Sites) (Van Liedekerke *et al.*, 2014). This indicator quantifies progress in the management of local contamination, identifies sectors with major contributions to soil contamination, classifies the major contaminants, and finally addresses issues of budgets spent for remediation. With this indicator, several activities causing soil pollution can be identified across Europe. The indicator also supports the implementation of existing legislative and regulatory frameworks (IPPC directive, Landfill Directive, and WFD) as they should further decrease soil contamination.

In urban environments, the priority has been to reduce emissions of contaminants. However, according to the JRC Report on Soil Threats in Europe (Stolte *et al.*, 2016), historically contaminated (brownfield) sites remain important contamination sources in many European cities. Ferber *et al.*, (2006) defined brownfields as 'sites that have been affected by the former uses of the site and surrounding land; are derelict and underused; may have real or perceived pollution problems; are mainly in developed urban areas; and require intervention to bring them back to beneficial use'. There is no EU regulation concerning brownfields, and few countries have developed national strategies to deal with them. Especially problematic brownfield types are smelter waste deposits that are usually barren due to the phytotoxicity of high-metal waste and therefore constitute secondary sources of pollution.

Some directives indirectly introduce elements to protect soil from contamination. These include The Sewage Sludge Directive (1986/278/EEC), which defines conditions for sewage sludge application to soils. The directive provides threshold trace metals contents in soil and sludge as well as allowed annual inputs of the following elements: Zn, Pb, Cd, Ni, Cu, and Hg.

The Urban Waste Water Treatment Directive (1991/271/EEC) is increasing the quantities of sewage sludge requiring adequate disposal and the number of households connected to sewers and increasing the level of treatment.

The purpose of the Directive on Environmental Liability with Regard to the Prevention and Remedying of Environmental Damage (2004/35/CE) is to establish a framework of environmental liability to prevent and remedy environmental damage. The directive aims at ensuring that the financial consequences of certain types of harm caused to the environment will be taken by the operator who caused this harm. This prevention instrument refers to various natural resources including protection against soil pollution.

The IPCC Directive (2010/75/EU) on Industrial Emissions aims to establish a general framework for the control of the main industrial activities, giving priority to intervention at the source, ensuring prudent management of natural resources and taking into account, when necessary, the economic situation and specific local characteristics of the place where the industrial activity is taking place.

The Landfill Directive (1999/31/EC) introduces stringent technical requirements for waste and landfills to prevent or reduce the adverse effects of landfill on the environment, in particular on surface water, groundwater, soil, air, and human health. It defines the different categories of waste (municipal waste, hazardous waste, non-hazardous waste, and inert waste) and applies to all landfills, defined as waste disposal sites.

The Directive on the Incineration of Waste (2000/76/EC) aims to prevent or reduce damage to the environment caused by the incineration and co-incineration of waste. The directive sets emission limit values and requires monitoring of pollutants to air such as dust, nitrogen oxides (NOx), sulphur dioxide (SO_2), hydrogen chloride (HCl), hydrogen fluoride (HF), heavy metals, dioxins, and furans. Most types of waste incineration plants fall within the scope of the WI Directive, with some exceptions such as those treating only biomass (e.g. vegetable waste from agriculture and forestry) and phytotoxicity of high-metal waste, and therefore constitute secondary sources of pollution.

With the steady increase in urban population and the growing awareness of contamination impacts, society is demanding an urban soil strategy which should include the establishment of a specific and common legal framework to protect urban soils from contamination. This legal framework should be integrated into the general strategy of soil protection of national and EU policies by strengthening the knowledge base and by taking into account the specific aspects of urban environments.

5.4.3 UK Soils Legislation

On 2 June 2016, The Commons Environmental Audit Committee (CEAC, 2016) warned that polluted soils are a potential health hazard in many urban areas because the UK Government no longer provides grants to decontaminate them. Previously there had been national funds to help local councils clean up polluted land, but this has now been closed. The government response is that planning policy sets a clear framework for the clean-up of land to be developed. But the CEAC Chair stated: 'Relying on the planning system to clean up contaminated land may be fine in areas with high land values, but it means that contamination in poorer areas will go untreated. Ministers must rethink their decision to phase out clean-up grants.' There were potentially now 300,000 contaminated sites in the United Kingdom, she added. The government has declared an objective of safeguarding all soils by 2030, but the MPs say there are no policies in place to deliver that promise.

5.4.3.1 Current UK Soils Legislation

In 2009, the UK Department for the Environment, Food and Rural Affairs (Defra) published 'Safeguarding our Soils – A Strategy for England' (Defra, 2009). This strategy was claimed to support the aims of the EU Thematic Strategy on Soil Protection. However, it arose due to a perceived need for *national* action to protect soils 'which is responsive to local circumstances' and, in effect, was a rejection of the harmonised European approach that had been proposed in the draft EU SFD.

The UK strategy envisaged that by 2030, 'all England's soils will be managed sustainably and degradation threats tackled successfully. This will improve the quality of England's soils and safeguard their ability to provide essential services for future generations,' the specific goals being that:

- Agricultural soils will be better managed, and threats to them will be addressed.
- Soils will play a greater role in the fight against climate change and in helping to manage its impacts.
- *Soils in urban areas will be valued during development, and construction practices will ensure that vital soil functions can be maintained.*
- *Pollution of soils is prevented, and our historic legacy of contaminated land is dealt with* (the two latter goals have been italicised to emphasise their particular relevance to urban soils).

There was also an undertaking to 'review the effectiveness of existing planning policy to protect important soils and consider whether there is a need to update it'. In addition there was a promise to 'publish new best practice guidance on decision-making later this year to help Local Authority officers make proportionate and robust decisions more confidently. It will also continue to encourage moves to more sustainable remediation practices that do not involve the wholesale removal and replacement of soil'.

UK Government policy on defining contaminated land takes a risk-based approach. Specifically, land is assessed in terms of whether contaminants pose a 'significant possibility of significant harm' (SPOSH) (Defra, 2008). As a consequence, current guidance considers that the amounts or concentrations of any contaminants are not the only factors to be taken into account. It is also necessary to consider to what extent the contaminants present may harm human health or the wider environment, that is, what is the risk caused by contaminants, and is that risk unacceptable? In short, the approach to be taken is one of risk management, not risk avoidance.

5.4.3.2 Planning Policy

Soil protection is mentioned in a number of Planning Policy Statements (PPSs), for example, PPS1 (Delivering Sustainable Development, Office of the Deputy Prime Minister (ODPM) 2005) and its Supplement on Climate Change and PPS11 (Regional Spatial Strategies), as well as in the Strategic Environmental Assessment and Environmental Impact Assessment directives. PPS1 recommends that development plan policies should take account of environmental issues including the conservation of soil quality. However, the document does not give any indication of the properties that characterise soil quality or any guidance on how those properties may be maintained or enhanced.

5.5 Conclusions

Sealing and contamination are the greatest threats to soils in urban areas. The major source of some pollutants in urban areas can be soils that are already contaminated due to historic pollution. Due to this contamination, some urban soils can be sources of pollution following soil disturbance. Estimating the potential risks to the environment and human health of contaminants present in urban soils can be difficult due to the heterogeneity and complexity of the matrix, the existence of multiple point and diffuse sources, and the presence of mixtures of contaminants. Soil sampling procedures developed for rural areas may not be appropriate for the assessment of the contamination of urban soils, due to the large spatial variability of contaminants, the fragmented distribution of soils, problems of accessibility, and rapid and unpredictable changes in land use. Metals continue to be emitted directly from traffic due to the combustion of fossil fuels, and indirectly due to the erosion of road surfaces and abrasion of vehicle components. Cadmium, Sb, and Pb continue to be added to urban soils as a result of traffic.

Metals retained by soils and subsequently dispersed may lead to ingestion by the human population long after the industrial activity that led to the original emissions has ceased. Soil PTEs from anthropogenic sources tend to be more available than those from natural sources. Lead is the PTE considered to pose the greatest threat to human health. Although there are concerns that soils are a permanent repository for Pb, concentrations in urban soils have decreased following the decrease in Pb emissions from traffic.

Due to their carcinogenic and/or mutagenic potential, PAHs are of concern. There may be up to $\leq\sim10$ times the concentrations of PAHs in urban soils than in rural soils.

Polluted soils are acknowledged to be a potential health hazard in many urban areas, yet the UK Government no longer provides grants to decontaminate them. Although the government has declared an objective of safeguarding all soils by 2030, there are no policies in place to deliver that promise.

References

Ajmone-Marsan, F. and Biasioli, M. (2010) Trace elements in soils of urban areas. *Water, Air and Soil Pollution*, 213, 121–143.

Alekseenko, V. and Alekseenko, A. (2014) The abundances of chemical elements in urban soils. *Journal of Geochemical Exploration*, 147, 245–249.

Bertazzi, P.A., Consonni, D., Bachetti, S., Rubagotti, M., Baccarelli, A., Zocchetti, C., and Pesatori, A.C. (2001) Health effects of dioxin exposure: A 20-year mortality study. *American Journal of Epidemiology*, 153, 1031–1044.

Biasioli, M., Barberis, R., and Ajmone-Marsan, F. (2006) The influence of a large city on some soil properties and metals content. *Science of the Total Environment*, 356, 154–164.

Blum, W.E.H. (1998) Soil degradation caused by industrialization and urbanization. In: Blume, H.-P., Eger, H., Fleischhauer, E., Hebel, A., Reij, C. and Steiner, K.G. (eds.), *Towards Sustainable Land Use*, Vol. I, Advances in Geoecology 31, Catena Verlag, Reiskirchen, pp. 755–766.

Brioschi, L., Steinmann, M., Lucot, E., Pierret, M.C., Stille, P., Prunier, J., and Badot, P.M. (2013) Transfer of rare earth elements (REE) from natural soil to plant systems: implications for the environmental availability of anthropogenic REE. *Plant Soil*, 366, 143–163.

Cachada, A., Ferreira da Silva, E., Duarte, A.C., and Pereira, R. (2016) Review – Risk assessment of urban soils contamination: The particular case of polycyclic aromatic hydrocarbons. *Science of the Total Environment*, 551–552, 271–284.

Cai, M., McBride, M.B., and Li, K. (2016) Bioaccessibility of Ba, Cu, Pb, and Zn in urban garden and orchard soils. *Environmental Pollution*, 208, 145–152.

Carpenter, D.O. (2006) Polychlorinated Biphenyls (PCBs): Routes of exposure and effects on human health. *Reviews on Environmental Health*, 21, 1–24.

Charlesworth, S., De Miguel, E., and Ordóñez, A. (2011) A review of the distribution of particulate trace elements in urban terrestrial environments and its application to considerations of risk. *Environmental Geochemistry and Health*, 33, 103–123.

Commons Environmental Audit Committee (CEAC). (2016).

Defra. (2008) Guidance on the Legal Definition of Contaminated Land. July 2008. https://www.gov.uk/government/collections/land-contamination-technical-guidance

Defra. (2009) Safeguarding Our Soils – A Strategy for England. http://webarchive.nationalarchives.gov.uk/20130123162956/http:/archive.defra.gov.uk/environment/quality/land/soil/documents/soil-strategy.pdf.

Diatta, J.B., Grzebisz, W., and Apolinarsk, K. (2003) Using dandelion (*Taraxacum officinale* Web) as a bioindicator. *Electronic Journal of Polish Agricultural Universities*, Environmental Development, 6.

Farago, M.E., Kavanagh, P., Blanks, R., Simpson, P., Kazantzis, G., and Thornton, I. (1995) Platinum Group Metals in the Environment: Their Use in Vehicle Exhaust Catalysts and Implications for Human Health in the UK. A report prepared for the UK Department of the Environment, 182 pp.

Farago, M.E., Kavanagh, P., Blanks, R., Kelly, J., Kazantzis, G., Thornton, I., Simpson, P.R., Cook, J.M., Parry, S., and Hall, G.M. (1996) Platinum metal concentrations in urban road dust and soil in the United Kingdom. *Fresenius Journal of Analytical Chemistry*, 354, 660–663.

Ferber, U., Grimski, D., Millar, K., and Nathanail, P. (2006) Sustainable Brownfield

Regeneration: CABERNET Network Report. The University of Nottingham.
Galušková, I., Mihaljevič, M., Borůvka, L., Drábek, O., Frühauf, M., and Němeček, K. (2014) Lead isotope composition and risk elements distribution in urban soils of historically different cities Ostrava and Prague, the Czech Republic. *Journal of Geochemical Exploration*, 147, 215–221.
Glüge J., Bogdal, C., Scheringer, M., and Hungerbühler, K. (2016) What determines PCB concentrations in soils in rural and urban areas? Insights from a multi-media fate model for Switzerland as a case study. *Science of the Total Environment*, 550, 1152–1162.
Hong, J., Wang, Y., McDermott, S., Cai, B., Aelion, C.M., and Lead, J. (2016) The use of a physiologically-based extraction test to assess relationships between bioaccessible metals in urban soil and neurodevelopmental conditions in children. *Environmental Pollution*, 212, 9–17.
JRC. (2014) Soil Themes: Soil Contamination. http://esdac.jrc.ec.europa.eu/themes/soil-contamination
Li, Z., Jia, M., Wu, L., Christie, P., and Luo, Y. (2016) Changes in metal availability, desorption kinetics and speciation in contaminated soils during repeated phytoextraction with the Zn/Cd hyperaccumulator Sedum plumbizincicola. *Environmental Pollution*, 209, 123–131.
Madrid, F., Reinoso, R., Florido, M.C., Díaz Barrientos, E., Ajmone-Marsan, F., Davidson, C.M., and Madrid, L. (2009) Estimating the extractability of potentially toxic metals in urban soils: A comparison of several extracting solutions. *Environmental Pollution*, 147, 713–722.
Mukwaturi, M. and Lin, L. (2015) Mobilization of heavy metals from urban contaminated soils under water inundation conditions. *Journal of Hazardous Materials*, 285, 445–452.
Nadal, M., Schuhmacher, M., and Domingo, J.L. (2004) Levels of PAHs in soil and vegetation samples from Tarragona County, Spain. *Environmental Pollution*, 132, 1–11.
ODPM. (2005) Planning Policy Statement 1: Delivering Sustainable Development. ISBN 0 11 753939 2.
Panagos, P., Van Liedekerke, M., Yigini, Y., and Montanarella, L. (2013) Contaminated sites in Europe: Review of the current situation based on data collected through a European network. *Journal of Environmental and Public Health*, Vol. 2013, Article ID 158764, 11 pg. 10.1155/2013/158764.
Payá Pérez, A.B., Van Lidekerke, M., and Jones, A. (2015) Soil Policy and Developments in the Management of Contaminated Sites in Europe EC–JRC Institute for the Environment and Sustainability, Ispra (VA) Italy.
Rodrigues, S.M., Cruz, N., Coelho, C., Henriques, B., Carvalho, L., Duarte, A.C., Pereira, E., and Römkens, P.F.A.M. (2013) Risk assessment for Cd, Cu, Pb and Zn in urban soils: Chemical availability as the central concept. Paul F.A.M. *Environmental Pollution*, 183, 234–242.
Stajic, J.M., Milenkovic, B., Pucarevic, M, Stojic, N., Vasiljevic, I., and Nikezic, D. (2016) Exposure of school children to polycyclic aromatic hydrocarbons, heavy metals and radionuclides in the urban soil of Kragujevac city, Central Serbia. *Chemosphere*, 146, 68–74.
Stolte, J., Tesfai, M., Øygarden, L., Kværnø, S., Keizer, J., Verheijen, F., Panagos, P., Ballabio, C., and Hessel, R. (2016) Soil Threats in Europe; EUR 27607 EN; doi:10.2788/488054 (print); doi:10.2788/828742 (online).
Urban, J.D., Wikoff, D.S., Bunch, A.T.G., Harris, M.A., and Haws, L.C. (2014) Review – A review of background dioxin concentrations in urban/suburban and rural soils across the United States: Implications for site assessments and the establishment of soil cleanup levels. *Science of the Total Environment*, 466–467, 586–597.
Van Camp, L., Bujarrabal, B., Gentile, A.R., Jones, R.J.A., Montanarella, L., Olazabal, C., and Selvaradjou, S.-K. (2004) Reports of the Technical Working Groups Established under the Thematic Strategy for Soil

Protection. EUR 21319 EN/4, 872 pp. Office for Official Publications of the European Communities, Luxembourg.
Van Liedekerke, M., Prokop, G., Rabl–Berger, S., Kibblewhite, M., and Louwagie, G. (2014) Progress in the Management of Contaminated Sites; JRC85913 EUR 26376 EN Office for Official Publications of the European Communities, Luxembourg.
Walraven, N., van Os, B.J.H., Klaver, G.Th., Middelburg, J.J., and Davies, G.R. (2014) The lead (Pb) isotope signature, behaviour and fate of traffic-related lead pollution in roadside soils in The Netherlands. *Science of the Total Environment*, 472, 888–900.
Webb, J., Fullen, M.A., and Blum, W.E.H. (2012) The pedological value of urban landscapes. In: C.A. Booth, F. Hammond, J.E. Lamond, and D.G. Proverbs (eds.), *Solutions for Climate Change Challenges of the Built Environment.* Wiley-Blackwell Publishing Limited, Oxford, pp. 113– 126.
Wiseman, C.L.S., Zereini, F., and Püttmann, W. (2015) Metal and metalloid accumulation in cultivated urban soils: A medium-term study of trends in Toronto, Canada. *Science of the Total Environment*, 538, 564–572.
Wong, C.S.C., Li, X., and Thornton, I. (2006) Urban environmental geochemistry of trace metals. *Environmental Pollution*, 142, 1–16.

6

Ground Gases in Urban Environments – Sources and Solutions

Andrew B. Shuttleworth

SEL Environmental Limited, Lancashire, United Kingdom

6.1 Introduction

This chapter will concentrate on ground gas issues arising from sources that commonly cause problems in urban areas where both the problems and solutions can be very different. Ground gases are considered to be all gaseous pollutants emanating from subsurface strata (other than those from leaking gas pipelines) with the potential to enter properties, or receptors, in sufficient quantities to possibly cause harm. While the global warming potential of these ground gases will be acknowledged, it will not be covered in detail. There are many forms of ground gas requiring action to be taken in urban areas, particularly during redevelopment; this commonly includes landfill gas. Ground gas problems are not confined to large, operational, or recently closed landfills, although this is the aspect with which most people are familiar. In fact, for many years, planning controls in developed societies have ensured that the large engineered landfills are established well away from urban areas and have gas controls built into their design and operation. In urban areas, in developed economies at least, it is unlikely to find the high methane generation rates capable of power generation, and thus this aspect of ground gas control will not be covered.

Natural gas emanation can also be problematic, and one in particular is radon, a naturally occurring radioactive gas generated as part of the uranium decay series (Public Health England, n.d.; Åkerblom *et al.*, 1984). Another form of naturally occurring ground gas is the variable mixture of methane and carbon dioxide generated from naturally occurring organic matter in alluvial drift deposits (Card *et al.*, 2012). Methane and carbon dioxide are also natural constituents of marsh gas; sometimes, populations live alongside marshes in the city centre (e.g. Colombo, Sri Lanka), although they are usually drained and filled, and then reclaimed for urban development. In the United Kingdom, one of the most notable examples is Victoria Embankment, which was reclaimed from marshes alongside the Thames in the nineteenth century (Thornbury, 1878); Manhattan, NYC, United States, is largely built on reclaimed marsh. Not only would the naturally occurring organic matter contribute to the generation of ground gas but fill material would often be unstable and contain organic waste.

6.1.1 Landfill Gas

The growing demand for development land has often resulted in the infilling of relatively small voids such as small quarries and

Urban Pollution: Science and Management, First Edition. Edited by Susanne M. Charlesworth and Colin A. Booth.

borrow pits, which often originated in rural areas, to provide materials for both urban and local rural needs. They were often left undeveloped until the towns and cities expanded, making them valuable building plots; they were then infilled and, in the absence of environmental controls, built on. Trees and other plants which had grown in the void over the years were often buried with various types of fill material and began to biodegrade. If construction was undertaken before tighter regulation came into force, it often proceeded with little concern for the inhabitants. However, when this land is targeted for redevelopment today, measured ground gas concentrations can give rise to regulatory demands for appropriate gas protection measures.

In addition to quarries and pits, linear features such as canals and railway cuttings can be infilled and reclaimed for building or used for agriculture. With age, ground gas production largely declines, but since lenders will not loan against properties with perceived environmental problems, even if regulators do not insist protection is put in place, it usually is where new properties are built on or close to such land. An example in the author's experience was the site of a former dry cleaning and workwear rental company which historically 'landfilled' oily rags and dirty clothing into a canal bed; this generated a flux of gas which also drove chlorinated solvent vapours toward a new development.

Such older domestic landfills in urban areas are not uncommon, but they are usually relatively small. The gradual creep of building toward previously isolated landfills was the cause of the most widely reported incident associated with ground gas in the United Kingdom, the Loscoe, Derbyshire, United Kingdom, methane explosion in March 1986 (Landfill Gas Expert, n.d.; Williams and Aitkenhead, 1991) which completely destroyed a house and badly injured three residents. Years before the explosion, there were signs of heating of the ground about 100 m outside the limit of a nearby landfill, but it was mistaken for a shallow-burning coal seam, and thus steps were not taken to protect the area (Williams and Aitkenhead, 1991). Close by, in Sandiacre, S. Derbyshire, United Kingdom, landfilling in the 1940s and 1960s was used to fill a former sand quarry subsequently built around with houses both during the quarrying and into the 1970s. After the Loscoe incident, the County Council surveyed the area, finding issues with gas emissions; as a consequence, this site is under constant surveillance, with a stand-by gas extraction scheme which is switched on if the methane or carbon dioxide levels reach prescribed limits.

6.1.2 Mine Workings

One of the most significant sources of methane-rich ground gas in some urban areas arises from old mine workings. Surface or near-surface coal mining operations in the United Kingdom are particularly prone to the generation of sufficient methane to cause problems in urbanised areas. Deeper mining can cause fracturing due to subsidence, which can release methane from unexploited shallow coal seams. An example of problems caused by the escape of mine gas is that of Arkwright Town, a NE Derbyshire mining village of 160 houses built in the late nineteenth century by a local mining company. Following nationalisation of the coal industry in 1947, it became the property of the National Coal Board (NCB) (later British Coal). In 1977, the village was sold off mainly to tenants, with the remainder being sold to a property company. As reported in *The Guardian* newspaper (2001), gas engineers were called to a suspected gas leak in 1988 when an inextinguishable fire occurred in the hearth of one of the properties. It was found that the gas had not escaped from the mains but was coal seam methane escaping from below the properties. As many as 110 people were subsequently evacuated for 2 weeks, and a methane drainage system was rapidly installed. Eventually, however, residents were relocated to a purpose-built new village some distance away, the old settlement was removed, and 3.2 million tons of coal was removed from under the village and surrounding areas by surface-mining the coal which vented the methane. The project took over 10 years (Bridgewater, n.d.).

Another source of ground gas is the degradation products of organic liquids released accidentally which then becomes trapped in the strata; the most common leakage is associated with petrol stations. The flux of gas can be sufficient to drive petrol vapours into surrounding properties.

6.1.3 Minor Sources of Ground Gas

An unusual source of ground gas is the release of carbon dioxide from the dissolution of carbonate rocks. While this constantly occurs in karstic areas, it is slow and so is not a problem; but if limestone or chalk is used as backfill in areas where the groundwater has a low pH, it can become an issue. For example, the author came across this when limestone pipe bedding was deposited against the outside of a cellar wall in an area where the groundwater was quite aggressive, resulting in the accumulation of carbon dioxide, which displaced oxygen in a basement, leading to potentially life-threatening effects on both the occupant of the property and the two firefighters who were sent to investigate. The response to these minor sources of ground gas is either to use the methods adopted for other sources or, where possible to combine this with source removal.

To protect against the effects of ground gas, it is first necessary to understand the sources and biochemical processes involved that generate the gases in the first instance.

6.2 The Biochemistry of Ground Gas Production

It is well known that historical waste tips are a likely source of potentially hazardous gases in the ground (Mor *et al.*, 2006). Other sources can be naturally occurring, for example, alluvial drift deposits from river alluvium and glacial till, peat layers, coal seams, carbonate strata, granites (radon emission from the radioactive decay of uranium), swamps, and wetlands or from artificial structures such as mine workings, sewage works, and slurry pits, burial grounds, service station fuel tanks, foundry sands, and abandoned wells.

6.2.1 Landfill Gas

Landfill gas is generated during the decomposition of degradable waste material by micro-organisms, principally producing carbon dioxide and methane as well as minor constituents such as hydrogen, hydrogen sulphide, and carbon monoxide. The decomposition process passes through five principle stages, each due to changes in physical and biochemical processes with time. Consequently, the composition and concentrations of the various gases in the landfill gas depends on the time since the waste was deposited (see Table 6.1).

Table 6.1 Phases in landfill gas generation (after Crowhurst and Manchester, 1993).

Phase		
I	Aerobic decomposition of biodegradable materials; entrained atmospheric oxygen is converted to carbon dioxide.	↓
II	Anaerobic decomposition commences as oxygen is used up; carbon dioxide concentration increases, and some hydrogen is produced; no methane is produced at this stage.	
III	Anaerobic methane production commences and rises to a peak; concentration of carbon dioxide declines; hydrogen production ceases.	Time
IV	Steady methane and carbon dioxide generation in proportions of between 50% and 70% and between 30% and 50%, respectively.	
V	Steady decline in generation of methane and carbon dioxide; gradual return to aerobic conditions.	

The ratio of carbon dioxide (CO_2) to methane (CH_4) affects the density and flammability of landfill gas. Gas with a CO_2:CH_4 ratio of greater than 3.5 is known as the 'limiting safe mixture' (LSM) and would not be flammable if mixed with air in any proportion. The CO_2:CH_4 ratio can be less than unity, but when greater than 0.87 it has neutral buoyancy and tends to settle, gathering close to the entry point of the affected receptors. Below 0.87, the mixture is less dense than air and tends to accumulate at higher levels in the available space. Methane with a concentration of between 5% and 15% by volume in air does form an explosive mixture. The potential volume of gas generated has been estimated at 130 m^3 per 1 m^3 of landfill material over a 13-year period, of which 80 m^3 will be methane (ICRCL, 1990). The rate of production of landfill gas is proportional to the rate of waste decomposition, which in turn is influenced by the following physical and environmental factors:

- Composition of the waste.
- Particle size and distribution of the waste material.
- Density, void ratio, and compaction state of the waste material.
- Depth and temperature of the waste.
- Landfilling techniques during waste deposition, for example, vertical distance and thickness of daily cover materials, historical locations of temporary haul roads across the landfill site, and associated sequence of filling.
- Moisture content; biological reactions can be retarded if moisture content drops below 40% and essentially stops when moisture content is below 20% (Mass.gov, n.d.)
- Capping material permeability, which in turn will influence the moisture content within the waste due to the rate of ingress of precipitation.

6.2.2 Natural Materials Producing Ground Gas

Naturally occurring biodegradable material in the strata beneath construction sites, mines, and waste tips can generate a mixture of gases similar to those from landfill, but these are usually released at lower levels. The gas may exceed the lower explosive limit, although fluctuations with groundwater levels can affect rates of emission due to the hydraulic pumping effect. Mine workings and colliery soil tips therefore produce methane, carbon dioxide, and small quantities of other hazardous ground gases generated by the anaerobic decomposition of ancient vegetation embedded in underlying rock; ethane, hydrogen, and helium may also be present. Dissolved methane can be present in groundwater that is released when the atmospheric pressure drops (in a similar way to opening a bottle of sparkling water). This phenomenon typically occurs at the surface of groundwater in underground artificial adits and constructions such as tunnels and mine workings (Crowhurst and Manchester, 1993). As explained in Section 6.1, carbon dioxide can also be released by the dissolution of carbonates from acid waters on limestone and chalk (Crowhurst and Manchester, 1993).

6.2.3 Radon

Radon is a naturally occurring radioactive gas with no taste, smell, or colour which forms from the radioactive decay of uranium found in many rocks such as granite as well as in some soils. It is by far the largest component of background radiation in the United Kingdom relating on average to 50% of the total contribution, equivalent to the sum of all the other contributing sources (POSTnote, 2001). Radon is measured in units of radioactivity using the SI derived becquerel per cubic metre, Bq m^{-3}; the average level across the United Kingdom is on the order of 20 Bq m^{-3}. Public Health England (http://www.ukradon.org/information/level) recommends an Action Level of 200 Bq m^{-3} in homes, at which remediation should take place to reduce it to a Target Level of 100 Bq m^{-3}. According to PHE (2009), radon increases the chances of developing lung cancer, and it was estimated that radon causes over 1,100 lung cancer deaths per year in the United Kingdom, this is in comparison with 28,000 deaths associated with smoking.

The Building Research Establishment's (BRE) guidance on radon protective measures for new buildings (Scivyer, 2015) gives a simple qualitative gas risk assessment procedure for protection against radon. It provides a decision from three basic levels of protection required; namely, no protection, basic protection, or full protection, ascertained by reading from a map showing the concentrations of radon affected areas in the United Kingdom.

As with other ground gases, radon enters a receptor by airflow from underground. Protection can be achieved with the use of a well-sealed radon-resistant barrier in the floor construction together with either sub-floor ventilation and/or the provision of a radon sump (POSTnote, 2001). The latter is a relatively small constructed void from which air is drawn using an extractor fan. This provides a collection zone for the radon, a preferential pathway and an extraction point back into the atmosphere. The Building Regulations 2000 (England and Wales), the Building (Scotland) Regulations 2004 and Building Regulations Northern Ireland (as amended 1990), supported by BRE reports BR211, BR376, and BR413, respectively, require that buildings and extensions (workplaces and dwellings) constructed after 2000 in Radon Affected Areas have appropriate radon protective measures installed during construction.

6.3 Ground Gas Monitoring and Risk Assessment

Assessing the risks associated with ground gas should be a phased approach with each one tailored on the basis of findings from the previous phase. The suggested approach includes: a desk study, an intrusive site investigation, and a risk assessment. These are detailed in the following sections, starting with the desk top study, which should be the first phase.

6.3.1 Desk Top Study

The purpose of the desk top study is to collate all available information to enable an informed decision to be made regarding the locations and depths of the phase 2 intrusive investigation. Typically, it would include as a minimum:

- Collection of all available historical and current Ordinance Survey maps of the area.
- A walk over survey of the site to be carried out, for visual evidence of pollution/contamination and to identify any potential receptors.
- Collection of anecdotal evidence through dialogue with local residents and questionnaires.
- Formulation of a plan clearly showing boundaries these require inspection.
- Gathering of photographic evidence of findings after appropriate permissions are in place.

From the information gathered, principal pollutant linkages should have been identified; the proposed locations and depths for phase 2 investigations can then be established. For further detailed information and recommendations for site investigations and associated case studies, see CIRIA C665 (Wilson *et al.*, 2007), BS10175, and BS 5930.

6.3.2 Intrusive Investigation

Phase 2 of the gas monitoring process usually involves the installation of intrusive gas monitoring points at strategic positions, designed independently to suit the individual strata and conditions at each location. In addition to monitoring locations centrally, peripheral monitoring points outside of the suspected areas of contamination should be installed to assess the potential for off-site migration. Shallow boreholes should be installed next to existing houses or any other potential receptors identified in phase 1. The possibility of monitoring internal spaces of existing dwellings should be considered to establish baseline levels of gas (if any) in the receptors. The monitoring point depth should be determined not only by the depth of the suspected source but also by consideration of the foundation arrangement and type of receptor. For example, ground conditions may dictate the use of deep piled foundations, in which case monitoring wells should extend the full depth of the proposed piling zone to intercept any potential pathways and pollutant linkages potentially breached during piling installation. Deep wells should be used for monitoring ground gases from deep sources, for example, coal measures or to extend into gas transmitting strata extending from the source beneath the proposed development.

The frequency and spacing of monitoring points would be established on the basis of site-specific conditions, found during phase 1, with cognisance of the proposed construction methods, end use of the development, and the sensitivity of all identified receptors. As a guide, monitoring points tend to be between 25 m and 75 m spacing depending on the nature of the source and the sensitivity of the end use. A closer monitoring point frequency, such as 5 m spacing, should be considered when investigating boundary gas migration.

6.3.2.1 Monitoring Gas in Wells

Standpipes are used to monitor the concentration of gas in well headspace (Wilson *et al.*, 2007). The response zone is the perforated length of standpipe where the gas can enter from the unsaturated zone. The gas will then settle in the upper unperforated section of the liner.

In order to ensure efficient gas monitoring of a site, the following points should be considered:

- PVC or HDPE liners are preferred; sometimes stainless steel may be more appropriate due to contamination.
- 19-, 25-, or 50-mm-diameter liners are preferable, but this depends on the combined use of the borehole and resultant annulus.
- Segmental 1-, 2- or 3-m-long perforated (preferably slotted) or plain liners are used, each section threaded to allow a mechanical connection. Gluing should be avoided due to possible presence of VOCs.
- Each monitoring point should be provided with a protected lockable cover.
- Unperforated liners at the top of the monitoring well should be between 0.5 and 1.0 m from the ground surface. It should be completely sealed with bentonite slurry to prevent both water ingress into the perforated section and dilution/mixing with atmospheric gases.
- One standard depth of response zone across the site is bad practice; the depths for each individual monitoring point should be determined on a site-specific basis reflecting the rock type, hydrogeology, and gas source. It needs to be sufficient to intercept any gassing source and any pathway; it may prove necessary for response zones sealed in different strata to eliminate other sources when multiple sources are present.
- Immediately following construction, gas concentrations will be similar to the overlying atmosphere; as ground gas diffuses into the standpipe and mixes with those in the pipe, an equilibrium will be reached; the time needed depends on the soil type, pore pressure, gas pressure gradient, and gas permeability.
- When drilling, care must be taken to avoid aquifers, response zones should avoid creating unexpected or unwanted pathways, and perforation of basal liners (either artificial or natural) should be avoided when determining the drill depth.
- Wherever possible, combining groundwater and gas monitoring wells should be avoided due to the potential effect of dissolved gases in groundwater; separate gas monitoring wells are preferred.

6.3.2.2 Monitoring Instruments

Infrared analysers are commonly used to detect carbon dioxide, methane, oxygen, and others, for example, hydrogen sulphide and carbon monoxide, but there are many instruments available on the market for detecting both the proportions of constituent gases and flow rates. It is not intended to list all the options in this chapter, but the operator needs to understand the differences. CIRIA C665 (Wilson *et al.*, 2007) and R131 (Table 5.1 in Crowhurst and Manchester, 1993) give more comprehensive information and tables of available monitoring instruments listing the factors and properties to be considered when selecting appropriate equipment.

In all cases, the following important points apply:

- All monitoring instruments and equipment should have regular, up-to-date accurate calibration by an accredited body.
- Consistent monitoring protocols should be adhered to. As well as flow rates and gas composition, the operator name, sampling point reference, date, time, weather conditions, and depth of water in well should all be recorded.
- Sampling methodologies and protocols should be in accordance with CIRIA 131 (Crowhurst and Manchester, 1993).

6.3.3 Risk Assessment Process

In general, the risk assessment needs to be sufficient to support an appropriate remedial solution or to justify that no actions are required. It is to help make legal, justified, transparent, and understandable decisions. A tiered risk-based decision process to deal with contamination is set out in EA (2004). The preparation of a gas risk assessment is a complex process requiring substantial knowledge of geology, physics, and chemistry. It should therefore be carried out by an appropriately qualified chartered and experienced professional person. While not intended to cover the whole process for producing a detailed risk assessment, the following section focuses on preparing gas risk assessments where methane and carbon dioxide are the significant contributors. The full procedure is well documented in CIRIA C665 (Wilson *et al.*, 2007) with examples in Sections 8.2 to 8.6.

6.3.4 Methane and Carbon Dioxide

The risk assessment process starts with the creation of a conceptual site model, which is used to develop a risk model to identify potential sources of gas and possible pathways toward receptors capable of causing harm. A qualitative risk assessment can then be performed on the basis of the likelihood of occurrence and severity of the consequence of the risks identified (see Rudland *et al.*, 2001). The risk model is then analysed to identify all potential pollutant linkages; if there are no linkages, there will be no risk. With the pollutant linkages established and an associated risk verified, the risk assessment divides into two avenues defined by the type of construction:

- Low-rise housing with ventilated underfloor void (min 150 mm)
- All other developments

6.3.4.1 Low-Rise Housing with Ventilated Underfloor Void (Min 150 mm)

Boyle and Witherington (2007) developed a simplified characterisation system specific to low-rise housing developments with a clear subfloor void min 150 mm deep. This is a risk-based approach using a 'traffic light' metaphor for the rate of gas emissions expressed in terms of gas screening values (GSVs) obtained using an empirical derivation from Wilson and Card (1999). The development is categorised as green, amber 1, amber 2, or red depending on the GSV; the prescribed gas protection measures required can then be sourced from Boyle and Witherington (2007).

6.3.4.2 All Other Developments

The classification of all other developments is given in CIRIA R149 (Card, 1995), in which characteristic situations 1–6 are defined depending on the GSVs, risk classification, additional factors, and the source of the gas generation. The typical scope of gas protection measures is advised depending on the characteristic situation and end use of the receptor.

6.5 Other Approaches

For some sites, the use of monitoring wells alone is insufficient to always accurately represent the gas regime. The quantity and frequency of gas monitoring boreholes are usually significantly lower than other exploratory boreholes, and consequently, the composition, concentration, pressure, and flow of gas measured in the well headspace may not be representative of the whole site. Other reasons for possible spurious monitoring results include:

- Wells may intercept horizontal pathways, misreading the vertical flow and diffusion.
- If there is little headspace in the well, low volumes of gas can cause artificially high concentrations.
- Differential gas pressures in strata can cause misleading flow rates and gas concentrations.
- Organic material or dissolved methane and carbon dioxide in groundwater can escape at a disproportionate rate from the unconfined groundwater surface in the well.
- Organic material can accumulate at the base of the well during construction, generating unexpected levels and flow rates of gas.
- Hydrocarbons from upper strata can collect in the well and degrade to methane and carbon dioxide over time.
- Methane can displace air in the well due to buoyancy, causing disproportionately elevated levels of methane.
- Methane can accumulate in the peat layers (sometimes up to 90% of the volume) and be intercepted by the gas monitoring wells. This may not necessarily be linked to high gas production rates but may just relieve gas pressure in the peat.

To achieve a more comprehensive risk assessment, it is sometimes possible to gain an understanding of the likely source of gas generation using data that can be collected more cheaply and quickly. This can be used in the first instance to establish if monitoring is absolutely necessary. For many sites with low organic content in the soils, the need for monitoring is less likely provided the risk is adequately characterised, and a robust conceptual model developed. This approach can also be used to refine the extent of any further monitoring that may be deemed necessary, reduce the period of monitoring, or avoid the need for additional monitoring if the results are not clear. The approach is to first differentiate between high-risk sites, for example, landfills and low-risk sites which may only comprise a thin layer of low organic content soils. According to Card *et al.* (2012), there are no recorded instances where ground gas from made ground or from natural soils with low organic content have caused gas ingress into buildings above ground level. Additionally, in modern construction, the requirements of Part L of the building regulations relating to the air tightness also leads to the need for well-sealed floor slabs (NHBC foundations, 2009) with damp-proof membranes and seals to pipe entry positions. In older construction, although the building offers less resistance to the penetration of ground gas, the inside of the building will have much greater natural ventilation and is therefore less likely to accumulate gases. There are sites, therefore, which are less likely to require gas monitoring, including:

- Natural soils which contain methane such as peat and alluvium; those with a high carbonate content including chalk and limestone
- Made ground, <5 m deep with low organic content
- Flooded mine workings where buildings are >20 m away from open shafts or adits
- Total organic carbon limits tabulated in Card *et al.* (2012)

And also sites more likely to require gas monitoring:

- Landfill with high organic concentrations
- Mine workings with large gas reservoirs which vent to ground level
- Made ground deeper than 5.0 m
- Where credible pathways exist from remote sources
- Total organic carbon limits tabulated in Card *et al.* (2012)

Where suspected, total organic carbon (TOC) content should be tested using Waste Acceptance Classification (WAC) testing as specified by the EA (2005). The limiting values of TOC are based on forensic descriptions, testing, and proportional representation of the soil fractions in the whole. For example, if 15% of the soil is a discrete zone with 30% TOC and the remaining 85% has a TOC of 0.6%, then the overall TOC will be 5%.

In all cases, a conceptual model is needed showing, as a minimum, site topography, and a drawn cross section to clearly identify the sources, pathways, and receptors. The information required to produce the conceptual model should be obtained from the exercise detailed in Section 6.3.

Where no credible sources of gas generation or pathways from external sources exist, low-risk sites situated on alluvial soils, lignite, or having buried layers of well-decomposed peat tend not to generate gas flows sufficient to exceed Characteristic Situation 2 under BS8485: 2007 (Wilson *et al.*, 2007); these sites therefore do not require monitoring. It is acceptable in these instances to install Characteristic Situation 2 protection measures such as passive or positive pressure ventilation layers with a correctly installed gas-resistant membrane. Passive ventilation layers or voids beneath slabs should be vented to atmosphere with a minimum open area of 1500 mm^2 per metre run of wall. It is also useful to fill cavity walls to below ground level. However, should monitoring cost less than potential savings on the gas protection design, it would make commercial sense to undertake it. Monitoring is still required for sites investigated as Part 2a of the Environmental Protection Act due to the need to investigate gas migration across boundaries; resistance to gas cannot be assumed in older, well-established dwellings.

6.6 Passive Barriers and Subfloor Ventilation

6.6.1 The Effect of Development on Ground Gases

Compaction of the ground, improvement techniques, and low-permeability surface layers, for example, asphalt on car parks, all reduce the potential for ground gases to diffuse into the atmosphere. They can lead to an accumulation of gas and increased pressure beneath the surface, known as the 'pressure cooker effect'. Figure 6.1 shows how these factors influence the pathways taken by ground gas, potentially leading to its ingress into buildings.

Should piled foundations be used, in particular vibrated stone columns, the ground gas pathways can be penetrated, providing alternative preferential routes for gases to enter the building. Foundation arrangements, especially perimeter slab thickenings, can form shapes that trap gas, causing accumulation over time beneath the structure. Service entry points, for example, sewers and incoming utilities, are prime targets for gas ingress directly into the building; at times, such features are often located in poorly ventilated service rooms that often remain closed for long periods of time, leading to increased risk of gas accumulation.

6.6.2 Primary Receptor Protection – Ventilation Layer

The purpose of a ventilation layer is to provide a continuous constructed void beneath the floor or slab of a building, giving a flow of fresh air to dilute and distribute the gases to a safe and acceptable concentration. There

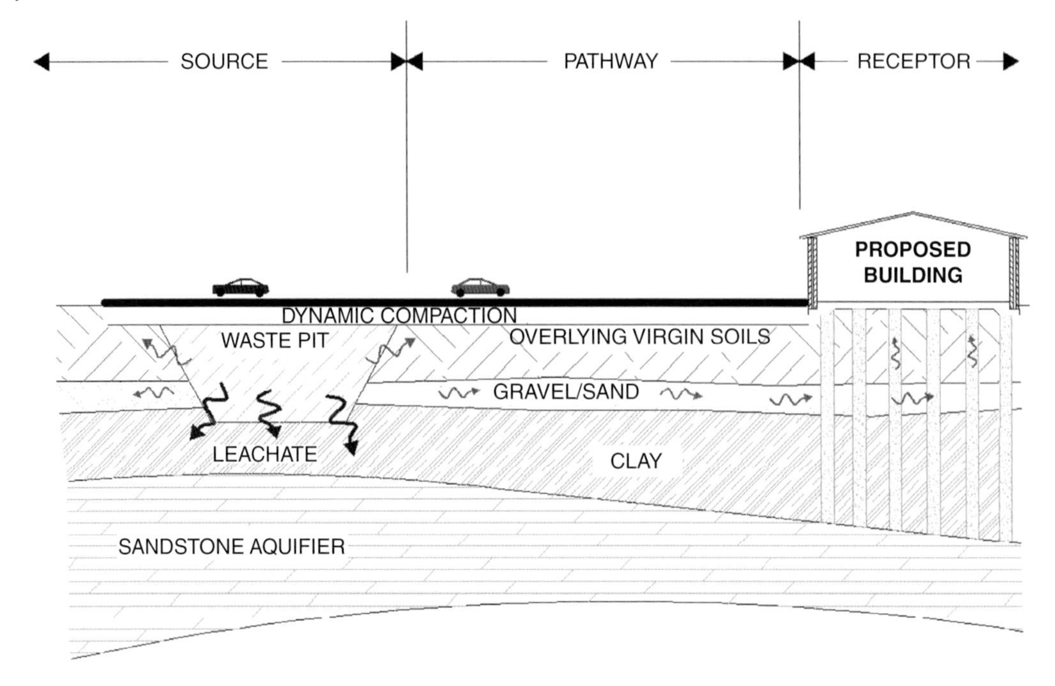

Figure 6.1 The influence of construction methods on preferential gas pathways.

are many methods, thicknesses, and various choices for forming the constructed void, ranging from traditional 'stilt-like' or 'sleeper-wall' types to proprietary plastic components installed permanently in the construction. The subfloor void should directly connect to the atmosphere, usually at the perimeter of the building to allow a free flow of air in or out of the void (passive ventilation), or it should connect to an appropriately sized fan (active ventilation).

Passive ventilation systems depend on the formation of a pressure differential between the inlets and outlets to create adequate air movement in order to vent the subfloor void. Some passive systems use the same ventilation termination type on all inlets and outlets, for example, cavity vents through masonry walls, relying on different wind speeds on each elevation of the structure to create the necessary pressure differential. Other passive systems use different types of inlet and outlet devices that create a pressure differential due to the variation in height, wind speed, and open area of each individual vent termination component (see Figure 6.2).

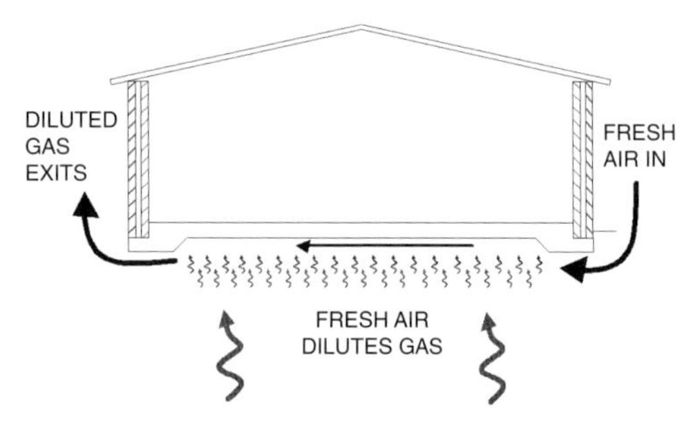

Figure 6.2 The principle of passive ventilation using subfloor voids.

Figure 6.3 illustrates two types of commercially available gas ventilation systems: on the left is a plastic-vacuum-formed structural void formerly known as a geocomposite which is rolled out on-site to create 25-mm-deep blankets of plastic void-forming 'egg-box-like' material. Typically, a concrete slab or foundation would be poured on top of the blanket to form a continuous void layer for ventilation beneath the slab. As is shown in Figure 6.3, sometimes it is installed in strips rather than a continuous blanket, particularly where lower levels of protection are required. On the right

Figure 6.3 Examples of proprietary ventilation systems.

of Figure 6.3, more robust, deeper (150 mm) injection- moulded plastic structural void-forming modules are shown, connected together to form a ventilation blanket beneath structures and roadways. This particular example is a proprietary material called Permavoid, which has a 92% void-to-plastic ratio and is strong enough to withstand loads greater than 70 tonnes per square metre.

Active ventilation can be used to either reduce the pressure in the void, causing fresh air to flow into the void from the atmosphere, or conversely increase pressure in the void to vent gases out from void into the atmosphere at safe and acceptable levels.

6.6.3 Secondary Protection – Low-Permeability Membranes

Low-permeability gas barriers act as an additional secondary protection in a passive gas protection system. The gas barrier should be correctly installed by suitably experienced and qualified professionals and in accordance with current best practice (CIRIA C735 2014) and an appropriate installation quality assurance protocol, for example, ISO 9001.

The main consideration in selecting gas-proof membranes is not their permeability but their durability during installation and continued intact operation after placement. When installing these membranes, care should be taken not to impose elongation beyond their yield capabilities, particularly where they pass movement joints in the structure and zones where differential settlement may be expected. Membranes containing aluminium foil in particular have very low coefficients of elasticity and will rupture with even slight elongation. Cavities in masonry need to be protected from potential gas ingress; sealed gas-resistant damp-proof courses should be used in these cases to avoid creation of slip planes during construction.

6.7 Practical Examples of Gas Protection Details

The following section shows examples of how gas protection details are shown on contract drawings and the corresponding photographs of each detail under construction. Figure 6.4 shows typical gas protection techniques required to adequately seal an isolated pile cap to prevent ingress of ground gas into the structure. Once the cast in situ pile has been constructed in the ground, the sacrificial pile top is broken down to the correct level, exposing the reinforcing steel tie bars. The top of the exposed surface of the pile is sealed with a liquid-based gas-resistant grout and the circumference of the pile sealed with gas-resistant self-adhesive flashing tape as shown in the lower photo. The upper photo

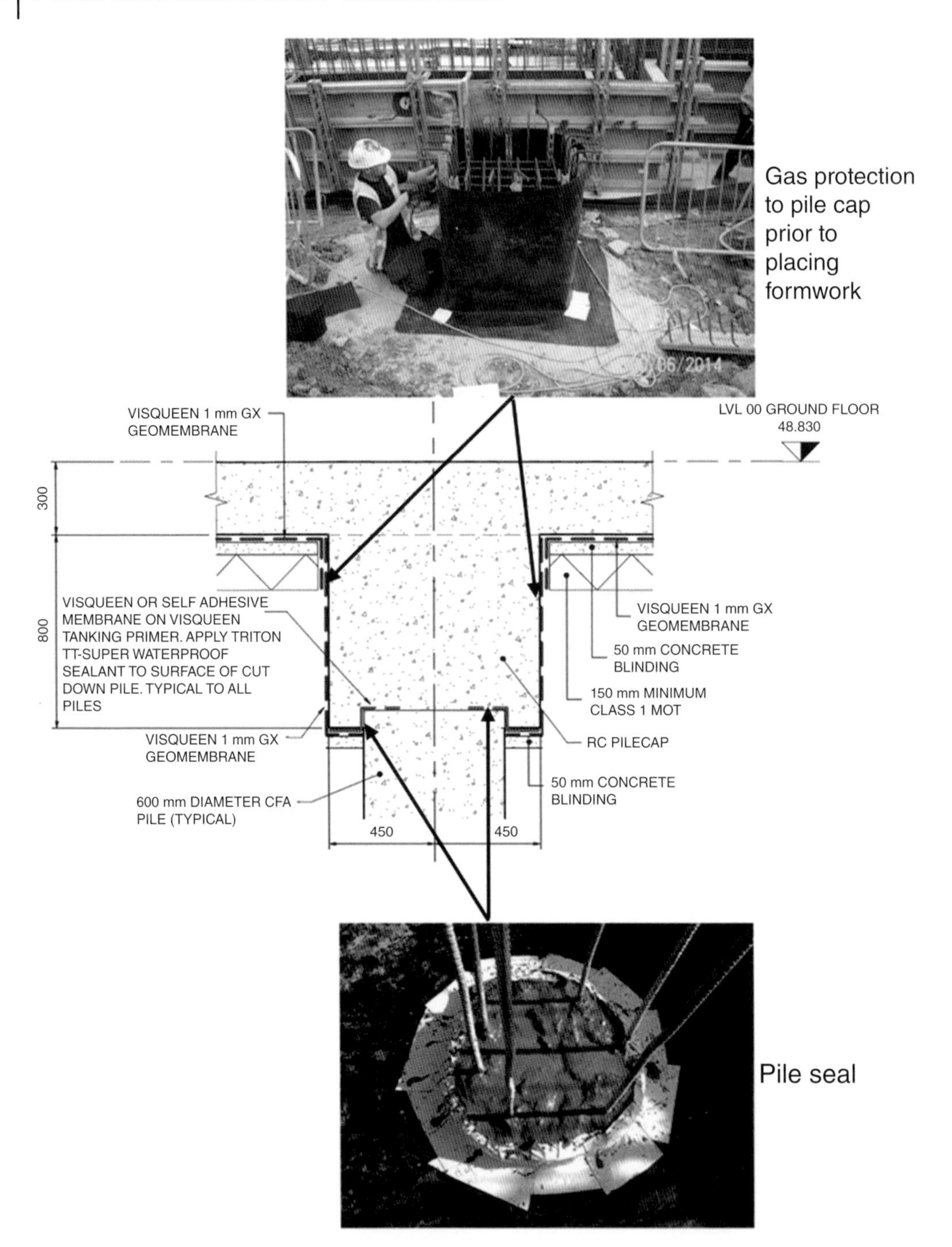

Figure 6.4 Isolated pile cap detail showing reinforcing steel bars.

shows the fully welded polyethylene gas-resistant liner around the isolated pile cap after the reinforcement steel has been fixed, but before the formwork is placed and concrete poured.

Figure 6.5 shows the gas protection membrane details of the pile caps containing multiple concrete cast in situ piles with horizontal and vertical fully welded polyethylene gas-resistant membrane prior to

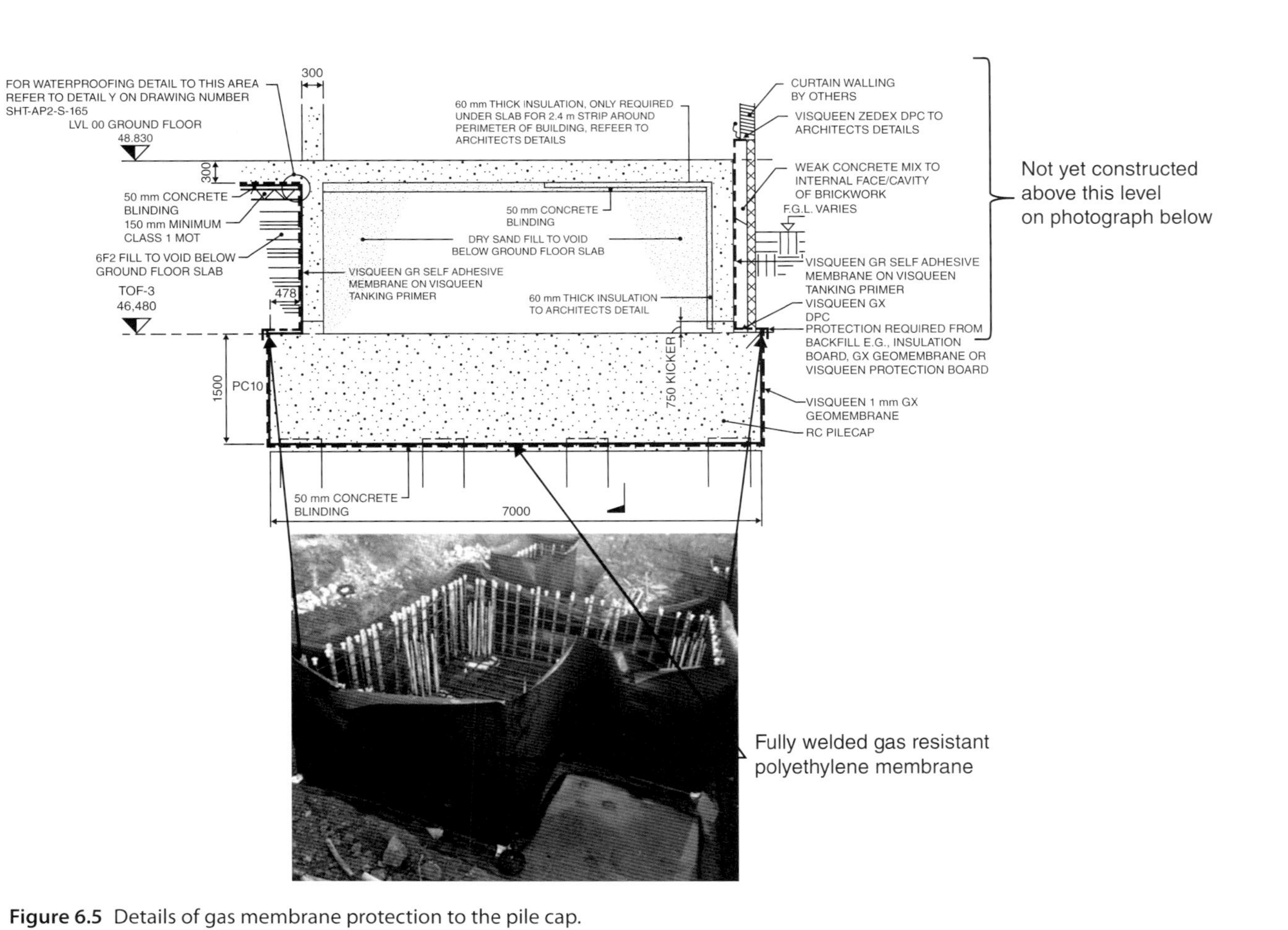

Figure 6.5 Details of gas membrane protection to the pile cap.

casting the reinforced concrete (RC) pile cap. Figure 6.5 shows the structures which would be installed above the pile cap.

6.7.1 Quality Assurance, Verification, and Integrity Testing

Verifying that a gas protection system works is as important as the quality of its design and installation in terms of the management of hazards posed by ground gases, and should be considered at the design stage. Following installation by appropriately qualified personnel, verification of the gas membranes should then be carried out by suitably competent, experienced, and trained persons, preferably independent of the installer. This verification comprises inspection of the ventilation layer, visual checking of compliance with the design, and confirmation there are no obstructions or impairments which might adversely affect its functioning. Vent terminations should undergo a final inspection to verify they are clear of any obstructions and that they are free flowing. Although integrity testing is not a substitute for well-designed and well-thought-out gas protection measures, a critical property of any gas membrane is its ability to survive the construction process, and therefore, integrity testing of installed gas membranes immediately prior to following trades is an important part of the verification process. Integrity testing is subdivided into two main elements:

Seam testing

- Pressurised air channel tests
- Mechanical point stress test
- Air lance test

In situ testing of large areas of membrane

- Tracer gas testing
- Dielectric porosity testing
- Smoke testing

For more detailed and comprehensive information and advice, Mallett *et al.* (2014) give authoritative guidance on both integrity testing and the documented verification procedure for gas protection systems. They also provide examples of good and bad practice.

6.8 Pathway Interception Systems for Receptor Protection and Control of Historic Sources

The risk management approach to evaluating ground gas protection regimes uses a source, pathway, receptor concept; each of these must exist in combination for a significant risk to human health to occur (see Figure 6.6).

As discussed earlier, the focus is generally on the receptor with protection provided in the form of substructure ventilation beneath an impermeable barrier as detailed in Section 6.4. This section concentrates on alternative solutions which intercept and disable the *pathway*, the advantage of which is that removal or treatment of the source is not necessary, nor is there any need to identify it as long as the migration route close to the receptor can be defined. Disabling the pathway eliminates the requirement of gas

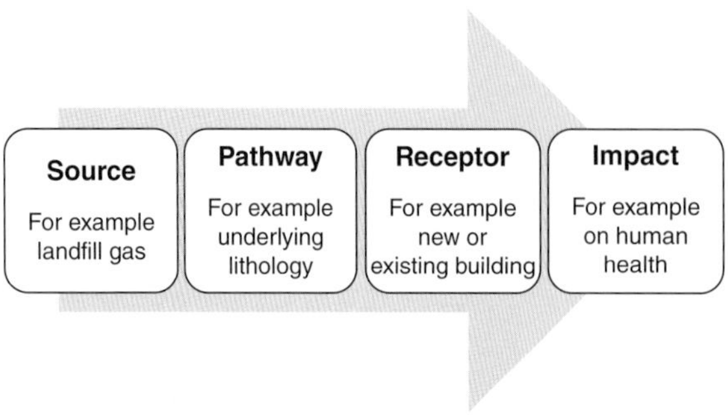

Source
For example landfill gas

Pathway
For example underlying lithology

Receptor
For example new or existing building

Impact
For example on human health

Figure 6.6 Source, Pathway, Receptor model.

protection measures for the receptor. Where it is appropriate to do so, pathway interception can be significantly more cost-effective than either source or receptor remedies; site operations associated with its installation tend to be remote from both source and receptor, so that construction times can be minimised. Under certain circumstances, the pathway interception approach can also permit development using receptor-centred protection otherwise deemed to be inadequate.

Pathway interception techniques need careful design considerations due to the inherent variability, depth, and extent of naturally occurring pathways. A comprehensive understanding of the site's gas regime together with a realistic conceptual model is required which, in addition to the depth and extent of the pathway, would consider the proximity of existing and planned development, hydrogeology, existence of services and utilities, temporary and permanent access roads as well as construction programme implications.

Traditional forms of pathway interception include 'cut-off trenches', where in-ground impermeable barriers such as bentonite slurry walls or welded impermeable geosynthetic liners are combined with ventilation trenches, filled with clean, single-sized aggregate enclosed with geotextiles to prevent migration of soil fines, to deal with the horizontal build-up of migrating ground gases. More recently, the aggregate part of the vent trenches have been replaced with more modern geosynthetic materials such as geocomposite void-forming blankets.

The procedures adopted have evolved considerably over time. For example, Figure 6.7 shows a scheme developed in 1997 whereby a 5.7-m-wide polyethylene membrane was vertically welded into the 7-m-deep trench. This approach was both dangerous, due to potential trench collapse and/or leachate contact on the welder, and it was also expensive. Modern-day health and safety regulations would prohibit this type of installation.

Figure 6.7 Vertical vent trench installation in 1997.

Figure 6.8 shows a section through a typical 'cut-off trench' formed using an 8.5-m-deep slurry wall (typically bentonite) with an immediately adjacent, shallow (approx. 4 m deep) gravel vent trench.

As is shown in Figure 6.8, the purpose of the cut-off trench is to intercept and 'cut off' the flow of ground gas emanating from the deeper area of waste and travelling along the sand lens pathways to the receptor. If a slurry wall is used, the construction plant required is substantial, carrying with it stability concerns which would typically prohibit construction of the trench any closer than 6 m from the site boundary as shown in Figure 6.8. One consequence of this is that a considerable volume of contaminated material would remain on the receptor side of the slurry wall, requiring extensive monitoring and investigation to demonstrate suitability for its retention, or its removal and replacement with inert material.

There are many other forms and variations of cut-off trenches or barrier system such as the installation of vertical impermeable

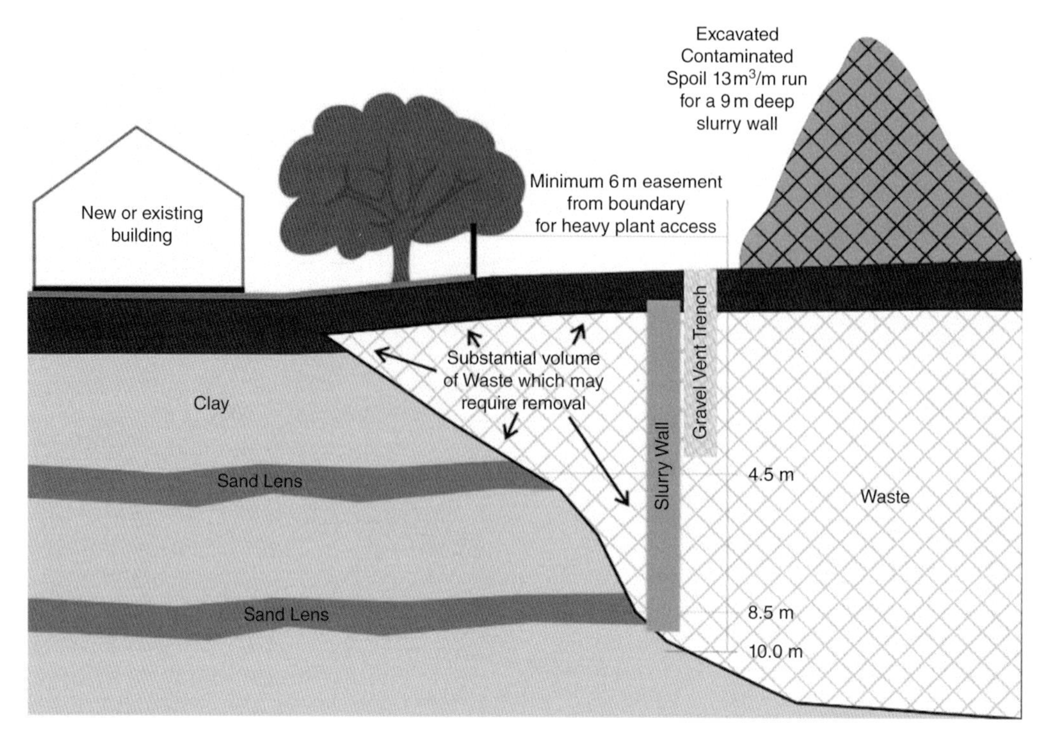

Figure 6.8 Section through typical 'cut-off' trench.

geomembrane liners in panels, with various types of proprietary jointing methods and installation techniques, as well as open cut 'V' trenches with geomembranes laid on the sloping sides of the excavation in combination with granular venting media or geosynthetic void formers, but all suffer from the same limitations associated with the need to excavate.

There is also a specific patented cut-off barrier system that requires little excavation and is installed at ground level with minimum risk to the workforce; it is particularly worthy of note due to the environmental and economic advantages that it provides. The system carries the trademark Virtual Curtain™ (see: http://www.virtual-curtain.com/) and is a unique no-dig system using a series of discreet vertically installed, highly voided, geosynthetic vent nodes, positioned at calculated spacings to create a virtual curtain in the ground, which depends on passive pressure differentials rather than barriers (see Figure 6.9). A zone of low pressure is created in the upper dilution duct which is situated beneath the ground surface; combining the pressure differentials with the vent pathways entrains the gases, passing them into the dilution duct. Venting is achieved through purpose-made vent stacks or bollards which together impart a flow of fresh air in the top ventilation duct for mixing with the ground gases, diluting them to safe levels prior to passively venting into the atmosphere.

The geosynthetic vents have an extremely high intrinsic permeability 'k' in excess of 1.75×10^{-5} m^2 (approximately 150 times greater than 20 mm single-sized gravel), which can accept inflows from both sides of the vertical vent nodes, reducing the potential for gas to accumulate. Each vent node has a zone of influence around it; the extent of this depends on local ground conditions, and thus the design is site specific. The vertical vent nodes are installed using high-frequency vibration equipment; as shown in Figure 6.10, they connect to a shallow upper horizontal

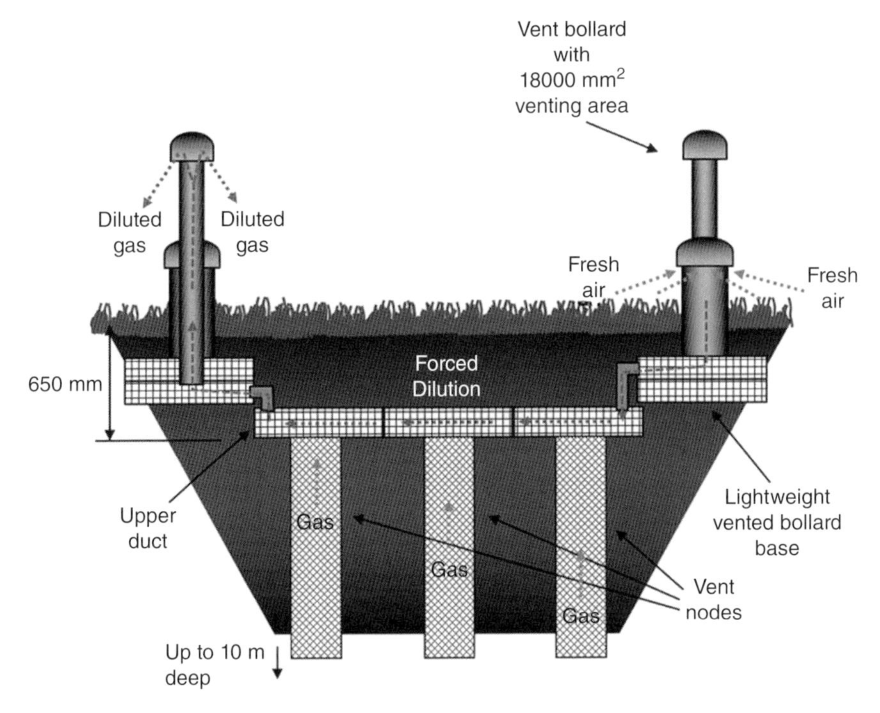

Figure 6.9 The principle of the Virtual Curtain™.

Figure 6.10 Insertion of void-forming vent nodes to construct the Virtual Curtain™.

gas collection and dilution duct, constructed in discrete sections with one end of each section attached to a proprietary vent bollard and the other to a vent stack. These operate together to induce a flow of fresh air through the duct to dilute and dissipate the ground gases. The vents can be offset from the virtual curtain itself, providing flexibility for subsequent construction.

An important characteristic of this approach is that it is not a physical barrier and will thus not affect the groundwater characteristics of a site, which will continue to behave normally, avoiding any potential issues with flooding, water abstraction, settlement, or heave.

Figure 6.11 demonstrates the potential advantages gained by selecting the Virtual Curtain™ in comparison with the same scenario as shown in Figure 6.8.

The Virtual Curtain system can be installed as close as possible (often less than 1 m) to the boundary due to the use of standard 360° tracked excavators for installation. This means there is a much smaller volume of waste material between the barrier and the boundary. Additionally, since it provides effective treatment from both sides, it is usually appropriate to leave a small amount of waste material against the site boundary in place. Only a small amount of excavation is required for the top dilution duct, which is generally such a small amount of inert material (approx. 0.65 m^3/m run) that it is usually unnecessary to remove from site and is not discernible once levelled out over the working area.

6.9 Examples of Other Techniques and Solutions: Biodegradation of Gaseous Pollutants

In urban situations, this branch of bioremediation has widest applications in odour control, such as where livestock need to be kept in urban environments (e.g. mounted police stables and urban educational farms). This air pollution control method often involves their removal using microorganisms on an engineered solid support, where the pollutants are retained long enough for the odorous, toxic, or otherwise harmful emissions to biodegrade. Figure 6.12 is a schematic of a simple system suitable for odour control in an urban animal housing situation.

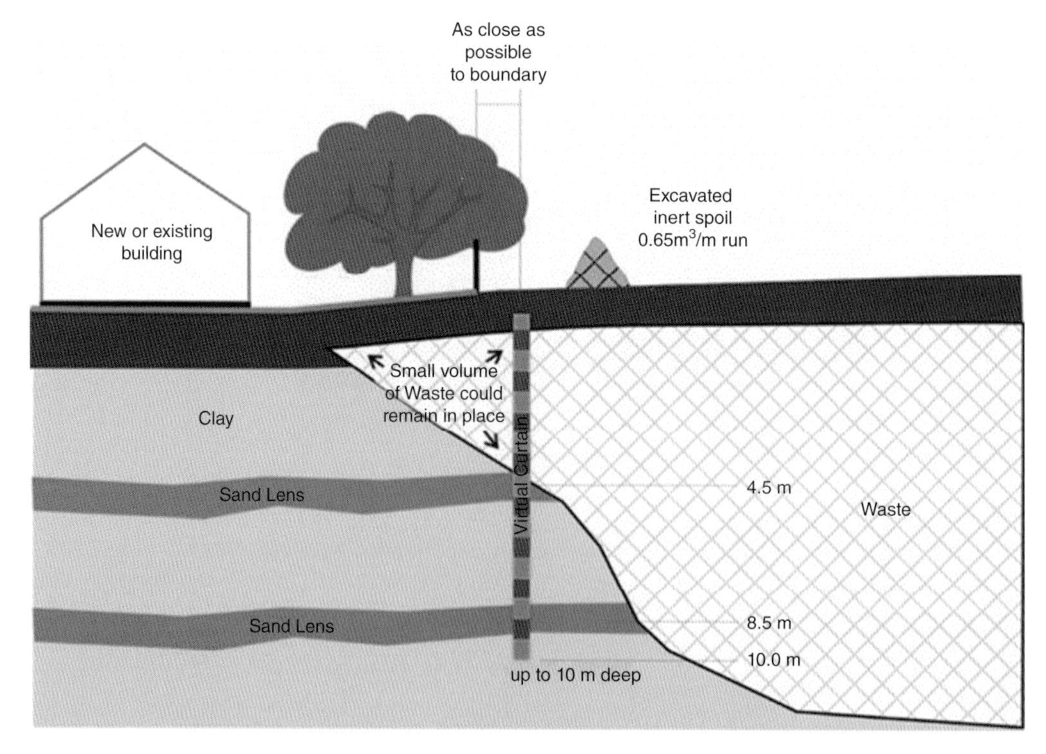

Figure 6.11 Section through typical Virtual Curtain™ installation.

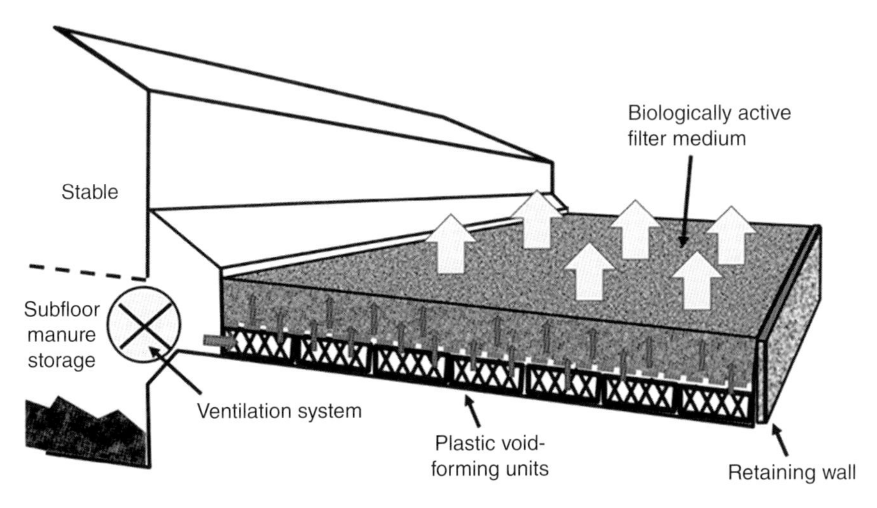

Figure 6.12 Schematic of a simple supported bed system suitable for odour control in an urban animal housing situation. Modified from Oliver (2017)

This scheme uses a simple biofilter such as an organic material which is porous and acts to filter the air; mulch, wood chips, and compost are examples of what can be used. Microbes live on the material as a biofilm, breaking chemicals such as hydrogen sulphide and ammonia to harmless products, removing up to 90% of the noxious smells from the air. They require proper design, construction, and maintenance but can be used equally efficiently on large-scale enterprises as well as smaller, urban-based ones.

6.10 Conclusion

This chapter has shown that there are many sources of ground gas and various solutions available to deal with the associated risks. Each solution should be developed to suit the specific requirements and needs of each individual project.

Several standards and best practice guidance documents are available containing an abundance of detailed information relating to each of the individual aspects associated with ground gas. It is therefore not the intention of this chapter to cover every aspect, but to provide a navigation route, from initial monitoring to final solution verification, through the issues, considerations, and necessary actions needed to provide robust and effective ground gas protection solutions to developments in urban areas.

Another purpose of this chapter is to highlight some of the less well-known aspects of ground gas, not merely the more unusual sources, for example, peat and carboniferous deposits, but also encompassing less obvious solutions like the pathway interception capabilities of the Virtual Curtain.

The author has over 25 years' experience with gas protection systems as a specialist installation contactor. Consequently, unusually and without apology, the chapter has a practical slant with sections devoted to the realities of actual installations. This is an immensely important aspect that is so often ignored in other publications.

References

Åkerblom, G., Andersson, P. and Clevensjö, B. (1984) Soil gas radon – A source for indoor radon daughters. *Radiation Protection Dosimetry*, 7, 1–4, 49–54.

Boyle, R. and Witherington, P. (2007) Guidance on Evaluation of Development Proposals on Sites Where Methane and Carbon Dioxide are Present, Incorporating 'Traffic Lights'. Report Ref 10627-R01-(02), National House Building Council. http://www.nhbc.co.uk/NHBCpublications/LiteratureLibrary/Technical/filedownload,29440,en.pdf

Bridgewater N. (n.d.) Awkright Town. http://www.oldminer.co.uk/arkwright-town.html

Card, G. (1995) Protecting Development from Methane. London: Construction Industry Research and Information Association.

Card, G., Wilson, S., and Mortimer S. (2012) A Pragmatic Approach to Ground Gas Risk Assessment. CL:AIRE Research Bulletin RB17. CL:AIRE, London, UK. ISSN 2047-6450. https://www.claire.co.uk/useful-government-legislation-and-guidance-by-country/212-assessing-risks-associated-with-gases-and-vapours-info-ra2-4

Crowhurst, D. and Manchester, S.J. (1993) The Measurement of Methane and Other Gases from the Ground CIRIA Report 131, CIRIA, London Assessment, Research Bulletin RB17, CL:AIRE .

Environment Agency (2004) Model Procedures for the Management of Land Contamination. Bristol: Environment Agency.

Guardian (2001) Deep Concern. https://www.theguardian.com/society/2001/jan/31/guardiansocietysupplement

ICRCL (1990) Notes on the Development and After-use of Landfill Sites. Guidance Note 17/78 – Interdepartmental Committee on the Redevelopment of Contaminated Land.

Landfill Gas Expert (xxxx n.d.) Notorious Landfill Gas Explosions During the 1980s in the UK and the US. http://landfill-gas.com/1980s-landfill-gas-explosions.html

Mallett, H. and Cox, L. (nee Taffel-Andureau), Wilson, S. and Corban, M. (2014) Good Practice on the Testing and Verification of Protection Systems for Buildings against Hazardous Ground Gases. CIRIA C735, London.

Mass.gov (n.d.) Appendix A Basics of Landfill Gas. http://www.mass.gov/eea/docs/dep/recycle/laws/lfgasapp.pdf

Mor, S., Ravindra, K., De Visscher, A., Dahiya, R.P. and Chandra, A. (2006) Municipal solid waste characterization and its assessment for potential methane generation: A case study. *Science of The Total Environment,* 371, 1–3, 1–10.

Oliver, J.P. (2017) Odor is More Than a Nuisance. http://smallfarms.cornell.edu/2017/01/09/odor-is-more/

POSTnote (2001) Reducing the Risk of Radon in the Home. Report 158. Parliamentary Office of Science and Technology. www.parliament.uk/post/home.htm

Public Health England (n.d.) Everything you need to know about radon http://www.ukradon.org/information/

Public Health England (PHE) (2009) Radon and Public Health. Report of an independent Advisory Group on Ionising Radiation. Chilton, Docs RCE 11. http://webarchive.nationalarchives.gov.uk/20101108170104/http://www.hpa.org.uk/Publications/Radiation/DocumentsOfTheHPA/RCE11RadonandPublicHealthRCE11/

Rudland, D. J., Lancefield, R.M. and Mayell, P.N. (2001) Contaminated Land Risk Assessment. A Guide to Good Practice (C552). CIRIA, 159pp.

Scivyer, C. (2015) Radon: Guidance on Protective Measures for New Buildings – Originally Introduced in 1991 and Amended in 1992, 1999 and 2007 Covering England and Wales. Report BR211. BRE Innovation Park, Watford, Bucknalls Lane, Watford. https://www.brebookshop.com/samples/327585.pdf

Thornbury, W. (1878) The Victoria Embankment. In: *Old and New London: Volume* 3, 322–329. *British History Online.* http://www.british-history.ac.uk/old-new-london/vol3/pp322-329

Williams, G.M. and Aitkenhead, N. (1991) Lessons from Loscoe: The uncontrolled migration of landfill gas. *Quarterly Journal of Engineering Geology and Hydrogeology,*

24, 191–207, 1 May 1991. 10.1144/GSL.QJEG.1991.024.02.03

Wilson, S. and Card, G. (1999) Reliability and risk in gas protection design. *Ground Engineering*, February 1999, 32–36.

Wilson, S., Oliver, S., Mallett, H., Hutchings, H. and Card, G. (2007) Assessing Risks Posed by Hazardous Ground Gases to Buildings. CIRIA Report C665.

7

Insights and Issues of Trace Elements Found in Street and Road Dusts

Susanne M. Charlesworth[1], *Eduardo De Miguel*[2], *Almudena Ordóñez*[3], *and Colin A. Booth*[4]

[1] *Centre for Agroecology, Water and Resilience, Coventry University, United Kingdom*
[2] *Laboratorio de Investigación e Ingeniería Geoquímica Ambiental (LI2GA) Environmental Geochemistry Research & Engineering Laboratory, Universidad Politécnica de Madrid, Spain*
[3] *Mining and Exploration Department, School of Mines, University of Oviedo, Spain*
[4] *Architecture and the Built Environment, University of the West of England, United Kingdom*

7.1 Introduction

Street and road dusts can accumulate in road gutters and gullypots (or catchpits), as well as adhering to road and pavement surfaces. There are various sources for these materials including diffuse and point sources, as well as from the atmosphere, surrounding soils, and the aquatic ecosystem. Land use (e.g. industrial, domestic, and vegetated) and design (e.g. road layout and traffic movements, distribution of buildings, and location of industry) have important consequences for the deposition and transport of these particulates around the urban environment. In their interactions with such activities, a combination of natural and anthropogenic chemical elements and compounds become associated with the dusts, which increases their potential to contaminate their surroundings (Foster and Charlesworth, 1996). This has consequences for human health, particularly since some of these particulates can become entrained into the air and subsequently inhaled, ingested, or adsorbed through the skin (Bowman *et al.*, 2003) and can also be tracked inside buildings on the feet of pedestrians and pets to be added to general house dust (Paustenbach *et al.*, 1997).

There has been much research to identify and characterise urban and non-urban sources of these contaminants both qualitatively, whereby elements are associated with their sources, and quantitatively, whereby relative contributions of different emission sources are compared to the total amount of a specified element in an identified urban material. Due to its toxicity, common occurrence in urban particulate materials, and the fact that its main urban source is well established (i.e. vehicles), Pb has been the most intensively investigated element in urban environments. However, it is now recognised that the introduction of unleaded fuel has reduced its further introduction into the environment, concentrating monitoring on Cu and Zn, as well as organic compounds (e.g. PAHs and TPHs) and new and emerging pollutants (e.g. nanoparticles). Studies, such as Charlesworth *et al.* (2003a), have shown that while historical Pb is still found in urban soils and street dusts, the amounts have reduced considerably. In a comparison of a study by Davis *et al.* (1987) in which the Pb concentrations in street dusts in Birmingham in 1976 were 1300

Urban Pollution: Science and Management, First Edition. Edited by Susanne M. Charlesworth and Colin A. Booth.

and 950 mg/kg in residential and industrial areas, respectively, they had reduced 11 years later to 791 and 527 mg/kg, respectively. Charlesworth *et al.* (2003a) found that Pb concentrations had reduced to 48.6 and 146.3 mg/kg across Birmingham in general, and specifically in the Balsall Heath area, respectively; the latter is the site of traffic calming measures, known for pollution associated with stop-start manoeuvres (Ewen *et al.*, 2009; Liu *et al.*, 2014).

This chapter considers the sources, pathways, deposits, and hazards associated with street and road dusts, and evaluates various management strategies to reduce their impacts in urban settings.

7.2 Sources of Street and Road Dusts

Sources of particulates can be classified into point and diffuse sources.

7.2.1 Characteristics of Atmospheric Diffuse Sources

Legislation to improve the quality of the air in cities in the United Kingdom was historically driven by the deaths of more than 4000 people due to the Great Smog of London in 1952. The introduced legislation mostly concentrated on reducing smoke, leading to the banning of open fires in the city. However, this did not have an impact on particles brought in from outside the city boundaries, which was combined with emissions from traffic, domestic heating systems, and industry, among others. Research sought to differentiate these sources by using the dusts' physico-chemical properties, an example of which is particle size distribution. It has been found (Van Dingenen *et al.*, 2004) that these airborne particulates have a bimodal particle size distribution, which differentiates the coarsest, or those of natural origin, which includes resuspended soil and mineral particles, from the finest (<2 μm), which are generally associated with anthropogenic activities, mainly combustion. Concentrating specifically on brake wear debris, Mosleh *et al.* (2004) also found this bimodal distribution; the fine fraction was at 350 nm. Research from different cities has determined that the dominant particle size is at the submicron size scale (Oberdörster *et al.*, 1995; Kasparian *et al.*, 1998).

In a review of the sources of particulates of non-vehicle exhaust origin, Thorpe and Harrison (2008) found reports that non-exhaust and exhaust sources were approximately equal in concentration, even on heavily trafficked roads and that the non-exhaust fraction was mostly represented by re-entrained road dust. They also state that in countries where studded tyres are used, 90% of the PM_{10} can be due to abrasion of the road surface. However, they do provide evidence that the proportion of brake wear debris in PM_{10}, $PM_{2.5}$, and $PM_{0.1}$ is 98, 39, and between 8% and 33%, respectively.

Viana *et al.* (2008) reviewed the various methods for elucidating the sources of particulate material across Europe, concluding that they fell into three main categories:

i) Monitored data, which involved numerical analysis of the resulting data to investigate trends, both spatially and temporally, comparing against background concentrations and undertaking correlations to establish associations.
ii) Analysis of emissions inventories or the construction of dispersion models. It was concluded, due to the lack of detailed emissions inventories and their general inaccuracy, that there were problems with this approach. However, it was felt that they were useful in scenario studies for the evaluation of emissions abatement strategies.
iii) The use of receptor models, whereby statistical analysis of the chemistry of particulate matter was used to identify sources of particulates in the aerosol. These models include chemical mass balance, which requires detailed information and multivariate models, which require far less.

They concluded that there was a lack of precise markers for specific emissions sources, giving the example of their inability to identify contamination due to shipping, and also that it was not possible to discriminate between fuels such as diesel and petrol. Loganathan *et al.* (2013) state that this kind of information is 'essential' for pollution control strategies and management approaches. While they list some of the more recent advances in physico-chemical and mineralogical analyses, the data from which have been used extensively in determining diagnostic ratios of elements, nevertheless they conclude that more research is needed in this area. Thorpe and Harrison (2008) go as far as to state that reliably tracing sources to road dusts 'remains problematic'.

Size and chemical composition determine the potential health effects of atmospheric particles. Particulate matter with a diameter below 10 μm (PM_{10}) is considered 'inhalable', while atmospheric particles with a diameter less than 2.5 μm ($PM_{2.5}$) are regarded as 'respirable'. According to Kappos *et al.* (2004), the latter fraction is associated with adverse impacts on human health; at its worst, it has been implicated in deaths, but it is also associated with conditions such as asthma. In terms of the urban environment, particulates cause ambient air quality issues, such as reduced visibility (Larson *et al.*, 1989; Lin and Tai, 2001). Consequently, research has mostly concentrated on the finest fraction, although Kappos *et al.* (2004) reported that there were difficulties in assessing the specific health effects of the ultrafine fraction (<0.5 μm). Some elements of toxicological concern that are found transported in the atmosphere include As, Cd, Cr, Hg, Mn, Ni, Pb, and V, exposure to which can cause conditions ranging from sinusitis, asthma, and chronic bronchitis to pneumonia, lung and brain haemorrhage, and lung cancer (Doadrio, 1984; Sadiq and Mian, 1994; Crosby, 1998). It is, however, difficult to be certain of these potential health effects in open, urban environments; nevertheless, their potential toxicity has led to research into the sources and levels of particulate trace elements in urban areas.

7.2.2 Point Sources of Particulates and Pollutants to Street and Road Dust

The aerodynamic diameter of the particle determines, in part, whether it stays airborne or settles out onto the pavement, soil, roof, window ledge, playground, and so on. Weather conditions are also important since the finest materials tend to stay suspended for longer or can be preferentially removed by wet deposition. Coarse particles, on the other hand, settle out by dry deposition (Jaffé *et al.*, 1993). Street dust can also incorporate large amounts of eroded soil, as well as particles that are not suspended after emission.

The residence time of street dust is typically short; in a coastal town in NW England, Allott *et al.* (1990) used ^{137}Cs as a tracer, finding that its half-life was between 190 and 370 days. Consequently, the temporal variability in the concentration of associated trace elements in street dust is high (Duggan, 1984), although monitoring is usually too short to be evaluated. However, in a study in NW Spain, Ordoñez *et al.* (2015) monitored the trace element concentration in street dusts every 5 years for a period of 15 years, representing four sampling campaigns, probably one of the longest such studies carried out on street dusts. They found that PM_{10} had decreased over the years, but that industrial sources such as Zn and Cd were still clearly identifiable, with elements such as As, Cd, Cr, Cu, Ni, and Zn exhibiting a net increase overall (Figure 7.1).

There are several point sources contributing to the load of road and street dust (Harrison, 1979; Hopke *et al.*, 1980; Schwar *et al.*, 1988), the most relevant of which is vehicular traffic, followed by emissions from domestic heating and fossil fuel burning, re-entrained polluted particulates arising from polluted soils and dusts, and other urban sources such as emissions from construction sites.

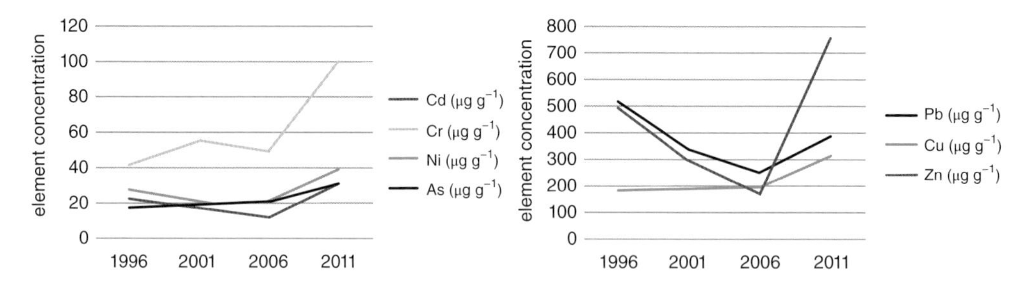

Figure 7.1 Increased concentrations of trace elements over time in a city in NW Spain (after Ordoñez *et al.*, 2015) (*Note:* Zn concentration × 10).

7.2.2.1 Sources Associated with Traffic

It is well known that vehicle exhaust emissions and car part wear are responsible for increased concentrations of Pb, Zn, Cd, Cu, and Ba near to roads. Lead was historically associated with the use of leaded petrol in internal combustion engines (Archer and Barrat, 1976), with Kowalczyk *et al.* (1978) concluding that the absolute concentration of Pb ranged from about 40% if there was minimal contribution from diesel traffic, to 4% with a significant diesel contribution due to the diluting effect of the carbonaceous particles emitted by diesel vehicles. The introduction of unleaded petrol in 1985 across Europe, reducing the content from 0.4 to 0.15 g^{-1}, led to a halving in atmospheric levels of Pb 'in a matter of weeks' according to Williams (2004, p 19). De Miguel *et al.* (1997) found the load of Pb in street dusts to be significantly reduced. Furthermore, a study by Charlesworth *et al.* (2003a) in Birmingham, West Midlands, United Kingdom, revealed that mean Pb concentrations in street dusts had reduced from 1300 mg kg^{-1} in residential streets in 1976 (Davies *et al.*, 1987) to 48 mg kg^{-1} by 2003. While Pb has reduced in street dusts and the urban aerosol, other studies monitoring the concentration of Pb in urban soils and deposited dust have shown that there remain significant levels of historical Pb stored in urban soils (e.g. Charlesworth *et al.*, 2003a).

The focus of research has subsequently switched from Pb to Zn and Cu, with Wong *et al.* (2006) suggesting that to be able to identify and quantify their sources, it would be useful to inventory their isotopic signatures, as has been carried out for Pb. Studies using radioactive isotope ratios, for instance, Pb and C (Widory *et al.*, 2004; Chen *et al.*, 2005a), enabled finer source discrimination. Widory *et al.* (2004) differentiated between diesel and fuel oil emissions using C isotopes, although they did admit that this was 'subject to debate' (p. 959). To further evaluate the relative contribution of traffic to the trace element load in urban particulate materials, the ratios of Ba/Pb, Br/Pb and, using factor analysis, by the scores on a factor that includes Pb, Cu, Ba, and Zn (Kowalczyk *et al.*, 1982; Sturges and Harrison, 1986; Boni *et al.*, 1988; Cornille *et al.*, 1990; Paterson *et al.*, 1996; De Miguel *et al.*, 1999; Viana *et al.*, 2006) have been used. Mazzei *et al.* (2008) modified factor analysis and positive matrix factorization with multiple sources, which they stated gave 'significant information on anthropogenic sources' (p. 87).

While high concentrations of Zn and Cd are usually linked to tyre wear (Thorpe and Harrison, 2008), Zn compounds, as well as Ba dispersions, are used as antioxidants and detergent or dispersant improvers in lubricating oils (Drew, 1975) and as corrosion inhibitors and smoke suppressants in diesel and other combustion engines. Oxidation of these lubricating oils by their exposure at high temperatures in air results in the formation of organic acids, alcohols, ketones, aldehydes, and other compounds, which are corrosive to metal. On contact with the oil, such metals wear, and since they are generally composed of Zn, Cu, and Cd alloys

(Drew, 1975), or of Ni, Cu, and Mo in the case of sinterised materials used in the automobiles' oil pump, they are released into the environment and can potentially deposit in association with street dust (De Miguel *et al.*, 1997).

Particle size distributions of exhaust emissions are particularly affected by driving patterns; for example, on the motorway, exhaust particulates have median diameters close to 0.1 μm; under normal driving conditions, they are usually about <1 μm; whereas urban driving causes a distinct shift towards coarser particle sizes of about >5 μm. However, this distribution can change with so-called stop-go and acceleration-deceleration activities, commonly found near traffic control lights, pedestrian crossings, or at road junctions (Charlesworth *et al.*, 2003a). This can cause resuspension of already deposited material, as well as the emission of larger particles from an exhaust system. These particulates then agglomerate, resulting in material that has a larger size than the constituent parts (Dongarrà *et al.*, 2003). Thus, due to the increase in particle size, most trace elements emitted in association with automobile exhausts do not travel far from their source; instead, they are deposited on pavement surfaces, to become street dust, or on local soils (Raunemaa *et al.*, 1986; Warren and Birch, 1987).

A further source of particulates to street dusts are those released during the construction of buildings, their subsequent weathering, renovation, and redecoration. In particular, galvanised street furniture, such as crash barriers, roofs, balconies, and window ledges can contribute substantial amounts of Zn and Cd to street dust (Fergusson and Kim, 1991). For example, raised concentrations of both elements to 44000 $\mu g\ g^{-1}$ and 20 $\mu g\ g^{-1}$ of Zn and Cd, respectively, were found in the material collected underneath metal ledges and balconies of old buildings (De Miguel *et al.*, 1997). Raised concentrations of Pb and Cd have been reported from many studies of paint flakes from weathered building facades or recently redecorated walls (Rundle and Duggan, 1986; Davies *et al.*, 1987; Schwar *et al.*, 1988; Fergusson and Kim, 1991). High levels of Ca are released in cement dust, especially when the ratio Mg/Ca is close to 0.14 (Kowalczyk *et al.*, 1982).

Catalytic converters were introduced in the United States in the mid-1970s and mid-1980s in Europe, whereupon it was realised that platinum group elements (PGEs) or platinum group metals (PGMs) (Pt, Pd, Rh, Ru, Ir, and Os; Ravindra *et al.*, 2004) were accumulating in the environment. Barbante *et al.* (2001) estimated that catalytic converters alone could release up to 1.4 ton Pt yr^{-1} on a global scale, with Schäfer *et al.* (1999) finding that the deposition rate for a typical urban area could be as much as 23 ng m^{-2} per day. Palacios *et al.* (2000) found that PGEs bioaccumulate and are transported at ultrafine particle sizes, generally <0.39 μm; these are the sizes generally considered to be inhalable and of greatest concern to human health (Kappos *et al.*, 2004). Ravindra *et al.* (2004) reviewed PGE levels in environmental materials, including the urban aerosol, concluding that their health effects were not well understood.

7.2.2.2 Domestic Heating, Coal, and Oil Combustion

Fuel that is used to heat domestic buildings varies, and therefore, its emissions profile also varies noticeably. The use of coal, for instance, leads to the release of Mn, Cr, Cu, Co, As, and Se depending on the type of coal used. In fact, trace elements found in resuspended soil can be similar to those of coal emissions, which can confuse the profile found and, therefore, the ability to estimate the relative contributions of coal to the atmosphere (Kowalczyk *et al.*, 1978; Tomza, 1984). Both of the latter authors emphasise the relative enrichment of As and Se and depletion of Mn in coal fly ash; while other studies (Pacyna, 1991; Rose *et al.*, 1994) have used Al, Si, and Ti as tracers of coal combustion found in the atmospheric environment.

Vanadium and, to a lesser extent, Ni and S have been used as tracers for oil combustion (e.g. Kowalczyk *et al.*, 1982; Boni *et al.*, 1988; Cornille *et al.*, 1990; Sadiq and Mian, 1994). However, it has been found that up to 40% of the V in the arid atmosphere has been due to a shale-like soil resuspension (Cornille *et al.*, 1990). Similar to coal combustion, it is difficult to assign an exact contribution of emissions due to oil combustion to the urban atmosphere as it depends on the particular origin of the oil used (Kowalczyk *et al.*, 1978).

7.2.2.3 Resuspension of Soil and Street Dust Particles

Thorpe and Harrison (2008) state that resuspension of material deposited on the road surface is an important source of particulates; both soil and street dust can be re-entrained by the wind where they can represent a significant proportion of the coarsest fraction (Harrison *et al.*, 1974). Other important means of resuspension include both the passage of vehicular and pedestrian traffic (e.g. Kupiainen, 2007; Patra *et al.*, 2008; Thorpe and Harrison, 2008), as well as activities associated with agriculture, construction, and street sweeping (Yuan *et al.*, 2003). Amato *et al.* (2010) reviewed the effectiveness of street sweeping, finding that vacuum-assisted and regenerative-air sweepers were more efficient than mechanical ones for the removal of fines, with the latter more efficient for removal of the coarser particles. It was, therefore, suggested that a combination of both types of sweeper would be more efficient overall. It has, however, been found that most forms of street sweeping are associated with the resuspension of the finer, more polluted fraction, leading to the use of dust suppressants, which simply adds to the mixture of chemicals on urban surfaces, and also damping down dusts using water, which is reliable, but not suitable where water-use restrictions apply (Amato *et al.*, 2010). Soil re-entrainment is probably mainly associated with K, Mg, and Mn, although as stated above, this is difficult to differentiate from coal combustion. Coal and soil may, therefore, provide significant sources of other elements such as Al, Ca, Ce, Cr, Fe, La, Sc, Sr, Ti, and Th (Kowalczyk *et al.*, 1978; Boni *et al.*, 1988).

Resuspension of street dusts can significantly add to the inhalable trace element load and, hence, to human health impacts. Laidlaw and Filippelli (2008) studied health risks, particularly to the young, from resuspended soil that was contaminated with Pb, both inside and outside their home. For children, the study found blood lead levels (BLLs) in excess of 10 μm dL^{-1} in some US cities, whereas in Cairo, Egypt, Sharaf *et al.* (2008) found children's BLLs up to 14.3 μm dL^{-1} by heavily trafficked roads. In comparison, the CDC (2007) advisory level is 10 μm dL^{-1}.

Particulate matter can originate outside cities, the finest fraction of which can constitute the 'natural' background resulting from crustal erosion, which can travel long distances (Cornille *et al.*, 1990). Soil outside urban areas can provide background values against which the chemical fingerprint of the street or road dust can be compared and enable an estimate of the enrichment to be made. Loganathan *et al.* (2013) reviewed the various ways in which enrichment could be represented, dividing the approaches into four general categories:

i) The geoaccumulation index, I_{geo}, has been used by several authors (e.g. Faiz *et al.*, 2009; Lu *et al.*, 2009; Wei *et al.*, 2009).

$$I_{geo} = \log_2(C_n/1.5B_n)$$

Where C_n = road dust pollutant concentration

B_n = background concentration of the element in soil

Level of contamination (after Nowrouzi, and Pourkhabbaz, 2014)

0	Uncontaminated
0–1	Uncontaminated to moderate contamination
1–2	Moderate contamination
2–3	Moderate to strongly contaminated
3–4	Strongly contaminated
4–5	Strongly to extremely contaminated
>5	Extremely contaminated

ii) The pollution index, PI, or contamination ratio or contamination factor (used by Faiz *et al.*, 2009; Christoforidis and Stamatis, 2009; Duong and Lee, 2011) and is represented by:

$PI = Cn/B_n$

Level of contamination (after Chen *et al.*, 2015)	
<1	Low
1–3	Middle
>5	High

iii) The integrated pollution index, IPI (used by Faiz *et al.*, 2009; Wei and Yang, 2010), is the mean PI of all contaminants. According to Wei and Yang (2010), the categorisation of the IPI is as follows: <1 = low; 1–2 = moderate; 2–5 = high, and >5 = extreme levels of pollution. In a study of street dusts in Iran, sampled from different land uses, Kamani (2015) reported IPIs of 3.65 for commercial areas, 2.76 for highly trafficked, 1.68 for industrial, 1.53 for urban parks, and 1.23 for residential, illustrating that while residential areas were the lowest, they were still moderately polluted, or above background, and that commercial land was highly contaminated.
iv) The enrichment factor, EF, is used widely and compares the concentration of the element of concern with a specific reference. This could be a published background value, perhaps of a global mean (e.g. Macklin, 1992), although if a local value is available, that is a better comparator.

$EF = (C_x/C_{ref})_{sample}$

$(C_x/Cr_{ef})_{background}$

Where C_x = pollutant concentration

C_{ref} = reference concentration

Level of enrichment (after Nowrouzi and Pourkhabbaz, 2014)	
<1	Not enriched
1–3	Minor
3–5	Moderate
5–10	Moderate–severe
10–25	Severe
25–50	Very severe
>50	Extremely severe

Table 7.1 summarises some of the results obtained in urban areas worldwide using the various enrichment factors when applied to urban solid matrices.

7.2.2.4 Other Urban Sources

Other sources, in addition to those detailed above, include those from specific industrial activities, incineration, construction activities, road weathering, and maintenance. The combustion-of-waste emission profile produces an profile dependent on various factors including the composition of the waste itself, the design of the chamber where the combustion is carried out, as well as the efficiency of the filters and scrubbers on the outlet chimney. Kowalczyk *et al.* (1978, 1982) and Pacyna (1983) report that such combustion is a major atmospheric source of Zn, Cd, and Sb, while Wadge *et al.* (1986) found high levels of Pb and Cd in the finest fraction of fly ash from waste combustion.

Building construction and renovation, and the weathering of building materials have impacts on the urban environment, mainly due to the release of Ca in cement dust, which Gatz (1975) was able to use as a tracer. Kowalczyk *et al.* (1982) categorised airborne cement particles according to their concentration ratios, whereby the K:Ca was about 0.006 and the Mg:Ca was about 0.16. According to Charlesworth and Foster (2005), relatively high activities of radioactive nuclides were found in some road gutter and street dusts in Coventry, United Kingdom. Possibly associated with pulses of construction activity, those with the highest activities exceeded the ICRP (1991) guidelines of 1 mSv yr^{-1} for members of the public.

In northern countries, during winter the use of spiked tyres results in attrition of the

Table 7.1 Examples of enrichment of trace metals in street dusts reported worldwide.

Method	Country reference	Elements				
		Pb	Zn	Cu	Cd	Ni
I_{geo}	Pakistan[1]	U–M	U	U–M	U	U
	India[2]	M–S	U	U	U	U
EF	Argentina[3]	Severe	M–Severe	Severe	n.d.	M–Severe
	China[4]	Severe	Severe	M	n.d.	Minor
PI	China[5]	Middle	Middle	Middle	Middle	Low
	Jordan[6]	High	High	Middle	Middle	Middle

Where n.d. = no data; U = uncontaminated; M = moderate contamination; S = strongly contaminated.
Sources: 1. Faiz *et al.* (2009); 2. Singh (2011); 3. Fujiwara *et al.* (2011); 4. Lu *et al.* (2009); 5. Chen *et al.* (2005b); 6. Al-Khashman (2013).

road surface, generating dust particles, which can become a major source of atmospheric particulate matter. In fact, Lindgren (1996) reported that an average of 24 g km^{-1} vehicle^{-1} of asphalt was worn off the road surface in Sweden during winter.

Other sources are site specific and include sea spray impacts for coastal cities. This can supply large amounts of Na to the atmosphere (Kowalczyk *et al.*, 1978), which needs to be accounted for to avoid misleading conclusions being drawn regarding the quality of the atmosphere (Pryor *et al.*, 2008).

Depending on their physical characteristics, airborne particulates can settle onto surfaces in urban areas – those accumulating outdoors are collectively referred to as 'street dust', while those found inside urban properties are commonly termed 'house dust'. While strictly beyond the scope of this chapter, nevertheless it is important to consider the contribution of street dusts to those accumulating inside buildings, and hence their risk to urban dwellers.

7.3 House Dust

Street dust can find its way inside a building on clothing and footwear, by the wind and adhere to the fur of domestic pets (Tong, 1998). As with street dust, there have been concerns with the potential of inhalation, ingestion, and dermal exposure to house dust, which has led to research being carried out on this material. It has been cited (Edwards *et al.*, 1998) as one of the main sources of pesticides and metals found indoors, particularly Pb in children.

In common with other studies, Turner and Simmonds (2006) found concerning enrichment of Cd, Cu, Pb, Sn, and Zn in dusts from four regions across the United Kingdom. While Tong and Kin Che Lam (2000) found that cleaning activities (such as floor sweeping and dusting) reduced the levels of metals indoors in Hong Kong, the type of paint used to decorate the house and its age were more important when sourcing indoor metal. In Australia, Chattopadhyay *et al.* (2003) found that concentrations of Pb in the outdoor urban atmosphere had reduced due to the introduction of unleaded petrol, but that of household dusts basically remained unchanged, due not only to the accumulation of Pb inside the house from the historical use of leaded paints, as well as the accumulation of decades of particulates associated with leaded petrol. Contaminated soil is also an important contributor to both street and indoor dusts because it acts as a storage repository for historical contamination (e.g. Charlesworth *et al.*, 2003a). The following section therefore considers levels and sources of soil contamination in urban areas.

7.4 Urban Soil

Urban soils can be both a source and a sink of polluted particulates (Chapter 5); concentrations are usually lower than those in street dust, but the soil can be enriched in trace elements relative to natural background (e.g. Charlesworth *et al.*, 2003a; Charlesworth and Foster, 2005; Biasioli *et al.*, 2007). The main sources and transport pathways of these trace elements include those already described for street dusts in previous sections. It is difficult to quantify the contributions of these sources, since their inputs are incorporated into the soil over time; however, they can retain a record of past pollution as they are a net accumulator of trace elements. Consequently, when re-entrained, they can be a source of additional particulates to street and house dusts. De Miguel *et al.* (1997) found that the highest concentrations of Pb in street dusts from Oslo were associated with nearby soils rather than with dense traffic, where Pb had accumulated from a smelter that had shut down several years earlier.

Urban soil is also disturbed regularly by processes including landscaping, construction, irrigation, and partial or complete replacement. However, atmospheric particle deposition is reasonably uniform across the city giving an idea of 'urban background', which is generally higher than that for natural soils. De Miguel *et al.* (1998), for example, found that the urban soil of Madrid was 2.3 (Zn), 2.6 (Cu), and 4.0 (Pb) times natural background levels. Industrial sources can considerably increase trace element soil concentrations: Ordóñez *et al.* (2003) found concentrations of Zn and Cd up to 2000 $\mu g\ g^{-1}$ and 8 $\mu g\ g^{-1}$, respectively, in soil downwind of a Zn smelter. It is also possible to identify so-called 'hot spots', whereby spatial distributions of certain elements can be plotted across a city. Charlesworth *et al.* (2007) plotted Zn, Ni, Cu, Cd, and Pb concentrations across the City of Coventry, United Kingdom, finding 'hot spots' associated with heavily trafficked main roads and industrial areas.

7.5 Urban Geochemical Cycles

In Charlesworth *et al.* (2000), the urban particulate environment was described as a 'cascade' in which sources included point sources, fluvial bed sediments, and polluted dusts. These particulates were transported as suspended sediment in water, storm sewers, rivers, and streams or as dusts via the atmosphere, eventually being deposited on the street surface, in gully pots or urban lakes. However, the urban environment is highly complex. Charlesworth *et al.* (2000) called it 'frustrating' (p. 356) with the physico-chemical characteristics of the particles being impacted by a wide variety of processes as they moved between the environmental compartments of the cascade. Figure 7.2 is based on De Miguel *et al.* (1998) and shows the production of street and house dusts from the various sources discussed in this chapter. In a fingerprinting exercise, Charlesworth *et al.* (2000) did not find many relationships between the geochemical and geophysical characteristics of the sediment taken from individual compartments of the cascade.

Trace elements are generally transported in association with particulates – or particulate-associated pollutants (PAPs). The finest fractions are particularly important for the following reasons:

i) Particles whose diameter is <100 μm can be re-entrained and easily transferred between soil, street dust, and atmospheric aerosol.
ii) Particles in the silt–clay size range have the highest binding capacity for trace elements, and therefore the highest capacity for transporting them.

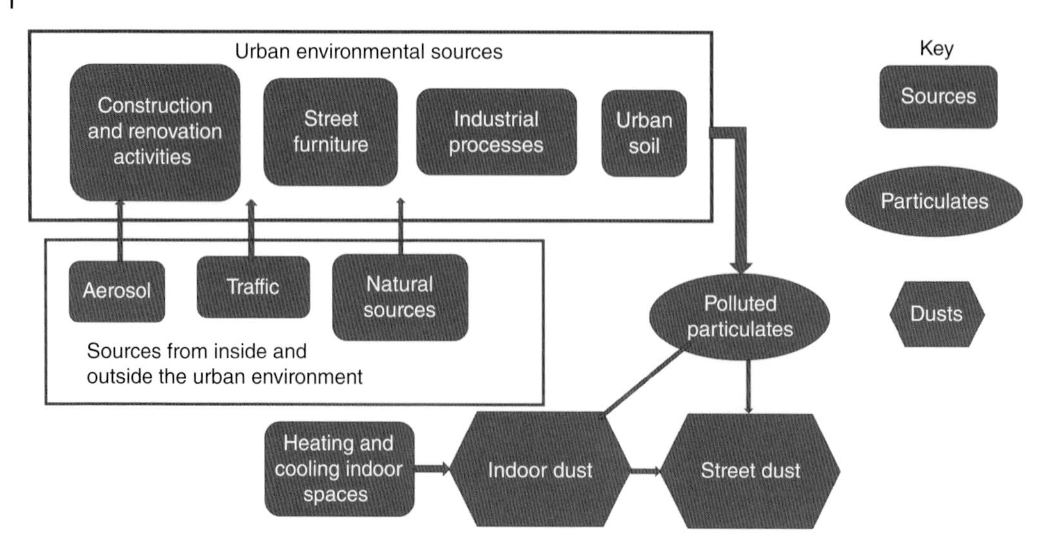

Figure 7.2 Sources from inside and outside of the urban environment associated with the production of house and street dusts.

As shown in Figure 7.2, the sources of material to street dusts always includes natural material, such as soil or airborne particulates from outside the urban environment. The chemistry of this material will be related to its geological origin, but may contain elements such as Ce, Ga, La, Th, and Y, which are found in urban deposits. These elements can provide a geochemical fingerprint since they can be found in cities worldwide, all through the urban particulate cycle, eventually appearing in urban sediments. De Miguel *et al. (*1999*)* were able to identify these natural trace elements in cities with quite different characteristics (e.g. Madrid, Oslo, and Ostrava).

De Miguel *et al.* (1998) were able to follow the pathway of Ag in the city of Madrid due to its incorporation in the compost applied by the local authority to public parks and gardens. Sources of Ag include X-ray plates, dental alloys, photographic film, or industrial materials in Ag–Zn and Ag–Cd batteries. These materials are eventually disposed of, releasing Ag into the urban aquatic system, and it finally passes through a wastewater treatment plant. The Ag can become concentrated in the sludge, reaching up to 45 $\mu g\ g^{-1}$; it is this sludge that is processed to become the compost widely used by authorities as a soil amendment in cities. Silver can be further concentrated in compost, reaching values of 50–70 $\mu g\ g^{-1}$, due to mass loss during the fermentation process as the sewage sludge matures. Not all the Ag makes it to the wastewater treatment plant; De Miguel *et al.* (2005) found concentrations of up to 16 $\mu g\ g^{-1}$ Ag in sediments of the river Manzanares, which runs through central Madrid, suggesting that some of the Ag is stored in the river sediments.

7.6 Conclusions

Urban population increase has seen cities become 'the most dominant human habitat in history' (Wong *et al.*, 2006, p. 12). However, associated with this concentration of human beings is the production of polluted materials, which accumulate and have the potential to have serious impacts on human health and quality of life. Fundamental changes have occurred, such as the gradual phasing out of leaded petrol after evidence of the harm caused to inhabitants of towns and cities came to light. However, once the main source of Pb had been addressed, it soon became evident that other pollutants in what Turner (1992) described as a 'multicomponent, multiphase' environment were also problematic, and research has subsequently focused on them. The urban environment is constantly changing, making the modelling of such complex mixtures difficult, but research is also evolving in terms of methodologies and analytical techniques. Knowledge of the pathways and processes leading to the accumulation of such pollutants in the urban environment is vital if they are to be managed, benefiting both environmental and human health.

References

Al-Khashman, O.A. (2013) Assessment of heavy metals contamination in deposited street dusts in different urbanized areas in the city of Ma'an, Jordan. *Environmental Earth Sciences,* 70, 2603–2612.

Allott, R.W., Hewitt, C.N., and Kelly, M.R. (1990) The environmental half–lives and mean residence times of contaminants in dust for an urban Environment: Barrow-in-Furness. *Science of the Total Environment,* 93, 403–410.

Archer, A. and Barrat, R.S. (1976) Lead levels in Birmingham dust. *Science of the Total Environment,* 6, 275–286.

Barbante, C., Veysseyre, A., Ferrari, C., Van de Velde, K., Morel, C., Capodaglio, G., Cescon, P., Scarponi, G., and Boutron, C. (2001) Greenland snow evidence of large scale atmospheric contamination for platinum, palladium and rhodium. *Environmental Science & Technology,* 35, 835–839.

Biasioli, M., Grčman, H., Kralj, T., Díaz-Barrientos, E., and Ajmone-Marsan, F. (2007). Potentially toxic elements contamination in urban soils: A comparison of three European cities. *Journal of Environmental Quality,* 36, 70–79.

Boni, C., Caruso, E., Cereda, E., Lombardo, G., Braga Marcazzan, G.M., and Redaelli, P. (1988) Particulate matter elemental characterization in urban areas: Pollution and source identification. *Journal of Aerosol Science,* 19, 1271–1274.

Bowman, C.A., Bobrowsky, P.T., and Selinus, O. (2003) Medical geology: New relevance in the earth sciences. *Episodes,* 270–278.

Boyd, H.B., Pedersen, F., Cohr, K.-H., Damborg, A., Jakobsen, B.M., Kristensen, P., and Samsøe-Petersen, L. (1999) Exposure scenarios and guidance values for urban soil pollutants. *Regulatory Toxicology and Pharmacology,* 30, 197–208.

CDC (Centers for Disease Control and Prevention) (2007) Advisory committee on childhood lead poisoning prevention. *MMWR Recommendations,* 56(RR–80), 1–16.

Charlesworth, S.M., Ormerod, L.M., and Lees, J.A. (2000) Tracing sediments within urban catchments using heavy metal, mineral magnetic and radionuclide signatures. In: Foster, I.D.L. (ed.), *Tracers in Geomorphology.* John Wiley & Sons.

Charlesworth, S.M., Everett, M., McCarthy, R., Ordóñez, A., and De Miguel, E. (2003a) A comparative study of heavy metal concentration and distribution in deposited street dusts in a large and a small urban area: Birmingham and Coventry, West Midlands, U.K. *Environment International,* 29/5, 563–573.

Charlesworth, S.M., Harker, E., and Rickard, S. (2003b) Sustainable Drainage Systems (SuDS): A soft option for hard drainage questions? *Geography,* 88 (2), 99–107.

Charlesworth, S.M. and Foster, I.D.L. (2005) Gamma-emitting radionuclide and metallic elements in urban dusts and sediments, Coventry, UK: Implications of dosages for dispersal and disposal. *Mineralogical Magazine,* 69 (5), 759–767.

Charlesworth, S.M., Booty, C., and Beasant, J. (2007) Monitoring the atmospheric deposition of particulate-associated urban contaminants, Coventry, UK. Morrison, G.M. and Rauch, S. (eds.), *Proceedings of the*

8th Highway and Urban Environment. Symposium Series: Alliance for Global Sustainability Bookseries, Vol. 12, 155–165.

Chattopadhyay, G., Lin, K.C.-P., and Feitz, A.J. (2003) Household dust metal levels in the Sydney metropolitan area. *Environmental Research*, 93, 301–307.

Chen, J., Tan, M., Li, Y., Zhang, Y., Lu, W., Tong, Y., Zhang, G., and Li, Y. (2005a). A lead isotope record of Shanghai atmospheric lead emissions in total suspended particles during the period of phasing out of leaded gasoline. *Atmospheric Environment*, 39, 1245–1253.

Chen, T.-B., Zheng, Y.M., Lei, M., Huang, Z.C., Wu, H.T., Chen, H., Fan, K.K., Yu, K., Wu, X., and Tian, Q.Z. (2005b) Assessment of heavy metal pollution in surface soils of urban parks in Beijing, China. *Chemosphere*, 60 (4), 542–51.

Cornille, P., Maenhaut, W., and Pacyna, J.M. (1990) Sources and characteristics of the atmospheric aerosol near Damascus, Syria. *Atmospheric Environment*, 24A, 1083–1093.

Crosby, D.G. (1998) *Environmental Toxicology and Chemistry*. Oxford University Press, pp. 336.

Davies, D.J.A., Watt, J.M., and Thornton, I. (1987) Lead levels in Birmingham dusts and soils. *Science of the Total Environment*, 67, 177–185.

De Miguel, E., Jiménez de Grado, M., Llamas, J.F., Martín-Dorado, A., and Mazadiego, L.F. (1998) The overlooked contribution of compost application to the trace element load in the urban soil of Madrid (Spain). *Science of the Total Environment*, 215, 113–122.

De Miguel, E., Llamas, J.F., Chacón, E., and Mazadiego, L.F. (1999) Sources and pathways of trace elements in urban environments: A multi-elemental qualitative approach. *Science of the Total Environment*, 235, 355–357.

De Miguel, E., Llamas, J.F., Chacón, E., Berg, T., Larssen, S., Røyset, O., and Vadset, M. (1997) Origin and patterns of distribution of trace elements in street dust. Unleaded petrol and urban lead. *Atmospheric Environment*, 31, 2733–2740.

Dongarrà, G., Sabatino, G., Triscari, M., and Varrica, D. (2003) The effects of anthropogenic particulate emissions on roadway dust and *Nerium oleander* leaves in Messina (Sicily, Italy). *Journal of Environmental Monitoring*, 5, 766 – 773.

Drew, H.M. (1975) *Metal-Based Lubricant Composition*. Noyes Data Corporation, 332 pp.

Duggan, M.J. (1984) Temporal and spatial variations of lead in air and surface dust – Implications for monitoring. *Science of the Total Environment*, 33, 37–48.

Duggan, M.J., Inskip, M.J., Rundle, S.A., and Moorcroft, J.S. (1985) Lead in playground dust and on the hands of schoolchildren. *Science of the Total Environment*, 44, 65–79.

Edwards, R.D., Yurkow, E.J., and Lioy, P.J. (1998) Seasonal deposition of housedusts onto household surfaces. *Science of the Total Environment*, 224, 69–80.

Ewen, C., Anagnostopoulou, M.A., and Ward, N.I. (2009) Monitoring of heavy metal levels in roadside dusts of Thessaloniki, Greece in relation to motor vehicle traffic density and flow. *Environmental Monitoring and Assessment*, 157, 483. doi:10.1007/s10661-008-0550-9.

Fergusson, J.E. and Kim, N.D. (1991) Trace elements in street and house dusts: Sources and speciation. *Science of the Total Environment*, 100, 125–150.

Ferreira-Baptista, L. and De Miguel, E. (2005) Geochemistry and risk assessment of street dust in Luanda, Angola: A tropical urban environment. *Atmospheric Environment*, 39, 4501–4512.

Foster, I.D.L. and Charlesworth, S.M. (1996) Heavy metals in the hydrological cycle: Trends and explanation. *Hydrological Processes*, 10, 227–261.

Gatz, D. (1975) Relative contribution of different sources of urban aerosols: Application of a new estimation method to multiple sites in Chicago. *Atmospheric Environment*, 9, 1–18.

Harrison, R.M. (1979) Toxic metals in street and household dusts. *Science of the Total Environment,* 11, 89–97.

Harrison, P.R., Draftz, R.G., and Murphy, W.H. (1974) Identification and impact of Chicago's ambient suspended dust. *Proceedings of 'Atmospheric–Surface Exchange of Particulate and Gaseous Pollutants (1974)',* Richland, Washington, 4–6 September 1974. Energy Research and Development Administration symposium series, CONF–740921. National Technical Information Service, U.S. Dept. of Commerce, pp. 557–570.

Hopke, P.K., Lamb, R.E., and Natusch, D.F.S. (1980) Multielemental characterisation of urban roadway dust. *Environmental Science & Technology,* 14, 164–172.

Jaffé, R., Carrero, H., Cabrera, A., and Alvarado, J. (1993) Organic compounds and heavy metals in the atmosphere of the city of Caracas, Venezuela – II: Atmospheric deposition. *Water, Air, & Soil Pollution,* 71, 315–329.

Kamani, H., Ashrafi, S.D., Isazadeh, S., Jaafari, J., Hoseini, M., Mostafapour, F.K., Bazrafshan, E., Nazmara, S., and Mahvi, A.H. (2015) Heavy metal contamination in street dusts with various land uses in Zahedan, Iran. *Bulletin of Environmental Contamination and Toxicology,* 94 (3), 382–386.

Kappos, A.D., Bruckmann, P., Eikmann, T., Englert, N., Heinrich, U., Höppe, P., Koch, E., Krause, G.H.M., Kreyling, W.G., Rauchfuss, K., Rombout, P., Schulz-Klempi, V., Thiel, W.R., and Wichmann, H.-E. (2004) Health effects of particles in ambient air. *International Journal of Hygiene and Environmental Health,* 207, 399–407.

Kasparian, J., Frejafon, E., Rambaldi, P., Yu, J., Vezin, B., Wolf, J.P., Ritter, P., and Viscardi, P. (1998) Characterization of urban aerosols using SEM–microscopy, X-ray analysis and Lidar measurements. *Atmospheric Environment,* 32, 2957–2967.

Kowalczyk, G.S., Choquette, C.E., and Gordon, G.E. (1978) Chemical element balances and identification of air pollution sources in Washington, D.C. *Atmospheric Environment,* 12, 1143–1153.

Kowalczyk, G.S., Gordon, G.E., and Rheingrover, S.W. (1982) Identification of atmospheric particulate sources in Washington, D.C., using chemical element balances. *Environmental Science & Technology,* 16, 79–90.

Kupiainen, K. (2007) *Road Dust from Pavement Wear and Traction Sanding.* Monograph of Boreal Environment Research, 26, 2007, Finnish Environment Institute, Helsinki.

Laidlaw, M.A.S. and Filippelli G.M. (2008) Resuspension of urban soils as a persistent source of lead poisoning in children: A review and new directions. *Applied Geochemistry,* 23, 2021–2039.

Larson, S.M., Cass, G.R., and Gray, H.A. (1989) Atmospheric carbon particles and the Los Angeles visibility problem. *Aerosol Science and Technology,* 10, 118–130.

Lin, J.J. and Tai, H.S. (2001) Concentrations and distributions of carbonaceous species in ambient particles in Kaohsiung City, Taiwan. *Atmospheric Environment,* 35, 2627–2636.

Lindgren, A. (1996). Asphalt wear and pollution transport. *Science of the Total Environment,* 189/190, 281–296.

Liu, E., Yan, T., Birch, G., and Zhu, Y. (2014) Pollution and health risk of potentially toxic metals in urban road dust in Nanjing, a mega-city of China. *Science of the Total Environment,* 476–477, 522–531.

Lu, X., Wang, L., Lei, K., Huang, J., and Zhai, Y. (2009). Contamination assessment of copper, lead, zinc, manganese and nickel in street dust of Baoji, NW China. *Journal of Hazardous Materials,* 161, 1058–1062.

Macklin, M. (1992). Metal pollution of soils and sediments. In M.D. Newson (ed.), *Managing the Human Impact on the Natural Environment: Patterns and Processes.* London, Belhaven Press, pp. 172–195.

Mazzei, F., D'Alessandro, A., Lucarelli, F., Nava, S., Prati, P., Valli, G., and Vecchi, R. (2008) Characterisation of particulate matter sources in an urban environment. *Science of the Total Environment*, 401, 81–89.

Nowrouzi, M. and Pourkhabbaz, A. (2014) Application of geoaccumulation index and enrichment factor for assessing metal contamination in the sediments of Hara Biosphere Reserve, Iran. *Chemical Speciation & Bioavailability*, 26 (2), 99–105.

Oberdörster, G., Gelein, R., Ferin, J., and Weiss, B. (1995) Association of particulate air pollution and acute mortality: involvement of ultrafine particles? *Inhalation Toxicology*, 7, 111–124.

Ordóñez, A., Loredo, J., De Miguel, E., and Charlesworth, S.M. (2003) Distribution of heavy metals in the street dusts and soils of an industrial city in Northern Spain. *Archives of Environmental Contamination and Toxicology*, 44, 160–170.

Ordoñez Alonso, M.A., Alvarez Garcia, R., Charlesworth, S.M., and De Miguel, E. (2015) Spatial and temporal variations of trace elements distribution in soils and street dust of an industrial town in NW Spain: 15 years of study. *Science of the Total Environment*, 524–525, 93–103.

Pacyna, J.M. (1983) Trace Element Emission from Anthropogenic Sources in Europe. Technical Report 10/82. Norsk Institutt for Luftforskning. Ref. 24781, 107 pp.

Pacyna, J.M. (1991) Chemical tracers of the origin of arctic air pollution. In: Sturges, W.T. (ed.), *Pollution of the Arctic Atmosphere*. Environmental Management Series (J. Cairns Jnr. and R.M. Harrison, eds.). Elsevier Science.

Palacios, M.A., Gómez, M., Moldovan, M., and Gómez, B. (2000) Assessment of environmental contamination risk by Pt, Rh and Pd from automobile catalyst. *Microchemical Journal*, 67, 105–113.

Paterson, E., Sanka, M., and Clark, L. (1996) Urban soils as pollutant sinks – A case study from Aberdeen, Scotland. *Applied Geochemistry*, 11, 129–131.

Patra, A., Colvile, R., Arnold, S., Bowen, E., Shallcross, D., Martin, D., Price, C., Tate, J., ApSimon, H., and Robins, A. (2008) On street observations of particulate matter movement and dispersion due to traffic on an urban road. *Atmospheric Environment*, 42, 3911–3926.

Paustenbach, D.J., Finley, B.L., and Long, T.F. (1997) The critical role of house dust in understanding the hazards posed by contaminated soil. *International Journal of Toxicology*, 16, 339–362.

Pryor, S.C., Barthelmie, R.J., Schoof, J.T., Binkowski, F.S., Delle Monache, L., and Stull, R. (2008) Modeling the impact of sea-spray on particle concentrations in a coastal city. *Science of the Total Environment*, 391, 132–142.

Raunemaa, T., Riihiluoma, V., and Kulmala, M. (1986) Particle emission from gasoline powered vehicles: Emission, deposition and re-emission under different traffic density situations. *Journal of Aerosol Science*, 17, 973–983.

Ravindra, K., Bencs, L., and Van Grieten, R. (2004) Platinum group elements in the environment and their health risk. *Science of the Total Environment*, 318, 1–43.

Rose, N., Juggins, S., Watt, J., and Battarbee, R. (1994) Fuel type characterization of spheroidal carbonaceous particles using surface chemistry. *Ambio*, 23, 296–299.

Rundle, S.A. and Duggan, M.J. (1986) Lead pollution from the external redecoration of old buildings. *Science of the Total Environment*, 57, 181–190.

Sadiq, M. and Mian, A.A. (1994) Nickel and vanadium in air particulates at Dharahn (Saudi Arabia) during and after the Kuwait oil fires. *Atmospheric Environment*, 28, 2249–2253.

Schäfer, J., Eckhardt, J.D., Berner, Z.A., and Stüben, D. (1999) Time-dependant increase of traffic-emitted platinum group elements (PGE) in different environmental compartments. *Environmental Science & Technology*, 33, 3166–3170.

Schwar, M.J.R., Moorcroft, J.S., Laxen, D.P.H., Thompson, M., and Armorgie, C. (1988) Baseline metal-in-dust concentrations in Greater London. *Science of the Total Environment*, 68, 25–43.

Sharaf, N.E., Abdel-Shakour, A., Amer, N.M., Abou-Donia, M.A., and Khatab, N. (2008) Evaluation of children's blood lead level in Cairo, Egypt. *American-Eurasian Journal of Agricultural & Environmental Sciences*, 3 (3), 414–419.

Sturges, W.T. and Harrison, R.M. (1986) Bromine: Lead ratios in airborne particles from urban and rural sites. *Atmospheric Environment*,, 20, 577–588.

Tomza, U. (1984) Trace Elements in the Atmospheric Aerosol at Katowice, Poland. Technical Report. Instituut voor Nucleaire Wetenschappen, Rijksuniversiteit Gent, Gent, Belgium.

Tong, S.T.Y. (1998) Indoor and outdoor household dust contamination in Cincinnati, Ohio, USA. *Environmental Geochemistry and Health*, 20 (3), 123–133.

Tong, S.T.Y. and Lam, K.C. (2000) Home sweet home? A case study of household dust contamination in Hong Kong. *Science of the Total Environment*, 256 (2–3), 115–123.

Turner, A. and Simmonds, L. (2006) Elemental concentrations and metal bioaccessibility in UK household dust. *Science of the Total Environment*, 371, 74–81.

Umoren, I.U., Udoh, A.P., and Udousoro, I.I. (2007) Concentration and chemical speciation for the determination of Cu, Zn, Ni, Pb and Cd from refuse dump soils using the optimized BCR sequential extraction procedure. *Environmentalist*, 27, 241–252.

Van Dingenen, R., Raes, F., Putaud, J.-P., Baltensperger, U., Charron, A., Facchini, M.-C., Decesari, S., Fuzzi, S., Gehrig, R., Hansson, H.-C., Harrison, R.M., Hüglin, C., Jones, A.M., Laj, P., Lorbeer, G., Maenhaut, W., Palmgren, F., Querol, X., Rodriguez, S., Schneider, J., ten Brink, H., Tunved, P., Tørseth, K., Wehner, B., Weingartner, E., Wiednsohler, A., and Wåhlin, P.A. (2004) European aerosol phenomenology – 1: Physical characteristics of particulate matter at kerbside, urban, rural and background sites in Europe. *Atmospheric Environment*, 38, 2561–2577.

Viana, M., Querol, X., Alastuey, A., Gil, J.I., and Menéndez, M. (2006) Identification of PM sources by principal component analysis (PCA) coupled with wind direction data. *Chemosphere*, 65, 2411–2418.

Wadge, A., Hutton, M., and Peterson, P.J. (1986) The concentrations and particle size relationships of selected trace elements in fly ashes from U.K. coal-fired power plants and a refuse incinerator. *Science of the Total Environment*, 13–27.

Warren, R.S. and Birch, P. (1987) Heavy metal levels in atmospheric particulates, roadside dust and soil along a major urban highway. *Science of the Total Environment*, 59, 253–256.

Wei, B. and Yang, L. (2010) A review of heavy metal contaminations in urban soils, urban road dusts and agricultural soils from China. *Microchemical Journal*, 94, 99–107.

Widory, D., Roy, S., Le Moullec, Y., Goupil, G., Cocherie, A., and Guerrot, C. (2004) The origin of atmospheric particles in Paris: A view through carbon and lead isotopes. *Atmospheric Environment*, 38, 953–961.

Williams, M. (2004) Air pollution and policy – 1952–2002. *Science of the Total Environment*, 334–335, 15–20.

Wong, C.S.C., Li, X., and Thornton, I. (2006) Urban environmental geochemistry of trace metals. *Environmental Pollution*, 142, 1–16.

Yuan, C.-S., Cheng, S.-W., Hung, C.-H., and Yu, T.-Y. (2003) Influence of operating parameters on the collection efficiency and size distribution of street dust during street scrubbing. *Aerosol and Air Quality Research*, 3 (1), 75–86.

8

Bioaccessibility of Trace Elements in Urban Environments

Eduardo De Miguel[1], Almudena Ordóñez[2], Fernando Barrio-Parra[1], Miguel Izquierdo-Díaz[1], Rodrigo Álvarez[2], Juan Mingot[1], and Susanne M. Charlesworth[3]

[1] *Laboratorio de Investigación e Ingeniería Geoquímica Ambiental (LI2GA), Environmental Geochemistry Research & Engineering Laboratory, Universidad Politécnica de Madrid, Spain*
[2] *Mining and Exploration Department, School of Mines, University of Oviedo, Spain*
[3] *Centre for Agroecology, Water and Resilience, Coventry University, United Kingdom*

8.1 Introduction

Contaminants in solid environmental matrices, that is, soil, settled and suspended particulate matter, and food, are bound to the matrix with strengths that vary with their physico-chemical mode of retention. Those contaminants will be selectively released from the solid matrix depending on the extracting agents and the extraction environment. Once released into a fluid medium, contaminants will be mobile and available for absorption into living organisms. For humans, exposure to chemical substances present in the surrounding environment takes place through three different routes: ingestion, inhalation, and dermal contact. Subsequent incorporation (and the possibility of distribution and absorption in organs and tissues) occurs if the contaminants are released from the original matrix and cross the respective interfaces separating the outer environment from the organism, that is, the intestinal epithelium, lung tissue, and the epidermis (DeSesso and Jacobson, 2001). The amount of a contaminant relative to its total (or pseudo-total, i.e., *aqua-regia* extractable) concentration that effectively penetrates one of those barriers and reaches the systemic circulation is commonly referred to as its bioavailable fraction.

Gastrointestinal bioavailability is estimated from the results of in vivo tests in which laboratory animals are fed or gavaged known amounts of the contaminant of interest under controlled conditions. Since these experiments are costly, time-consuming, and controversial (from a bioethical perspective), in vitro extraction tests have been developed to determine the oral bioaccessibility of a contaminant in a solid matrix, which can be defined as the fraction of that substance which is soluble in the gastrointestinal environment and is available for absorption (Ruby *et al.*, 1999). Analogously, inhalation bioaccessibility refers to the fraction of a contaminant that is soluble in the extracellular environment of the deep lung (Boisa *et al.*, 2014). Bioaccessibility can be regarded as a conservative proxy of bioavailability in that it should overestimate the latter, because only a fraction of the dissolved (i.e. bioaccessible) contaminant will cross the intestinal or lung barriers to become bioavailable. The following section discusses the analytical protocols that have been developed to assess the

Urban Pollution: Science and Management, First Edition. Edited by Susanne M. Charlesworth and Colin A. Booth.

bioaccessibility of contaminants potentially able to impact human health.

8.2 Analytical Protocols

In order to evaluate a contaminant's bioaccessibility, the physiological conditions of the gastrointestinal tract or lung environment must be reproduced in the laboratory (Dean, 2007). Since the early 1990s (Davis *et al.*, 1992; Ruby *et al.*, 1993), different extraction protocols have been developed to determine the oral bioaccessibility of trace elements in non-dietary solid matrices. The most comprehensive (and analytically complex) ones replicate separately both the gastric and the intestinal environments (inorganic and organic constituents, pH, transit times, and temperature) (Ruby *et al.*, 1996; Rodriguez *et al.*, 1999) and, in some cases, the conditions in the mouth cavity and the oesophagus as well (Hamel *et al.*, 1999; Sips *et al.*, 2001; Oomen *et al.*, 2002, 2003) (Figure 8.1). Simplified gastric-only extractions, on the other hand, attempt to simulate merely the dissolution process in the stomach phase of human digestion. These simplified protocols commonly consist of simple extractions with hydrochloric acid (Rasmussen *et al.*, 2008) or with a glycine/pepsin + hydrochloric acid/NaCl solution (Hamel *et al.*, 1998; Rodriguez *et al.*, 1999; Turner and Simmonds, 2006; Drexler and Brattin, 2007; Juhasz *et al.*, 2007). In spite of their simplicity, these extractions yield results that correlate with in vivo measures of bioavailability as well as or even better than those of complete gastrointestinal digestions, especially for elements and matrices for which the rate-controlling step in oral absorption is the dissolution/desorption process in the stomach (Rodriguez *et al.*, 1999). Similarly, correlations with in vivo studies have been reported to be greater for the gastric phase extracts than for the intestinal phase ones in complete gastrointestinal extraction protocols (Ruby *et al.*, 1996).

The lack of agreement between different studies on the average (oral and respiratory) bioaccessibility of trace elements in urban matrices and on the factors controlling it is due not only to the inherent heterogeneity of the solid materials analysed but also to the diversity of methodological approaches employed by different researchers (Ng *et al.*, 2010; Wiseman, 2015). These discrepancies mostly refer to analytical aspects (i.e. composition and pH of extraction fluids, solid-to-extractant ratios, mixing conditions) but also include the mathematical definition of 'average' bioaccessibility (Izquierdo *et al.*, 2015). Efforts are under way to harmonise those different analytical approaches. The most successful one to date is the Unified BARGE Method, UBM (Wragg *et al.*, 2009), a complete gastrointestinal protocol that is being promoted as the European standard. More recently, and adopting a similar perspective to that of the UBM, an Inhalation Bioaccessibility Method, IBM, was proposed which also attempts to standardise the inorganic and organic composition of the lung

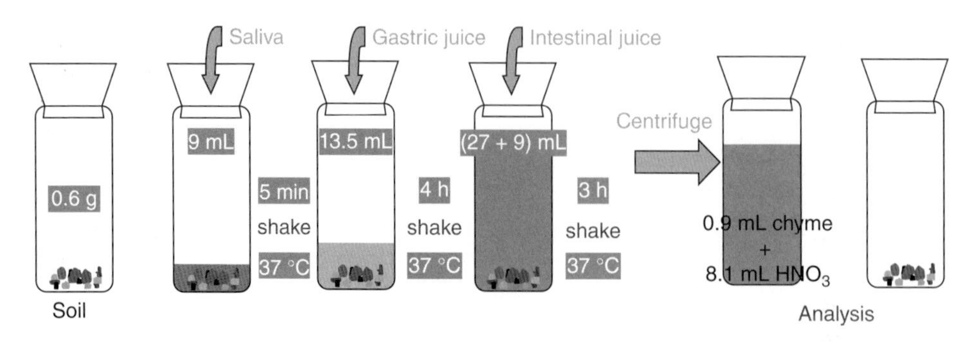

Figure 8.1 Example of a complete gastrointestinal extraction protocol (adapted from Oomen *et al.*, 2003).

fluid used for inhalation bioaccessibility studies (Boisa *et al.*, 2014).

8.3 Bioaccessibility and Urban Environments

Most research efforts in the bioaccessibility of trace elements have taken place in the context of urban environments, probably because the percentage of the global population living in cities is expected to keep growing (UN, 2015) and so is their exposure to those contaminants. Urban solid matrices pose some specific challenges in terms of the determination of bioaccessibility. Many bioaccessibility extraction protocols have been tested and/or validated in highly contaminated soils (Ruby *et al.*, 1993; Rodriguez *et al.*, 1999; Oomen *et al.*, 2003; Juhasz *et al.*, 2007; Roussel *et al.*, 2010; Wragg *et al.*, 2011). When applied to typical urban soils, which, with the exception of severely affected brownfields, present much lower concentrations, they result in extracts whose levels of contaminants of concern may lie near or below analytical quantification limits.

Urban environments include distinctive exposure scenarios such as urban gardens and orchards (Plate 8.1) or playgrounds (Plate 8.2) (Ljung *et al.*, 2007; Guney *et al.*, 2010; De Miguel *et al.*, 2012; Izquierdo *et al.*, 2015; Cai *et al.*, 2016). This is reflected in Table 8.1, which summarises some of the research undertaken in these environments and the various analytical techniques representing trace element bioaccessibility. These results show how variable specific urban environments and the substrates found there can be. Other solid exposure media vary in their geochemical nature (and consequently the bioaccessibility of trace elements in association with them) and can differ markedly from other environs, as in the case of indoor/household dust (Rieuwerts *et al.*, 2006; Yu *et al.*, 2006; Turner and Ip, 2007; Rasmussen *et al.*, 2008, 2011; Ibanez *et al.*, 2010; Le Bot *et al.*, 2010; Turner, 2011; Glorennec *et al.*, 2012; Reis *et al.*, 2015;), street and road dust (Carrizales *et al.*, 2006; Glorennec *et al.*, 2012; Potgieter-Vermaak *et al.*, 2012), and suspended particulate matter (Voutsa and Samara, 2002; Falta *et al.*, 2008; Varshney *et al.*, 2016). Among the differentiating characteristics of these materials in urban environments, the following can be mentioned: significant influence of emissions from traffic and domestic heating systems, and of building construction/renovation activities in their elemental composition; trace element concentrations generally higher than in peri-urban and rural particulate materials;

Plate 8.1 An example of an urban garden in Madrid (Spain).

Plate 8.2 An example of an urban playground (Madrid, Spain).

highly heterogeneous spatial distribution; and relatively high contents of organic matter and inorganic carbon (Rasmussen *et al.*, 2008; Charlesworth *et al.*, 2011).

8.3.1 Bioaccessibility in Urban Soil

Urban soil is the most thoroughly studied solid matrix in urban environments as regards the bioaccessibility of trace elements in its composition (Charlesworth *et al.*, 2011). Since the 1990s, efforts have been under way to discern the variables and mechanisms that control that bioaccessibility. Although there is a general agreement that the same soil properties that regulate the retention of trace elements on the solid phases of the soil, that is, soil organic matter, Fe oxyhydroxides, clay content, calcium carbonate, and pH (and mineralogy when the elements of concern are present, forming mineral phases of their own or as accessories in other crystalline structures), also determine the degree to which trace elements are released in the gastrointestinal tract, attempts to generalise their exact influence have met with little success. The disparity of results is a consequence of the different nature of urban soils in different regions, the diversity of grain sizes retained for analysis in different studies, the dissimilar extraction techniques employed by different research groups (Ng *et al.*, 2010), and the complexity and singularity of the interactions between individual trace elements and, on the one hand, the original solid phase and, on the other hand, the solution that serves as the extraction medium.

In very broad terms, metals that are retained through electrostatic bonds with negatively charged surface sites are expected to be released under the very acidic conditions of the gastric environment (assumed to be approximately pH 1.5 under fasting conditions) (Ruby *et al.*, 1996). The lowest bioaccessibilities are expected for elements which occur as part of crystalline mineral structures (i.e. in the 'residual' phase), like chromium, which in its Cr (III) oxidation state is mostly present in the form of mixed Fe–Cr oxides (Kabata-Pendias and Pendias, 1992; Poggio *et al.*, 2009). Some authors have suggested that a fraction of the metals released in the chloride-rich synthetic gastric fluid may be re-captured on positively charged mineral colloids as negatively charged metal–chloride complexes (De Miguel *et al.*, 2012) and thus present a lower gastric bioaccessibility than may have been expected.

Raising the pH of the extraction medium from very acidic, such as in a synthetic gastric juice, to approximately neutral in a simulated intestinal fluid should result in partial removal of metals from the solution through re-adsorption on inorganic colloids, enhanced complexation/absorption, and precipitation (Ellickson *et al.*, 2001; Grøn and Andersen, 2003; Ljung *et al.*, 2007; Roussel *et al.*, 2010) and consequently produce intestinal bioaccessibilities lower than their gastric equivalents. Several researchers, however, have reported results that contradict this assumption (Poggio *et al.*, 2009; De Miguel *et al.*, 2012; Cai *et al.*, 2016), citing as possible explanations the high affinity of some metals for organic compounds in the intestinal fluid or competition with previously released major ions for binding sites. Trace elements present as oxyanions, particularly arsenic, seem to exhibit similar (Ellickson *et al.*, 2001) or lower bioaccessibilities in intestinal fluid

Table 8.1 Bioaccessibility values (%) reported for soil in urban playgrounds and urban gardens/orchards.

Playgrounds

Reference	Method		As	Cd	Co	Cr	Cu	Mn	Ni	Pb	Sb	Sr	V	Zn
(De Miguel *et al.*, 2012)	RIVM		42		64	42	67			69				55
	SBET		63		26	6.2	47		18	59				56
	HCl		52		28	8.5	55		18	57				61
(Glorennec *et al.*, 2012)	HCl		< 20	> 75		< 20	> 50	> 50		> 50	< 20	> 75	< 25	
(Ljung *et al.*, 2007)	RIVM		9.7–29	13–27		1.9–4.5			3.7–7.1	2.8–15				
(Madrid *et al.*, 2008)	SBET	Torino				1–6	38–57		8–14	46–60				34–43
		Seville				4–16	13–24		34–86	30–51				32–83
(Luo *et al.*, 2012)	PBET						58			59				38

Urban Gardens

Reference	Method		Ba	Cd	Co	Cr	Cu	Mn	Ni	Pb	Zn
(Izquierdo *et al.*, 2015)	SBET				32	8.1	39	40	25	47	17
(Cai *et al.*, 2016)	PBET	G	< 25				< 20			< 30	25–60
		GI	< 10				30–70			< 15	< 20
(Pelfrêne *et al.*, 2012)	UBM	G		78						58	32
		GI		46						21	9
(Pelfrêne *et al.*, 2013)	UBM G			76						67	
(Li *et al.*, 2015)	IVG			3–13		2–3	49 –72		27–38	47– 68	

RIVM: Dutch National Institute of Public Health and the Environment; SBET: Simplified Bioaccessibility Extraction Test; HCl: Hydrochloric Acid extractable; PBET: Physiologically Based Extraction Test; UBM: Unified BARGE Method; IVG: In Vitro Gastrointestinal method; G: Gastric; GI: Gastrointestinal.

than in gastric juice (Rodriguez *et al.*, 1999; Mingot *et al.*, 2011; Wragg *et al.*, 2011), although some authors found slightly higher intestinal bioaccessibilities in soils near chromated copper arsenate (CCA)-treated wood structures (Girouard and Zagury, 2009).

8.3.2 Lung Bioaccessibility

Although it is commonly acknowledged that the highest exposure to contaminants in urban environments takes place through the route of ingestion (of food, dust, and soil) (Dudka and Miller, 1999; Turner, 2011; Potgieter-Vermaak *et al.*, 2012; Cao *et al.*, 2016), recent research demonstrates that the contribution of inhaled pollutants to the overall dose of trace elements received by urban residents is far from negligible (Boisa *et al.*, 2014; Cao *et al.*, 2016). There is also evidence that the potential occurrence of deleterious effects is more strongly correlated with the fraction of soluble toxic elements in the urban aerosol than with the total mass of suspended particles or the total concentration of contaminants in the inhaled material (Costa and Dreher, 1997).

Of all the inhaled particles, the finest among them penetrate deep into the respiratory tract, whereas larger inhaled particles can be removed via the tracheal mucocilia and transferred to the gastrointestinal tract, where they will be absorbed as if they had been ingested. For this latter case, the bioaccessibility of interest would be their gastric or gastrointestinal bioaccessibility, estimated as previously discussed for urban soil (Falta *et al.*, 2008). The finer fraction of the inhaled aerosol, in turn, reaches the alveoli and terminal bronchioles and the absorption of trace elements from this particulate material and their subsequent incorporation into the bloodstream (i.e. respiratory bioavailability) are strongly influenced by their solubility in lung fluids (i.e. their lung bioaccessibility).

Different extraction solutions have been used in the laboratory to estimate this solubility: distilled water (Costa and Dreher, 1997), hydrochloric acid (Costa and Dreher, 1997; Voutsa and Samara, 2002), artificial lysosomal fluid used to represent the intracellular environment of the macrophage (Potgieter-Vermaak *et al.*, 2012) and, most often, Gamble's solution, which is representative of the extracellular environment of the deep lung, and is used with slightly varying compositions (Zoitos *et al.*, 1997; Wragg and Klinck, 2007; Broadway *et al.*, 2010; Caboche *et al.*, 2011). Gamble's solution, however, does not accurately reproduce the effect on bioaccessibility of the mixture of lipids and proteins that act as a surfactant in the lung fluid (Goerke, 1998; Veldhuizen *et al.*, 1998). In order to address this shortcoming, simulated epithelial lung fluids that more accurately replicate the molecular composition of the original fluid have recently been developed and applied to the characterisation of lead bioaccessibility in PM_{10} in urban environments; for example, Boisa *et al.* (2014) simulated neutral pH conditions in the interstitial lung environment. The results of this study indicated that there was low potential inhalation bioaccessibility for lead in the samples of urban soils and mining wastes, but that the data would be useful for policymakers in any assessment of risk, particularly in contaminated areas. Although the analytical approach in Boisa *et al.*'s (2014) study has been proposed as an equivalent of UBM for respiratory bioaccessibility, there is no generally accepted and in vivo validated, standard protocol to evaluate respiratory bioaccessibility. Current practices differ not only in the composition of extraction fluids, but also in relevant analytical parameters such as solid/sample-to-liquid/reagents ratio, temperature, and extraction times (Mukhtar and Limbeck, 2013; Wiseman, 2015).

Rather than analysing environmental samples, Pelfrêne *et al.* (2017) used simulated lung fluids (phosphate-buffered saline, Gamble's solution, and artificial lysosomal fluid) and reference materials (road dust, soil, and urban atmospheric particulate matter) to assess metal respiratory bioaccessibility. They found that the levels of bioaccessibility obtained were highly dependent on the simulated lung fluid used and its pH.

8.4 Bioaccessibility and Human Health Risk Assessment

Harmonisation of gastrointestinal and lung bioaccessibility analytical protocols should be followed by unification of criteria for how their results are used and interpreted in human health risk assessments. If the bioaccessibility of the contaminant of concern in the delivered dose is different from its bioaccessibility in the material administered or studied to establish its oral Reference Dose (RfD) or Slope Factor (SF) (for example soil vs. feed or diet, as in the case of Pb and Zn (Budroe *et al.*, 2009; USEPA, 2016)), the results of risk characterisation will be skewed (Sips *et al.*, 2001; Grøn and Andersen, 2003). This problem is particularly acute when toxicity data are derived from animal experiments or epidemiological studies in which the contaminant is delivered in solution. For these substances, a direct comparison of their RfDs or SFs with environmental doses estimated from pseudo-total concentrations implies an assumption of 100% bioaccessibility and hence an overestimation of risk. Such is the case for As, for which the oral RfD published in the USEPA's IRIS database is derived from epidemiological studies in which human populations were exposed to As in drinking water (Tseng *et al.*, 1968; Tseng, 1977) and also Cr (VI), whose RfD is estimated from the results of a laboratory assay in rats supplied with drinking water containing K_2CrO_4 (MacKenzie *et al.*, 1958), Ba, and Sb.

The lowest oral bioaccessibility values of trace elements in solid urban matrices probably lie in the range of 0% to 5%–10%. Failure to consider this bioaccessibility in a risk assessment leads to numerical outcomes of risk that may overestimate the real value by more than one order of magnitude.

8.5 Conclusions

The realisation of this problem and its potential consequences, that is, inaccurate predictions of risk, has fuelled the development of numerous methodological approaches of varying complexity to estimate oral and lung bioaccessibilities as conservative proxies of bioavailability in the gastrointestinal and respiratory tracts. Still needed is a general consensus on a standard, practicable, and cost-effective protocol that can be universally implemented and that will yield harmonised results.

Other factors that introduce a potentially similar degree of uncertainty in human health risk assessments in urban environments must be addressed simultaneously. Two of these other critical sources of uncertainty are the estimates of toxic potency of individual substances and, especially of mixtures, and the quantitative characterisation of exposure variables, such as contact rates and exposure frequencies.

Lastly, assessments of bioaccessibility in urban environments need to be site specific: the complexity of the geochemical variables involved in the sorption–desorption and dissolution–precipitation equilibria in the gastric and intestinal environments and the highly heterogeneous nature of urban solid materials make the extrapolation of results from one medium or location to another very uncertain, even when exactly the same analytical protocols are employed.

References

Boisa, N., Elom, N., Dean, J.R., Deary, M.E., Bird, G., and Entwistle, J.A. (2014) Development and application of an inhalation bioaccessibility method (IBM) for lead in the PM10 size fraction of soil. Environment International, 70, 132–142. doi:10.1016/j.envint.2014.05.021

Broadway, A., Cave, M.R., Wragg, J., Fordyce, F.M., Bewley, R.J.F., Graham, M.C., Ngwenya, B.T., and Farmer, J.G. (2010) Determination of the bioaccessibility of chromium in Glasgow soil and the implications for human health risk assessment. Science of the Total Environment. doi:10.1016/j.scitotenv.2010.09.007.

Budroe, J.D., Brown, J.P., Collins, J.F., Marty, M.A., Salmon, A.G., Sandy, M.S., Claire Sherman, M.D., Tomar, R.S., and Zeise, L. (2009) Technical Support Document for Cancer Potency Factors. Appendix B. California Environmental Protection Agency.

Caboche, J., Perdrix, E., Malet, B., and Alleman, L.Y. (2011). Development of an in vitro method to estimate lung bioaccessibility of metals from atmospheric particles. Journal of Environmental Monitoring, 13, 621–30. doi:10.1039/c0em00439a

Cai, M., McBride, M.B., and Li, K. (2016) Bioaccessibility of Ba, Cu, Pb, and Zn in urban garden and orchard soils. Environmental Pollution, 208, 145–152. doi:10.1016/j.envpol.2015.09.050

Cao, S., Duan, X., Zhao, X., Chen, Y., Wang, B., Sun, C., Zheng, B., and Wei, F. (2016). Health risks of children's cumulative and aggregative exposure to metals and metalloids in a typical urban environment in China. Chemosphere, 147, 404–411. doi:10.1016/j.chemosphere.2015.12.134

Carrizales, L., Razo, I., Téllez-Hernández, J.I., Torres-Nerio, R., Batres, L.E., Cubillas, A.-C., and Dıaz-Barriga, F. (2006) Exposure to arsenic and lead of children living near a copper–smelter in San Luis Potosi , Mexico: Importance of soil contamination for exposure of children. Environmental Research, 101, 1–10. doi:10.1016/j.envres.2005.07.010

Charlesworth, S., de Miguel, E., and Ordóñez, A. (2011) A review of the distribution of particulate trace elements in urban terrestrial environments and its application to considerations of risk. Environmental Geochemistry and Health, 33, 103–123. doi:10.1007/s10653–010–9325–7

Costa, D.L., and Dreher, K.L. (1997) Bioavailable transition metals in particulate matter mediate cardiopulmonary injury in healthy and compromised animal models. Environmental Health Perspectives, 105 Suppl, 1053–1060. doi:10.2307/3433509

Davis, A., Ruby, M., and Bergstrom, P. (1992) Bioavailability of arsenic and lead in soils from the Butte, Montana, mining district. Environmental Science & Technology, 26, 461–468. doi:10.1021/es00027a002

De Miguel, E., Mingot, J., Chacón, E., and Charlesworth, S. (2012). The relationship between soil geochemistry and the bioaccessibility of trace elements in playground soil. Environmental Geochemistry and Health, 34, 677–687. doi:10.1007/s10653–012–9486–7

Dean, J.R. (2007) *Bioavailability, Bioaccessibility and Mobility of Environmental Contaminants*. John Wiley & Sons, Chichester, United Kingdom.

DeSesso, J.M. and Jacobson, C.F. (2001). Anatomical and physiological parameters affecting gastrointestinal absorption in humans and rats. Food and Chemical Toxicology, 39, 209–228. doi:10.1016/S0278–6915(00)00136–8

Drexler, J.W. and Brattin, W.J. (2007) An in vitro procedure for estimation of lead relative bioavailability: With validation. Human and Ecological Risk Assessment, An International Journal, 13, 383–401. doi:10.1080/10807030701226350

Dudka, S. and Miller, W.P. (1999) Permissible concentrations of arsenic and lead in soils based on risk assessment. Water, Air, and Soil Pollution, 113, 127–132. doi:10.1023/A:1005028905396

Ellickson, K.M., Meeker, R.J., Gallo, M.A., Buckley, B.T., and Lioy, P.J. (2001) Oral bioavailability of lead and arsenic from a NIST standard reference soil material. Archives of Environmental Contamination and Toxicology, 40, 128–135. doi:10.1007/s002440010155

Falta, T., Limbeck, A., Koellensperger, G., and Hann, S. (2008) Bioaccessibility of selected trace metals in urban PM2.5 and PM10 samples: a model study. Analytical and Bioanalytical Chemistry, 390, 1149–1157. doi:10.1007/s00216–007–1762–5

Girouard, E. and Zagury, G.J. (2009) Arsenic bioaccessibility in CCA–contaminated soils: Influence of soil properties, arsenic fractionation, and particle–size fraction. Science of the Total Environment, 407, 2576–2585. doi:10.1016/j.scitotenv.2008.12.019

Glorennec, P., Lucas, J.P., Mandin, C., and Le Bot, B. (2012) French children's exposure to metals via ingestion of indoor dust, outdoor playground dust and soil: Contamination data. Environment International, 45, 129–134. doi:10.1016/j.envint.2012.04.010

Goerke, J. (1998) Pulmonary surfactant: Functions and molecular composition. Biochimica et Biophysica Acta – Molecular Basis of Disease, 1408, 79–89. doi:10.1016/S0925–4439(98)00060–X

Grøn, C. and Andersen, L. (2003) Human Bioaccessibility of Heavy Metals and PAH from Soil, Environmental Project No. 840 2003, Technology Programme for Soil and Groundwater Contamination. Danish Environmental Protection Agency.

Guney, M., Zagury, G.J., Dogan, N., and Onay, T.T. (2010) Exposure assessment and risk characterization from trace elements following soil ingestion by children exposed to playgrounds, parks and picnic areas. Journal of Hazardous Materials, 182, 656–664. doi:10.1016/j.jhazmat.2010.06.082

Hamel, S.C., Buckley, B., and Lioy, P.J. (1998) Bioaccessibility of metals in soils for different liquid to solid ratios in synthetic gastric fluid. Environmental Science & Technology, 32, 358–362. doi:10.1021/es9701422

Hamel, S.C., Ellickson, K.M., and Lioy, P.J. (1999) The estimation of the bioaccessibility of heavy metals in soils using artificial biofluids by two novel methods: Mass-balance and soil recapture. Science of the Total Environment, 243–244, 273–283. doi:10.1016/S0048–9697(99)00402–7

Ibanez, Y., Le Bot, B., and Glorennec, P. (2010) House-dust metal content and bioaccessibility: A review. European Journal of Mineralogy, 22, 629–637. doi:10.1127/0935–1221/2010/0022–2010

Izquierdo, M., De Miguel, E., Ortega, M.F., and Mingot, J. (2015) Bioaccessibility of metals and human health risk assessment in community urban gardens. Chemosphere, 135, 312–318. doi:10.1016/j.chemosphere.2015.04.079

Juhasz, A.L., Smith, E., Weber, J., Rees, M., Rofe, A., Kuchel, T., Sansom, L., and Naidu, R. (2007) In vitro assessment of arsenic bioaccessibility in contaminated (anthropogenic and geogenic) soils. Chemosphere, 69, 69–78. doi:10.1016/j.chemosphere.2007.04.046

Kabata-Pendias, A. and Pendias, H. (1992) *Trace Elements in Soils and Plants*, 2nd edition. CRC Press.

Le Bot, B., Gilles, E., Durand, S., and Glorennec, P. (2010) Bioaccessible and quasi–total metals in soil and indoor dust. European Journal of Mineralogy, 22, 651–657. doi:10.1127/0935–1221/2010/0022–2052

Li, N., Kang, Y., Pan, W., Zeng, L., Zhang, Q., and Luo, J. (2015) Concentration and transportation of heavy metals in vegetables and risk assessment of human exposure to bioaccessible heavy metals in soil near a waste-incinerator site, South China. Science of the Total Environment, 521–522, 144–151. doi:10.1016/j.scitotenv.2015.03.081

Ljung, K., Oomen, A., Duits, M., Selinus, O., and Berglund, M. (2007) Bioaccessibility of metals in urban playground soils. Journal of Environmental Science and Health, Part A: Toxic/Hazardous Substances and Environmental Engineering, 42, 1241–1250. doi:10.1080/10934520701435684

Luo, X., Yu, S., and Li, X. (2012) The mobility, bioavailability, and human bioaccessibility of trace metals in urban soils of Hong Kong. Applied Geochemistry, 27, 995–1004. doi:10.1016/j.apgeochem.2011.07.001

MacKenzie, R.D., Byerrum, R.U., Decker, C.F., Hoppert, C.A., and Langham, R.F. (1958)

Chronic toxicity studies. II. Hexavalent and trivalent chromium administered in drinking water to rats. American Medical Association Archives of Industrial Health, 18, 232–234.

Madrid, F., Biasioli, M., and Ajmone-Marsan, F. (2008) Availability and bioaccessibility of metals in fine particles of some urban soils. Archives of Environmental Contamination and Toxicology, 55, 21–32. doi:10.1007/s00244–007–9086–1

Mingot, J., De Miguel, E., and Chacón, E. (2011) Assessment of oral bioaccessibility of arsenic in playground soil in Madrid (Spain): A three-method comparison and implications for risk assessment. Chemosphere, 84, 1386–1391. doi:10.1016/j.chemosphere.2011.05.001

Mukhtar, A. and Limbeck, A. (2013) Recent developments in assessment of bio–accessible trace metal fractions in airborne particulate matter: A review. Analytica Chimica Acta, 774, 11–25. doi:10.1016/j.aca.2013.02.008

Ng, J.C., Juhasz, A.L., Smith, E., and Naidu, R. (2010) Contaminant bioavailability and bioaccessibility. Part 1: A scientific and technical review. CRC CARE Technical Report no. 14, CRC for Contamination Assessment and Remediation of the Environment. Adelaide (Australia).

Oomen, A., Hack, A., Minekus, M., Zeijdner, E., Cornelis, C., Schoeters, G., Verstraete, W., Van De Wiele, T., Wragg, J., Rompelberg, C., Sips, A., and Van Wijnen, J.H. Van (2002) Comparison of five in vitro digestion models to study the bioaccessibility of soil contaminants. Environmental Science & Technology, 36, 3326–3334.

Oomen, A.G., Rompelberg, C.J.M., Bruil, M.A., Dobbe, C.J.G., Pereboom, D.P.K.H., and Sips, A.J.A.M. (2003) Development of an in vitro digestion model for estimating the bioaccessibility of soil contaminants. Archives of Environmental Contamination and Toxicology, 44, 281–287. doi:10.1007/s00244–002–1278–0

Pelfrêne, A., Cave, M. R. Wragg, J., and Douay F. (2017) In vitro investigations of human bioaccessibility from reference materials using simulated lung fluids. International Journal of Environmental Research and Public Health, 2017, 14, 112; doi:10.3390/ijerph14020112

Pelfrêne, A., Douay, F., Richard, A., Roussel, H., and Girondelot, B. (2013) Assessment of potential health risk for inhabitants living near a former lead smelter. Part 2: Site–specific human health risk assessment of Cd and Pb contamination in kitchen gardens. Environmental Monitoring and Assessment, 185, 2999–3012. doi:10.1007/s10661–012–2767–x

Pelfrêne, A., Waterlot, C., Mazzuca, M., Nisse, C., Cuny, D., Richard, A., Denys, S., Heyman, C., Roussel, H., Bidar, G., and Douay, F. (2012) Bioaccessibility of trace elements as affected by soil parameters in smelter-contaminated agricultural soils: A statistical modeling approach. Environmental Pollution, 160, 130–138. doi:10.1016/j.envpol.2011.09.008

Poggio, L., Vrscaj, B., Schulin, R., Hepperle, E., and Ajmone Marsan, F. (2009) Metals pollution and human bioaccessibility of topsoils in Grugliasco (Italy). Environmental Pollution, 157, 680–689. doi:10.1016/j.envpol.2008.08.009

Potgieter–Vermaak, S., Rotondo, G., Novakovic, V., Rollins, S., and van Grieken, R. (2012) Component-specific toxic concerns of the inhalable fraction of urban road dust. Environmental Geochemistry and Health, 34, 689–696. doi:10.1007/s10653–012–9488–5

Rasmussen, P.E., Beauchemin, S., Chénier, M., Levesque, C., MacLean, L.C.W., Marro, L., Jones–Otazo, H., Petrovic, S., McDonald, L.T., and Gardner, H.D. (2011) Canadian house dust study: Lead bioaccessibility and speciation. Environmental Science & Technology, 45, 4959–65. doi:10.1021/es104056m

Rasmussen, P.E., Beauchemin, S., Nugent, M., Dugandzic, R., Lanouette, M., and Chénier, M. (2008) Influence of matrix composition on the bioaccessibility of copper, zinc, and nickel in urban residential dust and soil.

Human and Ecological Risk Assessment, An International Journal, 14, 351–371. doi:10.1080/10807030801934960

Reis, A.P., Costa, S., Santos, I., Patinha, C., Noack, Y., Wragg, J., Cave, M., and Sousa, A.J. (2015) Investigating relationships between biomarkers of exposure and environmental copper and manganese levels in house dusts from a Portuguese industrial city. Environmental Geochemistry and Health, 37, 725–744. doi:10.1007/s10653–015–9724–x

Rieuwerts, J.S., Searle, P., and Buck, R. (2006) Bioaccessible arsenic in the home environment in southwest England. Science of the Total Environment, 371, 89–98. doi:10.1016/j.scitotenv.2006.08.039

Rodriguez, R.R., Basta, N.T., Casteel, S.W., and Pace, L.W. (1999) An in vitro gastrointestinal method to estimate bioavailable arsenic in contaminated soils and solid media. Environmental Science & Technology, 33, 642–649. doi:10.1021/es980631h

Roussel, H., Waterlot, C., Pelfrêne, A., Pruvot, C., Mazzuca, M., and Douay, F. (2010) Cd, Pb and Zn oral bioaccessibility of urban soils contaminated in the past by atmospheric emissions from two lead and zinc smelters. Archives of Environmental Contamination and Toxicology, 58, 945–954. doi:10.1007/s00244–009–9425–5

Ruby, M.V., Davis, A., Schoof, R., Eberle, S., and Sellstone, C.M. (1996) Estimation of lead and arsenic bioavailability using a physiologically based extraction test. Environmental Science & Technology, 30, 422–430. doi:10.1021/es950057z

Ruby, M.V., Schoof, R., Brattin, W., Goldade, M., Post, G., Harnois, M., Mosby, E., Casteel, S.W., Berti, W., Carpenter, M., Edwards, D., Cragin, D., and Chappell, W. (1999) Advances in evaluating the oral bioavailability of inorganics in soil for use in human health risk assessment. Environmental Science & Technology, 33, 3697–3705. doi:10.1021/es990479z

Ruby, M.V, Davis, A., Link, T.E., Schoof, R., Chaney, R.L., Freeman, G.B., and Bergstrom, P. (1993) Development of an in-vitro screening-test to evaluate the in-vivo bioaccessibility of ingested mine-waste lead. Environmental Science & Technology, 27, 2870–2877. doi:10.1021/es00049a030

Sips, A.J.A.M., Bruil, M.A., Dobbe, C.J.G., van de Kamp, E., Oomen, A.G., Pereboom, D.P.K.H., Rompelberg, C.J.M., and Zeilmaker, M.J. (2001) Bioaccessibility of contaminants from ingested soil in humans. RIVM report 711701012/2001.

Tseng, W.P. (1977) Effects and dose–response relationships of skin cancer and blackfoot disease with arsenic. Environmental Health Perspectives, 19, 109–119.

Tseng, W.P., Chu, H.M., How, S.W., Fong, J.M., Lin, C.S., and Yeh, S. (1968) Prevalence of skin cancer in an endemic area of chronic arsenicism in Taiwan. Journal of the National Cancer Institute, 40, 453–463.

Turner, A. (2011) Oral bioaccessibility of trace metals in household dust: A review. Environmental Geochemistry and Health, 33, 331–341. doi:10.1007/s10653–011–9386–2

Turner, A. and Ip, K.H. (2007) Bioaccessibility of metals in dust from the indoor environment: Application of a physiologically based extraction test. Environmental Science & Technology, 41, 7851–7856. doi:10.1021/es071194m

Turner, A. and Simmonds, L. (2006) Elemental concentrations and metal bioaccessibility in UK household dust. Science of the Total Environment, 371, 74–81. doi:10.1016/j.scitotenv.2006.08.011

UN (United Nations) (2015) World Urbanization Prospects: The 2014 Revision, (ST/ESA/SER.A/366). United Nations, Department of Economic and Social Affairs, Population Division. New York.

USEPA (2016) Zinc and Compounds; CASRN 7440–66–6. Integrated Risk Information System. https://cfpub.epa.gov/ncea/iris2/chemicalLanding.cfm?substance_nmbr=426.

Varshney, P., Saini, R., and Taneja, A. (2016) Trace element concentration in fine particulate matter (PM2.5) and their bioavailability in different microenvironments in Agra, India: A case study. Environmental Geochemistry and Health, 38, 593–605. doi:10.1007/s10653–015–9745–5

Veldhuizen, R., Nag, K., Orgeig, S., and Possmayer, F. (1998) The role of lipids in pulmonary surfactant. Biochimica et Biophysica Acta – Molecular Basis of Disease, 1408, 90–108. doi:10.1016/S0925–4439(98)00061–1

Voutsa, D. and Samara, C. (2002). Labile and bioaccessible fractions of heavy metals in the airborne particulate matter from urban and industrial areas. Atmospheric Environment, 36, 3583–3590. doi:10.1016/S1352–2310(02)00282–0

Wiseman, C.L.S. (2015) Analytical methods for assessing metal bioaccessibility in airborne particulate matter: A scoping review. Analytica Chimica Acta, 877, 9–18. doi:10.1016/j.aca.2015.01.024

Wragg, J., Cave, M., Basta, N., Brandon, E., Casteel, S., Denys, S., Grøn, C., Oomen, A., Reimer, K., Tack, K., and Van de Wiele, T. (2011) An inter-laboratory trial of the unified BARGE bioaccessibility method for arsenic, cadmium and lead in soil. Science of the Total Environment, 409, 4016–4030. doi:10.1016/j.scitotenv.2011.05.019

Wragg, J., Cave, M., Taylor, H., Basta, N., Brandon, E., Casteel, S., Grøn, C., Oomen, A., and Van de Wiele, T. (2009) Inter-Laboratory Trial of a Unified Bioaccessibility Testing Procedure. British Geological Survey, Open Report, OR/07/027. Keyworth, Nottingham.

Wragg, J. and Klinck, B. (2007) The bioaccessibility of lead from Welsh mine waste using a respiratory uptake test. Journal of Environmental Science and Health, Part A 42, 1223–1231. doi:10.1080/10934520701436054

Yu, C., Yiin, L. and Lioy, P. (2006) The bioaccessibility of lead (Pb) from vacuumed house dust on carpets in urban residences. Risk Analysis, 26, 125–134.

Zoitos, B.K., Meringo, A. De, Rouyer, E., Thelohan, S., Bauer, J., Law, B., Boymel, P.M., Olson, J.R., Christensen, V.R., Guldberg, M., Koenig, A.R. and Perander, M. (1997) In vitro measurement of fiber dissolution rate relevant to biopersistence at neutral pH: An interlaboratory round robin. Inhalation Toxicology, 9, 525–540. doi:10.1080/089583797198051

9

The Necessity for Urban Wastewater Collection, Treatment, and Disposal

Colin A. Booth[1], *David Oloke*[2], *Andrew Gooding*[3], *and Susanne M. Charlesworth*[4]

[1] *Architecture and the Built Environment, University of the West of England, United Kingdom*
[2] *Faculty of Science and Engineering, University of Wolverhampton, United Kingdom*
[3] *Wessex Water, Bath, United Kingdom*
[4] *Centre for Agroecology, Water and Resilience, Coventry University, United Kingdom*

9.1 Introduction

Faecal contaminated water puts people at risk of contracting dysentery, cholera, typhoid, schistosomiasis, trachoma, and intestinal worms (UNESCO, 2006). Deaths attributed to outbreaks of cholera (a potentially fatal bacterial infection) in Iraq (2007), Congo (2008), Zimbabwe (2008), Haiti (2010), and Sierra Leone (2012) are a shocking reminder of the importance of treating urban and rural wastewater (Mason, 2009; Piarroux *et al.*, 2011; Nguyen *et al.*, 2014). For instance, inadequate wastewater management (sanitation) is alleged to cause an estimated 280,000 diarrhoeal deaths annually (WHO, 2017).

Goal Six – Clean Water and Sanitation – of the 17 Sustainable Development Goals (SDGs) set out by the United Nations is an initiative with eight global targets, whose aims are to attempt to substantially reduce untreated wastewater and improve water quality by reducing pollution. Wastewater management is indifferent worldwide. Most towns and cities that have developed through a formal planning system have access to a sewage infrastructure system that manages their wastewater through a sewerage treatment plant to improve water quality and minimise public health risks. This is starkly different from many unregulated settlements, where the poorest minorities of urban households exist; they tend to lack access to sanitation and sewage infrastructure (Coker *et al.*, 2011). Sadly, open defecation or use of pit latrines is still prevalent in the urban areas of some developing countries. This is a public health risk and potential source of groundwater contamination.

Claimed by many as the first epidemiological study, Dr John Snow, a public health worker, investigated a severe cholera outbreak in Soho, London, in 1854, that was responsible for killing 616 people (Newsom, 2006; Tulodziecki, 2011). Around this time, London had already suffered incapacitating outbreaks of cholera (1832 and 1849), which had killed thousands of people. To unearth the evidence he needed, he set about mapping the location of those people who contracted the deadly disease, and was able to link their movements and activities to those who drank from a water well on Broad Street (now Broadwick Street). Dr Snow was convinced that contaminated water (and not foul air) was responsible for spreading the infectious disease. Conclusive by modern-day standards, it took the removal of the pump handle and abatement of the outbreak to convince some officials that the disease may be waterborne. In fact, further investigation

Urban Pollution: Science and Management, First Edition. Edited by Susanne M. Charlesworth and Colin A. Booth.

revealed the well was sited close to a cesspit that was discharging into groundwater and, by doing so, highlighted that faecal waste may somehow be responsible for the contamination of the drinking water and for the original cholera outbreak.

Without a doubt wastewater management is an essential requisite of all modern city infrastructures. Proper collection, appropriate treatment, and consented discharge of wastewater, together with suitable disposal of the resulting sludge, helps protect the quality of the natural environment and prevents public health issues. This chapter provides some insights into the collection, treatment, and disposal of urban wastewater systems that are utilised in many towns and cities, before it focuses attention on WASH (water, sanitation, and hygiene) in developing countries.

9.2 Wastewater Collection in Developed Countries

Urban wastewater, usually known as sewage, typically represents a combination of domestic wastewater from homes (e.g. greywater from baths, showers, sinks, washing machines, and dishwashers; blackwater from our toilets), mixed with wastewater from factories and industries and stormwater (from precipitation events) run-off from roads, pavements, and other hard surfaces. To remove wastewater, most towns and cities have traditionally used a single underground pipe or drain. The drains require a sloping gradient and careful design and installation because the water flows away by gravity and, with it, carry along small amounts of waste solids. Urban drains, in and around buildings, usually connect to a community drainage system so the wastewater is transferred to a large sewerage treatment plant.

Combined sewers have now been replaced. Nowadays, most housing developments are required to install both a public foul water sewer and public surface water sewer, to create separate drainage systems. While the costs of installing a double system are much higher, by separating the wastewaters, the stormwater run-off can be channelled directly to watercourses, because it usually requires no treatment, and the volume of foul wastewater requiring treatment at the sewerage plant is notably reduced, which saves both energy and costs. Furthermore, given increases in urban flash flood events occurring, having separate systems means there is a reduced risk of stormwater surges overburdening sewerage treatment plants.

Sewage treatment service providers are responsible for maintaining and improving public sewers. The purpose of drainage and sewage treatment systems is to minimise public health risk, protect the environment, convert wastewater into stable end-products, recover and recycle materials (if possible), dispose of the end-products in a safe manner, provide a reliable and regular service, and to comply with appropriate standards and legislation (e.g. Urban Wastewater Treatment Directive). In the United Kingdom, for instance, the EU Directive regulates the collection and treatment of wastewater from homes and from industry to protect the water environment from the adverse effects of urban wastewater discharges.

Without treatment, wastewater can impair the quality of the natural environment and create public health issues. Untreated sewage typically comprises mainly water (depending on time and local conditions), together with organic solids (proteins, carbohydrates, and fats) and inorganic solids (e.g. salts, metals, and grits). Whether accidental or intentional, failure to treat wastewater before it enters a stream or river can cause a pollution (point-source) event, which reduces the water quality of the receiving waterbody by lessening the dissolved oxygen levels and destroying wildlife. Table 9.1 shows some typical biochemical oxygen demand (BOD) levels.

Table 9.1 Typical biochemical oxygen demand (BOD) levels of various waterbodies.

	BOD mg/L
Farm slurry	50,000
Raw domestic sewage	300
Treated domestic sewage	30
Polluted river	20
Large lowland river	5
Upland stream	1

9.3 Wastewater Treatment and Disposal in Developed Countries

Many people flush the toilet with little regard to what happens after it leaves the pan. It is only when we stop and think about the number of times we each flush the toilet daily and then multiply this by the number of people in each town, city, or country that we realise the volume of wastewater that needs to be treated each day. In the United Kingdom, for instance, the ~347,000 km of sewers beneath our feet collect >11 billion litres of wastewater daily, which is then treated at ~9,000 sewage treatment works across the country. This is the case for ~96% of the population, who are connected to sewers leading to sewage treatment works; the remainder are served by small private treatment works, cesspits, or septic tanks (DEFRA, 2002).

Urban infrastructure includes a wastewater/sewerage treatment plant. This is a combination of stepwise treatment components that are used to carry out particular treatment processes. The exact combination of treatment components typically depends on the characteristics of the wastewater input and the consented effluent quality that the treated wastewater needs to meet before it can be discharged to the receiving waterbody.

Most urban sewage treatment plants follow a similar layout/design (Figure 9.1).

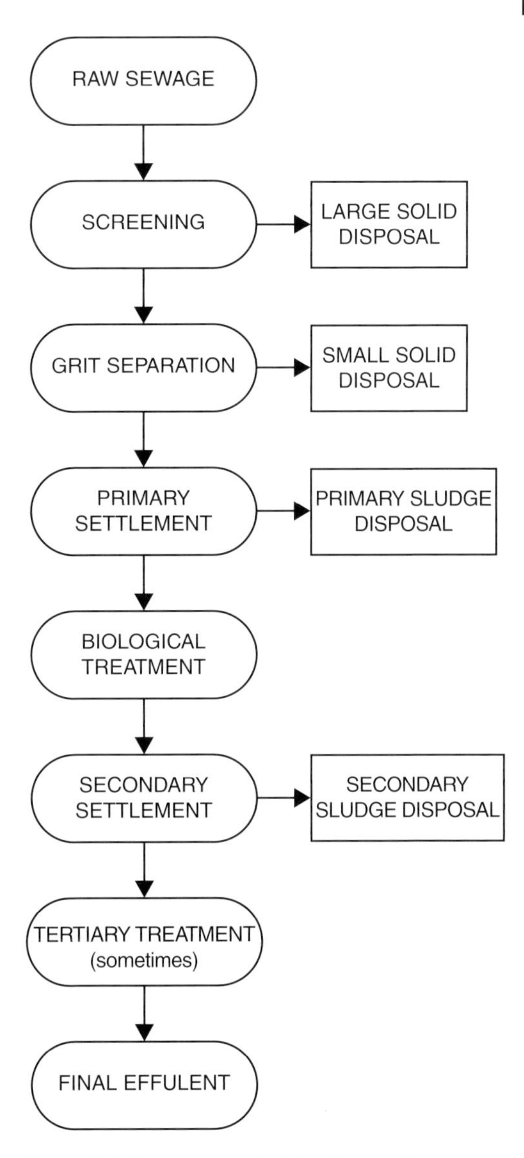

Figure 9.1 Processes undertaken in an urban sewage treatment plant.

The inlet stage (Figure 9.2a) controls the volume of wastewater entering the treatment plant. When excessive amounts of wastewater arrive at the inlet, it is diverted to storm tanks for storage until the system is able to treat the wastewater. Once wastewater has entered the treatment system, it is screened to remove large solids and grit (Figure 9.2b), which may cause blockages and damage other parts of the treatment system. Steel bar screens (~25 mm apart) are commonly used to remove large solids (e.g. rags, nappies, sanitary products, paper, wood, plastic, among others), while steel wire mesh screens (~10 mm apart) are commonly used to remove smaller solids (e.g. stones, glass, peas, sweet corn, among others). The solid wastes are collected in a skip and transferred to a landfill (Figure 9.2c and 9.2d). Meanwhile, the wastewater is transferred to the next treatment component.

Once screened, the wastewater enters the primary treatment stage. This is where large sedimentation tanks receive the wastewater in their centre (Figure 9.2e). The wastewater is slowly circulated at a very low velocity, which allows the fine solids to fall out of suspension by gravity. These solids form a grit–sludge at the base of the tank that is scraped to an outlet and transferred for separate treatment. Primary sedimentation normally takes ~2 hours, and typically removes ~55% of the suspended solids; most importantly, it reduces the BOD by about 35%. The surface of the wastewater commonly gathers a scum layer (formed of oils and grease), so a

Figure 9.2 Wastewater treatment stages (for an explanation of 9a–g, see Section 9.3).

Figure 9.2 *(Continued)*

rotating scraping blade is used to skim off the scum, before the 'partially cleaner' wastewater now spills over the tank sides (Figure 9.2f) and is transferred to the next treatment component.

The secondary treatment stage (activated sludge) is to remove the remaining matter, by exposing the wastewater to biochemical reactions under aerobic conditions. Purification is carried out (for ~12 h) by aerobic bacteria and microorganisms (protozoa), which need air to live, and, in turn, they convert the impurities in the wastewater to settled sewage (food) to multiply. Aeration lanes are continuously fed with wastewater, which flows around a transverse of lanes. Each lane is constantly mixed and aerated by mechanical blowers, which bubble up oxygen-rich air from the tank base (Figure 9.2g). It is critical to keep a balanced system so that the wastewater is treated to the regulatory requirements. For instance, insufficient oxygen will kill the aerobic bacteria and microorganisms and cause the wastewater to pass through the system untreated.

Before the effluent can be discharged to a neighbouring waterbody, the wastewater is first transferred to the final settlement tanks. These are designed (similar to the primary settlement tanks) to enable the collection of microorganisms produced during the treatment to be reused for 'seeding or activating' the next incoming sewage. Depending on the permitted discharge levels, consented by the appropriate regulatory body (e.g. the Environment Agency in the United Kingdom), the effluent is now released from the treatment plant. Permitted effluent water quality levels at this point should now be ideally similar to the river that is receiving the effluent, where it is dispersed downstream in the flow of the river (Table 9.2). Occasionally, where the water quality of the receiving river is exceptional high or the waterbody has no

Table 9.2 Typical water quality characteristics of wastewater before, during, and after treatment (after Weiner and Matthews, 2003).

	Untreated wastewater	After primary treatment	After secondary treatment
BOD (mg/L)	250	175	15
SS (mg/L)	220	60	15
P (mg/L)	8	7	6

obvious flow (e.g. a canal), the regulatory body can request that the water quality levels are further improved. Where this is the case, the wastewater is usually processed through a tertiary treatment stage. There are various ways of improving and purifying the water. This can be done by sand filtration or carbon adsorption to remove residual suspended matter and toxins or by disinfection using ozone, chlorine, ultraviolet light, or sodium hypochlorite.

Where single or isolated groups of buildings or communities are too far away (economically) from a sewage treatment plant to have the sewage piped away or they have chosen to live 'off-grid' and want to use more natural treatment methods, alternative means are utilised to manage the wastewater. These can include (1) cesspools, (2) septic tanks, (3) stabilisation ponds, and (4) reed beds, among others.

9.4 Sludge Treatment and Disposal in Developed Countries

Sludge is an unavoidable semisolid by-product of the sewage treatment process, and its treatment/disposal poses an environmental challenge. Historically, until the introduction of the Urban Waste Water Directive, 1998, ~25% of sludge was either dumped at sea or discharged to surface waters (DEFRA, 2002) with disregard for the environmental consequences this caused. Fortunately, with improved environmental regulations in place, this practice is no longer permitted throughout Europe and many nations beyond.

It was estimated in 1999 that in the United Kingdom alone produced 1,130,066 tonnes of sludge dry solids, which equates to ~20 kg generated by each person in the country (DEFRA, 2002). Since then, it is estimated that the total volume of sludge requiring disposal has sizably increased due to national population growth, more households being connected to the sewerage system, and more stringent requirements for the treatment of sewage effluent. As such, this has exacerbated the disposal challenge for the national water and sewage companies.

Solids in the sludge contain nutrients and humus-like material that is of value to plants, as it improves the capacity of poor soils to hold water and air. However, the characteristics of sludge can vary dramatically, which means it is not always cost-effective to treat sludge for application on land. This is because toxic materials (including heavy metals) can be introduced from industrial, household wastes, and urban run-off. Human waste also contains harmful organisms, disease-causing bacteria, viruses, and parasites. Therefore, before sludge can be disposed, it needs to be treated.

Raw sludge may be treated (thickened) to remove some of its water (which is returned to the wastewater treatment process) and then further processed to produce biosolids. This additional processing may include sludge stabilisation, dewatering, and utilisation (Spellman and Drinan, 2003). In sludge stabilisation, raw sludge is allowed to decompose in a digester. In the dewatering process, most of the remaining liquid is removed by using filters, drying beds, among other methods. Sludge utilisation involves the application of biosolids to be disposed by land application, composting, incinerating, or landfilling.

Once treated, the environmentally favoured disposal option (regulated under the Sludge (use in Agriculture) Regulations

1989 (as amended)) is for the sewage sludge to be applied to agricultural land (by spreading, spraying, or injection onto or below the land surface) as soil conditioner and fertiliser to improve the structure of the soil (Garg, 2009). In fact, recycling to farmland is the current main disposal route (73%, up from 59% in 2000), followed by incineration (18% down from 22%). Disposal to landfill has significantly decreased in recent years (6% down to <1%), and use in land restoration has increased (6% from 0%).

The primary focus of sludge treatment is to decrease its weight and volume to reduce disposal costs, and to diminish any potential health risks attributed to the disposal options. Water removal is the main means of decreasing its weight and volume, while the removal of pathogens is usually accomplished through digestion or incineration. Much sludge is treated using a variety of digestion techniques (anaerobic digestion and advanced anaerobic digestion), the purpose of which is to reduce the amount of organic matter and the number of disease-causing microorganisms present in the solids.

As technological advancements have progressed, energy can now be recovered from sludge through gas production during anaerobic digestion or through incineration of dried sludge. As a consequence, the Anaerobic Digestion Strategy and Action Plan in the United Kingdom for producing energy from waste was introduced in 2011 (DEFRA, 2011) to increase the use of anaerobic digestion for human, animal, and food waste to produce renewable energy, reduce landfill, and lower greenhouse gas emissions, while creating a fuel source from an unavoidable waste product. The UK water industry has been using anaerobic digestion to produce biogas. The process of anaerobic digestion of sewage begins with the sludge being contained in digestion tanks and degrading it (by thermal hydrolysis or enzymic hydrolysis) to give off biogas. These gases are then extracted and stored in tanks, where it combusts to provide heat and electricity or is fed into the national networks (Gooding and Booth, 2017).

9.5 WASH in Developing Countries

Globally, not all wastewater is collected and treated to the standards described earlier in this chapter. In many developing countries, the picture is very different; this means many people are at risk of contracting water-related diseases. This section will now focus on WASH.

Universal access to safe drinking water, sanitation, and hygiene has been a long-standing sustainable development goal and the linkages between improvements in WASH and the achievement of targets relating to poverty, health, nutrition, education, gender equality, sanitation, and hygiene, and sustainable economic growth are well established (United Nations, 2012). Adequate access to good WASH practices, facilities, and infrastructures have been strongly credited for considerable reductions in acute respiratory infections and reduced infant mortality (Luby *et al.*, 2005; Jefferson *et al.*, 2009). Cairncross and Valdmanis (2006) established that the spread of water-washed diseases can be contained by improved sanitation and hygiene. WASH practices, such as hand washing, sanitation, water treatment, and safe drinking water storage, have each been proven to reduce diarrhoea rates by 30%–40% (Curtis and Cairncross, 2003; Clasen *et al.*, 2007).

The WHO (2012) report stresses that enabling universal access to basic WASH and a safe faecal disposal system can protect entire communities from disease exposure and significantly reduce diarrhoeal diseases up to 70%. Similarly, the UNDP (2010) noted that child mortality was high among households with poor access to clean water and sanitation facilities. It has been estimated that adequate and safe WASH could decrease the number of diarrhoea cases by one third. It is generally accepted that lack of potable water and basic sanitation services remains one of the world's most urgent health issues.

Prüss-Üstün *et al.* (2008) highlights that unsafe water, inadequate sanitation, and insufficient hygiene account for an estimated 9.1 % of the global burden of disease and 6.3 % of all deaths. Nearly half of all people in developing countries have infections or diseases associated with inadequate water supply and sanitation. Barbara (2014) argues that inadequate access to sufficient good-quality water and appropriate sanitation exacerbates already difficult situations for people being treated for chronic or potentially fatal illnesses such as HIV/AIDS. They further affirmed that very few interventions would have a greater impact on the lives of the world's poorest and most marginalised people than reducing the time spent collecting water and addressing the health problems caused by poor water supply, sanitation, and hygiene.

The time saved by children, particularly the females, that would otherwise be spent searching for water and distances travelled in the bush or in unsafe places to defecate, can lead to improved school attendance and may result in higher percentage of school completion for girls. Access to adequate WASH facilities will mean fewer days lost in the home, at school, or work due to sickness, greater comfort, privacy, and safety especially for women, children, the elderly, and people living with disabilities, which enhances dignity and general well-being (WaterAid, 2013).

According to the WHO (2012) study *Global Costs and Benefits of Drinking-Water Supply and Sanitation Interventions to Reach the MDG Target and Universal Coverage*, it was estimated that an overall gain of 1.5% of global GDP and a $4.3 return would be gained for every US$1 invested in WASH services. This should translate to reduced healthcare costs for individuals and society. Furthermore, in an earlier study, Hutton and Haller (2004) estimated that the global economic benefits of investing in WASH come in several forms: (1) Healthcare savings of US$7 billion a year for health agencies and US$340 million for individuals; (2) 320 million productive days would be gained each year in the 15–59 year age group, an extra 272 million school attendance days a year, and an added 1.5 billion healthy days for children under five years of age, together representing productivity gains of US$9.9 billion a year; (3) Time savings resulting from more convenient drinking water and sanitation services, totalling 20 billion working days a year, giving a productivity payback of some US$63 billion a year; and (iv) values of deaths averted, based on discounted future earnings, amounting to US$3.6 billion a year.

Globalisation and interdependence among the world's economies mean that a growth crisis in one region could have a subsequent effect on the developed world. Economic growth seems to be dependent on high levels of access to WASH infrastructure; a decline in access to WASH facilities is expected to result in the following: a higher disease burden, lower education levels, lower worker productivity, higher labour costs, and slower economic growth. The Institute of Medicine (IOM, 2009), however, highlight that while poverty has been a major barrier to gaining access to adequate sanitation in many parts of the developing world, access to and the availability of clean water is a prerequisite to the sustainable growth and development of communities around the world.

WASH interventions have been shown to be highly effective in reducing the environmental exposure to diseases (Campbell *et al.*, 2014). Adequate treatment and disposal of excreta and both household and industrial wastewater creates less pressure on freshwater resources (Hesselbarth, 2005). According to Pickford (1991), poor environmental conditions arising from unhygienic disposal of excreta and sullage, plus accumulation of solid wastes, contributes to the spread of disease.

Esrey *et al.* (1991) and Soares-Magalhães *et al.* (2011) claim that locking excreta away in an enclosed receptacle (such as a pit) will bring about immediate benefit to the environment such as noticeable visual improvement, reduction in foul odour and flies, and improvements to the quality of surface water from a reduction of excreta polluting local water courses. Moreover, there are essential elements of sustainable WASH delivery that must continue to be contextually implemented in order to ensure that all WASH benefits are achieved in developing countries.

9.6 Conclusions

Turning on the tap in our homes to fill a glass with fresh water without worrying about its quality and/or being able to flush the toilet without regard to what happens to it when it leaves the pan means you are probably living in a developed country, where both the drinking water has been purified and the wastewater is treated to protect the natural environment and minimise public health risks. This chapter has described and detailed the processes involved in the collection and complex stepwise treatment of wastewater in a developed-country setting, together with the energy benefits extracted from the residual sludge before its disposal as an agricultural fertiliser or, if too contaminated, its disposal in landfill. It has also proffered insights of what life is like for people in developing countries, where public health risks from WASH are a daily risk/decision for the poorest in society; whereby, child mortality is highest among those households with poor access to clean water and sanitation facilities. Where there is a breakdown in safe faecal disposal, entire communities are exposed to life-threatening illnesses, particularly those with diarrhoeal diseases (e.g. cholera). Hence, the lack (or breakdown) of basic sanitation services remains one of the world's most urgent health issues.

References

Barbara, B. (2014) Nigeria Review African Economic Outlook. [Online] [Accessed 12 May 2015] Available at: <outlookhttp:// www.africaneconomicoutlook.org/

Cairncross, S. and Valdmanis, V. (2006) Water supply, sanitation, and hygiene promotion. In: Jamison, D.T., Breman, J.G., Measham, A.R., Alleyne, G., Claeson, M., Evans, D.B., Jha, P., Mills, A., and Musgrove, P. (eds.), *Disease Control Priorities in Developing Countries*. 2nd edition. Washington (DC), World Bank. pp. 771–792.

Campbell, S.J., Savage, G.B., Gray, D.J., Atkinson, J.-A.M., Soares-Magalhães, R.J., Nery, S.V., McCarthy, J.S., Velleman, Y., Wicken, J.H., Traub, R.J., Williams, G.M., Andrews, R.M., and Clements, A.C.A. (2014) Water, sanitation, and hygiene (WASH): A critical component for sustainable soil-transmitted helminth and schistosomiasis control. *PLoS Neglected Tropical Diseases*, 8 (4), e2651. https://doi.org/10.1371/journal.pntd.0002651

Clasen, T., Schmidt, W., Rabie, T., Roberts, I., and Cairncross, S. (2007). Interventions to improve water quality for preventing diarrhoea: Systematic review and meta-analysis. *British Medical Journal*, 334 (7597), 782.

Coker, A.O., Oluremi, J.R., Adeshiyan, R.A., Sridhar, M.K., Coker, M.E., Booth, C.A., Millington, J.A., and Khatib, J.M. (2011) Wastewater management in a Nigerian leper colony. *Journal of Environmental Engineering and Landscape Management*, 19, 260–269.

Curtis, V. and Cairncross, S. (2003) Effect of washing hands with soap on diarrhoea risk in the community: A systematic review. *Lancet Infectious Diseases*, 3, 275–281.

DEFRA (2002) *Sewage Treatment in the UK: UK Implementation of the EC Urban Waste Water Treatment Directive*. DEFRA, London.

DEFRA (2011) *Anaerobic Digestion Strategy and Action Plan: A Commitment to Increasing Energy from Waste through Anaerobic Digestion*. HM Government, London.

Esrey, S.A., Potash, J.B., Roberts, L., and Shiff, C. (1991) Effects of improved water supply and sanitation on ascariasis, diarrhoea, dracunculiasis, hookworm infection, schistosomiasis, and trachoma. *Bulletin of the World Health Organization*, 69, 609–621.

Garg, N.K. (2009) Multicriteria Assessment of Alternative Sludge Disposal Methods. Unpublished MSc Thesis, University of Strathclyde, UK.

Gooding, A. and Booth, C.A. (2017) Insights and issues into the uptake and development of advanced anaerobic digestion within the UK water industry. In: Brebbia, C.A., Longhurst, J., Marco, E. and Booth, C.A. (eds.), *Sustainable Development & Planning* IX (in press).

Hesselbarth, S. (2005) *Socio-Economic Impacts of Water Supply and Sanitation Projects.* KfW Entwicklungsbank, Frankfurt.

Hutton, G. and Haller, L. (2004) *Evaluation of the costs and benefits of water and sanitation improvements at the global level.* Geneva, World Health Organization.

Institute of Medicine (IOM) (2009) *Global Issues in Water, Sanitation, and Health.* Washington, DC, The National Academies Press.

Jefferson, T., Del Mar, C., Dooley, L., Ferroni, E., Al-Ansary, L.A., Bawazeer, G.A., van Driel, M.L., Foxlee, R., and Rivetti, A. (2009) Physical interventions to interrupt or reduce the spread of respiratory viruses: systematic review. *British Medical Journal*, 339 (b3675).

Luby, S.P., Agboatwalla, M., Feikin, D.R., Painter, J., Billhimer, W., Altaf, A., and Hoekstra, R.M. (2005) Effect of handwashing on child health: A randomized controlled trial. *The Lancet*, 366 (9481), 225–233.

Mason, P.R. (2009) Zimbabwe experiences the worst epidemic of cholera in Africa. *The Journal of Infection in Developing Countries*, 3, 148–151.

Newsom, S.W.B. (2006) Pioneers in infection control: John Snow, Henry Whitehead, the Broad Street pump, and the beginnings of geographical epidemiology. *Journal of Hospital Infection*, 64, 210–216.

Nguyen, V.D., Sreenivasan, N., Lam, E., Ayers, T., Kargbo, D., Dafae, F., Jambai, A., Alemu, W., Kamara, A., Sirajul Islam, M., Stroika, S., Bopp, C., Quick, R., Mintz, E.D., and Brunkard, J.M. (2014) Cholera epidemic associated with consumption of unsafe drinking water and street-vended water – Eastern Freetown, Sierra Leone, 2012. *American Journal of Tropical Medicine and Hygiene.* 90, 518–523.

Piarroux, R., Barrais, R., Faucher, B., Haus, R., Piarroux, M., Gaudart, J., Magloire, R., and Raoult, D. (2011) Understanding the cholera epidemic, Haiti. *Emerging Infectious Diseases*, 17, 1161–1168.

Pickford, J. (1991) *The Worth of Water: Technical Briefs on Health, Water and Sanitation.* Intermediate Technology Publications, London, 140pp.

Prüss-Üstün, A., Bos, R., Gore, F., and Bartram, J. (2008) *Safer Water, Better Health: Costs, Benefits and Sustainability of Interventions to Protect and Promote Health.* World Health Organization (WHO), Geneva

Soares-Magalhães, R.J., Barnett, A.G., and Clements, A.C. (2011) Geographical analysis of the role of water supply and sanitation in the risk of helminth infections of children in West Africa. *Proceedings of the National Academy of Sciences of the United States of America*, 108, 20084–20089. doi: 10.1073/pnas.1106784108

Spellman, F.R. and Drinan, J. (2003) *Wastewater Treatment Plant Operations Made Easy.* DEStech Publications, p. 232.

Tulodziecki, D. (2011) A case study in explanatory power: John Snow's conclusions about the pathology and transmission of cholera. *Studies in History and Philosophy of Science Part C: Studies in History and Philosophy of Biological and Biomedical Sciences*, 42, 306–316.

United Nations Educational, Scientific and Cultural Organization (UNESCO) (2006) *Water: A Shared Responsibility.* Berghahn Books, New York, USA. pp. 203–240.

United Nations (2012) On the Right Track: Good Practices in Realising the Rights to Water and Sanitation. United Nations Special Rapporteur. Available at http://www.ohchr.org/Documents/Issues/Water/BookonGoodPractices_en.pdf

Urban Waste Water Directive (1998) http://ec.europa.eu/environment/water/water-urbanwaste/legislation/directive_en.htm

WaterAid (2013) *Everyone, Everywhere: A Vision for Water, Sanitation and Hygiene Post-2015*. WaterAid, London, UK.

Weiner, R.F. and Matthews, R.A. (2003) *Environmental Engineering*. 4th edition. Butterworth-Heinemann, Elsevier.

WHO (2012) *Global Costs and Benefits of Drinking-Water Supply and Sanitation Interventions to Reach the MDG Target and Universal Coverage*. Geneva, World Health Organization. Available at http://www.who.int/water_sanitation_health/publications/2012/globalcosts.pdf

WHO (2017) Sanitation Fact Sheet. http://www.who.int/mediacentre/factsheets/fs392/en/

10

Living Green Roofs

Sara Wilkinson[1] *and Fraser Torpy*[2]

[1] *Faculty of Design Architecture and Building, University of Technology Sydney, Australia*
[2] *School of Life Sciences, University of Technology Sydney, Australia*

10.1 Introduction

Along with the well-documented ecosystem services provided by urban plants such as run-off abatement, provision of biodiversity, and mitigation of the urban heat island effect, there is a growing body of evidence that plants can have a major influence on the quality of city air, both positively, and in some cases, negatively. A thorough understanding of the ecosystem services and disservices provided by various forms of greenspace is thus essential during the planning stage of urban developments, to ensure that the implementation of greenspace provides the sought-after functions. Furthermore, retrofit and adaptation should take advantage of the opportunity to increase the amount of urban greenspace.

There are increasing challenges for urban settlements in both developed and developing countries. These challenges can include rapid urbanisation, poorly legislated and/or controlled development, undocumented overoccupation of residential buildings particularly where housing affordability is an issue, the increasing impact of climate change, and rising temperatures. For example, issues of stormwater surges, sewer overflows, and consequent pollution are driving legislation and programmes with regard to green roofs and infrastructure in Nashville, Tennessee (NRDC, 2015). Individually or concurrently, such challenges can add to urban pollution, thereby reducing liveability and negatively impacting on human health and well-being. This chapter explores the relationship between urban development, pollution, and green roofs as a valuable, contributory means of mitigating the effects of urban pollution using examples from two major Australian cities, Melbourne and Sydney.

10.2 Increasing Urbanisation: Urban Growth

Over hundreds of thousands of years, the global population grew to 1 billion; then in another 200 years, it grew seven times (UN, 2015). The total global population had reached 7 billion in 2011, and by 2015, it had increased to around 7.3 billion. Furthermore, according to the UN (UN DESA, 2015), this figure is predicted to reach 8.5 billion by 2030, 9.7 billion by 2050, and 11.2 billion by 2100. This growth is predicated by a number of factors, including the greater numbers of people surviving to reproductive and older age, together with considerable changes in fertility rates, increasing urbanisation, and accelerating migration. These trends will have far-reaching implications

Urban Pollution: Science and Management, First Edition. Edited by Susanne M. Charlesworth and Colin A. Booth.

for generations to come and one of those implications is increased pressures on existing and rapidly expanding urban settlements (UNPF, 2015).

Currently the world is undergoing its most rapid and largest upsurge of urban growth in history. In 2015, more than half of the world's population now reside in towns and cities, with the prediction that by 2030 this figure will increase to about 5 billion (UNFPA, 2015). Further, by 2050, it is estimated that 66% of the total population, or 6.4 billion people, will be urbanised (RICS, 2015). Although much of this urbanisation will unfold in Africa and Asia, bringing significant social, economic, and environmental transformations, all countries and cities will be affected.

Potentially, urbanisation could usher in a new era of well-being, resource efficiency, and economic growth. However, many cities accommodate large concentrations of inequality and poverty, and wealthy communities can coexist alongside, and separate from, slums and informal settlements. So cities will grow, in many cases faster than ever before. It follows that planning and governance are needed to deliver transitions from the building scale to the city scale, ensuring that infrastructure can support growing populations and changing land uses. Construction of the built environment and its associated infrastructure causes considerable resource consumption as well as disruption, with the potential for soil, water, air, and noise pollution for local inhabitants. As existing areas expand and new areas are developed to accommodate greater numbers of people, predominant land uses undergo some change; consideration of sustainable development is needed ensure that urban pollution, in all its forms, is minimised.

10.3 Increasing Urbanisation: Soil, Water, and Air Pollution

Pollution is an integral outcome of most human activities. While agricultural pollution is a major cause of environmental degradation worldwide, the dramatically increasing urbanisation of human settlement and the demographic transition under way in many regions of the world results in pollution in urban areas and cities having the greatest effect on human health. Urban pollution may take many forms and degrees of severity, depending on the population density, level of development, topography, climate, and past and current major activities of an area. Pollution in urban environments can affect soil, water, and air, with the latter having the greatest effect on human health, well-being, and longevity.

10.3.1 Soil Pollution

The chemical contamination of urban soils results from both a broad range of industrial activities and vehicular traffic emissions. While most urban soils have at least some contaminants present (Mitchell *et al.*, 2014), depending on the usage history of an area, pollutant concentrations in soils range from fairly safe to dangerously high (Luo *et al.*, 2012). Many forms of soil contamination are of significant concern for human health, especially when a high level of contact with soil is commonplace, for example, when food crops are grown on polluted soils (Wiseman *et al.*, 2013; Galal and Shehata, 2015).

Vehicular exhaust emissions are the primary cause of traffic-sourced soil pollutants, which are made up of various gases along with suspended particulate matter, composed of carbon, a range of non-volatile hydrocarbons, and metal compounds (Laschober *et al.*, 2004). These materials may linger in the air proximal to roadways, or become deposited in nearby soils, where they may accumulate, be absorbed by plants (Jozic *et al.*, 2009), or contribute to pollution in stormwater run-off. Wear products from brakes and tyres are also a source of heavy metal (Zereini *et al.*, 2012). Concentrations of heavy metals including lead, copper, zinc, and palladium in both urban soil and proximal vegetation have been linked to increases in road traffic density (Onder and Dursun,

2006, Cicchella *et al.*, 2008; Farmaki and Thomaidis, 2008; Alyemenia and Almohisen, 2014). Lead is the traffic-sourced heavy metal of most concern in urban soils (Lu *et al.*, 2015) and is found at high levels in cities worldwide (Laidlaw and Taylor, 2011; Lu *et al.*, 2015). While the use of lead in motor vehicle fuels has ceased, millions of tonnes of this pollutant remain in soils, often at concentrations that are dangerous to health (Mielke *et al.*, 2011). Unless actively removed, leached into the groundwater or eroded away, lead and other heavy metals are cycled through the biosphere and remain in the local food chain indefinitely.

Certain industries can also lead to significant soil pollution in local soils. Islam *et al.* (2015) showed that Bangladeshi soils contaminated with tannery waste, metalwork industry pollution, and electrical wastes were heavily contaminated by nickel, copper, arsenic, cadmium, and lead. Peña-Fernández *et al.* (2015) found high levels of arsenic, nickel, and beryllium in soil from Alcalá de Henares (Spain), originating from local ceramic, jewellery, and electronic manufacturing industries. Pollutant levels in these studies often exceeded safe levels, and heavy metal translocation from soil to plants has been demonstrated (Finster *et al.*, 2004), indicating potential health concerns if food plants are grown in contaminated urban gardens (Wiseman *et al.*, 2013; Galal and Shehata 2015). Metal contamination may thus be a concern if plants grown in roof gardens are used for food. In such instances, it would be prudent to understand the types of pollutants taken up by different crop species (Hu *et al.*, 2013; McBride, 2013), replace the cropping soils with those from an uncontaminated source, or utilise an effective remediation technique (e.g. see Mitchell *et al.*, 2014; Angotti, 2015; Wilkinson and Torpy, 2016) to ensure safety of the produce grown. The use of pollutant hyper-accumulator plant species (Jambhulkar and Juwarkar, 2009) in roof gardens could both de-contaminate soil prior to the planting of soil crops, and prevent contaminants from accumulating in the garden soil and subsequently migrating to stormwater, and could become a focus of future research.

Despite the ubiquity of urban soil contamination, and localised high concentrations, it is generally not believed be of major concern in most areas, especially if efforts are made to use less contaminated soils for cropping (e.g. Sipter *et al.*, 2008; McBride *et al.*, 2014). Further, relatively few studies have detected dangerous levels of contamination in food products grown in urban areas (e.g. Hu *et al.*, 2013; Ferri *et al.*, 2015). In any case, increasing the consumption of fresh fruit and vegetables should provide far greater health benefits than any deleterious effects resulting from the consumption of lightly contaminated produce (Guitart *et al.*, 2014).

10.3.2 Water Pollution

Many urban waterways suffer from elevated concentrations of nitrogen and phosphorus (Nidzgorski and Hobbie, 2016), sourced primarily from agricultural, recreational facilities (e.g. golf courses) and domestic garden fertiliser use. High levels of these nutrients can accelerate microbial growth, leading to algal blooms, the subsequent starvation of oxygen from a body of water, the release of offensive odours, and production of microbial toxins. Urban trees have been shown to be capable of substantially reducing the leaching of these pollutants to groundwater, resulting in large potential savings over the installation of traditional stormwater infrastructure to achieve the same performance (Nidzgorski and Hobbie, 2016). Similarly, Teemusk and Mander (2011) demonstrated reduced levels of nitrogen and phosphorus in stormwater run-off when urban vegetation in the form of green roofs were present, relative to hard surfaces. Given the difficulty in dealing with nutrient rich urban run-off using any other means, the development of urban greenspace in areas where stormwater accumulates has become a growing practice in many developed and developing countries (Charlesworth, 2010; Ezema and Oluwatayo, 2014).

Precipitated air pollutants such as heavy metals also lead to significant pollution concerns in urban run-off (Miguntanna *et al.*, 2010). Studies by Mendez *et al.* (2011) and Vijayaraghavan *et al.* (2012) have both provided proof-of-concept models that roof gardens reduce the transfer of metal pollutants to the stormwater system, thus contributing to improved waterway quality. These pollutants, however, will accumulate in the plant growth substrate, and become subject to release in extreme storm events, along with high levels of suspended solids if erodible substrates are used (Gnecco *et al.*, 2013). The careful design of green infrastructure systems will thus be required to ensure the best performance, especially noting the roof pitch, expected volume of stormwater, and frequency of high-rainfall events.

With appropriate design, much of the water pollution remediation capacity of urban greenspace will be associated with reduction in these occasional, high-level storm flows. Soils retain water effectively, and once in the substrate, it is either utilised by vegetation, or subject to high evaporation rates due to the innate porosity of plant growth media (Bliss *et al.*, 2009). The roofs of buildings, being a major contributor to the total impervious surface area of urban areas, are perhaps the ideal location for green space aimed at storm flow mitigation. This has received some study, with roof gardens observed to reduce gross rainwater discharge from buildings by 29%–55% (Graham and Kim, 2003; Spolek, 2008; Berndtsson, 2010).

10.3.3 Air Pollution

Air pollution is a universal component of urban environments (Begg *et al.*, 2007). Poor urban air quality has thus become a major worldwide health problem, and has been convincingly linked to the deaths of ~8 million people yearly (WHO, 2014), with global healthcare and associated costs estimated at US$1.7 trillion (OECD, 2014). In Australia, the total cost burden of air pollution has been estimated to be as high as 1% of the gross domestic product (Brindle *et al.*, 1999) or AUD$12 billion yearly, from both direct medical expenses and lost productivity (Environment Australia, 2003). Exposure to even low levels of air pollution can lead to a very wide range of respiratory diseases, especially in infants (Saravia *et al.*, 2013), with some research even linking poor air quality to autism (Volk *et al.*, 2013). As much as 90% of urban air pollution is due to fossil fuel emissions, comprising a mixture of carbon dioxide, carbon monoxide, nitrogen oxides, sulphur oxides, volatile organic compounds, ozone, and particulate matter (Thurston, 2008).

Emerging evidence indicates that most forms of urban greenspace can mitigate high airborne pollutant concentrations (Chen and Jim, 2008; Yang *et al.*, 2008; Rowe 2011; Zheng *et al.*, 2013), including particulate matter (Sæbø *et al.*, 2012; Blanusa *et al.*, 2015; Irga *et al.*, 2015; Janhäll, 2015) and a range of gases (Vos *et al.*, 2013, Janhäll, 2015) including ozone, nitrogen dioxide (Fantozzi *et al.*, 2015), and sulphur dioxide (Yin *et al.*, 2011). Additionally, vegetation has been shown to bind aerosol heavy metals in the foliage (Blanusa *et al.*, 2015) and the surrounding soil (Fantozzi *et al.*, 2013). Dense areas of plants can also form effective barriers against pollutant intrusion (Salmond *et al.*, 2013; Janhäll, 2015), and there is even evidence of reduced illegal dumping of garbage when green space is present (Yoo and Kwon, 2015).

While not all studies have detected significant effects, depending on the location of the study (e.g. Setälä *et al.*, 2013), there remain insufficient empirical studies to allow accurate quantification of the air pollutant mitigation capacity of green plants (Setälä *et al.*, 2013), there remains convincing evidence that plants have a positive effect on air quality, and that the expansion of urban greenspace could be a key component of future air quality improvement measures. Given the growing land values and space constraints in cities worldwide, integration of the required vegetation into the existing and future built environment is necessary, and thus green roofs, green walls, and other forms of integrated green infrastructure should become a key part of future city buildings.

Table 10.1 Pollution types, sources, and impacts attenuated by green roofs and infrastructure.

Pollution	Type	Sources	Impacts
Soil	Chemical	Industrial activities	Human health if ingested via food grown in contaminated soils
	Chemicals (including lead)	Vehicle traffic emissions	Human health if ingested via food grown in soils contaminated by leaded fuels. Also inhalation of polluted particulates
Water	Chemical (nitrogen and phosphorous)	Agricultural run-off, commercial and domestic garden fertilisers	Odours and microbial toxins
	Chemical	Precipitated air pollutants, e.g. metals	Can contaminate and effect water quality
	Chemical	Storm surges overloading sewers	Discharge of contaminated water and solids
Air	Poor air quality	Industrial processes	Human health, respiratory disease, and possibly autism
		Vehicle exhaust fumes	Human health, respiratory disease, and possibly autism

Source: Wilkinson and Torpy.

Thus, there is compelling evidence that urban vegetation provides significant positive ecosystem services associated with several forms pollution abatement (see Table 10.1). However, several studies have quantified disservices from urban plants, including the release of allergenic materials and gases which contribute to ozone formation, along with the largely unknown rates and processes by which trapped particulate matter may become resuspended (Setälä *et al.*, 2013). To these may be added the minor risks associated with the consumption of food crops with high pollutant levels. While further study is clearly required to determine the balance of outcomes resulting from the expansion of urban vegetation, currently, it appears that the positive effects or urban greenspace greatly outweigh the negatives.

10.4 Urban Heat Islands and Human Health

All cities are characterised by intense energy-using activities and heat-retaining materials such as concrete and bitumen. These factors combine to create the 'urban heat island' effect, wherein cities have a substantially greater ambient temperature than surrounding suburban, rural, and undeveloped areas (Cheng *et al.*, 2010). In warm climates, these elevated temperatures lead to a greater load on the cooling component of air conditioning systems, and thus increase the energy use of urban areas as a whole. As air conditioning energy usage constitutes the greatest energy load in buildings, and between 10% and 20% of the total energy consumption in developed countries (Perez-Lombard *et al.*, 2011), these effects can be substantial.

Added to these human-induced contributions to overall temperature and heat in cities, global temperatures are increasing owing to climate change. As such, some cites are more vulnerable than others to extreme heat. Melbourne in Victoria, Australia, is one such city, where some areas are more vulnerable than others. Here, vulnerability is defined as a function of exposure, sensitivity of the exposed population, and capacity to adapt (Melbourne, 2015). Urban green spaces, including green roofs and walls, play a key role in cooling the city and reducing exposure to heatwaves; however, the distribution of green spaces varies considerably, with some areas having among the lowest rates of tree canopy cover in Australia (Melbourne, 2015). In the Central Business District (CBD) especially, there is great potential to increase green space provision in the form of green roofs, and Plates 10.1 and 10.2 show typical roofscapes in Melbourne.

Between 14 and 17 January 2014, Victoria experienced the hottest four-day period on record, with daily temperatures exceeding 41°C (Victoria Health, 2015). Maximum temperatures were 12°C or more above average for much of the state, with the intensity and duration of the heatwave greater than those recorded in January 2009, January 1939, and January 1908 (BOM, 2014). The consequent health impacts due to the 2014 heatwave were assessed by examining the following health service use and mortality data and included a 7% increase in public hospital emergency department presentations, although people

Plate 10.1 Typical roof designs in Melbourne CBD. (*Source*: Wilkinson)

Plate 10.2 The Melbourne CBD. (*Source*: Wilkinson)

aged 75 years plus increased presentations by 23%. There were 621 heat-related presentations during the week of the heatwave, higher than the 105 expected (a fivefold increase), with 45% for people aged over 75 years. There was a 25% increase in the Ambulance Victoria emergency caseload in the metropolitan region, with numbers of emergency dispatches peaking on 16–17 January 2014. Numbers of emergency incidents continued to be high for a few days following the heatwave (Victoria Health, 2015). In addition there was a 56% increase in after-hours doctor consultations during the heatwave and a threefold increase in heat-related calls to the on-call nurse service. There were 228 deaths reported to the Coroners Court of Victoria for investigation during the heatwave, over twice the 105 expected. Finally, the 858 deaths, compared to the 691 expected during the week of the heatwave, represented a 24% increase in mortality (Victoria Health, 2015).

Therefore, health impacts in high-density urban environments can be significant, and furthermore with an ageing population it is set to increase if similar conditions are experienced. Green infrastructure alone cannot provide the whole solution. Part of the response is also social in nature, and Melbourne identified the need to strengthen social cohesion and the community, citing a Chicago study during a 1995 heatwave (Melbourne, 2015). The mortality and health impacts in two Chicago neighbourhoods with almost identical demographic profiles were compared, and the neighbourhood with a culture of checking on the elderly, vulnerable, and infirm residents suffered much lower rates of mortality than the other which had no such culture; 3 deaths per 100,000 were recorded compared to 33 per 100,000. With Melbourne bracing for a future of more frequent and intense extreme heat events, social cohesion as well as green infrastructure are vital components in the ability to withstand these conditions.

As well as roadside vegetation, roof gardens can reduce the ambient temperature of buildings through shading and evaporative cooling from both evaporation of substrate water and plant transpiration (Taha *et al.*, 1997). Chen *et al.* (2013) performed a simulation study that estimated mean city temperature decreases of up to 0.7°C by doubling the quantity of vegetation in Melbourne. While not a great magnitude overall, temperature reductions in individual buildings would be much greater than this exterior value. Such effects could lead to a substantial reduction in the thermal regulation requirements of the affected buildings, and consequent reductions in the building carbon footprint across cities (Norton *et al.*, 2015). In combination with the well-researched air cleaning capacities of indoor plants (Torpy *et al.*, 2015) leading to reductions in air conditioning ventilation cycles, much more sustainable cities could result, with very low expenditure compared to the mechanical infrastructure required to achieve the same effects. Furthermore, if green walls are adopted, greater benefits could be realised because the potential area of facades is greater than roof space (see Plate 10.3, Central Park Sydney). Wilkinson and Reed's (2009) analysis of the Melbourne CBD rooftop estimated that 17% of office roofs could be retrofitted with green roofs without any additional structural strengthening work being required. If adopted on a wide scale, green roofs, and walls could attenuate the urban heat island generally and mitigate the effects of heat waves.

10.5 Green Roof Options

Green roofs are versatile and can be used on flat and pitched roof designs. They can be new build or retrofitted to existing structures and to high-rise, high-density, and low-rise, low-density buildings as shown in Plates 10.3, 10.4, and 10.5. In Plate 10.3, a former brickworks site in inner Sydney has been converted to a retail and start-up commercial space in the former brickworks buildings. Conversion involved remediation of residual contamination on the former industrial site. Part of the development

Plate 10.3 The Grounds Alexandria Sydney in 2014. (*Source*: Wilkinson)

Plate 10.4 The pitched green roof at The Grounds Alexandria Sydney August 2016. (*Source*: Wilkinson)

Plate 10.5 Pitched green roofs on residential property Newtown Sydney. (*Source*: Wilkinson)

comprises a new build 'shed' with a pitched green roof, and is shown in 2014 when it was initially planted. Plate 10.4 shows the same roof two years later in full bloom. Not only is the roof attractive, it attracts biodiversity, attenuates stormwater run-off and improves air quality. Located in the inner city, it also absorbs pollution from vehicle emissions.

It is possible to design low-rise residential buildings with pitched green roofs as shown in Plate 10.5. Here, the terraces of three new buildings constructed in 2015 have been planted out in an inner-city Sydney suburb. In this example, the roofs will provide high levels of thermal insulation, as well as insulating against noise pollution. This suburb is site close to the domestic and international Sydney airport, and aircraft noise is a local issue. Stormwater run-off is also attenuated with the design, and air quality improved. The plants will also absorb some chemical airborne and precipitated air pollutants; however, currently these community-wide pollution abatement benefits are not recognised in the regulations, measured, or quantified, and there may be a case for incentivising building owners to adopt green roofs to deliver pollution abatement as cities continue to accommodate higher-density populations.

Green roofs have multiple benefits collectively or individually; for example, a green roof designed as a food-producing garden will have thermal performance, biodiversity, stormwater attenuation, air quality benefits, and social amenity benefits, as well as producing food with low carbon miles for people to consume. In areas where air or soil pollution is high, there is also the opportunity to specify plant species known to have a higher pollutant absorption capacity.

10.6 Case Study: University of Technology, Sydney, Food-Producing Roof and Urban Pollution

The authors performed a preliminary trial of growing plants and vegetables on three different rooftops at the University of Technology, Sydney (UTS), New South Wales, Australia, in 2013 and 2014. Three different types of rooftop garden beds were used to provide three illustrative case studies.

A plastic wicking bed was built on the northwest-facing side of the seven-storey Science Faulty building roof. The structure consisted of a 320 L plastic pond, with a reservoir created by laying a 100 mm agricultural drainage pipe across the bottom of the pond, with a vertical tube to allow filling of the reservoir, and an overflow pipe 100 mm from the base. A layer of geotextile fabric was placed on top of the agricultural pipe to allow capillary action to irrigate the garden bed. The bed was filled with ~200 mm soil and compost and fertilised with seaweed fertiliser, and planted with a mix of aubergine, courgette, basil, carrots, beetroot, lettuce, chilli, capsicum, silverbeet, celery, rocket, mizuna, and marigolds.

Paired beds were also built on the roof of a nine-storey student housing structure, facing southeast. The garden beds chosen for this site consisted of 594 L recycled plastic raised beds supported on timber frame trestles. A food-grade waterproofing membrane was placed over the trestles as weather protection, before the plastic pods were sealed in place on the membrane with silicone. A layer of 20 mm drainage cell sheeting was placed across the floor of the beds covered with a layer of geotextile fabric, a 20 mm layer of fine gravel, followed by another layer of geotextile fabric. One bed was filled with composted cow manure and garden soil, the other a lightweight engineered substrate with a mixture of coir bark, perlite P400, Canadian peat, composted pine bark, trace elements, calcium nitrate, coarse granular dolomite, gypsum, superphosphate, zeolite, and magrilime. Both beds were fertilised and planted as given for the Science Faculty roof above.

The third bed was built on the level 5 roof of another university building, facing northwest. This bed was shaded by the surrounding buildings, and experienced minimal direct sun in the winter months. The garden beds used on this site consisted of a timber frame and a large base bed with a series of metal-framed horizontal trays, which were self-watered with an electric irrigation system. These beds were planted with a variety of herbs including basil, mint, and oregano and vegetables such as lettuce, eggplants, silverbeet, and capsicums. Larger root vegetables such as carrots and beetroots were grown in the base bed.

There was considerable interaction and interest from staff where the beds were sited around the university, eliciting self-organised groups to water and maintain the beds. Students in the housing building took a keen interest in the gardens, journalism students wrote about them, and industrial design students used the gardens and rooftop agriculture in their major design projects. Overall, the social interactions were high and positive and were analysed by Ghosh *et al.* (cited in Wilkinson and Dixon, 2016). This is a good example of social benefits coexisting with the environmental benefits of reducing urban pollution and improving air quality in a green roof setting.

While the economic returns from the gardens would take many years to repay the costs of the beds, the plants grew successfully despite challenging weather conditions and an exceptionally hot summer; subsequent crops continued to grow well in all the substrates used. In terms of building performance, all green roofs were expected to reduce air conditioning system cooling loads to some degree, although this could not be recorded in this pilot trial due to any effects in the buildings being masked by increased air conditioning loads. Nonetheless, the temperature of the roofs under the beds was markedly cooler than the surrounding areas, especially the wicking bed due to its evaporative effects. Water absorption by all the beds noticeably reduced stormwater run-off from the roofs of all the buildings on which they were built, and several species of birds and insects were attracted to the beds, which were in areas otherwise bereft of observable biodiversity.

Testing was performed for heavy metals in the original growth substrates, and again for subsequent crops and the vegetables grown. While some lead was detected in all the original potting substrates, this was taken up by the first crop, after which very low levels were detected. While the lead levels in the first crops were detectable, they were not unsafe for human consumption — it may nonetheless be wise to grow an initial, sacrificial crop to bioremediate metals in any food crops grown in commercial substrates.

10.7 Conclusions and Next Steps

It is thus well known that urban green space provides a range of valuable ecosystem services to the urban environment; reducing the urban heat island effect, mitigating storm water run-off, and providing a base for the development of biodiversity in city areas, along with the generally underestimated effect of promoting well-being and happiness in city residents (Wilson, 1984). Population growth, climate change, and increasing urbanisation are putting even more pressure on providing and/or maintaining liveable urban settlements. As the need for methods to provide clean air in cities becomes more critical, is clear that plants in urban areas, especially taller shrubs and trees, can have a major effect on air quality.

The heatwave of 2014 in Melbourne is used as an example to show how acute events can have negative social, environmental and economic impacts on the environment and on the health of large numbers of people. However, a widespread adoption of green infrastructure including trees, green walls, and green roofs could mitigate the effects of these intense temperatures substantially. Some cities, such as Nashville, Tennessee, United States, are enacting legislation to mitigate the impacts of pollution caused by stormwater surges, and others such as Toronto, Canada, have mandated green roofs for multiple reasons including air quality, pollution abatement, and some of the community-wide benefits green roofs can deliver. Monitoring and evaluation of such approaches would be useful to examine whether cases can be made for other cities to adopt such approaches.

Green roofs are suited to pitched and flat-roofed buildings, to low-rise and high-rise buildings across a broad ranges of land uses. Moreover, they can be designed into new buildings and retrofitted to existing buildings. Quantifying the multiple benefits of green roofs is imperative to establish all the concurrent benefits that arise when they are installed. Urban green infrastructure is already performing effective roles in mitigating stormwater run-off, along with removing toxins and high nutrient loads in run-off streams. This initiative needs to continue and grow.

References

Alyemenia, M.N. and Almohisen, I.A.A. (2014) Traffic and industrial activities around Riyadh cause the accumulation of heavy metals in legumes: A case study. *Saudi Journal of Biological Sciences*, 21, 167–172.

Angotti, T. (2015) Urban agriculture: Long-term strategy or impossible dream? Lessons from Prospect Farm in Brooklyn, New York. *Public Health*, 129, 336–341.

Begg, S., Vos, T., Barker, B., Stevenson, C., Stanley, L., and Lopez, A. (2007) The Burden of Disease and Injury in Australia 2003, Cat. no. PHE 82 ed. Australian Institute of Health and Welfare (AIHW), Canberra, Australia.

Berndtsson, J. (2010) Green roof performance towards management of runoff water quantity and quality: A review. *Ecological Engineering*, 36 (4), 351–360.

Blanusa, T., Fantozzi, F., Monaci, F., and Bargagli, R. (2015) Leaf trapping and retention of particles by holm oak and other common tree species in Mediterranean urban environments. *Urban Forestry and Urban Greening*, 14, 1095–1101.

Bliss, J., Neufeld, D., and Ries, J. (2009) Storm water runoff mitigation using a green roof. *Environmental Engineering Science*, 26 (2), 407–418.

Brindle, R., Houghton, N., and Sheridan, G. (1999) Transport generated air pollution and its health impacts. A source document for local government. In: ARRB Transport Research. Vermont South, Victoria.

Bureau of Meteorology (2014) Special Climate Statement 48 – One of Southeast Australia's Most Significant Heatwaves, Bureau of Meteorology, Melbourne.

Bureau of Meteorology and Commonwealth Scientific and Industrial Research Organisation 2012, State of the climate 2012, <http://www.csiro.au/Outcomes/Climate/Understanding/State-of-the-Climate-2012.aspx>.

Charlesworth, S. (2010) A review of the adaptation and mitigation of global climate change using sustainable drainage in cities. *Journal of Water and Climate Change*, 1 (3), 165–180.

Chen, W. and Jim, C.Y. (2008) Assessment and valuation of the ecosystem services provided by urban forests. In: Carreiro, M., Song, Y.-C., and Wu, J. (eds.), *Ecology, Planning, and Management of Urban Forests*. Springer, New York. pp. 53–83.

Chen, D., Wang, X., Khoo, Y., Thatcher, M., Lin, B., Ren, Z., Wang, C., and Barnett, G. (2013) Assessment of urban heat island and mitigation by urban green coverage. *Mitigating Climate Change*, 34 (2), 247–257.

Cheng, C.Y., Cheung, K.K.S., and Chu, L.M. (2010) Thermal performance of a vegetated cladding system on facade walls. *Building and Environment*, 45 (8), 1779–1787.

Cicchella, D., De Vivo, B., Lima, A., Albanese, S., McGill, R.A.R., and Parrish, R.R. (2008). Heavy metal pollution and Pb isotopes in urban soils of Napoli, Italy. *Geochemistry:*

Exploration, Environment, Analysis, 8 (1), 103–112.

Environment Australia (2003) BTEX Personal Exposure Monitoring in Four Australian Cities. Technical Paper No. 6. Canberra, Australia, Environment Australia.

Ezema, I.C. and Oluwatayo, A. (2014) Densification as sustainable urban policy: The case of Ikoyi, Lagos, Nigeria. *Proceedings, International Council for Research and Innovation in Building and Construction (CIB) Conference*, University of Lagos, 28–30 January 2014.

Fantozzi, F., Monaci, F., Blanusa, T. and Bargagli, R. (2013) Holm Oak (*Quercus ilex* L.) canopy as interceptor of airborne trace elements and their accumulation in the litter and topsoil. *Environmental Pollution*, 183, 89–95

Fantozzi, F., Monaci, F., Blanusa, T., and Bargagli, R. (2015) Spatio-temporal variations of ozone and nitrogen dioxide concentrations under urban trees and in a nearby open area. *Urban Climate*, 12, 119–127

Farmaki, E.G. and Thomaidis, N.S. (2008) Current status of the metal pollution of the environment of Greece—A review. *Global NEST Journal*, 10, 366–375.

Ferri, R., Hashim, D., Smith, D.R., Guazzetti, S., Donna, F., Ferretti, E., Curatolo, M., Moneta, C., Beone, G.M., and Lucchini, R.G. (2015) Metal contamination of home garden soils and cultivated vegetables in the province of Brescia, Italy: Implications for human exposure. *Science of the Total Environment*, 518–519, 507–517.

Finster, M.E., Gray, A.K., and Binns, H. (2004) Lead levels of edibles grown in contaminated residential soils: A field survey. *Science of the Total Environment*, 320, 245–257.

Galal, T.M. and Shehata, H.S. (2015) Bioaccumulation and translocation of heavy metals by *Plantago major* L. grown in contaminated soils under the effect of traffic pollution. *Ecological Indicators*, 48, 244–251.

Gnecco, I., Palla, A., Lanza, L.G., and La Barbera, P. (2013) The role of green roofs as a source/sink of pollutants in storm water outflows. *Water Resources Management*, 27 (14), 4715–4730.

Graham, P. and Kim, M. (2003) Evaluating the stormwater management benefits of green roofs through water balance modelling. In: *Proceedings of 1st North American Green Roof Conference: Greening Rooftops for Sustainable Communities*, Chicago. Cardinal Group, Toronto.

Guitart, D.A., Pickering, C.M. and Byrne, J.A. (2014) Color me healthy: Food diversity in school community gardens in two rapidly urbanising Australian cities. *Health and Place*, 26, 110–117.

Hu, J., Wu, F., Wu, S., Sun, X., Lin, X., and Wong, M.H. (2013). Phytoavailability and phytovariety codetermine the bioaccumulation risk of heavy metal from soils, focusing on Cd-contaminated vegetable farms around the Pearl River Delta, China. *Ecotoxicology and Environmental Safety*, 91, 18–24.

Intergovernmental Panel on Climate Change (2014) *Climate Change 2013:* The Physical Science Basis, contribution of working group I to the fifth assessment report of the Intergovernmental Panel on Climate Change, Cambridge University Press, Cambridge, United Kingdom.

Irga, P.J., Burchett, M.D., and Torpy, F.R. (2015) Does urban forestry have a quantitative effect on ambient air quality in an urban environment? *Atmospheric Environment*, 32 (2), 171–185.

Islam, M.S., Ahmed, M.K., Habibullah-Al-Mamun, M., and Masunaga, S. (2015) Potential ecological risk of hazardous elements in different land-use urban soils of Bangladesh. *Science of the Total Environment*, 512–513, 94–102.

Jambhulkar, H.P. and Juwarkar, A.A. (2009) Assessment of bioaccumulation of heavy metals by different plant species grown on fly ash dump. *Ecotoxicology and Environmental Safety*, 72, 1122–1128.

Janhäll, S. (2015) Review on urban vegetation and particle air pollution – Deposition and dispersion. *Atmospheric Environment*, 105, 130–137.

Jozic, M., Peer, T., and Turk, R. (2009) The impact of the tunnel exhausts in terms of heavy metals to the surrounding ecosystem. *Environmental Monitoring and Assessment*, 150, 261–271.

Laidlaw, M.A.S. and Taylor, M.P. (2011) Potential for childhood lead poisoning in the inner cities of Australia due to exposure to lead in soil dust. *Environmental Pollution*, 159, 1–9.

Laschober, C., Limbeck, A., Rendl, J., and Puxbaum, H. (2004) Particulate emissions from on-road vehicles in the Kaisermühlen-tunnel (Vienna, Austria). *Atmospheric Environment*, 38, 2187–2195.

Lu, Y., Song, S., Wang, R., Liu, Z., Meng, J., Sweetman, A.J., Jenkins, J., Ferrier, R.C., Li, H., Luo, W., and Wang, T. (2015) Impacts of soil and water pollution on food safety and health risks in China. *Environment International*, 77, 5–15.

Luo, X-S., Yu, S., Zhu, Y-G., and Li, X.-D. (2012). Trace metal contamination in urban soils of China. *Science of the Total Environment*, 421–422, 17–30.

McBride, M.B., Shayler, H.A., Spliethoff, H.M., Mitchell, R.G., Marquez-Bravo, L.G., Ferenz, G.S., Russell-Anelli, J.M., Casey, L., and Bachman, S. (2014) Concentrations of lead, cadmium and barium in urban garden-grown vegetables: The impact of soil variables. *Environmental Pollution*, 194, 254–261.

McBride, M.B., Simon, T., Tam, G., and Wharton, S. (2013) Lead and arsenic uptake by leafy vegetables grown on contaminated soils: Effects of mineral and organic amendments. *Water, Air and Soil Pollution*, 224, 1378.

Mendez, B., Klenzendorf, B., Afshar, R., Simmons, T., Barrett, E., Kinney, A., and Kirisits, J. (2011). The effect of roofing material on the quality of harvested rainwater. *Water Research*, 45 (5), 2049–2059.

Mielke, H.W., Laidlaw, M.A.S., and Gonzales, C.R. (2012) Estimation of leaded (Pb) gasoline's continuing material and health impacts on 90 US urbanized areas. *Environment International*, 37, 248–257.

Miguntanna, N.S., Egodawatta, P., and Goonetilleke, A. (2010) Pollutant characteristics on roof surfaces for evaluation as a stormwater harvesting catchment. *Desalination and Water Treatment*, 19 (1–3), 205–211.

Mitchell, R.G., Spliethoff, H.M., Ribaudo, L.N., Lopp, D.M., Shayler, H.A., Marquez-Bravo, L.G., Lambert, V.T., Ferenz, G.S., Russell-Anelli, J.M., Stone, E.B., and McBride, M.B. (2014) Lead (Pb) and other metals in New York city community garden soils: factors influencing contaminant distributions. *Environmental Pollution*, 187, 162–169.

NDRC (2015) Rooftops to Rivers II: Green Strategies for Controlling Stormwater and Combined Sewer Overflows. https://www.nrdc.org/resources/rooftops-rivers-ii-green-strategies-controlling-stormwater-and-combined-sewer-overflows

Nidzgorski, D.A. and Hobbie, S.E. (2016) Urban trees reduce nutrient leaching to groundwater. *Ecological Applications*, 26 (5), 1566–1580.

Norton, B.A., Coutts, A.M., Livesley, S.J., Harris, R.J., Hunter, A.M., and Williams, N.S.G. (2015). Planning for cooler cities: A framework to prioritise green infrastructure to mitigate high temperatures in urban landscapes. *Landscape and Urban Planning*, 134 (0), 127–38.

OECD. (2014) *The Cost of Air Pollution Health Impacts of Road Transport*. OECD Publishing.

Onder, S. and Dursun, S. (2006) Air borne heavy metal pollution of Cedrus libani (A. Rich) in the city centre of Konya (Turkey). *Atmospheric Environment*, 40, 1122–1133.

Peña-Fernández, A., Lobo-Bedmar, M.C., and González-Muñoz, M.J. (2015) Annual and seasonal variability of metals and metalloids in urban and industrial soils in Alcaláde Henares (Spain). *Environmental Research*, 136, 40–46.

Perez-Lombard L., Ortiz J., and Maestre I.R. (2011) The map of energy flow in HVAC systems. *Applied Energy*, 88, 5020-31.

RICS (2015) Our Changing World: Lets Be Ready. http://www.rics.org/au/knowledge/research/insights/futuresour-changing-world/

Rowe, D.B. (2011). Green roofs as a means of pollution abatement. *Environmental Pollution*, 159, 2100–2110.

Salmond, J.A., Williams, D.E., Laing, G., Kingham, S., Dirks, K., Longley, I., and Henshaw, G.S. (2013). The influence of vegetation on the horizontal and vertical distribution of pollutants in a street canyon. *Science of The Total Environment*, 443, 287–298.

Saravia, J., Lee, G.I., Lomnicki, S., Dellinger, B., and Cormeir, S.A. (2013) Particulate matter containing environmentally persistent free radicals and adverse infant respiratory health effects: A review. *Journal of Biochemical and Molecular Toxicology*, 27, 56–68.

Sæbø, A., Popek, R., Nawrot, B., Hanslin, H.M., Gawronska, H., and Gawronski, S.W. (2012) Plant species differences in particulate matter accumulation on leaf surfaces. *Science of The Total Environment*, 427–428, 347–354.

Setälä, H., Viippola, V., Rantalainen, A.-L., Pennanen, A., and Yli-Pelkonen, V. (2013) Does urban vegetation mitigate air pollution in northern conditions? *Environmental Pollution*, 183, 104–112.

Sipter, E., Rozsa, E., Gruiz, K., Tatrai, E., and Morvai, V. (2008) Site-specific risk assessment in contaminated vegetable gardens. *Chemosphere*, 71, 1301–1307.

Spolek, G. (2008). Performance monitoring of three ecoroofs in Portland Oregon. *Urban Ecosystems*, 1 (11), 349–359.

Taha, H., Douglas, S., and Haney, J. (1997). Mesoscale meteorological and air quality impacts of increased urban albedo and vegetation. *Energy and Buildings*, 25 (2), 169–177.

Teemusk, A. and Mander, A. (2007) Rainwater runoff quantity and quality performance from a greenroof: The effects of short-term events. *Ecological Engineering*, 30 (3), 271.

Thurston, G.D. (2008) Outdoor air pollution: Sources, atmospheric transport, and human health effects. In: Kris, H. (ed.), *International Encyclopedia of Public Health*. Academic Press, Oxford. pp. 700–712.

Torpy, F.R., Irga, P.J., and Burchett, M.D. (2015) Reducing indoor air pollutants through biotechnology. In: Pacheco Torgal, F., Labrincha, J.A., Diamanti, M.V., Yu, C.P., and Lee, H.K. (eds.), *Biotechnologies and Biomimetics for Civil Engineering*. Springer International Publishing. pp 181–210.

UN (2015) Sustainable Development Transforming Our World: The 2030 Agenda for Sustainable Development. Resolution adopted by the General Assembly on 25 September 2015. https://sustainabledevelopment.un.org/post2015/summit

UN Department of Economic and Social Affairs (DESA) (2015) World Population Prospects: The 2015 Revision. http://www.un.org/en/development/desa/news/population/2015-report.html

UN Populations Fund. World Population Trends (2016) http://www.unfpa.org/world-population-trends#sthash.oMfXUZJO.dpuf

Victoria Department of Health (2015) The Health Impacts of the January 2014 Heatwave in Victoria. www.health.vic.gov.au/cheifhealthofficer

Vijayaraghavan, K., Joshi, U.M., and Balasubramanian, R. (2012). A field study to evaluate runoff quality from green roofs. *Water Research*, 46 (4), 1337.

Volk, H.E., Lurman, F., Penfold, B., Herz-Picciotto, I., and McConnell, R. (2013) Traffic-related air pollution, particulate matter and autism. *JAMA Psychiatry*, 70, 71–76.

Vos, P.E.J., Maiheu, B., Vankerkom, J., and Janssen, S. (2013) Improving local air quality in cities: To tree or not to tree? *Environmental Pollution*, 183, 113–122.

Wilkinson, S. and Torpy, F.R. (2016) Urban food production on retrofitted rooftops. In:

Wilkinson, S.J. and Dixon, T. (eds.), *Green Roof Retrofit: Building Urban Resilience.* Chapter 9. Wiley-Blackwell; New Jersey, USA. pp. 158–183.

Wilson, E.O. (1984) *Biophilia: The Human Bond with Other Species.* Harvard University Press.

Wiseman, C.L.S., Zereini, F., and Püttmann, W. (2013) Traffic-related trace element fate and uptake by plants cultivated in roadside soils in Toronto, Canada. *Science of the Total Environment,* 442, 86–95.

World Health Organisation (2008) Improving Public Health Responses to Extreme Weather/Heat-Waves EuroHEAT, World Health Organisation, Copenhagen, Denmark.

World Health Organisation (WHO) (2014) Burden of Disease from Ambient Air Pollution for 2012. The World Health Organisation, Geneva, Switzerland.

Yang, J., Yu, Q., and Gong, P. (2008) Quantifying air pollution removal by green roofs in Chicago. *Atmospheric Environment,* 42, 7266–7273.

Yin, S., Shen, Z., Zhou, P., Zou, X., Che, S., and Wang, W. (2011) Quantifying air pollution attenuation within urban parks: An experimental approach in Shanghai, China. *Environmental Pollution,* 159, 2155–2163.

Yoo, Y. and Kwon, Y. (2015) Urban street greenery as a prevention against illegal dumping of household garbage – A case in Suwon, South Korea. *Urban Forestry and Urban Greening,* 14, 1088–1094.

Zereini, F., Alsenz, H., Wiseman, C.L.S., Püttmann, W., Reimer, E., Schleyer, R. et al. (2012) Platinum group elements (Pt, Pd, Rh) in airborne particulate matter in rural vs. urban areas of Germany: Concentrations and spatial patterns of distribution. *Science of the Total Environment,* 416, 261–268.

11

Light Pollution

Fabio Falchi

ISTIL - Light Pollution Science and Technology Institute, Italy

11.1 Introduction

The Oxford dictionary defines pollution as 'the presence in or introduction into the environment of a substance which has harmful or poisonous effects'. More precisely, pollution can be defined as an alteration and contamination of the natural conditions in the environment due, for example, to physical, chemical, or biological factors.

Anthropogenic changes to the natural environment are sources of pollution, whose effects vary depending on the magnitude of the contamination. Sometimes it is not even necessary to have a contaminant for pollution to occur. In fact, raising the temperature of the water of a river to cool an electric power station is thermal pollution that changes the temperature of the river without contaminating it with any substance (Hollan, 2009).

Artificial light at night (ALAN) can be considered to be physical pollution, from the fact it operates against the natural characteristics of light/dark cycles which reflect the rotation of the Earth on its axis, the basis for temporal organisation to which life has adapted through evolution. The natural light levels and spectral composition in a given geographical location change with these daily cycles. Compared to natural nights, dimly lit by natural sources of light, mainly the Moon, the natural airglow, the stars, the Milky Way, and zodiacal light, the lighting levels after the introduction of ALAN are higher by several orders of magnitude.

The light coming from the night sky produces an illumination, a light flux per unit of surface, that is measured in lux (i.e. lumen per square metre). The illuminance due to natural sources is always very low, from a few tenths of millilux (on a moonless overcast night), to about a millilux on a clear moonless starry night, to a few hundreds of lux with the quarter Moon, to a maximum of a quarter of lux when the full Moon is high in the sky. In comparison, roads are lit typically in the 5 to 30 lux range, 100,000 times natural night-time light levels.

It is important to distinguish light pollution from its many negative effects. The most evident is, of course, the sky glow arising from cities and the loss of the starry night. Other negative consequences of light pollution are connected with ecology, human health, and energy and money wastage. Light pollution is thus a global issue that needs to be confronted from different perspectives and competences to be successfully solved. Ecologists, biologists, and physicians need to join astronomers in this undertaking (Settele, 2009). Scientific research has already shown how to limit the negative consequences of the use of ALAN (Falchi *et al.*, 2011). The main practical problem is how to implement rules as effective laws and norms that apply

Urban Pollution: Science and Management, First Edition. Edited by Susanne M. Charlesworth and Colin A. Booth.

everywhere. Moreover, there is a cultural limitation that makes it difficult to think of ALAN as a pollutant. A lot of work is needed to explain that a positive and healthy factor during the daytime may turn into a negative and dangerous factor at the wrong time, during the night.

The main sources of light pollution are inside or near cities, somewhat correlated with population density and the country's development, with significant differences between countries of comparable per capita income (Kyba *et al.*, 2015). Figures 11.1 and 11.2 show how light pollution affects the population and territories of the G20 countries. The sources of outdoor ALAN emerge mainly from road and area lighting, commercial centres, lit buildings, indoor lights escaping from windows, advertising signs, monitors and billboards, sport facilities, private lighting (e.g. home, gardens, and, in USA, flag lighting), transportation, industry plants, train stations, harbours, and airports. These lights can pollute directly by intruding on the environment, in private properties and in windows (e.g. bedrooms) or indirectly via reflection over lit surfaces such as roads, and diffusion by particulates, aerosols, and gases in the atmosphere. In particular, skyglow can be generated by light escaping directly up on the horizon plane from incompletely shielded sources and light reflected by lit areas such as roads, football fields, and parking lots.

Not all the light pollutes the sky in the same way, since light travelling at low angles above the horizon, such as that typically emitted by partially shielded fixtures, will pollute much more than that going straight up, such as that reflected by lit surfaces (Cinzano and Castro, 2000). Luginbuhl *et al.* (2009) showed that at 100 km from sources, a direct light flux of only 3% (i.e. 3% up to the sky at low angles, 97% down towards the road) pollutes almost three times more than the same flux emitted by a fully shielded source. To compare the relative importance in producing skyglow of the direct light emitted above the horizon plane from the lighting fixtures and the indirect light reflected

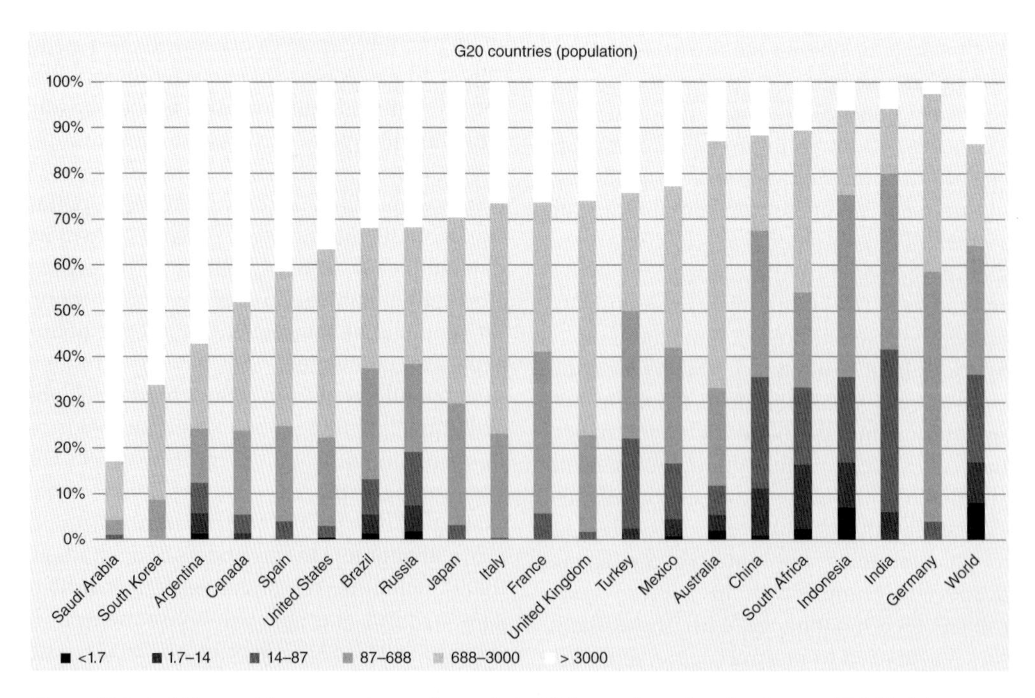

Figure 11.1 In G20 countries, percentages of population living under a sky of a given artificial brightness, listed from left to right in order of pollution. Saudi Arabia has more than 80% of its population living under the white level, with a light pollution so high that it prevents human eyes from adapting to darkness, maintaining the cones active. Levels in μcd m^{-2}. Adapted from Falchi *et al.*, 2016. (key given below figure).

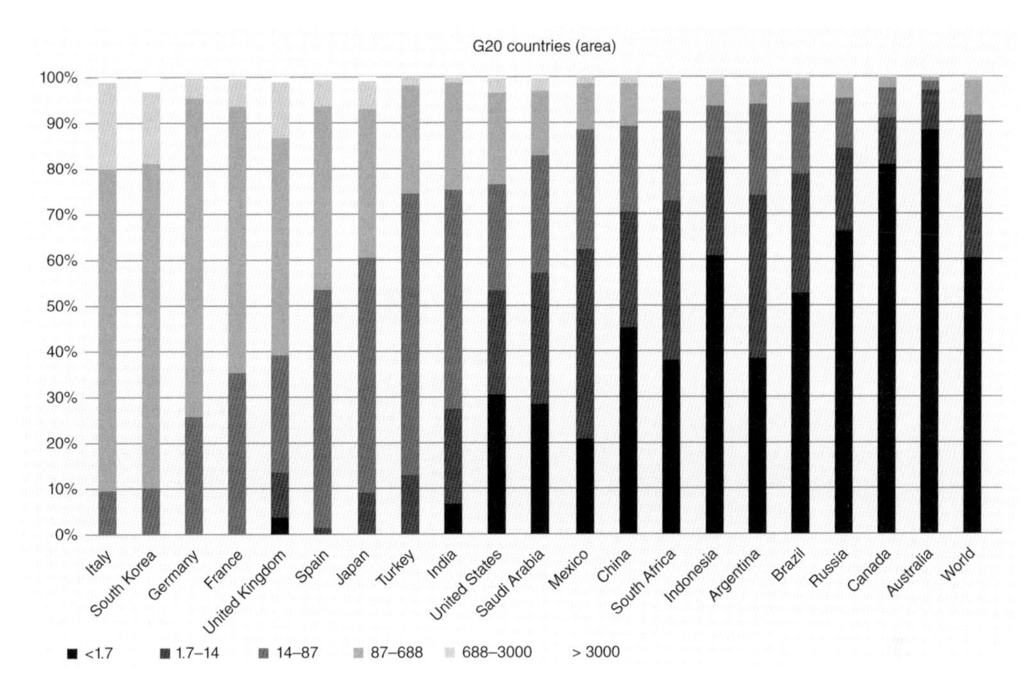

Figure 11.2 In G20 countries, percentages of territory under a sky of a given artificial brightness, listed from left to right in order of pollution. Italy and South Korea have 90% of their territories at the highest levels of pollution (key given below figure). Levels in μcd m^{-2}. Adapted from Falchi *et al.*, 2016.

by roads, parking lots, and lighted areas in general, a study by Falchi (2010) compared the zenith sky brightness measured with snow coverage to the brightness measured in the same places with no snow. This comparison calculated the ratio of the pollution of the sky produced by direct light to that produced by reflected light. In fact, the direct flux emitted by the fixtures remained unchanged, while the reflected flux increased because of snow coverage. In all the six sites analysed, the main cause of the artificial night sky brightness was direct light. This means that by cutting all direct upward flux – maintaining the downward flux unchanged – skyglow could be lowered substantially.

11.2 Environmental and Health Effects of Light Pollution

As seen above, in the natural night environment, animals and plants are exposed to lighting levels that vary from a few tenth of millilux of the overcast sky to a maximum of 0.1–0.3 lux in the week around the full Moon. It is not surprising that artificial lights thousands and thousands times brighter have strong environmental effects on animals, affecting their foraging, mating, orientation, migration, communication, competition, and predation habits as well as plant phenology. A lit road, even if not built with fences, will fragment the ecosystem it crosses, effectively becoming an insurmountable obstacle for nocturnal species that rely on dark nights for their activities. Light pollution may be the most pervasive and, at the same time, neglected form of pollution. As Longcore and Rich, 2004 state: 'What if we woke up one morning to realize that all conservation planning of the last 30 years told only half the story – the daytime story?' For reviews and examples of the ecological consequences of light pollution, see in bibliography: Falchi *et al.* (2011); Gaston *et al.* (2012 and 2013); Grubisic et al. (2018); Hölker *et al.* (2010); Kempenaers *et al.* (2010); Knop *et al.* (2017); Kyba *et al.* (2011); Longcore and Rich (2004); Longcore (2010); Massetti (2018); Navara and Nelson (2007).

Besides behavioural consequences, exposure to artificial light during the night hours affects animal physiology. One of the best studied effects is its suppression of pineal gland melatonin production and secretion, which starts during darkness. The lowering of melatonin production depends on several factors: the light spectrum, its intensity, timing, and exposure duration. It was almost 40 years ago that Lewy *et al.* (1980) discovered the suppression of human melatonin production due to exposure to ALAN. Now it is known that the light intensity levels affecting human physiology are very low; illuminance of the order of only one lux can affect circadian rhythms (Glickman *et al.*, 2002; Wright *et al.*, 2001). Typical evening bedroom illumination, such as that used to read, can reduce and delay melatonin production (Gooley *et al.*, 2011). Intrusive light entering through windows from street lighting, sport fields, advertising signs, and giant monitor screens may lead to chronic exposure to ALAN.

The discovery of a new photoreceptor, the non-image-forming photoreceptor (NIFP), some 20 years ago, opened the way to a better understanding of how light affects human beings and other mammalian temporal organisation. Beside light intensity, it was revealed that the spectrum of the light is of primary importance in determining responses to light (Brainard *et al.*, 2001; Thapan *et al.*, 2001). Cajochen *et al.* (2005) studied how the spectrum of light varied in its impact on humans by measuring melatonin concentration, alertness, thermoregulation, and heart rate. They noted that 2 h of exposure to blue light (spectrum peak at 460 nm wavelength) in the late evening lowered melatonin secretion, while it did not happen using green light (peaking at a wavelength of 550 nm) of the same intensity, timing, and duration. Since melatonin may act as an oncostatic agent for breast and prostate cancers, lowering its concentration in blood by exposure to ALAN may promote (compared to normal night melatonin levels in blood) the growth of some type of cancers (Bullough *et al.*, 2006; Garcia-Saenz *et al.*, 2018; Glickman *et al.*, 2002; Haim *et al.*, 2010; Haim and Portnov, 2013; James *et al.*, 2017; Keshet-Sitton *et al.*, 2016; Kloog *et al.*, 2008; Kloog *et al.*, 2009; Schwimmer *et al.*, 2013; Stevens *et al.*, 2007; Stevens, 2009).

Apart from the direct action on physiology by ALAN, there are also indirect effects due to sleep disturbance and deprivation arising from chronic exposure. Sleep disorders have negative effects on several diseases such as diabetes, obesity, and others (Bray and Young, 2012; Bass and Turek, 2005; Haus and Smolensky, 2006; Reiter *et al.*, 2011; Smolensky *et al.*, 2016). The International Agency for Research on Cancer classifies shiftwork that involves circadian disruption in their group 2A, a list of factors probably carcinogenic to humans (Straif *et al.*, 2007).

It has been shown above that exposure to ALAN induces circadian disruption. ALAN is thus a rising public health issue, as shown by medical research (Cho *et al.*, 2015; Fonken and Nelson, 2011, Hatori *et al.*, 2017; Pauley, 2004; Stevens, 2009). On the basis of this, in 2009, the AMA (American Medical Association; AMA, 2009) issued a resolution affirming that light pollution is a public health hazard.

The continued increase in the number and flux of outdoor lighting installations (Kyba *et al.* 2017) combined with the race by municipalities to be the pioneers of light emitting diodes (LED) will serve to increase light pollution and its subsequent negative impacts on human health and the night environment. In fact, today the primary attention is given to energy efficiency, without considering any other environmental factor. It is like aiming for the highest possible fuel efficiency in cars by removing all the air pollution abatement devices. The engine will be surely more efficient, but also much more pollutant. A similar situation is happening with outdoor lighting where the high efficient white LEDs, which have strong short-wavelength emissions (blue light content in the light emitted) are preferred over less efficient but of lower environmental impact warm white and amber LEDs. To help authorities make rational choices, AMA released a report (AMA, 2016) which advised the use of LED sources with the lowest possible emissions in the blue part of the spectrum, to pay attention to glare, to shield lights completely in order to minimise the light trespass into homes alongside roads, and to limit the light shining into the sky.

The strong evidence of the adverse effects of ALAN on animals and on human health should be balanced against the assumed (but yet to be proven) positive effects on safety and security (Marchant, 2004). A multi-year survey in England and Wales showed no effect of four street lighting modifications (switch off, part-night lighting, dimming, and white light) on road casualties and crime (Steinbach *et al.*, 2015; Steinbach *et al.*, 2016). Road lighting can surely increase the visibility of obstacles, animals, pedestrians, and bikers on the carriageway, provided that the lights do not cause glare, in which case reduction of visibility can be expected instead. However, an increase in visibility does not automatically mean an increase in safety. In fact, higher lighting levels may increase the *perception* of safety, eventually inducing drivers to speed up and consequently increase the numbers and the danger of incidents. Another aspect to consider is that the lighting poles are dangerous obstacles that are always present, 24 hours a day, and their existence actually reduces road safety.

Research in this field should follow the example of pharmaceutical research in using randomised controlled trials. Also, public registration of protocols and trials should be implemented to lower the problem of publication bias (Copas, 2005) by ensuring that 'against lighting' results remain as visible as 'pro lighting' ones.

11.3 How to Reduce Light Pollution

To obtain more robust results in order to protect the night-time environment both inside and outside of cities, action is needed at all levels, from the simple action of the single citizen who turns off or dims their garden lights to the municipality that chooses fixtures and sources that pollute the least, to the province/region/national authorities that must take action now. Each new installation typically operates and, consequently, pollutes the environment for at least 20 years. Therefore, each new installation that does not follow best practices to keep light pollution to the absolute minimum is a waste of money and energy, and its environmental impact is likely to be felt for a long time.

Lowering pollution generated by lighting installations will benefit the community and the urban, rural, and wild environment. Light pollution generated in cities has the potential to pollute far from the cities themselves. The artificial glow of Honolulu is clearly visible from the summit of Mauna Kea, 300 km away. Pollution from Las Vegas propagates in every direction, as shown in Figure 11.3. Therefore, limiting light pollution inside cities will improve the night environment all around.

The following lists actions required in order to mitigate the light pollution problem, based on the Starlight Initiative (Falchi and Marin, 2012), CieloBuio, and ISTIL, the Light Pollution Science and Technology Institute:

The first step towards efficiency is to limit unnecessary light:

a) Develop standard rules to encourage substantial reductions in the levels currently used in outdoor lighting, guiding the market of LED and OLED technologies towards more environmentally friendly lighting.
b) Investigate whether there is a relationship between light, lowered levels of crime, and/or traffic accidents.
c) Weigh any benefits (if found under point c) against those of different actions having the same cost for the community (e.g. more control by police, video surveillance, and lower speed limits).
d) Consider the health problems of using light at night, especially with blue content, and so limit its blue content at night, indoors, and outdoors.
e) Accept that specific light frequencies affect many wildlife species and alter

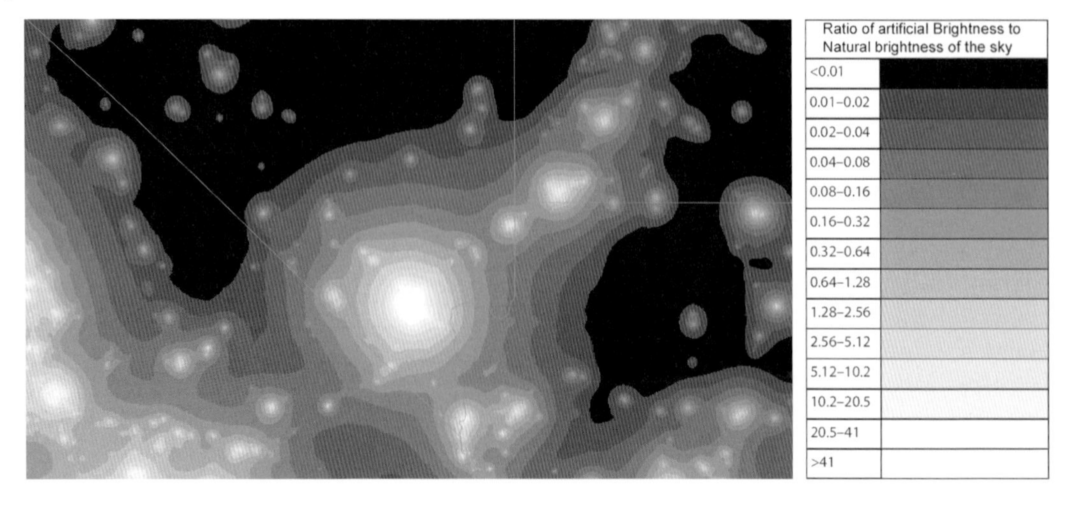

Figure 11.3 Las Vegas lights pollutes the sky well over 100 miles away. Adapted from Falchi *et al.*, 2016. (*See colour plate section for the colour representation of this figure.*)

ecosystem function in urban environments and beyond.

f) Properly define the concept of eco-friendly light technologies, apart from energy efficiency, that should take account of the control of light pollution in all its aspects by giving the following minimum rules for all outdoor sources of light:

- Do not allow luminaires and lighting installations to send any light directly at and above the horizontal, nor into private properties and windows.
- Keep lighting levels to the absolute minimum necessary for the task.
- Do not waste downward light flux outside the area to be lit.
- Shut off lights when the area is not in use.
- Enforce zero growth in total installed flux, promptly followed by a decrease (as is happening with most other pollutants).
- Strongly limit the short-wavelength 'blue' light, both indoors (particularly during the night) and outdoors.

In respect to points e) and f), the following should also be noted: In order to reduce the adverse health effects caused by decreased melatonin production and circadian rhythm disruption in humans and animals, due to the relationship between light wavelength and melatonin suppression, a total ban on night-time artificial light at wavelengths shorter than 540 nm is recommended. The relatively low emissions of high-pressure sodium lamps in this spectral range could be set as the upper limit on what can be initially acceptable in terms of short-wavelength light emissions. So, the following rule should be observed:

> The wavelength range of the visible light spectrum under 540 nm, due to the high sensitivity of the melatonin suppression action spectrum, should be established as a protected range. Lamps that emit an energy flux in the protected range larger than that emitted by the standard high-pressure sodium lamp on the basis of equal photopic output should not be installed for night-time use.

It is to be noted that the correlated colour temperature (CCT) of a light source is often used as a proxy of its blue content, and usually the higher the colour temperature, the higher the blue content of the source. So a 4000 or 5000 kelvin white-bluish LED has a several-fold higher blue content than the once standard yellowish high-pressure sodium lamp. Nevertheless, the correct characterisation of the blue content of a light source should be obtained from its spectrum.

Enforcing all the above points will reduce light pollution by an order of magnitude or more. This can be done without cost, if it is accepted that the current levels of light pollution will continue for at least 30 years, simply by using these rules for new installations only. Of course, the process could – and should – be speeded up by changing existing installations, starting from the most polluting ones. The proposed rules will save on energy expenditure with a payback time of 10 years at most. As an example, Italy spends about a billion euro each year on outdoor public lighting energy (and a similar amount on private lighting). Most of this money could be saved by adopting all the rules proposed above. Taking into account the total cost of ownership of public installations, the annual cost is much higher. Other hidden costs are associated with other environmental damage due to light pollution (e.g. reduction of biodiversity, loss of species, and health consequences of exposure to light at night).

Evidence of light pollution's negative consequences on the environment and human health has been quickly accumulating in the last few years of scientific research. Light pollution is no longer, as is commonly believed, just a problem for astronomers. It is a hazard to environmental and human health.

11.4 The example of the Italian Regional Laws

The main impacts of associations which protect the night environment, mainly IDA Italian Section and CieloBuio, have resulted in the following: 18 out of 20 Italian regions have enforced laws against light pollution so that most of the territory and population are, more or less, protected from this contamination. Thirteen of these prescribe the use of fully shielded fixtures in both private and public new lighting installations. All the laws enforced after 2002 prescribe fully shielded fixtures. Other measures are used combined with the full shielding, like the maximum allowed luminance or illuminance, the shutting off or dimming of lights at curfew, the use of low-impacting lamp spectra, depending on the region. Special rules are laid down for particular lighting (e.g. monuments and historic buildings). Unfortunately, some of the older regional laws, such as Valle d'Aosta, and Basilicata, are not effective, because their technical prescriptions are too weak.

The laws only temporarily halted the constant increase of light pollution experienced previously. Additional measures should now be implemented to lower the light pollution impact in the long term, in particular, a limit on newly installed light flux, followed by a lowering of the total installed light flux, and a limit on the blue wavelengths of light produced.

Italian regional laws against light pollution are shown in Figure 11.4, along with the year of adoption. The laws that allow for zero direct upward flux at and above the horizontal plane (i.e. lights are fully shielded, and no light can escape in an upward direction from road and area lighting) are shown in darker grey along the east coast and northern regions. Laws with zero direct upward flux with some limited exceptions (e.g. decorative and historic lighting) are shown in mid grey along the western coast. Laws with even higher limits or no technical prescriptions (light grey) are located in the northwest (Valle d'Aosta) and south (Basilicata). Regions without any law are shown in white (Calabria and Sicily). Veneto, the first region to adopt a law in 1997 that allowed a 3% direct upward flux, modified its law in 2009, upgrading to zero upward flux. The upgrade was driven by the evidence that the 3% upward flux limit was ineffective in stopping light pollution. In fact, while this could limit the worst lighting installations such as those with spheres, nevertheless it permitted the use of other very polluting and low-efficient lighting fixtures such as those that use indirect light. Moreover, the light emitted at low angles above the horizontal plane, such as that

Figure 11.4 Italian regional laws with year of enforcement. The letter 'S' indicates the position of the Osservatorio Astronomico di San Benedetto Po, where sky brightness has been monitored since 1998.

escaping from partially shielded fixtures, is disproportionately effective at polluting the night environment, especially far from sources. Piemonte too updated its old law towards a zero upward one in 2018.

Most of the regional laws have a similar core of technical demands, as follows.

The main rules for roads, large areas, and buildings (not historic) are:

a) The light intensity of the fixture, as installed, should not exceed 0.49 cd/klm at and above the horizontal plane.
b) The luminance of the road or the illuminance on the lighted surface cannot exceed the minimum value suggested by safety norms. A flux reduction is mandatory with low traffic volumes.

In some regions, additional rules are enforced:

c) A limit is imposed on the CCT of light sources.
d) The ratio of the distance between the poles to the pole height should be greater than 3.7 (this rule is to minimise the proliferation of poles).

These rules are enforced across the whole territory of each region, not just in specific areas around astronomical observatories or natural parks. This is because protecting only relatively small areas is ineffective since pollution from light can contaminate areas tens of kilometres or more from its sources.

These rules are effective for all the new installations (public and private alike), while for the old installations the timing of upgrades to the new less polluting fixtures varies from region to region. Sometimes there is no such time schedule, so that the old polluting installations work until their natural end of life.

The full shielding requirement has exceptions for certain types of lighting only:

Monuments and historical buildings can be lit upwards, with restrictions on luminance/illuminance and a specified curfew.

Very small installations (e.g. home garden lighting) with lamps of less than 1500 lumens each can have more than zero upwards flux, with a total for each installation of no more than 2250 lumens above the horizon plane (this translates to a maximum of three globes, for example).

Luminous signs should not exceed 4500 lumens of total flux and will be subject to a curfew, usually starting at 11 p.m.

The success in enforcing these rules prescribed by law is not equal. The most important of all (fully shielding, rule a) above) is very successful and is normally used even in regions where it is not imposed. The only sectors where this rule is not enforced well are football fields (and in Italy *every* small village has such a field!) and the illumination of industrial sheds and surroundings. The reasons behind this are not technical, but mainly this stems from a lack of culture in the installations and in the lack of controls on private property by local authorities.

The results of the adoption of the regional laws in terms of sky brightness have been measured as a time series since 1998 at the San Benedetto Po Observatory, near the centre of the group of 'fully shielded law' regions, northeast of Italy. The letter 'S' in Figure 11.4 indicates the position of the site. The measurements show that night sky light pollution is much lower than it would have been without the laws, but it has not decreased in absolute terms. Now the sky brightness has the same value as it had in 1998 notwithstanding the almost doubling of the total light flux produced by outdoor lighting installations. The increase in the produced flux is mainly due to two decades of new installations and the increase in average light efficiency of light sources. This demonstrates that one fundamental rule is missing in the laws: enforcing a decrease in the installed flux over time. This rule in necessary to counteract the rebound effect of increasing light efficiency that drives the Jevons paradox (Blake, 2005). Historically, the increase in light efficiency had always reflected an increase in consumption, derived by the lower cost of maintaining more efficient lights and new applications of lights being enabled by their low cost. This has always happened, from oil lamps to gas lamps to incandescent bulbs, to high-intensity discharge lamps, and now to LEDs (Kyba *et al.*, 2017). For this reason, enforcing a progressive reduction in the total produced light flux (and not simply a reduction of the energy consumed for lighting) is mandatory to abate light pollution in the long term. This may be easier to achieve as new technologies in other sectors arise, such as driverless cars that do not need public lighting to be safe.

11.5 Conclusions

This chapter has shown that light pollution is not just an irritant impacting astronomers wishing to have a clearer view of the night sky. While this is an important consideration, nevertheless it has been found that, particularly in the blue wavelengths of the light spectrum, light pollution can adversely affect human and environmental health. There has been a suggested connection between cancer and light pollution, for example, and biodiversity is affected, impacting animal and plant breeding. Reducing the effects of lighting will also reduce energy use, and hence has positive financial implications for the individual householder, and also at the local authority level.

Suitable legislation could quite easily reduce lighting and its orientation in new built estates, and examples across Italy have shown how this can work. However, there is work yet to be done to address light pollution from the infrastructure already in existence. Worldwide, as populations grow, urban and industrial areas expand with their associated requirements for lighting. This is therefore not a problem that is going to go away soon, and in the meantime, the impacts will increase. Great strides have been made in addressing chemical pollution in urban environments; the problem of light pollution needs a similar effort as a matter of some urgency.

References

American Medical Association. (2009) House of Delegates, Resolution 516 – Advocating and Support for Light Pollution Control Efforts and Glare Reduction for both Public Safety and Energy Savings.

American Medical Association. (2016) Council on Science and Public Health, Report 2-A-16, Human and Environmental Effects of Light Emitting Diode (LED) Community Lighting.

Bass, J. and Turek, F.W. (2005) Sleepless in America: A pathway to obesity and the metabolic syndrome? *Archives of Internal Medicine*, 165, 15–16.

Blake, A. (2005) Jevons' paradox. *Ecological Economics*, 54 (1), 9–21. doi:10.1016/j.ecolecon.2005.03.020

Brainard, G.C., Hanifin, J.P., Greeson, J.M., Byrne, B., Glickman, G., Gerner, E. and Rollag, M.D. (2001) Action spectrum for melatonin regulation in humans: Evidence for a novel circadian photoreceptor. *Journal of Neuroscience*, 21 (16), 6405–6412.

Bray, M. and Young, M. (2012) Chronobiological effects on obesity. *Current Obesity Reports*. doi: 10.1007/s13679-011-0005-4

Bullough, J.D., Rea, M.S. and Figueiro, M.G. (2006) Of mice and women: Light as a circadian stimulus in breast cancer research. *Cancer Causes Control*, 17, 375–383.

Cajochen, C., Munch, M., Kobialka, S., Krauchi, K., Steiner, R., Oelhafen, P., Orgul, S. and Wirz-Justice, A. (2005) High sensitivity of human melatonin, alertness, thermoregulation, and heart rate to short wavelength light. *Journal of Clinical Endocrinology and Metabolism*, 90, 1311–1316.

Cho, Y., Ryu, S.H., Lee, B.R., Kim, K.H., Lee, E. and Choi, J. (2015) Effects of artificial light at night on human health: A literature review of observational and experimental studies applied to exposure assessment. *Chronobiology International*, 32 (9), 1294–1310. doi: 10.3109/07420528.2015.1073158

Cinzano, P., Diaz Castro, F.J. (2000) The artificial sky luminance and the emission angles of the upward light flux, in *Measuring and Modelling Light Pollution*, ed. P. Cinzano, *Mem. Soc. Astron. Ita.*, 71, 1, 251–256 (ISSN: 0037-8720).

Copas, J. (2005) The downside of publication. *Significance*, 2 (4), 154. doi: 10.1111/j.1740-9713.2005.00127.x

Falchi, F. and Marin, C. (2012) There are several ways of Lighting the Future, Starlight Initiative-CieloBuio-ISTIL (2012) http://cielobuio.org/wp-content/uploads/supporto/download/LightGreenPaperComments_FINAL.pdf

Falchi, F., Cinzano, P., Elvidge, C.D., Keith, D.M. and Haim, A. (2011) Limiting the impact of light pollution on human health, environment and stellar visibility. *Journal of Environmental Management*, 92, 2714–2722.

Falchi, F. (2010) Campaign of sky brightness and extinction measurements using a portable CCD camera. *Monthly Notices of the Royal Astronomical Society*, 412, 33–48.

Falchi, F., Cinzano, P., Duriscoe, D., Kyba, C.M., Elvidge, C.D., Baugh, K., Portnov, B.A., Rybnikova, N.A. and Furgoni, R., (2016) The new world atlas of artificial night sky brightness. *Science Advances*, 2 (6), e1600377.

Fonken, L.K. and Nelson, R.J. (2011) Illuminating the deleterious effects of light at night. *F1000 Medical Reports*, 3, 18.

Garcia-Saenz, A. and 23 others (2018) Evaluating the association between artificial light-at-night exposure and breast and prostate cancer risk in Spain (MCC-Spain Study). *Environ Health Perspectives*. doi:10.1289/EHP1837

Gaston, K.J., Davies, T.W., Bennie, J. and Hopkins, J. (2012) Reducing the ecological consequences of night-time light pollution: Options and developments. *Journal of Applied Ecology*, 49, 1256–1266. doi: 10.1111/j.1365-2664.2012.02212.x

Gaston, K.J., Davies, T.W., Bennie, J. and Hopkins. (2013) The ecological impacts of nighttime light pollution: A mechanistic appraisal *Biological Reviews*. doi: 0.1111/ brv.12036

Glickman, G., Levin, R. and Brainard, G.C. (2002) Ocular input for human melatonin regulation: relevance to breast cancer. *Neuroendocrinology Letters*, 23 (Suppl 2), 17–22.

Gooley, J.J., Chamberlain, K., Smith, K.A., Khalsa, S.B., Rajaratnam, S.M., Van Reen, E., Zeitzer, J.M., Czeisler, C.A. and Lockley, S.W. (2011) Expo- sure to room light before bedtime suppresses melatonin onset and shortens melatonin duration in humans. *Journal of Clinical Endocrinology and Metabolism*, 96, 463–472.

Grubisic, M., Grunsven, R., Kyba, C., Manfrin, A. and Hölker, F. (2018) Insect declines and agroecosystems: does light pollution matter?. *Annals of Applied Biology*, doi:10.1111/aab.12440.

Haim, A., Yukler, A., Harel, O., Schwimmer, H. and Fares, F., (2010) Effects of chronobiology on prostate cancer cells growth in vivo. *Sleep Science*, 3 (1), 32–35.

Haim, A. and Portnov, B.A. (2013) *Light Pollution as a New Risk Factor for Human Breast and Prostate Cancers*. Springer, ISBN 978-94-007-6220-6.

Hatori, M., Gronfier, C., Van Gelder, R.N., Bernstein, P.S., Carreras, J., Panda, S., Marks, F., Sliney, D., Hunt, C.E., Hirota, T., Furukawa, T., and Tsubota, K. (2017) Global rise of potential health hazards caused by blue light induced circadian disruption in modern aging societies. *npj Aging and Mechanisms of Disease*, 3 (9). doi:10.1038/ s41514-017-0010-2

Haus, E. and Smolensky, M. (2006) Biological clocks and shift work: Circadian dysregulation and potential long-term effects. *Cancer Causes Control*, 17, 489–500.

Hölker, F., Wolter, C., Perkin, E.K. and Tockner, K. (2010) Light pollution as a biodiversity threat. *Trends in Ecology & Evolution*, 25, 681–682.

Hollan, J. (2009), What is Light Pollution, and How Do We Quantify It? http://amper.ped.muni.cz/jenik/light/lp_what_is.pdf

James, P., K. A. Bertrand, J. E. Hart, E. Schernhammer, R. M. Tamimi, and F. Laden. (2017) Outdoor Light at Night and Breast Cancer Incidence in the Nurses' Health Study II. https://ehp.niehs.nih.gov/doi/10.1289/EHP935

Kempenaers, B., Borgström, P., Loës, P., Schlicht, E. and Valcu, M. (2010) Artificial night lighting affects dawn song, extra-pair siring success, and lay date in songbirds. *Current Biology*, 20 (19), 1735–1739.

Keshet-Sitton, A., Or-Chen, K., Huber, E. and Haim, A. (2016) Illuminating a risk for breast cancer: A preliminary ecological study on the association between streetlight and breast cancer. *Integrative Cancer Therapy*. doi: 10.1177/1534735416678983

Kloog, I., Haim, A., Stevens, R.G., Barchana, M. and Portnov, B.A. (2008) Light at night co-distributes with incident breast but not lung cancer in the female population of Israel. *Chronobiology International*, 25 (1), 65–81.

Kloog, I., Haim, A., Stevens, R.G. and Portnov, B.A. (2009) Global co-distribution of light at night (LAN) and cancers of prostate, colon, and lung in men. *Chronobiology International*, 26 (1), 108–125.

Knop, E., Zoller, L., Ryser, R., Gerpe, C., Hörler, M., and Fontaine, C. (2017) Artificial light at night as a new threat to pollination. *Nature*, 548, 206–209.

Kyba, C.C.M., Garz, S., Kuechly, H., Sanchez de Miguel, A., Zamorano, J., Fischer, J. and Hölker, F. (2015) High-resolution imagery of earth at night: New sources, opportunities and challenges. *Remote Sensing*, 7 (1), 1–23; doi:10.3390/rs70100001

Kyba, C.C.M., Ruhtz, T., Fischer, J. and Hölker, F. (2011) Cloud coverage acts as an amplifier for ecological light pollution in urban ecosystems. *PLoS ONE*, 6 (3), e17307. doi:10.1371/journal. pone.0017307

Kyba, C.C.M., Kuester, T., Sánchez de Miguel, A., Baugh, K., Jechow, A.,Hölker, F., Bennie, J., Elvidge, C.D., Gaston, K.J., and Guanter, L. (2017) Artificially lit surface of Earth at night increasing in radiance and extent. *Science Advances* 22 Nov 2017: Vol. 3, no. 11, e1701528 DOI:10.1126/sciadv.1701528

Lewy, A.J., Wehr, T.A., Goodwin, F.K., Newsome, D.A. and Markey, S.P. (1980) Light suppresses melatonin secretion in humans. *Science*, 210, 1267–1269.

Longcore, T. and Rich, C. (2004) Ecological light pollution. *Frontiers in Ecology and the Environment*, 2 (4), 191–198

Longcore, T. (2010) Sensory ecology: Night lights alter reproductive behavior of blue tits, *Current Biology*, 20 (20), R893–R895.

Luginbuhl, C.B., Walker, C.E. and Wainscoat, R.J. (2009) Lighting and astronomy. *Physics Today*, 62 (12), 32.

Marchant, P.R. (2004) A demonstration that the claim that brighter lighting reduces crime is unfounded. *British Journal of Criminology*, 44, 441–447.

Massetti, M. (2018) Assessing the impact of street lighting on Platanus x acerifolia phenology. *Urban Forestry & Urban Greening*, 34, 71–77; doi:10.1016/j. ufug.2018.05.015.

Navara, K.J. and Nelson, R.J. (2007) The dark side of light at night: physiological, epidemiological, and ecological consequences. *Journal of Pineal Research*, 43, 215–224.

Pauley, S.M. (2004) Lighting for the human circadian clock: Recent research indicates that lighting has become a public health issue. *Medical Hypotheses*, 63, 588–596.

Reiter, R., Tan, D., Korkmaz, A. and Ma, S. (2011) Obesity and metabolic syndrome: Association with chronodisruption, sleep deprivation, and melatonin suppression. *Annals of Medicine*. doi:10.3109/07853890.586365

Reiter, R., Tan, D., Sanchez-Barcelo, E., Mediavilla, M., Gitto, E. and Korkmaz, A. (2011) Circadian mechanisms in the regulation of melatonin synthesis: Disruption with light at night and the pathophysiological consequences. *Journal of Experimental and Integrative Medicine*, 1, 13–22. doi: 10.5455/ jeim.101210.ir.001

Schwimmer, H., Metzer, A., Pilosof, Y., Szyf, M., Machnes, Z.M., Fares, F., Harel1, O. and Haim, A. (2013) Light at night and melatonin have opposite effects on breast cancer tumors in mice assessed by growth rates and global DNA methylation. *Chronobiology International*. ISSN: 0742-0528 print / 1525-6073 online 1-7 (short communication).

Settele, J. (2009) Ecologists should join astronomers to oppose light pollution. *Nature*, 576, 379.

Smolensky, M.H., Hermida, R.C., Reinberg, A., Sackett-Lundeen, L., and Portaluppi, F. (2016) Circadian disruption: New clinical perspective of disease pathology and basis for chronotherapeutic intervention. *Chronobiology International*, 33 (8), 1101–1119. doi: 10.1080/07420528.2016.1184678

Steinbach, R., Perkins, C., Tompson, L., Johnson, S., Armstrong, B., Green, J., Grundy, C., Wilkinson, P. and Edwards, P. (2015) The effect of reduced street lighting on road casualties and crime in England and Wales: Controlled interrupted time series analysis. *Journal of Epidemiology and Community Health*, 69, 1118–1124.

Steinbach, R., Perkins, C., Tompson, L., Johnson, S., Armstrong, B., Green, J., Grundy, C., Wilkinson, P. and Edwards, P. (2016) The effect of reduced street-lighting on road collisions in England and Wales 2000–2013. *Injury Prevention*, 22, A89. doi:10.1136/injuryprev-2016-042156.243

Stevens, R.G., Blask, E.D., Brainard, C.G., Hansen, J., Lockley, S.W., Provencio, I., Rea, M.S. and Reinlib, L. (2007) Meeting report: The role of environmental lighting and circadian disruption in cancer and other diseases. *Environmental Health Perspectives*, 115 (9), 1357–1362.

Stevens, R.G. (2009) Light-at-night, circadian disruption and breast cancer: Assessment of existing evidence. *International Journal of Epidemiology*. doi:10.1093/ije/dyp178

Straif, K., Baan, R., Grosse, Y., Secretan, B., El Ghissassi, F., Bouvard, V., Altieri, A., Benbrahim-Tallaa, L. and Cogliano, V. (2007) Carcinogenicity of shift-work, painting, and fire-fighting. *Lancet Oncology*, 8 (12), 1065–1066. PubMed PMID: 19271347.

Thapan, K., Arendt, J. and Skene, D.J. (2001) An action spectrum for melatonin suppression: Evidence for a novel non-rod, non-cone photoreceptor system in humans. *Journal of Physiology*, 535, 261–267.

Wright, K.P., Jr., Hughes, R.J., Kronauer, R.E., Dijk, D.J. and Czeisler, C.A. (2001) Intrinsic near-24-h pacemaker period determines limits of circadian entrainment to a weak synchronizer in humans. *Proceedings of the National Academy of Sciences USA*, 98 (24), 14027–14032.

12

The Role of Forensic Science in the Investigation and Control of Urban Pollution

Kenneth Pye

Kenneth Pye Associates Ltd, Blythe Valley Innovation Centre, United Kingdom

12.1 Introduction

Forensic science is concerned with the application of scientific methods and techniques to matters under examination by a court of law. The word *forensic* is derived from the Latin *forensis*, meaning pertaining to the *forum*, a public place used for judicial and other purposes. Forensic investigation methods can be applied to any matter of enquiry, including urban pollution. In many such cases, forensic scientific investigation is concerned primarily with identifying the nature and source of a pollutant (contaminant), while the court is often concerned with establishing legal responsibility for the pollution, and potentially its remediation or removal. Interest in these issues is as old as urbanisation itself, and a wide range of types of evidence have been used over the centuries, including oral testimony, indicators of personal identity in deposited waste, photographic evidence, and, more recently, scientific evidence. Although a distinction is sometimes made today between 'environmental forensics' and 'criminal forensics', there is in fact considerable overlap between the two since problems of environmental pollution frequently lead to prosecutions in the criminal courts, as well as hearings in the civil courts.

This chapter provides a brief overview of some of the main contributions that can be made by forensic science to the investigation of the causes, sources, and control of urban pollution. Further, more detailed information about different types of urban and wider environmental pollution, and the techniques used to investigate them, can be obtained from a number of textbooks and review articles including Murray and Tedrow (1992), Morrison (2000), Pye and Croft (2004), Morrison and Murphy (2006), Pye (2007) Ruffell and McKinley (2005, 2008), Hester and Harrison (2008), Ritz *et al.* (2009), Bergslien (2012), Pirrie *et al.* (2013), Petrisor (2014), and Murphy and Morrison (2015).

12.2 Types of Urban Pollutants

Five main aspects of the urban environment may be exposed to pollution: air, soil, surface water, the ground surface, and groundwater; some example pollution types associated with each are shown in Figure 12.1. In practice, many pollutants affect more than one type of environment; for example, run-off or groundwater movement from an area of contaminated soil may affect surface water courses and/ or soil outside the originally contaminated area. In this context, a basic distinction can be made between 'diffuse pollutants', usually affecting a wide area and derived from multiple sources (e.g. background air

Urban Pollution: Science and Management, First Edition. Edited by Susanne M. Charlesworth and Colin A. Booth.

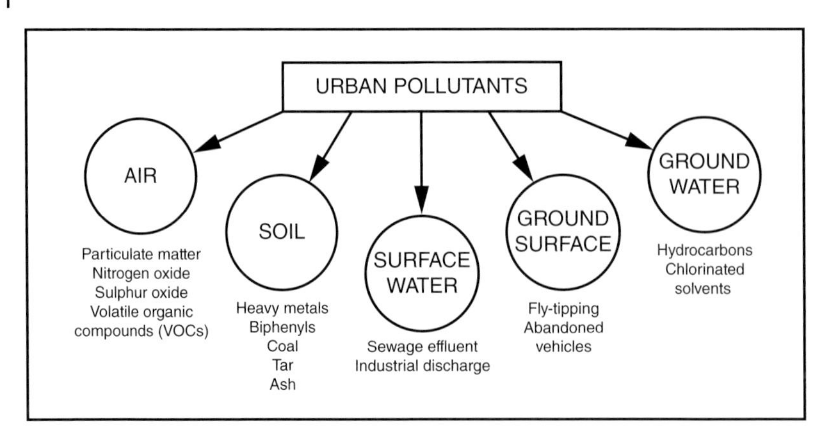

Figure 12.1 Aspects of urban pollution and examples of major pollutant types.

pollution), and pollutants derived from point sources (e.g. an effluent discharge pipe entering a river), which, often, although not always, affect a more restricted geographical area. There is a vast array of potentially contaminating materials and substances, ranging from large individual items such as abandoned cars, through composite fly-tipped waste and industrial spoil heaps to micro-scale particles, suspensions, solutions, and gases.

Official statistics compiled by the UK Department of Environment, Food and Rural Affairs (DEFRA, 2017) show that in 2007–2008 there were over 1.2 million reported fly-tipping incidents in England. The figure fell to a minimum of approximately 720,000 in 2012–2013 but has since increased again to circa 900,000 in 2015–2016 as charges for commercial waste disposal have increased, and tighter regulations have been imposed on domestic waste disposal (Figure 12.2). The majority of individual fly-tipping incidents are relatively small (Figure 12.3) but result in significant expense both for local councils and private landowners on whose land the waste is deposited. The waste also often poses a major health hazard, containing toxic chemicals, carcinogens, and medical waste. Some recent fly-tipping cases have also been on a near-industrial scale, carried out by well-organised groups/gangs, including one near Biddulph, North Staffordshire, in February 2017 (Stoke Sentinel, 2017). Analysis of monthly waste tipping reports by local authority responsibility area identified that certain urban areas in London, the North West, and the North East of England are particular 'hot spots' (Webb *et al.*, 2006).

Many urban areas in the United Kingdom, and elsewhere, have been affected by a range of industrial activities over decades and even centuries, resulting in the creation of numerous waste tips and more diffuse contamination to soils, surface waters, and groundwater. Industrial waste tips can contain a wide range of materials including fly ash, slag, and industrial process waste products, which may contain potentially hazardous substances such as asbestos, heavy metals, polyaromatic hydrocarbons (PAHs), polychlorinated biphenyls (PCBs), and pathogens (Morrison, 2000; Morrison *et al.*, 2009). Numerous domestic landfill sites have also been used over the centuries, mostly now abandoned and many unmanaged. Leakage of ammonium into groundwater and rivers, and its subsequent breakdown to nitrogen, can be a major cause of nutrient enrichment and algal blooms (e.g. Gooddy *et al.*, 2014). Surface run-off and wind action from landfill sites and waste tips can lead to pollution of the surrounding area by contaminated sediment, and escape of gaseous methane, carbon dioxide, and other gases sites can have a significant impact on local air quality and health (Macklin *et al.*,

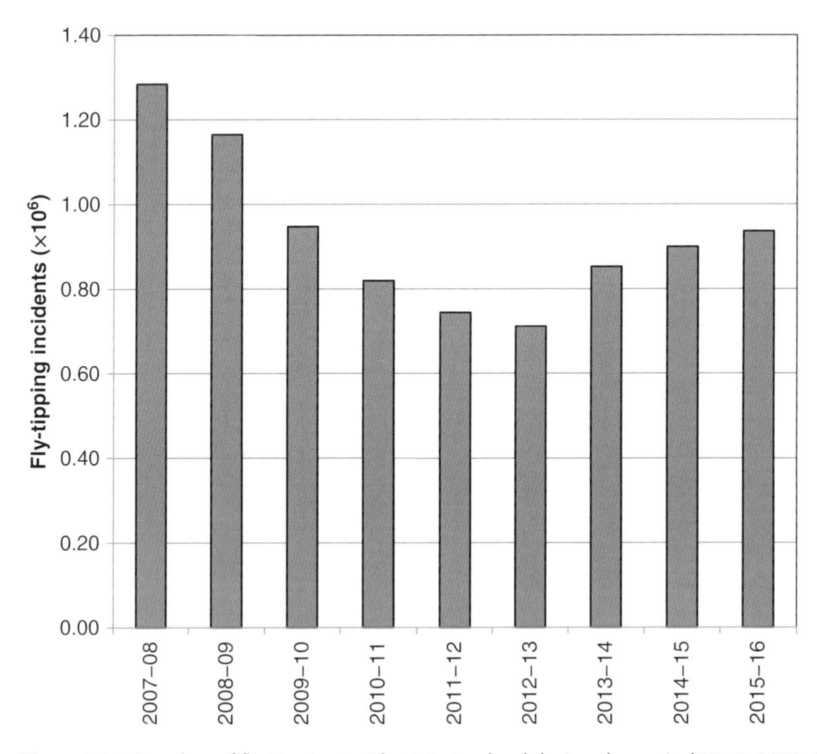

Figure 12.2 Number of fly-tipping incidents in England during the period 2007–2016. *Data Source*: DEFRA.

Figure 12.3 Small-scale fly tipping adjacent to an urban road.

2011). Emissions from industrial chimneys and power stations have also caused widespread diffuse contamination of surface soils and waterways with fly ash, organic compounds, and heavy metals, of which Cu, Pb, Zn, and Pb are of widest concern (Alloway, 1995). A further source of air and soil pollution is provided by wind erosion of dust from industrial stockpiles and construction sites (e.g. King, 1996).

Controlled or uncontrolled release discharges of wastewater from water treatment works and seepage from septic tanks can have a major effect on nutrient levels and oxygen demand within urban and peri-urban water courses, and may have an adverse impact on local drinking water sources (Schaider *et al.*, 2016). Although regulations relating to wastewater and sewage discharges have been tightened considerably in recent decades, in accordance with national UK legislation and European directives, very large-scale water pollution episodes remain all too frequent. For instance, in March 2017, Thames Water was prosecuted by the Environment Agency and was fined £20.3M following huge leaks of sewage into the River Thames and its tributaries, which had serious impacts on fish, wildlife, farmers, and residents (BBC News Online, 2017).

Another major source of air, water, and soil pollution in urban environments is provided by motor vehicles. Several different types of pollutants arise from this source, including metallic and paint fragments from bodywork, rubber particles containing metals (especially Zn) from tyre wear, particulate matter from engine emissions (especially diesel engines), incompletely burnt hydrocarbons, a range of volatile organic compounds (VOCs), and various gases including carbon monoxide (CO), carbon dioxide, (CO_2), nitrogen oxides (NOx), and sulphur dioxide (SO_2). The effects of these pollutants on people, plants, animals, buildings, and climate (local, regional, and potentially global) can be significant. Particulate matter smaller than 2.5 μm in size ($PM_{2.5}$) and 10 μm (PM_{10}), NOx, and VOC are particularly significant in terms of human health (RCP and RCPCH, 2016), and CO_2 and SO_2/SO_3 are particularly important for the weathering of stone and other building materials (Figure 12.4). NOx is also important as a cause of nutrient enrichment in soils and surface waters, and emissions of CO_2, methane (CH_4), and water vapour (e.g. from cooling towers) are important in terms of their impact on both global climate and local microclimate. Many of these aspects are considered more fully in other chapters of this book.

Figure 12.4 Deterioration of stonework due to atmospheric pollution, Pembroke College, Cambridge.

12.3 Stages in the Forensic Investigation of Urban Pollution

Forensic science can have an important part to play in the identification of pollution source contributions and in the development of remedial strategies. The solution to a pollution problem can rarely be achieved if the nature of the pollution source is unknown or is subject to high uncertainty. Regardless of pollutant type, several, sometimes overlapping, stages are involved in any environmental, and potentially criminal, scientific forensic investigation (Figure 12.5). The first stage involves field investigation to locate and prescribe the limits of a questioned material. The second stage involves laboratory analysis to identify the nature and potential source(s) of the material. The third stage involves evaluation of the scientific data alongside other lines of evidence to determine whether potential liability can be determined and a case presented to a relevant court or other authority. This stage is principally the responsibility of statutory regulators, lawyers, and/ or the police but usually involves the preparation of reports by a range of technical experts, including scientists. Finally, a fourth stage may involve court proceedings to establish legal responsibility and any subsequent remedial action, compensation, or fines.

12.4 Methods Used to Identify Sources of Pollutants

The most appropriate forensic methods and techniques to use will depend on the specific problem being addressed, and in many instances analytical scientific methods will be used alongside other lines of enquiry. For example, incidents of fly tipping are usually investigated using a range of methods including CCTV, eyewitness reports, tyre track analysis, examination of the deposited waste for documentary and photographic information which may identify the source, and DNA and fingerprint analysis to identify the originators and/or those responsible for the tipping. Since 2004, householders in England have been required by law to check that anyone removing waste from their property is authorised to do so, usually in the form of a Waste Carrier licence issued by the Environment Agency, and councils can fine householders who hire a company to take material away that ends up being fly-tipped. The problem of illegal dumping is also

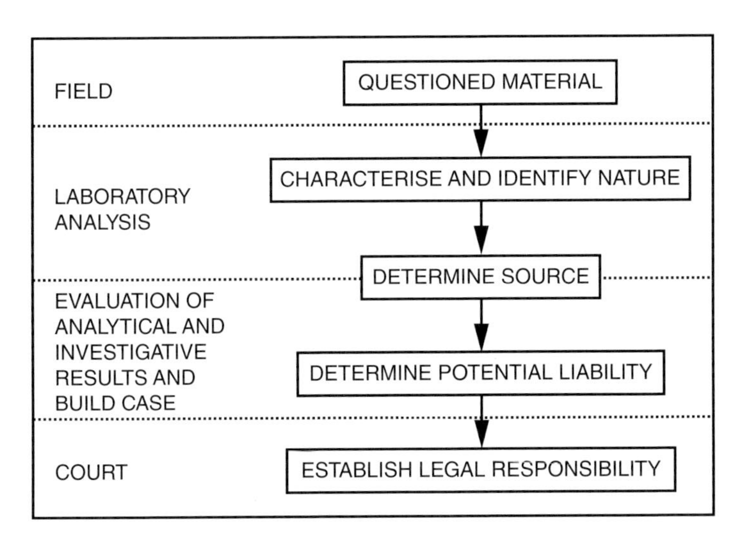

Figure 12.5 Stages in the forensic investigation of urban pollution, their impacts, and potential liability.

widespread across Europe and other parts of the world and has prompted a recent upsurge in the development of techniques to identify sources and perpetrators (e.g. Lega *et al.*, 2012).

A Pollution Crime Forensic Investigation Manual has recently been published by Interpol (2014) in response to recognition that environmental crime is a growing international crime area. In 2010, the Interpol Environmental Crime Committee working group set up and maintain a network of environmental, technical, and forensic experts to promote best practice in environmental forensics, and to capture this in the form of a published manual. Its purpose is to provide a guide through the investigation process from initial receipt of information of a potential violation, through planning and implementation of the evidence-gathering process, to preparation and presentation of data and evidence in a prosecution brief. Volume I of the Manual describes a series of common environmental investigation scenarios relating to surface water, groundwater, soil, and air pollution and summarises, in each case, how evidence is collected through sampling analysis and subsequently presented. Volume II contains sections that describe commonly used field chemistry tests, sampling equipment, and methods suitable for surface waters, soils, air and gases, airborne deposits, biota, and hazardous wastes. The final section describes the procedures necessary for the shipment of legal samples, and the appendices contain example documentation for chain of custody.

In cases where an individual or vehicle suspected of fly tipping has been apprehended, soil and geological trace evidence on footwear, types, vehicle wheel arches, and so on, may be used to test a possible connection with the tipping site. The techniques used in this process are identical to those used in other types of forensic geoscience investigation, as described by Murray and Tedrow (1992), Pye and Croft (2004), Pye *et al.* (2006), Pye (2007), and Bergslein (2013). A wide range of bulk properties and individual particle properties can be used to characterise and compare soil and sediment samples, including size, shape, surface texture, colour, mineral composition, elemental composition, isotopic composition, and biological / biochemical characteristics (Table 12.1). Ruffell *et al.* (2013) specifically discuss trace evidence in urban environments and its potential applications, pointing out that urban environments frequently contain a wider range of particle types, many of anthropogenic origin, than those found in typical rural soils, and samples taken from urban 'soils' often contain 'unusual' particle assemblages, which may be specific to one, or only a small number, of locations/source areas.

In some investigations, only one or two analytical techniques have been used to characterise and compare the materials under investigation, but best practice requires that three or more independent techniques are used wherever possible. Where individual particles are the focus of interest, for example, industrial fly ash or coal dust, optical microscopy and scanning electron microscopy combined with X-ray chemical micro analysis are widely used for initial assessment (e.g. Pye, 2004; Tishmack and Burns, 2004; Kutchko and Kim, 2006; Isphording, 2013; Williamson *et al.*, 2013). This may be followed by quantitative electron microprobe analysis and laser ablation mass spectrometry to provide further information about trace element composition and isotopic signatures.

Physical and chemical 'fingerprinting' methods may be used in a similar manner to compare questioned deposited waste, or any other contaminating material, with a number of potential source materials. In the context of environment forensics, 'fingerprinting' is a generic term that includes methods aimed at identifying specific correlations within the variability of experimental data based on which patterns are established that could be linked to sources and/or release times (Petrisor, 2005). A wide range of natural and synthetic materials may be analysed,

Table 12.1 Bulk sediment and individual particle properties used frequently in forensic comparison.

Properties	Commonly used analysis methods
Bulk sediment properties	
Particle size distribution	Sieving, laser diffraction, electro-sensing size analysis
Particle shape	Image analysis
Surface area	Nitrogen gas absorption
Colour	Spectrophotometry
Major and trace element composition	Inductively coupled plasma spectrometry, X-ray fluorescence, neutron activation
Mineralogy	Optical microscopy, quantitative scanning electron microscopy and electron microprobe analysis, X-ray diffraction
Particle typology	Visual examination, microscopy
Magnetic properties	Magnetometry
Stable isotope ratios	Mass spectrometry
Radiogenic isotope ratios	Thermal ionization mass spectrometry, inductively coupled plasma mass spectrometry
Lipid biomarkers	Gas chromatography mass spectrometry
Pollen assemblage	Optical microscopy
Diatom assemblage	Optical microscopy
Individual particle properties	
Particle size	Physical measurement, microscopy
Particle shape	Physical measurement, microscopy
Surface texture	Optical microscopy, SEM (scanning electron microscopy)
Colour	Spectro-photometry, optical microscopy
Internal structure	X-ray radiography, X-ray tomography
Major and trace element composition	—Laser ablation mass spectrometry, qualitative and semi-quantitative energy-dispersive and wavelength dispersive chemical microanalysis, quantitative electron microprobe analysis
Mineralogy	Optical microscopy, electron microprobe analysis X-ray crystallography
Stable isotope ratios	—Laser ablation mass spectrometry
Radiogenic isotope ratios	—Inductively coupled plasma mass spectrometry, thermal ionization mass spectrometry
Spectroscopic characteristics	Fourier transform infra-red, ultra-violet, laser Raman spectroscopy
Nano-structures	X-ray diffraction, transmission electron microscopy, field emission SEM, scanning acoustic microscopy, scanning tunnelling microscopy

including petroleum products, the constituents of chemical effluents, organic wastes, gases, and micro-particulates (Morrison, 2000; Morrison and Murphy, 2006). Some of the methods most commonly used to characterise hydrocarbons, solvents, and other VOCs include gas chromatography (GC), high performance liquid chromatography (HPLC), infra-red spectroscopy (IRS), and isotope ratio mass spectrometry (IRMS) (see Petrisor (2014) and Murphy and Morrison (2015) for further information).

Any forensic soil or other environmental forensic investigation will frequently proceed

Table 12.2 Stages in a typical forensic soil investigation.

1	Attend briefing session, and read relevant background information.
2	Carry out laboratory examination of suspect items and obtain questioned soil samples.
3	Visit scene and other relevant locations to collect control samples.
4	Undertake laboratory analysis.
5	Compare results for questioned samples and controls using graphical and statistical methods.
6	If necessary, collect additional control samples.
7	Undertake further laboratory analysis on additional samples.
8	Compare data for questioned samples and controls with background database information.
9	Synthesise results, formulate opinion, and write report/witness statement.

in a series of steps, starting with project briefing and culminating in the preparation of an expert report which summarises the results of the test performed and the interpretations made (Table 12.2). It is always important that the questioned samples are compared with a suitable range of 'control' samples from known locations, or reference samples from known sources/product batches. A suitable field sampling programme must be devised early in the investigation, and all relevant background information collected to develop hypotheses and inform the analytical strategy. Figure 12.6 illustrates an example where stable sulphur isotopes were used to investigate the potential sources of sulphur present in degraded structural concrete and render, and where sulphate attack had been identified as the principle deterioration mechanism. The initial working hypothesis in this case was that sulphate in highway de-icing salts had migrated into adjacent structures and gradually built up over a period of years to the point where calcium silicate and calcium hydroxide bonding agents within the concrete and render were broken down into secondary mineral phases, including ettringite and thaumasite, which have minimal or no bonding strength. A range of materials containing potential sulphur sources was investigated and the range of sulphur isotope values determined. Through this process, it proved possible to demonstrate that de-icing salts were highly unlikely to be a major contributor to the sulphate present within the deteriorated concrete, and that atmospheric pollution was a more likely source (Pye and Schiavon, 1989).

12.5 Conclusions

Forensic investigation of pollution events in urban and other settings has become increasingly common and technically more sophisticated over the past 40 years in response to increased public concern about the impacts of pollution on human populations and the wider environment, including water and air quality, climate, and biota. Defined protocols for the environmental forensic investigation have been developed and their adoption across international boundaries is being actively promoted by the International Criminal Police Organization (INTERPOL). The increase in sophistication of environmental forensic investigations has also been facilitated by improvements in environmental monitoring in surveillance, sampling methodologies, analytical techniques, and data interrogation and presentation. The technical capability now exists to determine the sources of pollution, including the identity of perpetrators, with much greater speed and certainty than ever before. However, the application of such procedures remains hampered in many countries by financial constraints and, in some parts of the world, by endemic high-level corruption. In the future, there are likely to be further refinements in methods of environmental monitoring/surveillance and material 'fingerprinting', leading to better chances of early detection, reduced investigative costs, and increased likelihood of prosecution.

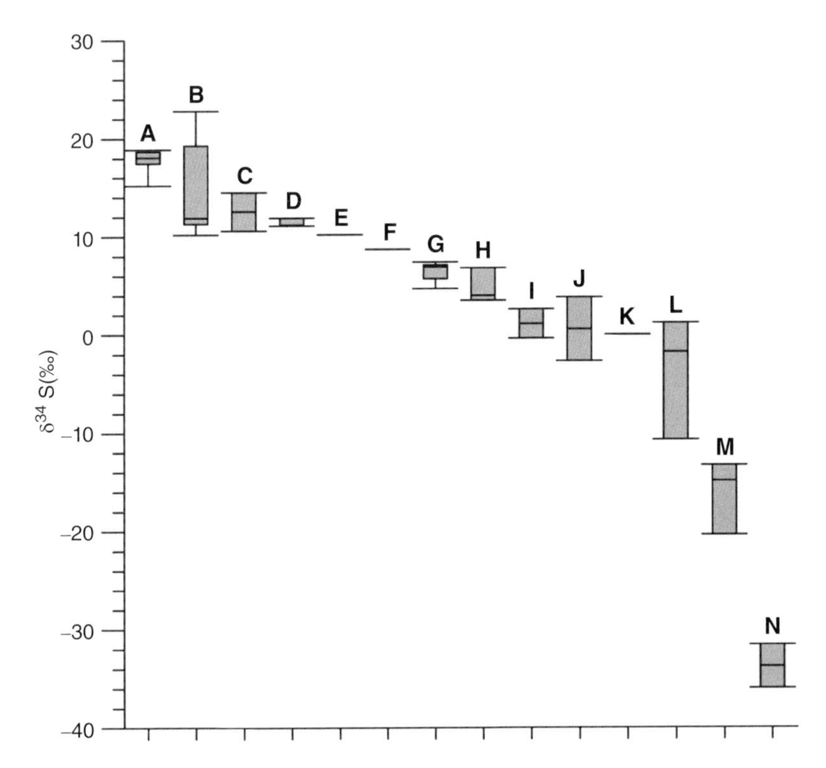

Figure 12.6 An example of the use of sulphur isotope analysis to determine the origin of sulphur compounds in degraded urban construction materials: $\delta^{34}S$ values for: A, Cheshire de-icing salt; B, other UK and overseas de-icing salt; C, un-hydrated Ordinary Portland Cement (OPC); D, Pristine concrete, Peterborough; E, Carbonated concrete pillar, Huntingdon; F, Sewer concrete affected by sulphates and sulphuric acid; G, Deteriorated concrete and render affected by atmospheric sulphate; H, Gypsum crusts on limestone buildings; I, Industrial ash fill beneath concrete floor slabs, Birmingham; J, Render attached by sulphates from brickwork, Cambridge; K, Concrete floor slab attacked by sulphate from underlying clay soil, Birmingham; L, Oxford Clay containing pyrite; M, Fletton bricks, Bedfordshire; N, Edale Shale, Derbyshire. *Source*: Data from Pye and Schiavon (1989).

References

Alloway, B.J. (ed.) (1995). *Heavy Metals in Soils*. 2nd edition. Blackie, Glasgow and London, 368 pp.

BBC News On-line (2017) Thames Water Fined £20M for Sewage Spill. March 22, 2017. Available at: http://www.bbc.co.uk/news/uk-england-39352755 (last accessed April 13, 2017).

Bergslien, E. (2012). *An Introduction to Forensic Geoscience*. John Wiley & Sons, Chichester, 514 pp.

DEFRA (2017) *Statistical Data Set. Env24 – Fly Tipping Incidents and Actions Taken in England*. Part of DEFRA Waste and Recycling Statistics, Updated March 2, 2017. Department of Environment, Food and Rural Affairs, London, available at: https://www.gov.uk/government/statistical-data-sets/env24-fly-tipping-incidents-and-actions-taken-in-england (last accessed April 13, 2017).

Gooddy, D.C., Macdonald, D.M.J., Lapworth, D.J., Bennett, S.A., and Griffiths, K.J. (2014) Nitrogen sources, transport and processing in peri-urban floodplains. *Science of the Total Environment*, 494–495, 28–38.

INTERPOL (2014) *Pollution Crime Forensic Investigation Manual*. INTERPOL

Environmental Security Sub-Directorate, Lyon, Volume I 183 pp and Volume II 178 pp.

Isphording, W.C. (2013). Using soil mineral signatures to confirm sources of industrial contaminant trespass. In: Pirrie, D., Ruffell, A., and Dawson, L.A. (eds.),*Environmental and Criminal Geoforensics*. Geological Society Special Publication 384, The Geological Society, London, pp. 81–86.

King, A. (1996). *Reduction of Fugitive Dust from Coal Stockpiles*. Contract No. 72220–EA/002, Final Report to the European Commission by the British Coal Corporation, Cheltenham, 32 pp.

Kutchko, B.G. and Kim, A.N. (2006) Fly ash characterization by SEM–EDS. *Fuel*, 85, 2537–2544.

Lega, M., Ceglie, D., Persechino, G., Ferrara, C., and Napoli, R.M.A. (2012) Illegal dumping investigation: A new challenge for forensic environmental engineering. In: *Waste Management and the Environment VI. WIT Transactions on Ecology and The Environment*, Vol. 163, Wessex Institute Press, Southampton: doi: 10.2495/WM120011.

Macklin, Y., Kibble, A., and Pollitt, F. (2011).*Impact on Health of Emissions from Landfill Sites. Advice from the Health Protection Agency*. Documents of the Health Protection Agency, Radiation, Chemical and Environmental Hazards, July 2011, 27 pp.

Morrison, A., McColl, S., Dawson, L.A., and Brewer, M. (2009) Characterization and discrimination of urban soils: Preliminary results from the soils Forensics University Network. In: Ritz, K., Dawson, L.A., and Miller, D. (eds.), *Criminal and Environmental Soil Forensics*. Springer, Dordrecht, pp. 75–86.

Morrison, R.D. (2000) *Environmental Forensics Principles and Applications*. CRC Press LLC, Boca Raton, Florida, 351 pp.

Morrison, R.D. and Murphy, B.L. (2006) *Environmental Forensics: Contaminant Specific Guide*. Academic Press, New York, 576 pp.

Murphy, B.L. and Morrison, R.D. (2015) *Introduction to Environmental Forensics*. 3rd edition. Academic Press, Oxford and San Diego, 719 pp.

Murray, R.C. and Tedrow, J.C.F. (1992) *Forensic Geology*. Prentice Hall, Englewood Cliffs, New Jersey, 203 pp.

Petrisor, I.G. (2005) Fingerprinting in environmental forensics. *Environmental Forensics*, 6, 101–102.

Petrisor, I.G. (2014) *Environmental Forensics Fundamentals: A Practical Guide*. CRC Press, Boca Raton, Florida, 445 pp.

Pirrie, D., Ruffell, A., and Dawson, L.A. (eds.) (2013) *Environmental and Criminal Geoforensics*. Geological Society Special Publication 384, The Geological Society, London, 273 pp.

Pye, K. (2004) Forensic examination of rocks, sediments, soils and dusts using scanning electron microscopy and X-ray chemical microanalysis. In: Pye, K. and Croft, D.J. (eds.), *Forensic Geoscience Principles. Techniques and Applications*. Geological Society Special Publication 232, The Geological Society, London, pp. 103–121.

Pye, K. (2007) *Geological and Soil Evidence: Forensic Applications*. CRC Press, Boca Raton, Florida, 335 p.

Pye, K. and Croft, D. (eds.) (2004) *Forensic Geoscience: Principles, Techniques and Applications*. Geological Society Special Publication 232, The Geological society, London, 318 pp.

Pye, K. and Schiavon, N. (1989) Cause of sulphate attack on concrete, render and stone indicate by sulphur isotope ratios. *Nature*, 342, 663–664.

Pye, K., Blott, S.J., Croft, D.J., and Carter, J.F. (2006) Forensic comparison of soil samples: Assessment of small-scale spatial variability in elemental composition, carbon and nitrogen isotope ratios, colour and particle size distribution. *Forensic Science International*, 163, 59–80.

RCP and RCPCH (2016) *Every Breath We Take: The Lifelong Impact of Air Pollution. Report of a Working Party, February 2016*. Royal College of Physicians and the Royal

College of Paediatrics and Child Health, *London*, 160 pp.

Ritz, K., Dawson, L.A., and Miller, D. (2009) *Criminal and Environmental Soil Forensics*. Springer, Dordrecht, 519 pp.

Ruffell, A. and McKinley, J. (2005) Forensic Geoscience: Applications of geology, geomorphology and geophysics to criminal investigations. *Earth Science Reviews*, 69, 235–247.

Ruffell, A. and McKinley, J. (2008) *Geoforensics*. John Wiley & Sons, Chichester, 332 pp.

Ruffell, A., Pirrie, D., and Power, M.R. (2013) Issues and opportunities in urban forensic geology. In: Pirrie, D., Ruffell, A., and Dawson, L.A. (eds.), *Environmental and Criminal Geoforensics*. Geological Society Special Publication 384, The Geological Society, London, pp. 147–162.

Schaider, L.A., Ackerman, J.M., and Rudel, R.A. (2016) Septic systems as sources of organic wastewater compounds in domestic drinking water wells in a shallow sand and gravel aquifer. *Science of the Total Environment*, 547, 470–481.

Stoke Sentinel (2017) Big Issue: Who is Behind the Mountains of Rubbish Appearing in North Staffordshire. *February* 04, 2017. Available at: http://www.stokesentinel.co.uk/big-issue-who-is-behind-the-mountains-of-rubbish-appearing-in-north-staffordshire/story-30109772-detail/story.html (last accessed April 13, 2017).

Tishmack, J.K. and Burns, P.E. (2004) The chemistry and mineralogy of coal and coal combustion products. In: Giere, R. and Stille, P. (eds.), *Energy, Waste, and the Environment: A Geochemical Perspective*. Geological Society Special Publication 236, The Geological Society, London, pp. 223–246.

Webb, B., Marshal, B., Czarnomski, S., and Tilley, N. (2006) *Fly Tipping: Causes, Incentives and Solutions*. Jill Dando Institute of Crime Science, University College London, 71 pp.

Williamson, B.J., Rollinson, G., and Pirrie, D. (2013). Automated mineralogical analysis of PM10: new parameters for assessing PM toxicity. *Environmental Science and Technology*, 47, 5570–5577.

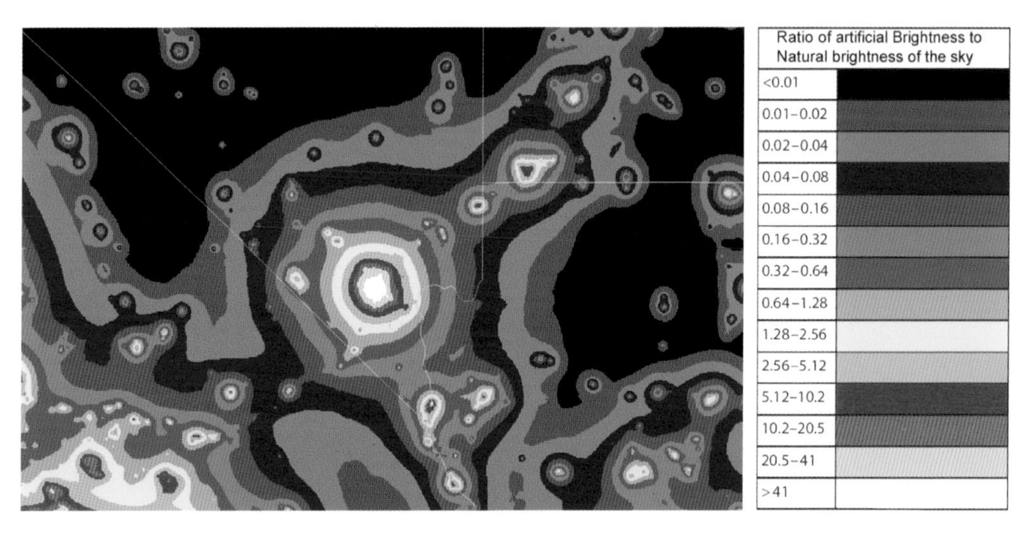

Figure 11.3 Las Vegas lights pollutes the sky well over 100 miles away. Adapted from Falchi *et al.*, 2016.

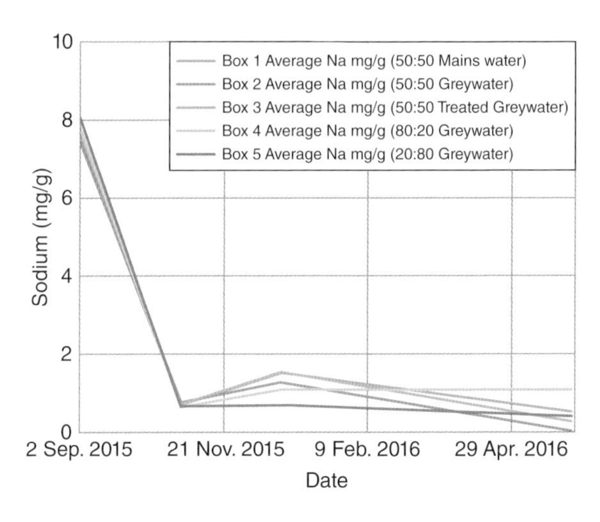

Figure 16.1 Longitudinal changes in plant leaf sodium concentrations for differing watering regimes; '50:50' refers to a plant substrate of 50 parts growing medium to 50 parts compost; '80:20' refers to a plant substrate of 80 parts growing medium to 20 parts compost; and '20:80' to a plant substrate of 20 parts growing medium to 80 parts compost.

Urban Pollution: Science and Management, First Edition. Edited by Susanne M. Charlesworth and Colin A. Booth.

13

River Ecology and Urban Pollution

Martin Fenn

Environment Agency, United Kingdom

13.1 Introduction

Rivers and streams are a very important resource, used by humankind for numerous purposes, the majority of which often cause deterioration in river quality. Towns and cities have tended to grow next to rivers due to their importance, but it is these large population densities that have caused the most damage. It is envisaged that population growth is likely to increase mainly in urban areas during the twenty-first century (Walsh, 2000), and this will further add to the pressures on our already stressed rivers and streams. However, rivers and streams are also the home to a complex ecosystem made up of a myriad of flora and fauna. Deteriorations in water quality or habitat will impact this ecosystem, altering the natural equilibrium. The exploitation of river water by man, and the associated impact on ecology, can be placed into five broad categories:

1) **Supply**. River water is often a main source of water and is used for public water supply and industrial processes. Too high a demand for water reduces the flow of the watercourse. This causes problems for the ecology of the river as fine sediments can be deposited covering substrate, flora, and fauna. As flows decrease dissolved oxygen levels can be reduced significantly, especially if tied in with an increase in water temperature. Another consequence of less water in the system is the reduction in dilution of pollutants, and where extreme water loss occurs, sections of rivers can dry up. Also, water can be transferred between catchments, increasing the risk of spread of invasive non-native species (INNS).
2) **Disposal of waste**. Sewage and industrial wastes are often pumped into rivers as an 'easy' method of disposing of unwanted waste. These are often described as 'point source' pollution types and can be extremely harmful to river life. Raw sewage can lower dissolved oxygen levels in the river (essential element for life), increase ammonia levels (toxic to some organisms), and increase turbidity (impacting light dispersal and covering substrate). Modern industrial discharges can be a concoction of hazardous chemicals that can be directly toxic to the flora and fauna living in our watercourses, while some chemicals can bioaccumulate in organisms (e.g. some pesticides) or bioconcentrate through the food chain.
3) **Transport**. This is perhaps not used to as great an extent in developed countries, but still has an impact on some of the larger rivers. Impacts include physical disturbance through boat traffic by stirring up sediment and physical modification of rivers with navigation weirs,

Urban Pollution: Science and Management, First Edition. Edited by Susanne M. Charlesworth and Colin A. Booth.

changing the hydrodynamics of the water and physically obstructing fish migration. There is always the potential for pollution either from the boat itself or its cargo. Boats moving between catchments can also increase the risk of spreading INNS.

4) **Leisure**. Rivers and streams are an important leisure resource, especially in urban areas. This can be just for somewhere to walk along or an active use such as fishing or canoeing. Recreational pressure on rivers can cause problems of bank side erosion, disturbance, and general pollution (Haslam, 1995). INNS can also be distributed unknowingly by leisure activities if biosecurity is not undertaken.
5) **Physical alteration**. In towns and cities, the dynamics of the fluvial system are far removed from the natural system. Large hardstanding areas prevent water soaking into the ground. This limits the replenishment of groundwater, but also causes large flow fluctuations in urban areas with large storm peaks due to run-off caused by heavy rain. This creates serious problems for the ecology in the rivers, which is required to adapt between times of low flow and raging torrent. In addition, run-off is often contaminated with a wide array of chemicals and sediment. Also, within the urban environment, rivers are often hidden away in culverts and diverted/straightened to move water away quickly. Light is an essential element in primary production in rivers, and those rivers flowing underground are flowing in total darkness. Culverts can also be impassable creating barriers to fish migration. In an attempt to prevent flooding, large flood defence schemes have been built in urban areas, removing essential riparian habitat for the ecology and just passing the problem of flooding downstream.

Monitoring techniques have been developed to be able to quantify these impacts and clearly inform policymakers on the state of our waterways. This has been a relatively recent phenomenon, and it is worth considering how and why these techniques were created.

13.2 History of River Ecology Monitoring

The increase in human population in Europe during the nineteenth century was accompanied by major water quality problems. These mainly manifested in urban areas, which saw the greatest increases in population densities and exerted the greatest pressures on our rivers. It has been well documented (Hynes, 1960; Holmes and Raven, 2014) how in the early nineteenth century the introduction of sewers and the carriage of untreated excrement in water caused widespread river pollution. Cholera outbreaks in large cities such as London were common, and loss of fish such as Atlantic salmon in the Thames and Mersey were recorded in the early nineteenth century. Finally, governments began to tackle the problem of river pollution through the late nineteenth century and the beginning of the twentieth century, initially restricting discharges in watercourses and then through advances in treatment of sewage (Mason, 1996). Gradually, water quality has improved, but a method was required to quantify improvement or deterioration in water quality. To begin with, water chemistry samples were collected and tested to assess the quality of our rivers and streams. However, this only tells part of the story. There are so many different concoctions of chemicals requiring measurement that testing for every single one is impossible. Also, chemical sampling usually involves the collection of a water sample that is then sent to a laboratory. Even if this sample is taken daily, it is only a snapshot of what is entering the watercourse and can miss discharges or other events. Finally, some chemicals can be toxic at very low levels which can be expensive and very difficult to detect in water samples. Although chemical sampling has a place in water monitoring, it does not tell the whole story. Monitoring the ecology of the river is able to overcome all these issues and is used throughout the world to assess the quality of our rivers and stream (Table 13.1).

13.2.1 Macroinvertebrates

The most commonly utilised group of organisms used to assess water quality are the macroinvertebrates. These are a diverse group of organisms that mainly stay in one place, show a reaction to different conditions within the watercourse, and are relatively easy to collect and identify (Chapman, 1996). On the whole, these organisms in the past have been used to measure the impact of organic pollution, but as we learn more about them we are able to assess impacts from low flows, acidification, and sedimentation (Extence *et al.*, 1999; Extence *et al.*, 2017). To understand why we can use these organisms to measure the impact of organic pollution, it is essential to understand the underlying mechanisms by which organic pollution impacts macroinvertebrates. Microorganisms within the river start to break down organic matter. This microbial activity begins to rapidly use up the available dissolved oxygen in the water and produces an oxygen sag. This activity can be measured as biochemical oxygen demand (BOD). This test measures the amount of oxygen used up by life in the water at 20°C over five days. Self-purification of the river occurs as the organic matter is used up, and water movement assists in the re-aeration process (Figure 13.1). With plenty of 'food' available, those macroinvertebrates that can survive in low dissolved oxygen levels dominate in large numbers. Tubificid worms and *Chironomus riparius* (bloodworms) contain the pigment haemoglobin, which has a high affinity for oxygen. Therefore, even at low dissolved oxygen levels, these species can survive. As self-purification occurs and dissolved

Table 13.1 List of ecological monitoring undertaken to assess pressures in rivers for Water Framework Directive.

Organism	Pressure
Macroinvertebrates	Organic enrichment
	Abstraction of water
	Sedimentation
	Acidification
	Pollution by toxic chemicals
Macrophytes	Nutrient enrichment (eutrophication)
	Morphological alterations
Phytobenthos (diatoms)	Nutrient enrichment (eutrophication)
Fish	Morphological alterations
	Abstraction of water
	Sedimentation
	Hazardous chemicals

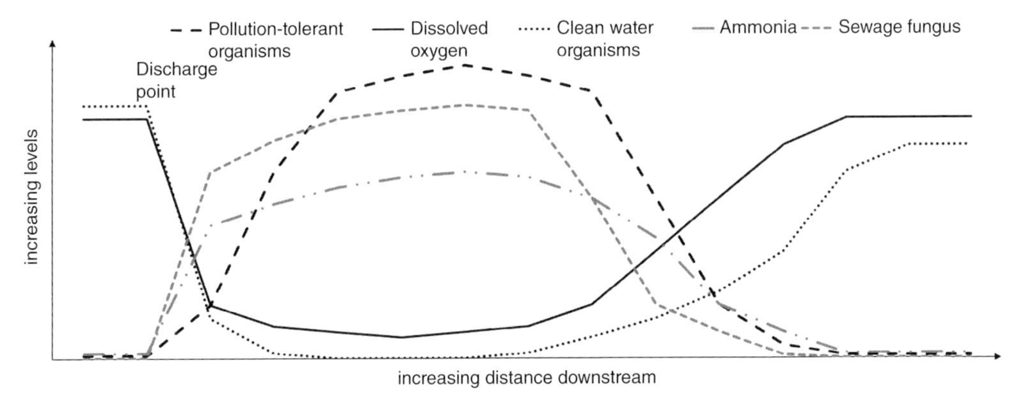

Figure 13.1 The impact of an organic discharge into a river showing the self-purification process (adapted from Hynes 1960).

oxygen levels begin to rise, species such as the water hog-louse (*Asellus aquaticus*) and leeches become more prevalent. These then give way to freshwater shrimps (*Gammarus pulex*) and mayflies (Ephemeroptera). Stoneflies (Plecoptera) are generally the most sensitive species to organic pollution (although tolerant of metal pollution) and are only found in unpolluted waters. This is because of how this group of organisms has evolved to thrive in cool fast-flowing waters with plenty of dissolved oxygen. Knowing the response of these macroinvertebrate groups makes it possible to assess water quality in respect to organic pollution. Biotic indices were created to describe these responses to non-ecologists (Mason, 1996).

The earliest biotic index created to measure water quality was developed in the early part of the twentieth century in continental Europe. The Saprobien system recognised four stages in the breakdown of organic matter in rivers: polysaprobic (grossly polluted), α-mesosaprobic, β-mesosaprobic, and oligosaprobic (recovered river) (Mason, 1996). Each of these stages, or zones, would be characterised by specific indicator organisms. The main criticism of this system was that the taxa used were not always in the correct category; hence, the system was not popular in English-speaking countries (Metcalfe, 1989).

In the United Kingdom, biotic indices were developed much later to assist in the assessment of the response of macroinvertebrates to organic pollution. These indices give a score depending on what macroinvertebrates are in the sample, showing a response to water quality. Certain groups of taxa (e.g. Plecoptera (stoneflies)) are sensitive to the presence of organic pollution, while other groups (Chironomidae (non-biting midge)) are generally tolerant of organic pollution. One of the first indices to be developed was the Trent Biotic Index in 1964, which is simply based on the presence or absence of certain groups of taxa. The total number of groups present is then cross-referenced against a hierarchy of indicator taxa ranging from stoneflies down to Oligochaetes (worms). A score between 0 (heavily polluted) to X (unpolluted water quality) is then given to the sample.

In 1979, the Biological Monitoring Working Party (BMWP) score was developed and continued to be used within the United Kingdom for the next 35 years. The BMWP index is primarily used to monitor the impact of organic water pollution, but will also show responses to toxic pollution. It is based on a standard monitoring approach of a 3-min kick sample and additional 1-min hand search. A numeric figure is given to macroinvertebrates depending on their tolerance to organic pollution. Macroinvertebrate families are given a score between 1 and 10. An overall score (BMWP score) is derived for a particular sample by summing all the family scores. Generally, the higher the overall score, the higher the biological quality of the sample and the cleaner the watercourse.

A score of 1 is given if the macroinvertebrate is very tolerant to organic pollution, for example, Oligochaetes, and a score of 10 if it is very sensitive to organic pollution, for example, certain stonefly families. Abundances of the animals present are not accounted for. The ASPT (average score per taxon) is derived by dividing the BMWP score by the total number of scoring families, and it is used as an indicator of organic pollution, also removing any bias from the sample size. For example, larger samples are likely to include a higher number of families, which will, therefore, increase the BMWP score, while not necessarily being composed of sensitive taxa. The higher the ASPT, the more sensitive the population is to pollution; lower scores are indicative of invertebrate communities suffering from stress due to reduced water quality. The number of scoring taxa can be used as a simple diversity index and can highlight issues with habitat or toxic pollution. The River Stour in the West Midlands has historically been a heavily polluted river suffering from industrial discharges and draining a highly populated urban area. The lower Stour was also the victim of the successful Kidderminster carpet industry. The river would often turn different colours from the dyes used on carpets, and the macroinvertebrate fauna was impoverished by the widespread use of pesticides to 'moth-proof' carpets. However, in 2011, it was named as one of the most improved rivers in the United Kingdom. Figure 13.2 demonstrates the improvement in the macroinvertebrate population with large increases in the BMWP score and number of scoring taxa. This came about in response to the reduction of inputs from the carpet industry. The ASPT has shown an increase, but it is far less impressive as organic pollution is still of concern within this river.

To be able to assess/classify different types of waterbodies, it is important to try and predict what macroinvertebrates should be at a site. The 16th Report of the Royal Commission of Environmental Pollution (1992) recommended that a classification scheme using biological data should be created for reporting river quality. In the United Kingdom, RIVPACS (River InVertebrate Prediction and Classification System) was developed by the Institute of Freshwater Ecology and was used in 1995 to produce the General Quality Assessment (GQA) of rivers in England and Wales. This system produces expected macroinvertebrate fauna for a river on the basis of environmental parameters and information from a series of sites with high biological quality. This can then be compared to observed BMWP scores and a ratio calculated (observed/expected), where a value of 1 indicates a macroinvertebrate fauna of high biological quality. A value less than 1 indicates that the macroinvertebrate population is under stress. Wright *et al.* (2000) discuss the development of RIVPACS and its use in the GQA of rivers, along with other assessment tools.

Waterbodies were then classified on this basis for the Water Framework Directive (WFD) within England and Wales for the first river basin cycle. For classification of waterbodies using macroinvertebrates, the Web-based programme RICT (River Invertebrate Classification Tool) is now used throughout the United Kingdom. This tool incorporates the RIVPACS models to produce an ecological quality ratio, which is then used to give a classification of High, Good, Moderate, Poor, or Bad (the expectation is for all waterbodies to achieve a minimum of Good

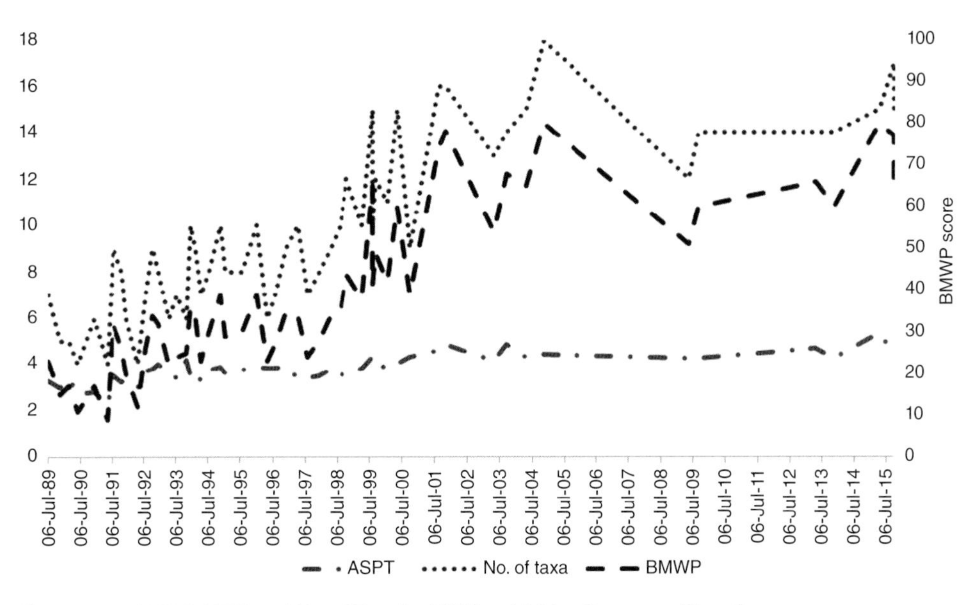

Figure 13.2 BMWP, ASPT, and No. of Taxa for Mill Road Bridge Stourport, River Stour.

status by 2027). For a waterbody to be classified, a spring and an autumn sample (from the same year) need to be collected and analysed. At present within England, macroinvertebrates are collected triennially, once every three years, for WFD classification purposes. However, a failing of the BMWP system has been the fact that a score is attained for a family if 1 or 1000 individuals are present. The abundance of certain macroinvertebrates can be very informative. For example, large numbers of pollution-tolerant leeches is indicative of pollution, while even in cleaner waters it is possible for a leech to be found in a sample. The Walley Hawkes Paisley Trigg (WHPT) score overcomes this issue by taking account of abundances. WHPT is also more accurate than BMWP because it was derived from an analysis of a very large set of real observations (more than 100,000 standard samples) from across the United Kingdom rather than on expert judgement sometimes based on the limited knowledge available in the late 1970s.

ASPT $_{(WHPT)}$ responds to the same environmental pressures as ASPT$_{(BMWP)}$. This includes urban organic discharges and the pressures associated with them, such as increases in organic loading and the concentrations of nutrients, ammonia, and suspended solids; and reduction in oxygen concentration and saturation and habitat degradation, including reduced habitat diversity and increased siltation. It, therefore, responds to activities that cause these pressures, including industrial discharges, reductions in flow, habitat degradation, and eutrophication (all of which are associated with urban areas). Unlike ASPT $_{(BMWP)}$, ASPT $_{(WHPT)}$ will respond to activities that affect the abundance of different invertebrates, which should improve its ability to distinguish finer alterations in water quality.

The number of taxa for WHPT responds to the same environmental pressures as that for BMWP. It responds to most environmental pressures including organic pollution, habitat degradation, acidification, and toxic pollution from a wide range of pollutants including metals. Whereas BMWP is based on analysis of 82 taxa, WHPT is based on 106 taxa, so its sensitivity is slightly different. WHPT better reflects the quality of the invertebrate community because of this. In particular, WHPT includes more families of Diptera, and some families that were grouped together in BMWP (known as BMWP composite taxa) are considered separately in WHPT.

However, macroinvertebrates can also be used to assess the impacts of low flows and sedimentation, both of which are often linked with urban areas. The Lotic-invertebrate Index for Flow Evaluation (LIFE) has been developed by Extence *et al.* (1999) to show the response of macroinvertebrates to low flows. A scoring system has been devised for both family- and species-level responses to low flows. This index gives ecologists a tool to measure the impact of low flows caused by abstractions and drought. The use of RICT to produce an expected LIFE score enables ecologists to produce an observed/expected ratio, with 0.94 used as the usual threshold for demonstrating the impact of low flows (chalk and sandy rivers are an exception and have a threshold of 1 as RIVPACS tends to under-predict for these river types). Sedimentation is categorised by using Proportion of Sediment-sensitive Invertebrates (PSI) index (Extence *et al.*, 2011). Fine sediment deposition is often linked with low flows, and sources of sediment can originate from agricultural or urban areas. Sedimentation can clog up the gills of macroinvertebrates, smother river substrate, and alter macrophyte and algae composition. The index scores family and species of macroinvertebrate taxa on tested response to fine sediments, and again using RICT can produce an observed/expected ratio. Within the Environment Agency Hydro-Ecological Validation (HEV), plots (Figure 13.3) are produced to demonstrate links to flows, water quality, and sedimentation. The graphs display flow against the observed/expected ratios for LIFE, PSI, ASPT (BMWP), and number of taxa (BMWP). It is important to look at all these graphs together with local knowledge of the sites to ensure

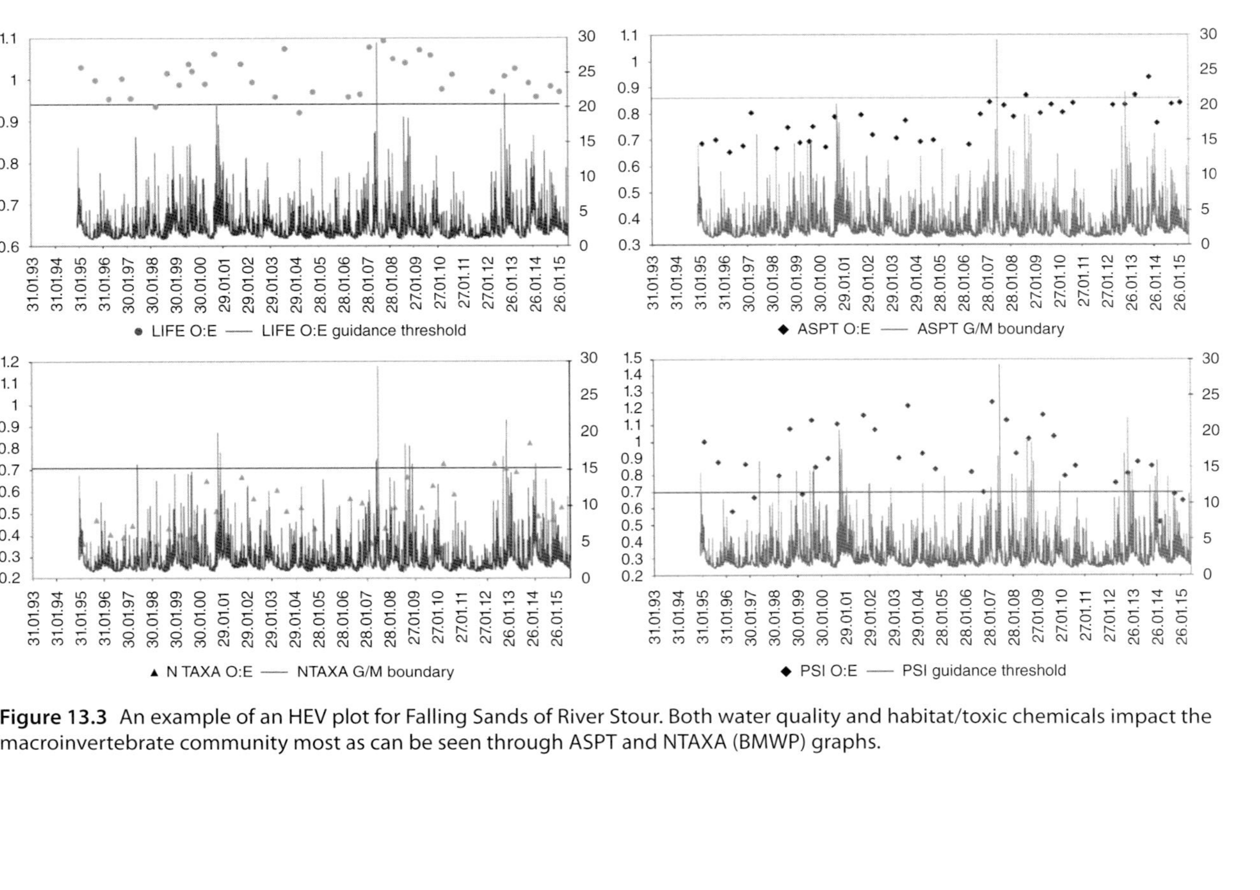

Figure 13.3 An example of an HEV plot for Falling Sands of River Stour. Both water quality and habitat/toxic chemicals impact the macroinvertebrate community most as can be seen through ASPT and NTAXA (BMWP) graphs.

that flow is impacting the site in question. Local knowledge is essential as certain INNS such as the signal crayfish (*Pacifasticus leniusculus*) can artificially heighten LIFE scores. This occurs as the crayfish will mainly predate on slow-moving soft-bodied macroinvertebrates which have lower LIFE scores.

Although macroinvertebrates have seen the greatest adoption in measuring impacts within rivers in urban areas, other organisms can be used to assess different pressures, as can be seen in Table 13.1.

13.2.2 Macrophytes

Eutrophication of rivers is caused by increased nutrient loading. The main nutrients in question are phosphorous and nitrogen, which are important chemicals for primary production. The term eutrophic has been used to classify the nutrient status of waterbodies for nearly a hundred years and is defined as a waterbody with an excessively high nutrient content. These excessive nutrients cause an imbalance in the river ecology, and those organisms that can make the most of the excess nutrients will thrive. Aquatic macrophytes (plants) and phytobenthos (diatoms – small algae with a siliceous skeleton) are used to measure the impact of eutrophication (Table 13.1) in rivers, while phytoplankton (algae) is also used in lakes. A great deal of research has gone into the impacts of nutrient levels on both of these groups of organisms (Kelly *et al.*, 2008; Wilby *et al.*, 2012). They are fairly easy to survey and identification is possible with training. The most common inputs of nutrients within the urban environment are from domestic sewage, industrial wastes, and storm water runoff. In developed countries (especially the United Kingdom), phosphorus levels in domestic sewage have become heightened by the use of detergents that contain sodium tripolyphosphate, which often makes up over half the total phosphorus in the effluents (Mason, 1996). Due to these high nutrient levels within sewage discharges, the Urban Waste Water Treatment Directive (UWWTD) was produced in 1991 by the European Union to ensure minimum treatment of urban wastewater.

For macrophytes, this led to the continued development of the Mean Trophic Rank (MTR) scoring system during the 1990s. MTR scores plants depending on their sensitivity to nutrients and also their abundance at a site (100 m stretch). For example, certain algae tolerant of high nutrients *Cladophora agg.* (blanketweed) only score 1 (Species Trophic Rank), while a large number of mosses and liverworts score high; for example, *Dicranella palustris* scores 10. This score is then combined with their Species Cover Value over a 100 m section of river.

0 =absent
1 =<0.1%
2 =0.1%–1%
3 =1%–2.5%
4 =2.5%–5%
5 =5%–10%
6 =10%–25%
7 =25%–50%
8 =50%–75%
9 =>75%

The Species Trophic rank is then multiplied with the Species Cover Value (SCV) to produce a Cover Value Score (CSV) for each species present in the survey. To calculate the MTR, the sum of all the CVSs is divided by the sum of all the SCVs and multiplied by 10:

$$\text{MTR} = \left(\text{sum of CVS} / \text{sum of SCV}\right) \times 10$$

This was used to designate sensitive areas to eutrophication under the UWWTD. Aquatic macrophytes were monitored upstream and downstream of large sewage treatment works (qualifying discharges are those with a greater loading than 10,000 population equivalent).

This was then superseded within the United Kingdom by the production of the classification tool for WFD, LEAFPACS. Here, the River Macrophyte Nutrient Index (RMNI) and per cent filamentous algae metrics dominate the classification as they best demonstrate macrophytes' reaction to high nutrient levels. Macrophytes are used solely in waterbodies where alkalinity is high (>200 mg/L $CaCO_3$), while in low alkalinity (<75 mg/L $CaCO_3$) waterbodies' diatoms show the best response to nutrient enrichment. In between these two extremes, both methods are used. One problem with macrophyte surveys is that shading can have a large impact on the growth of aquatic plants, so densely shaded sites will not necessarily show a response to high nutrient levels.

13.2.3 Diatoms

For diatoms, the Trophic Diatom Index (TDI) was developed by Kelly and Whitton (1995) to assist in investigations required for UWWTD. The method uses benthic (bottom-dwelling) diatom communities and relies on known responses to nutrient loading in watercourses. The TDI score produced ranges from 0 (low nutrient loading) to 100 (very high nutrient loading). To bring diatom results in line with MTR scores for macrophytes and other biotic indices, the Diatom Quality Index (DQI) was created. This is simply produced by subtracting the TDI score from 100; therefore, a low DQI score shows high nutrient loading (poor water quality). It is also useful to look at the percentage of motile valves (diatoms) to check if any other non-nutrient factor might be influencing the diatom population. The standard format for viewing this data can be seen in Figure 13.4.

The TDI has also been used to help classify waterbodies for WFD by using DARLEQ. The acronym DARLEQ describes the outputs from two projects – DARES (Diatom for Assessing Rivers Ecological Status) and DALES (Diatoms for Assessing Lake Ecological Status). The DARLEQ programme calculates an expected TDI based on alkalinity, which is then compared to observed scores over two seasons (spring and autumn) to produce an ecological quality ratio.

13.2.4 Fish

Within the urban environment, physical modification is a major factor that impacts the river ecology (Cowx and Welcomme, 1998). This can be in the form of weirs or modification of the banks in an attempt to prevent erosion, culverting, and the building of flood defences. Fish are ideal indicators for assessing these impacts. The loss of bankside habitat for fish is a common problem within urban areas. For different life stages, this habitat has varying significance. Fish fry, especially those of coarse fish, require sheltered bays during the summer where water warms up quickly, there is ample food, and they are out of the main flow of the river and protected from predatory fish (Everard, 2015). As they get older, the fish require trees to give cover from predators, especially birds. Certain species also need to lay their eggs on submerged vegetation (phytophiles), for example, roach (*Rutilus rutilius*).

Weirs are barriers to migration, especially for diadromous fish that require to migrate between freshwater and the sea to complete their life cycle. The Atlantic salmon is the most iconic of these mainly due to their efforts seen jumping at weirs, often being unsuccessful. There are a number of other species that also need to travel between freshwater and the sea to complete their life cycle that are unable to proceed over the majority of barriers and need passes to be built (Twaite shad (*Alosa fallax*), European eel (*Anguilla anguilla*), and sea lamprey (*Petromyzon marinus*), to name but a few). However, even those fish that spend their whole lives in rivers can migrate large distances to feed or breed. For example, studies have shown that barbel (*Barbus barbus*) can travel well over 20 km, especially during spawning (Britton and Pegg, 2011). Weirs also change the flow dynamics of a watercourse. By its very nature, a weir will hold back water, but will also hold back sediment and with the loss in velocity fine sediment deposition is common.

The Environment Agency undertook a study of the fish population on the River Severn within the town of Shrewsbury. Boom boat electric fishing was used (Figure 13.5) for three sections of the river surveying from the Showground to English Bridge (Figure 13.6). The topmost section from the Showground

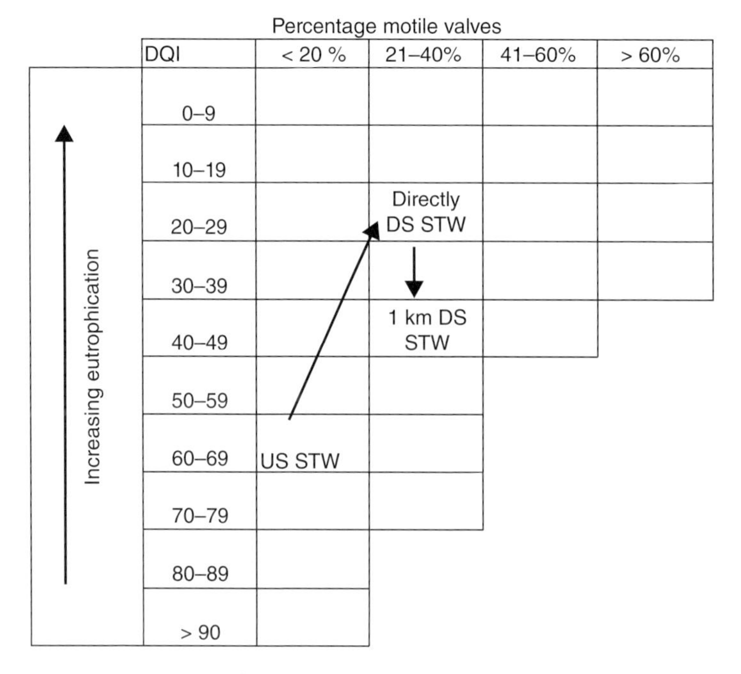

Figure 13.4 Diagram illustrating changes in DQI (eutrophication) and percentage motile valves caused by a sewage treatment works (STW).

Figure 13.5 Hard bank reinforcements on the River Severn in Shrewsbury, Shropshire. Environment Agency staff undertaking boom boat electric fishing survey to assess fish population within town centre.

to Welsh Bridge was more natural with partial mature tree cover and some varied habitat. The next two sections were generally more degraded with fewer trees, extensive bank modifications, and a dredged channel. The size of the river in Shrewsbury is too large to fish quantitatively, and so the margins were mainly fished using catch per unit effort time;

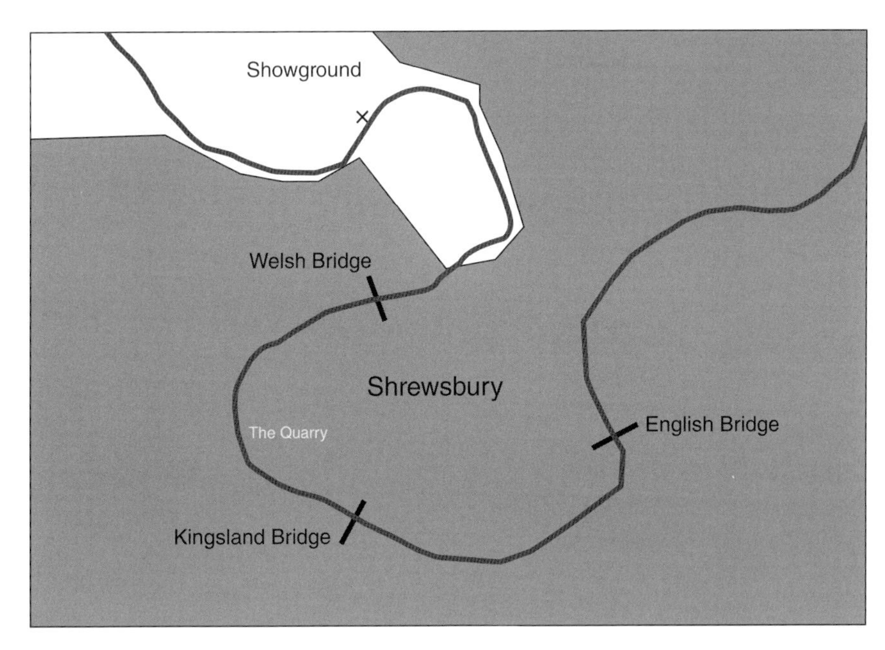

Figure 13.6 Sections surveyed by boom boat electric fishing on the River Severn in Shrewsbury. Grey area is main urban conurbation of Shrewsbury.

the results are shown in Table 13.2. There were more fish and greater species diversity caught in the furthest upstream section from the Showground to Welsh Bridge. Very few fish were caught in the areas of the river where the impacts of urbanisation were the greatest,

Table 13.2 Numbers of fish caught per minute within the three sections boom boat electric fished on the River Severn in Shrewsbury 2013.

Species	Showground to Welsh Bridge	Kingsland Br to English Bridge	Welsh Bridge to Kingsland Bridge
Atlantic salmon [*Salmo salar*]	0.3	0.04	0
Bullhead [*Cottus gobio*]	0	0.04	0
Chub [*Leuciscus cephalus*]	0.07	0	0.04
Dace [*Leuciscus leuciscus*]	0.78	0.09	0
European eel [*Anguilla anguilla*]	0.07	0.09	0.04
Gudgeon [*Gobio gobio*]	0.04	0	0.42
Minnow [*Phoxinus phoxinus*]	0.04	0	0
Perch [*Perca fluviatilis*]	0.11	0	0.15
Pike [*Esox lucius*]	0.11	0.13	0.19
Roach [*Rutilus rutilus*]	1.37	0.04	0.04
Ruffe [*Gymnocephalus cernuus*]	0.04	0.04	0.04
Stone loach [*Barbatula barbatula*]	0.04	0	0.04
Total	**2.96**	**0.48**	**0.96**

The Environmental Quality Standard Directive (EQSD) is linked with the WFD and focuses on some of the hazardous substances that would be difficult to detect with standard chemical monitoring due to the low levels required in the water column before they become hazardous. These chemicals bioaccumulate in certain freshwater organisms, causing problems for top predators and humans. These substances include industrial chemicals, metals, and pesticides and can originate from a number of sources:

- Sewage and trade discharges
- Water passing through contaminated land or abandoned mines
- Rain water run-off from roads or industrial sites

The Environment Agency in England have devised a monitoring programme in key locations around the country to sample species that are considered to be good biomonitors for contaminants and can be caught in sufficient numbers. These include fish species brown trout (*Salmo trutta*), roach (*Rutilus rutilus*) or chub (*Leuciscus cephalus*), and the signal crayfish (*Pacifasticus leniusculus*), for those chemicals fish do not bioaccumulate.

13.3 Success Stories

13.3.1 Case Study: Love Your Rivers Telford

Telford was created in the 1960s and 1970s incorporating a number of small towns as well as industrial and agricultural land. Major development occurred throughout the 1960s, 1970s, and 1980s, in which large numbers of housing estates and industrial units were created. However, minimal thought was given to the rivers and streams of the area at the time, and they were often culverted to hide them away. Urban run-off, misconnections, and industrial pollution have seriously degraded some of these rivers. The Environment Agency has worked closely with three other government organisations, four nongovernment organisations, one water company, one university, 16 community groups, and an industry-led environmental group to set up Love Your Rivers Telford to start to tackle some of these issues.

The project has worked to include the whole community and formed 'clean stream teams'. This is led by two paid-for members of staff (one from the Environment Agency and one from the water company), but they also receive support from volunteer groups, the local community, business community, the stakeholder organisations, and schools. While standard water quality monitoring techniques have been used, volunteer groups have also been trained up by experts in macroinvertebrate and aquatic macrophyte surveys, and the Shropshire Wildlife Trust has been running its River Rangers course at the primary school in the area. A system similar to the Riverfly angling project, where anglers have been used to monitor certain indicator macroinvertebrate groups in rivers around the country, has been created in Telford. The volunteer groups have sites allocated to them that they check regularly, and any deteriorations noted are reported to the clean stream team. This project has been very successful and has seen water quality and physical improvement to large sections of urban waterbodies.

13.3.2 River Wandle, South West London

Back in the 1960s, this river was declared 'a sewer' due to its polluted and degraded state. This river is a tributary of the River Thames and flows through Croyden, Mitcham, Morden, Wimbledon, and Wandsworth in South West London. As populations expanded rapidly during the late nineteenth century, parts of this river were culverted, straightened, and canalised as water quality deteriorated. The river was often running different colours depending on which colours the local tanneries were using at the time and had 90 mills along its length.

During the latter part of the twentieth century, dramatic improvements in water quality have been achieved, and the Wandle was listed as one of ten of the most improved rivers in the United Kingdom by the Environment Agency in 2011. The Wandle Trust is a charity dedicated to restoring and maintaining the health of this river. With the help of volunteers and the local water company, river clean-up days are run on the second Sunday of every month. Volunteer pollution patrol teams are also working with the Environment Agency to regularly check on the Wandle, and children in local schools have been heavily involved. The fish population has significantly improved to become one of the area's most important urban coarse fisheries. This has not just been due to improvement in water quality, but also due to re-naturalising the river in certain sections and creating passages through many of the weirs that has enabled European eel (*Anguilla anguilla*) to migrate upstream once again.

13.4 Conclusions

The Wandle and Love Your Rivers Telford are just a few examples of where we have seen large-scale improvements within our urban rivers. However, it also highlights the need to involve the local communities to give them some ownership of their rivers and to educate them about the importance of the river ecology.

The ecosystem services approach is key to improving river ecology, especially in urban areas. The natural benefits of interacting with the rivers are of huge value to people's mental and physical health. A clean and healthy river will also provide benefits through better public water supply, irrigation for crops, natural flood protection (slow the flow in upland areas), and, especially in developing countries, a food supply. A healthy ecosystem will also buffer any potential impacts from climate change that we can expect. Freshwater systems are seen as one of the most vulnerable ecosystems to climate change (Edwards *et al.*, 2012), and with population expansion in urban areas, there is potentially increased pressure on rivers and the ecology within them.

Ecological monitoring programmes are an essential piece of the jigsaw puzzle when assessing river water quality. Macroinvertebrates will continue to be the main group of organisms used, but the use of plants, diatoms, and fish is also invaluable. Monitoring programmes, though, should be tailored towards the risk being assessed and delivered by a SMART approach (RRC, 2011):

- Specific (concrete, detailed, well defined)
- Measurable (quantity, comparison)
- Achievable (feasible, actionable)
- Realistic (considering resources)
- Time (a defined timeline)

One thing is certain, there will be continued expansion of our urban areas, the impacts on the river ecology will need to be ameliorated.

References

Britton, J. R. and Pegg, J. (2011) Ecology of European barbel *Barbus barbus*: Implications for river, fishery, and conservation management. *Reviews in Fisheries Science*, 19 (4), 321–330.

Cowx, I. and Welcomme, R.L. (1998) *Rehabilitation of Rivers for Fish*. Fishing News Books.

Everard, M. (2015) *River Habitats for Coarse Fish: How Fish Use Rivers and How We Can Help Them!* Old Pond Publishing.

Extence, C.A., Balbi, D.M., and Chadd, R.P. (1999) River flow indexing using British benthic macroinvertebrates: A framework for setting hydroecological objectives. *Regulated Rivers: Research & Management*, 15, 543–574.

Extence, C.A., Chadd, R.P., England, J., and Naura, M. (2017) Application of the proportion of sediment-sensitive invertebrates (PSI) biomonitoring index at local and national scales. River Research and Applications, 33 (10), 1596–1605.

Haslam, S.M. (1995) *River Pollution: An Ecological Perspective.* Belhaven Press

Holmes, N. and Raven, P. (2014) *River: A Natural and Not-So-Natural History.* British Wildlife Publishing.

Hynes, H.B.N. (1960) *The Biology of Polluted Waters.* Liverpool University Press.

Kelly, M.G., Juggins, S., Bennion, H., Burgess, A., Yallop, M., Hirst, H., King, L., Jamieson, B.J., Guthrie, R., and Rippey, B. (2008) Use of Diatoms for Evaluating Ecological Status in UK Freshwaters. Science Report: SC030103/SR4. Environment Agency.

Kelly, M.G. and Whitton, B.A. (1995) The Trophic Diatom Index: A new index for monitoring eutrophication in rivers. *Journal of Applied Phycology*, 7, 433–444.

Mason, C. F. (1996) *Biology of Freshwater Pollution.* Longman.

Metcalfe, J.L. (1989) Biological water quality assessment of running water based on macroinvertebrate communities: History and present status in Europe. *Environmental Pollution*, 60 (1–2), 101–139.

River Restoration Centre (RRC) (2011) Practical River Restoration Appraisal Guidance for Monitoring Options (PRAGMO). Guidance document on suitable monitoring for river and floodplain restoration projects. The River Restoration Centre (RRC).

Walsh, C.J. (2000) Urban impacts on the ecology of receiving waters: A framework for assessment, conservation and restoration. *Hydrobiologia*, 431 (2), 107–114.

Wilby, N., Pitt, J.A., and Phillips, G. (2012) The Ecological Classification of UK Rivers Using Aquatic Macrophytes. Report – SC010080/R1. Environment Agency

Wright, J.F., Sutcliffe, D.W., and Furse, M.T. (2000) *Assessing the Biological Quality of Freshwaters: RIVPACS and Other Techniques*. Freshwater Biological Association.

14

Urban Meadows on Brownfield Land

Lynn Besenyei

Faculty of Science and Engineering, School of Sciences, University of Wolverhampton, United Kingdom

14.1 Introduction

A large number of traditionally managed semi-natural hay meadows have been lost from the British countryside since 1945 as a consequence of agricultural improvement (Blackstock *et al.*, 1999). They have been ploughed, fertilised, drained, or reseeded, in order to make the land more productive, or have been lost to development for housing and industry. These species-rich meadows have high conservation value, and their loss as a habitat has resulted in declines in associated plant and animal species, and their isolation within large expanses of species-poor agricultural land. Areas of species-rich grassland vegetation created in urban conurbations, and known as urban meadows, can help to address this loss from the countryside and increase the biodiversity in towns and cities. These can particularly be created on brownfield land, with a previous history in urban areas, with some careful planning and management and with the introduction of species, which are found in these semi-natural meadows.

Only a small number of these hay meadows remain and are hotspots of British biodiversity with up to 30 plant species per square metre. Some of the best examples are preserved as National Nature Reserves (NNRs) and protected by European legislation (e.g. Foster's Green Meadows in Worcestershire, Draycote Meadows in Warwickshire, and Mottey Meadows in Staffordshire) or remain in private ownership. They are still maintained by traditional management. In 2013, some of Britain's best meadows were selected for conservation by Prince Charles as 'Coronation Meadows' – he wanted to save these meadows (in celebration of the anniversary of the coronation of the Queen) before more were lost (www.coronationmeadows.org.uk).

In urban areas, people attempted to create replicas of these meadows in public open spaces (such as verges, roundabouts, and school grounds) by creating 'political meadows' (Baines, 1989). Such meadows provide specified mixtures of flower species associated with the countryside which have been planted from seed mixtures harvested from the wild (Figure 14.1). These urban meadows bring much needed splashes of colour into urban areas and contain examples of species found in meadows in the countryside (e.g. ox-eye daisy *Leucanthemum vulgare*) as well as species often known as 'agricultural weeds' and associated with cornfields, many of which are now very rare (e.g. corn cockle *Agrostemma githago* and corn marigold *Glebionis segetum*). Some superb examples of 'pictorial' meadows, collections of wild flowers from various different habitats – and

Urban Pollution: Science and Management, First Edition. Edited by Susanne M. Charlesworth and Colin A. Booth.

Figure 14.1 Pictorial meadow in the centre of Coventry.

not always natives – were showcased at Queen Elizabeth Park (www.nigeldunnett.info/PictorialMeadows) for the London Olympics, and visitors gloried in their grand expanses of colour. These were all created from wild flower seed sown on an old brownfield site in East London. These meadows followed earlier examples created by Landlife in Liverpool (Ash *et al.*, 1992; Luscombe and Scott, 1994)

In the late 1980s, Chris Baines (Baines and Smart, 1984; Baines, 1986) encouraged gardeners and park wardens to plant native species in urban areas. The benefits of this are twofold: first, to preserve native flower species, but also to provide a natural habitat for native animal species. Some plant ecologists took this one step further by attempting to create ecological replicas of plant communities in rural areas by creating meadows on ex-arable land (Wells *et al.*, 1986; Wells, 1987; Wells *et al.*, 1989), by restoring existing meadows to their previous condition where species diversity had declined (www.ydmt.org/hay-time) and also through habitat creation (Trueman and Millett, 2003) on brownfield sites in urban areas of known plant communities (Gilbert and Anderson, 1998).

Brownfield land, formally defined as "any land that has previously been used, or developed, and is not currently fully in use and may be vacant, derelict or contaminated" (Alker *et al.*, 2000), provides a wide array of opportunities for meadow creation projects on a wide variety of different land types. Once subjected to creation work, this land not only benefits from increased plant diversity but also facilitates animal diversity for creatures using these meadows as habitats. This is carried out either by sowing seeds in proportion to their species constituents in natural communities or by introducing the species (Walker *et al.*, 2004) in other ways, such as the use of turf or plug plants, or by introducing dry hay, or more naturally, by utilising species-rich green hay cut from a semi-natural hay meadow and spread across the land to be seeded (Trueman and Millett, 2003).

In the most successful cases, it has been possible to create almost exact replicas of the semi-natural MG5 crested dog's-tail – black knapweed *Cynosurus cristatus – Centaurea nigra* hay meadow community (Rodwell, 1992) in the English Midlands by utilising freshly strewn green hay (Trueman and Millett, 2003) on various brownfield sites. The range and extent of these sites vary but include quarries and quarries turned into landfill sites and then further re-profiled into amenity public open space (Figure 14.2), building demolition areas such as old factory sites (Figure 14.3), old coal mines, old aerodromes, school fields, and covered reservoirs.

The presence of biodiversity is seen as an important success criterion of brownfield greening projects and the UK government sees this as a prime tool for regional economic regeneration, neighbourhood renewal, and achieving its international biodiversity commitments (Doick *et al.*, 2009). Brownfield sites can be used as a component of green infrastructure development for sustainable cities. The promotion of sustainable land use by developing these kinds of sites is preferable to the loss of greenfield sites from the countryside and adding to existing urban sprawl. However, brownfield sites which have been neglected and left undisturbed are often tranquil havens for biodiversity, providing opportunities for wildlife and localised species diversity.

The halt of species decline in recent years in urban areas is often due to the colonisation of such sites and their resultant classification as nature reserves for specific species conservation.

14.2 Creating Flower-Rich Meadows

Very often, brownfield sites are harsh environments. This is due to a number of reasons, such as previous toxicity, high daytime temperatures (in sheltered aspects in direct sunlight), poor or non-existent soils, poor or excessive drainage as a consequence of compaction from heavy machinery, and uneven substrates with large pore sizes in the sediments, for example, large bricks and rubble. However, brownfield sites are "neutral" in terms of existing seed banks – initially they have no seed bank but in order to become

Figure 14.2 Green-winged orchids *Anacamptis morio* on a landfill site, Kitchen Lane, Wolverhampton. Photo credit: Alison Wilkes.

Figure 14.3 Cowslips *Primula veris* in vegetation created in an old aircraft factory site, Castle Vale, Birmingham. Photo credit: Alison Wilkes.

colonised, they are dependent on seeds blowing in on the wind or being brought there by animals. Almost always these are sites lacking in high levels of available nutrients, unlike much of the British countryside, which has been heavily fertilised owing to its use for agriculture.

Species-rich meadows can be created on brownfield land by killing, or cutting short, the existing stand of vegetation. Killing is recommended where vigorous weeds are present, but it is only necessary to cut short the existing stand of vegetation where this contains desirable species (Grime *et al.*, 1988). Then the topsoil should be prepared by raking it into a fine tilth to prepare a seed bed. All bricks and other large debris should be removed from the site in the early stages of creation to prevent damage to cutting machinery at a later date. Hay should be cut from a species-rich source early in the morning during mid-July, ideally baled (for convenience), and transported to the site immediately after cutting while still green. On arrival, the bales should be distributed evenly around the site, and the hay should be evenly spread as soon as possible using machinery such as a tedder (Gamble, 2015) or a muck spreader, or by hand with the help of volunteers if such machinery is unavailable (Trueman and Millett, 2003). The hay should then be left in situ to shed its seed and to act as a mulch for germinating seedlings. Depending on the intervening weather, or vernalisation requirements, some seeds may germinate in the first autumn, or next spring, but others may take longer to break their dormancy. The site may result initially in the growth of weed species owing to the presence of airborne seeds which have germinated in the fine tilth, or were already present prior to creation.

A sward with reasonable species diversity can be achieved by sowing seeds harvested from species-rich grasslands by the use of a forage harvester. Such seed mixtures can now be easily obtained from commercial seed merchants, although better species diversity (more similar to existing native semi-natural grasslands) (Figure 14.4) can be obtained from the green hay. A study by Jones *et al.* (1995) found that good ground preparation and spreading technique as well as the use of appropriate management techniques led to species diversity as much as 90% similar to the source meadow. However, a number of species have proved too difficult to transfer for various reasons, including early or late seed set outside normal harvesting time.

At an old industrial estate, previously a wartime airfield, vegetation rich in flowering dicotyledonous forbs was obtained by Besenyei, 2000 within two years of hay being spread on an area equivalent to a third the size of the source site using hay from a suitable flower-rich MG5 crested dog's-tail – black knapweed *Cynosurus cristatus–Centaurea nigra* hay meadow. Such good results on brownfield sites of the creation of a species-rich sward within five years can be achieved with green hay at a spreading ratio of between 4:1 and 1:4 depending on site fertility and provided that good meadow management can be maintained.

Some brownfield sites are known to be ecologically interesting for their assemblages of plant species that have arisen spontaneously through the germination of seeds carried to the site by the wind or by animal vectors such as birds. These seeds have managed to germinate despite the often harsh conditions on some sites. Over

Figure 14.4 Created meadow vegetation detail showing yellow rattle *Rhinanthus minor*. Photo credit: Alison Wilkes.

time, succession has taken place – often resulting in a variety of vegetation assemblages at these sites. This type of open mosaic habitat on previously developed land (Rodwell, 2000) is a priority for conservation and is protected by the United Kingdom's Biodiversity Action Plan (BAP) and as such is a legally protected habitat. Preserving this kind of vegetation on existing brownfield sites is important for town and city planners as it helps with international biodiversity commitments (Buglife, 2009). Where spontaneous open habitats have developed, further meadow creation is unnecessary, although there may be some opportunities where the site is large enough and if the vegetation is less developed in some places.

Under the National Pollinator Strategy (Defra, 2014), bees and hoverflies require support across towns and cities as well as on agricultural areas. High plant species diversity in urban meadows, and open habitats on brownfield land, will lead to an associated diversity of invertebrate species, particularly pollinators. Brownfield sites can thus become refuges for key invertebrate species in urban areas (Buglife, 2009), and improving floral diversity at such sites will provide additional habitats and sources of food for a wide variety of insects. Around 30 species of butterfly and many species of moths are associated with brownfield sites and can be considered to be good indicators of brownfield and urban habitat health, responding rapidly to any changes in condition of these sites (Butterfly Conservation, 2006). Butterflies in particular have a preference for the open, varied habitats of brownfield land, often with sparsely scattered patches of vegetation which they can colonise naturally.

14.3 Brownfield Soils for Meadow Creation

Soils at brownfield sites are often composed of various types of substrates which have arisen from the component materials from the site's previous use (this frequently includes building rubble with little actual topsoil). Soils on such sites can take many years to form and result from the decomposition of invading plants, which leads to a slow accumulation of organic material over time. Such sites can be considered a blank canvas because they lack a seed bank and are dependent on plant species coming in from elsewhere. Initially such sites are invaded by lichens, mosses, and liverworts and later by grasses and herbaceous species, which can be slow to colonise, particularly in areas of high toxicity. The first of the flowering plants to colonise sites are ruderal species such as the common nettle (*Urtica dioica*), thistle (*Cirsium* spp.), and rosebay willowherb (*Chamerion angustifolium*), which are often ephemeral, having short life cycles and quickly exploiting available resources. Plants will exploit localised resources as these become available, and some species will even tolerate the relatively high toxicities and extremes of high or low pH levels associated with brownfield land. Such sites can be very variable in their water-holding capacities, with localised water accumulation where soils are heavily compacted, or very well drained in locations where substrate sizes are large and water can percolate through rapidly. Interesting assemblages of aquatic plant species can quickly colonise open areas of ponds with concomitant colonisation by amphibians and aquatic invertebrates.

High soil fertility can be a major constraint in the creation of grasslands and has led to investigations into the relationships between soil characteristics and site species richness. McCrea *et al.* (2001a) studied soil nutrient content in two created meadows on brownfield land in Wolverhampton, West Midlands, United Kingdom, in an attempt to identify optimal nutrient ranges when evaluating soils for meadow creation. One of their study meadows was created on the site of an old grain mill and its dam, where the mill had been demolished, rubble from the mill was used to fill the dam, and the site regraded resulting in sandy and loamy soils. The other meadow was on land restored to amenity use on public open space which had previously been associated with an old quarry and had a rubbly substrate. Of the range of soil parameters studied, they found that phosphorus and potassium concentrations separately accounted for the most variation in species composition. High levels of phosphorus were found to inhibit species richness, but intermediate levels of potassium were found to be beneficial for diversity. They recommend that soils containing less than 7 mg of extractable phosphorus per 100 g, and between 10 and 30 mg of extractable potassium per 100 g, would yield optimal plant species diversity (results are based on Truog's method of analysis for soil phosphorus and on ammonium acetate extraction for soil potassium). Thus, on sites where soil nutrients are found to be higher than these levels, meadow creation is not recommended. Soils can be tested at ADAS-accredited laboratories around the country before any detailed planning commences.

Further work by McCrea *et al.* (2004) investigated the relationship between soil fertility and species diversity in eight meadows created on urban soils (including the two mentioned above) and 28 semi-natural meadows. The soils had pHs of between 4.5 and 6.5 classified as mesotrophic grasslands by Rodwell (1992) in the phytosociology of British plant communities. The main differences between the soils of the created and semi-natural meadows were that the created ones contained more extractable phosphorus and less mineralisable nitrogen and organic matter. High phosphorus concentrations were thought to be the main reason few species colonise or survive in grasslands on many urban soils. The urban soils also had higher pHs, which Gilbert (1989) attributed to the presence of mortar rubble characteristic of brownfield sites. Gilbert (1989) also considered that high phosphorus levels in urban soils might be released from constituent clay minerals derived from brick rubble in the soil. Thus, McCrea *et al.* (2004) proposed phosphorus as the main edaphic factor controlling the establishment of created meadows in many urban areas.

Where nutrient levels in soils are high, vigorous growth of undesirable species is likely to occur, and thus it is best to avoid spending a lot of time and effort creating meadows on sites with unsuitable soils. However, depending on the site, it might be worth considering depleting nutrient levels by growing crops for a season, or two, in order to attempt meadow creation when nutrient levels have fallen to the desired range (McCrea *et al.*, 2001b). Where the top soil is high in nutrients, another option might be to remove it off site and create the meadow in the subsoil. In such cases, seed germination will still take place, but, overall, plant height will be appreciably less than normal with vegetation development taking longer until organic matter builds up sufficiently to retain nutrients. At a factory site in Bridgnorth, Shropshire, a swale was created as part of a Sustainable Drainage System by removing the topsoil. A species-rich green hay was then spread onto the remaining subsoil, and a sparse covering of a mix of ruderal species and meadow species was obtained the following year, with the meadow species increasing in abundance in subsequent years (author's own observations).

14.4 Management of Created Meadows

Semi-natural meadows have been managed traditionally for hay since Saxon times. The hay meadow was the most important field on the manor because this was where the food to over-winter livestock, such as oxen used for ploughing, was grown. Meadows would have been cut between mid-July and mid-August with the hay being left on the ground for about two weeks and turned occasionally to dry. When dry, the hay would have been collected, baled, and stored for use during winter. Any aftergrowth would have been grazed throughout the autumn until the ground was poached. This would allow for seeds fallen onto the ground during hay making to have some open soil to germinate into. Hay meadows would then have been closed up to allow germination, new growth, and flowering before being cut again.

Meadow creation is more successful if the traditional management of semi-natural meadows is carried out. One of the main problems of habitat creation is the necessity to manage them once created, and many schemes fail as a consequence of not addressing the necessary management regime (Parker, 1995). Often, this can be as simple as installing a gate to allow access to machinery. The very minimum requirement for created meadows is cutting and removal of the hay, ideally in late summer. If this is not achieved, the diversity of the vegetation can deteriorate within two seasons. It is thus imperative to facilitate access to sites for a tractor-mounted flail mower to cut the vegetation. Very small sites can be cut by hand with a strimmer, but this is hard work at larger sites and makes the hay difficult to collect and remove once cut. Where possible, after mowing, the site should be grazed, ideally with young beef cattle (where available) or less preferably sheep or horses, until the vegetation stops growing when the temperature drops in autumn. This is unlikely to be feasible in urban areas due to security issues with livestock, so the effects of grazing should be simulated by damaging the sward in some way, for example, by scarifying, to expose the soil in places to facilitate regeneration gaps for seeds to germinate in. Failure to cut for hay quickly results in taller-growing species outcompeting the shorter species. Grazing prevents the development of a thick thatch which would otherwise preclude seeds from reaching the soil and germinating, resulting very quickly in the loss of annuals from the sward. In addition, failure to remove the vegetation will result in the unchecked growth of competitive species, such as cock's-foot *Dactylis glomerata* and hogweed *Heracleum sphondylium*, which continue to grow throughout the autumn and rapidly out-compete the shorter-growing species and seedlings of annual species attempting to germinate.

It is often difficult to decide when to initiate cutting after creation; it is not recommended to do so after the first season's growth. Much of the germination after creation can be weed species that are not members of the required vegetation community. Ideally, these should be eliminated by spot spraying or weed-wiping with herbicide in the first season before they set seed. At the end of the second season, depending on how much resulting growth of the required species there has been, the sward can be cut and removed from the site. If there has been little growth, due to lack of nutrients in the soil, then cutting may need to be delayed until after the third season. New species will continue to germinate despite the cutting process, but these should be facilitated by keeping the growth of more vigorous species in check.

On brownfield sites, growth of the new vegetation might be slow owing to localised stressing due to lack of nutrients in the soil, or lack of a top soil, or localised toxicity preventing, or retarding, vegetation growth. Germination and growth of species in the first season is often disappointing due to disturbance, which often results in the germination of weed species which can grow more vigorously than the desired species and because many species do not flower for a couple of seasons after establishment. However, orchids, particularly the common spotted orchid *Dactylorrhiza fuchsii* and the green-winged orchid *Orchis morio*, have been found flowering on brownfield sites in the second season after seeding, but, in general, these take longer to establish (author's own observations).

Brownfield sites should also be managed for pollinators, including bees, hoverflies, and butterflies. This can be done after a site evaluation for its value for wildlife (Defra, 2014). There is a need for advice for land managers to provide essential resources and habitats for pollinators (www.beesneeds.org.uk), which require pollen and nectar as food sources between February and October, so plants need to be allowed to flower continuously throughout the season, possibly requiring some areas to be left uncut. They also require places to nest and shelter, and these areas should not be disturbed during the hibernation period (Defra, 2014). These requirements can be made available in urban meadows, but in addition, some patches of land need to be left to grow wild to provide other sources of food such as leaves for caterpillars, or hibernating insects. To prevent sparse vegetation for butterflies from disappearing from brownfield sites, they need occasional disturbance and scrub clearance (Butterfly Conservation, 2006). The use of pesticides should be avoided across nature conservation sites.

14.5 Opportunities for Urban Meadow Creation

Sir John Lawton made an appeal to make space for nature in his review of England's wildlife sites (Lawton *et al.*, 2010). The report said 'there is compelling evidence that England's collection of wildlife sites are generally too small and too isolated, leading to declines in many of England's characteristic species' and made 24 recommendations to benefit wildlife and people in order to try to rebuild nature. It called for targeted conservation efforts to improve existing natural sites in conjunction with linking to adjacent sites in surrounding areas to strengthen and expand areas available to wildlife. Brownfield land offers an opportunity to provide additional unpopulated land, and thus, facilitating its re-colonisation by nature will extend the amount of land available for wildlife and strengthen ecological networks.

The National Brownfield Strategy (English Partnerships, 2003) (the Government's policy framework for brownfield development at the time) set out policy measures necessary to ensure the sustainable use of brownfield land and acknowledged that available land should be set aside for its most appropriate use depending on the site's characteristics. This need not be housing, or industrial use, but could quite easily be for nature and amenity use, urging a better use of these finite resources. In many cases, redeveloping such land would miss an opportunity for nature conservation by allowing land that has naturally re-colonised to persist as a nature reserve by habitat creation and ecological restoration techniques. This could allow these areas to be rewoven into the natural fabric of the land and enable the development of natural habitats in urban areas. In this way, large areas of land which currently remains unused will provide habitats for nature (Buglife, 2009).

The UK Government are also trying to provide opportunities to widen access to nature to a greater variety of people in order for them to enjoy being in the natural world and thus, additionally, to enable conservation by providing funding for local ecological creation and management schemes. The Government's Access to Nature grant programme was funded by the Big Lottery Fund and run by Natural England. It funded 115 projects which ranged from local-community-based schemes to national initiatives with the aim to 'bring lasting change to people's awareness of, access to and engagement with the natural environment' (Icarus, 2014). It sought to bring awareness for people about the natural places around them and found that as a result of the project there were an estimated 2.93 billion visits to the natural environment by the adult population of England between March 2013 and February 2014, and 45% of visits "to nature" were to a green space in an urban location a mile from home. Two thirds (68%) of visits to the natural environment were within two miles of the respondent's home, or other starting point, and a quarter (27%) included a visit to a park in a town, or city (Natural England 2015). The outcomes of the project suggest a link to increased health and well-being from visiting a green space. The combination of increased nature conservation value of brownfield sites and increased human access to them should engender benefits to both human society and its well-being and to the natural environment.

14.6 Conclusions

There are many possibilities for providing beautiful flower-rich vegetation on brownfield land, as a habitat for wildlife, which can also be enjoyed by local people in urban areas close to their homes. The wide range of brownfield sites, resulting from a variety of previous uses, can offer opportunities for influencing the design of development in favour of the local ecology by facilitating the creation of species-rich urban meadows containing very similar species assemblages to the remaining examples in the countryside and at the same allowing the preservation of the floral species which make up this community. An initial site assessment should first assess any existing vegetation that has already developed over the intervening years since abandonment, as often the edaphic features at these sites can allow the development of open flower-rich habitat mosaics which can be excellent habitat for invertebrates. These should not be damaged or destroyed and should remain at the site without the need for creation work, or should be woven into any ecological enhancement.

Created urban meadows need soils which have a pH between 5 and 7, which should be free draining, and ideally have some organic matter content. High soil nutrient levels are not required; in fact, low levels of available nutrients (nitrates, phosphates, and potassium) are preferred in order to prevent the dominance of fast-growing competitive species. Access should be available to facilitate machinery for cutting the hay, which should be cut at the end of the flowering season, and still in summer, once the majority of seeds have set. The cut hay should be removed from the site, so as not to smother shorter-growing species and not return nutrients to the site after decomposition. Where possible, the meadow should be grazed after mowing during autumn. If this is not possible, any new growth should be kept short, and the sward should be disturbed in some way to allow seeds to set.

The Government is currently working to legislate for statutory registers of brownfield land suitable for housing to be kept by councils. This will allow for quicker planning proposals on such land and might well mean that land currently ecologically valuable might well be lost to development. It is hoped that where meadows have already become established, with existing vegetation that has high biodiversity value, they will be allowed to persist and be kept well managed for the long-term enjoyment of the public and as habitat for wildlife.

References

Alker, S. Joy, V., Roberts, P., and Smith, N. (2000) The definition of Brownfield. *Journal of Environmental Planning and Management*, 43 (1), 49–69.

Ash, H.J., Bennett, R., and Scott, R. (1992) *Flowers in the Grass: Creating and Managing Grasslands with Wild Flowers*. English Nature, Peterborough.

Baines, C. (1986) *The Wild Side of Town*. BBC publications and Elm Tree Books, London.

Baines, C. (1989) Choices in habitat re-creation. In Buckley, G.P. (ed.), *Biological Habitat Reconstruction*. Belhaven Press, London.

Baines, C. and Smart, J. (1984) *A Guide to Habitat Creation*. Ecological Handbook No.2, Greater London Council.

Besenyei, L. (2000). The Management of Artificially Created Species-rich Meadows in Urban Landscaping Schemes. Unpublished PhD thesis, University of Wolverhampton.

Blackstock, T.H., Rimes, C.A., Stevens, D.P., Jefferson, R.G., Robertson, H.J., Mackintosh,

J., and Hopkins, J.J. (1999) The extent of semi-natural grassland communities in lowland England and Wales: A review of conservation surveys 1978–96. *Grass and Forage Science*, 54, 1–18.

Buglife (2009) *Planning for Brownfield Biodiversity: A Best Practice Guide.* Buglife – The Invertebrate Conservation Trust, Peterborough.

Butterfly Conservation (2006). *Brownfields for Butterflies.* British Butterfly Conservation Society, East Lulworth.

Department for Environment, Farming and Rural Affairs (2014) The National Pollinator Strategy: For Bees and Other Pollinators in England. (www.gov.uk/government/publications/national–pollinator–strategy–for–bees–and–other–pollinators–in–england)

Doick, K.J., Sellers, G., Castan–Broto, V., and Silverthorne, T. (2009) Understanding success in the context of brownfield greening projects: The requirement for outcome evaluation of success assessment. *Urban Forestry and Urban Greening*, 8, 163–178.

English Partnerships (2003) *Towards a National Brownfields Strategy.* Research Findings for the Deputy Prime Minister.

Gamble, D. (2015) Hay time meadow restoration. *Field Studies* (www.fsj.field–studies–council.org)

Gilbert, O.L. (1989) *The Ecology of Urban Habitats.* Chapman and Hall, London.

Gilbert, O.L. and Anderson, P. (1998) *Habitat Creation and Repair.* Oxford University Press, Oxford.

Grime, J.P., Hodgson, J.G., and Hunt, R. (1988) *Comparative Plant Ecology: A Functional Approach to Common British Species.* Unwin Hyman, London.

Icarus (2014) Access to Nature: Inspiring People to Engage with Their Natural Environment. Final Evaluation Report English Nature Report, Worcester.

Lawton, J.H., Brotherton, P.N.M., Brown, V.K., Elphick, C., Fitter, A.H., Forshaw, J., Haddow, R.W., Hilborne, S., Leafe, R.N., Mace, G.M., Southgate, M.P., Sutherland, W.J., Tew, T.E., Varley, J., and Wynne, G.R. (2010) Making Space for Nature: A Review of England's Wildlife Sites and Ecological Network. Report to Defra.

Luscombe, G. and Scott, R. (1994) *Wildflowers Work – A Technical Guide to Creating and Managing Wildflower Landscapes.* Landlife, Liverpool.

McCrea, A.R., Trueman, I.C., Fullen, M.A., Atkinson, M.D., and Besenyei, L. (2001a) Relationships between soil characteristics and species richness in two botanically heterogeneous created meadows in the urban English West Midlands. *Biological Conservation*, 97, 171–180.

McCrea, A.R., Trueman, I.C., and Fullen, M.A. (2001b) A comparison of the effects of four arable crops on the fertility depletion of a sandy silt loam destined for grassland habitat creation. *Biological Conservation,* 97, 181–187.

McCrea, A.R., Trueman, I.C., and Fullen, M.A. (2004) Factors relating to soil fertility and species diversity in both semi-natural and created meadows in the West Midlands of England. *European Journal of Soil Science,* 55, 335–348.

Natural England (2015) Monitor of Engagement with the Natural Environment: The National Survey on People and the Natural Environment – Annual Report from the 2013–14 survey. Natural England, Peterborough.

Parker, D.M. (1995) *Habitat Creation a Critical Guide.* English Nature, Peterborough.

Rodwell, J.S. (ed.) (1992) *British Plant Communities Volume 3.* Grassland and Montane Communities. Cambridge University Press, Cambridge.

Rodwell, J.S. (2000) *British Plant Communities Volume 5.* Maritime Communities and Vegetation of Open Habitats. Cambridge University Press, Cambridge.

Trueman, I.C. and Millett, P. (2003) Creating wild-flower meadows by strewing green hay. *British Wildlife,* 15 (1), 37–44.

Walker, K.J., Stevens, P.A., Stevens, D.P., Mountford, J.O., Manchester, S.J., and

Pywell, R.F. (2004) The restoration and re-creation of species-rich lowland grassland on land formerly managed for intensive agriculture in the UK. *Biological Conservation*, 119, 1–18.
Wells, T.C.E. (1987) The Establishment of Floral Grasslands. *Acta Horticulturae*, 195, 59–69.
Wells, T. C. E., Frost, A., and Bell, S. (1986) Wild Flower Grasslands from Crop-grown Seed and Hay-bales. Focus on Nature Conservation No.15. Nature Conservancy Council, Peterborough.
Wells, T.C.E., Cox, R., and Frost, A. (1989) The Establishment and Management of Wildflower Meadows. Focus on Nature Conservation No.21. Nature Conservancy Council, Peterborough.

15

Urban Pollution and Ecosystem Services

Rebecca Wade

Urban Water Technology Centre, School of Science Engineering and Technology, Abertay University, Scotland, United Kingdom

15.1 Introduction

Urban pollutants can degrade and inhibit ecological functions and processes. Those natural processes provide vital benefits and services to humans. The 'services' range from food and water provision, to aesthetics, cultural benefits, health and recreation opportunities, and also climate regulation (including water and air quality regulation and flood regulation). These services are referred to collectively as ecosystem services (ES). To understand the impacts of urban pollution, and the opportunities for mitigating its effects, it is important to explore the relationship between pollution and ES. This can lead to better decision-making for urban infrastructure and spaces and can provide increased opportunities for gaining multiple ES from urban environments.

15.2 Ecosystem Services (ES), the Ecosystem Approach, and Ecosystem Service Valuation

Ecosystem Services are the benefits provided to humans by natural systems. The use of the term *ecosystem services* became widespread after the publication of the Millennium Ecosystem Assessment (MA) in 2005. The focus of the MA was to assess the consequences of ecosystem change for human well-being; it provided a state-of-the-art scientific appraisal of the condition and trends in the world's ecosystems and the services they provide (MA 2005). In 2011, the UK National Ecosystem Assessment (UK NEA) provided the first analysis of the United Kingdom's natural environment in terms of the benefits it provides to society and economic prosperity in the United Kingdom. An ecosystems approach was adopted; this was designed to provide a framework for examining whole ecosystems in decision-making, and for valuing the ES they provided. The aim of this approach is to ensure that society can maintain a healthy and resilient natural environment now and for future generations (DEFRA, 2013).

There are several different definitions and classifications of ES. Both the Millennium Ecosystem Assessment (MA, 2005), and the UK National Ecosystem Assessment (UKNEA 2011) classify ES along functional lines into the four categories of provisioning, regulating, supporting, and cultural services (MA, 2005; UK NEA, 2011). Ecosystem functions determine the capacity of a natural resource system to sustain ES that are fundamental to human well-being. Examples of ES include products such as food and water; regulation of floods, soil erosion, and disease outbreaks; noise reduction; and non-material

Urban Pollution: Science and Management, First Edition. Edited by Susanne M. Charlesworth and Colin A. Booth.

benefits such as recreational, educational, cultural, and spiritual benefits in natural areas (UKNEA, 2011; URBES, 2013).

The term 'services' is usually used to encompass the tangible and intangible benefits that humans obtain from ecosystems, which are sometimes separated into 'goods' and 'services' (UKNEA, 2011). It is clear from the findings of the MA (2005) and UKNEA (2011) that damage to the environment is seriously degrading the ability of natural ecosystem functions to deliver these services. It is now accepted that this degradation will have direct economic implications for the management of functions such as pollution regulation, flood protection, regulation of the chemical composition of the atmosphere, and pollination, on which human well-being directly depend (POSTnote 281, 2007). The multiple values of ES (both monetary and non-monetary) can be used to identify and assess the importance of these services for society (URBES, 2013). Table 15.1 identifies some of the different values that can be attributed to urban ES; these values, however, are not often recognised in urban planning and decision-making and, consequently, the impact of their loss remains invisible.

The monetary and health values which ES provide can be exemplified as follows: urban vegetation moderates the local temperature and buffers noise pollution. Loss of urban vegetation can lead to economic costs due to energy demand for heating and cooling buildings, healthcare expenses related to respiratory diseases, and maintenance of expensive infrastructures to abate noise and pollution (URBES, 2013; Säumel *et al.*, 2016). Restoring urban green spaces can provide substantial long-term benefits to cities as well as delivering non-monetary or indirect ES values (such as species diversity and richness of habitat). Among the sociocultural values, there is increased community cohesion and local ecological knowledge. Enhancing the amenity value of urban areas can also positively influence public health by promoting physical activity (Säumel *et al.*, 2016), and increased physical and mental

Table 15.1 Example of the multiple values that can be attributed to urban ecosystem services (ES) (adapted from URBES, 2013)[a].

Value-type	Relevance to urban ES
Economic values	Direct or indirect monetary values provided by urban ecosystems, e.g. saved costs for air pollution reduction by technical solutions or property damage by natural barriers to environmental extremes.
Ecological values	Environmental outputs, which have value for humans, e.g. air purification, carbon storage and sequestration, water filtration, and genetic diversity.
Sociocultural values	Moral, spiritual, aesthetic, and ethical values associated with urban biodiversity and ES, including emotional, affective, and symbolic views attached to urban nature, as well as local ecological knowledge.
Health values	Health benefits obtained from urban green spaces, consisting of reduction in air pollution, improved water quality, enhanced recreation potential, physical and mental health benefits of spending time in nature.
Insurance values	The contribution of green infrastructure and ES to increased resilience and reduced vulnerability to shocks, such as flooding and landslides.

a) Several of these services directly reference mitigation of air and water pollution or avoidance of cost related to pollution effects in urban areas.

well-being are certainly considered as health values (URBES, 2013).

It should be noted that ecosystem *dis*services are also recognised in the literature; Säumel *et al.* (2016) have published a comprehensive table listing ecosystem services and disservices provided by roadside vegetation, in an article on liveable and healthy urban streets. They found that some plants *decreased* air quality by emitting biological aerosols (volatile organic compounds, aeroallergens), and indeed Rogers *et al.* (2006)

reported that plant stressors such as air pollution and increased temperatures can induce increased levels of allergenic proteins in the pollen – and thereby related health problems. Therefore, the greater the impacts of urban pollution on urban ecosystems, the greater the incidence of ecosystem disservices (at least in some cases).

More information on ES can be accessed via the UKNEA (2011), in particular, Chapter 10 (Urban), which provides information, key messages, examples, and valuations relevant to urban ES.

15.3 Urban Impacts on ES

Urbanisation can have significant impacts on biodiversity and ES. Population growth, increased industry and commercialisation, and expanded provision for new housing and transport, all greatly contribute to the increasing extent of impermeable surfaces. With urban development comes increased surface run-off and pollution, associated with which is the loss of regulating and provisioning services and important cultural benefits (Davies *et al.*, 2011). All urban areas will be subject to some of the disadvantages of populated areas, such as loss and isolation of habitats (Kong *et al.*, 2010) and increased water and air pollution (Davies *et al.*, 2011; Lundy and Wade, 2011); these impacts will be greater in more densely populated areas.

Urban areas can contain rich flora that contribute significantly to biodiversity, and thus they also provide many ES benefits to humans. Urban ecosystem goods and services will differ according to urban population size, boundary, location, and surroundings. The provision of ES in urban settings depends in large part on the quality and quantity of urban green infrastructure (URBES, 2013) and also on the connectivity of green spaces. The more integrated the green infrastructure, the greater its potential to provide ES benefits. Figures 15.1 and 15.2

Figure 15.1 Trianon Park, central São Paulo, Brazil. The park occupies two city blocks, it is surrounded by the concrete streets and skyscrapers (concrete forest) of Brazils' largest city. However, inside the park the air temperature is lower, the traffic fumes disappear, and even the traffic noise subsides. Trianon Park was created in 1892 and is now the only remaining piece of native Atlantic Forest (Mata Atlântica) in the area. (Photo credit: Rebecca Wade, 2015).

Figure 15.2 Princes Street Gardens, a public park in the centre of Edinburgh, Scotland, United Kingdom, in the shadow of Edinburgh Castle. The Gardens were created in the 1770s and 1820s. The prominent central location, and large size of the gardens in this busy tourist-destination capital city have ensured that the gardens are Edinburghs' best known parks with the highest visitor figures for both residents and visitors. Photo credit: Rebecca Wade, 2017.

provide examples of urban green space in cities.

As with ES, green infrastructure has different definitions. Broadly, green infrastructure includes parks, gardens, urban allotments, green routes or paths, urban forests, wetlands, lakes, and ponds in cities, but also natural areas – such as forests, mountains, and wetlands – surrounding urban spaces. It can also be described from a planning perspective as the design elements in urban settings that contribute to the delivery of the Green Network, brought together in a placemaking masterplan (GCV GNP, no date). Green infrastructure as defined by Wise (2008) is 'the interconnected network of open spaces and natural areas – such as greenways, wetlands, parks, forest reserves, and native plant vegetation – that naturally manages stormwater, reduces the risk of floods, captures pollution, and improves water quality'. This definition specifically identifies pollution reduction as a feature or design element of green infrastructure.

Urban populations draw heavily on external resources (outside the city or town boundary) for provisioning services (food and water, fuel, and fibre) and other ES. They export waste and release pollution emissions to air, water, and land that extend far beyond the urban boundary (Luck *et al.*, 2001). The ecological footprint of urban areas is widely recognised as being much larger than the geographical area of the city or town. Urban areas also export visitors to other habitats, giving rise to associated transport pressures (pollution and infrastructure) (Davies *et al.*, 2011) and adding pressures to ES in other areas.

15.4 ES and Urban Pollution in the UK Legislative Context

In the last decade, in the United Kingdom and elsewhere, it has become increasingly possible to deliver green infrastructure elements and thereby to enhance some ES with multiple benefits through existing legislative frameworks and requirements. For example, there is an obligation to meet surface water management legislation sustainably, from both a water quantity (flooding and drought) aspect and a water quality (pollution) perspective. In Europe, the EU Water Framework Directive has been in place for more than a decade and provides an overarching piece of legislation that aims to harmonise existing water policy and to improve water quality in all of Europe's aquatic environments (Kaika and Page, 2003). The Water Framework Directive (European Commission, 2000) affects 27 countries and marks an important trend towards an ecosystem-based approach to water policy and water resource management (Kallis and Butler, 2001).

However, reporting for the Royal Commission on Environmental Pollution, Goode (2006) concluded that cross-cutting guidance was needed, that it should be based firmly on the multifunctional green space network, and ensure that it provides the necessary advice for sustainability. More recently, POSTnote 376 (2011) recognised that natural resources are being impacted by a range of pressures, such as biodiversity loss and climate change. They stated that 'if governments do not monitor effectively the use or degradation of natural resource systems in national account frameworks ("environmental accounting"), the risks of incurring costs to future economic productivity are not taken into account, nor are impacts on human wellbeing' (POSTnote 376, 2011). While there is increased recognition that biodiversity and ES can contribute greatly to improve quality of life in cities, and that their maintenance will deliver economic benefit, their multiple values are usually not fully taken into account in urban policymaking (URBES, 2013).

The UK government provide guidance for policymakers and decision-makers on using an ecosystems approach and on valuing ES (DEFRA, 2013). The Scottish government have conducted research and mapping of ES for use in decision-making across land use policy priorities (such as a low carbon economy, sustainable food production, and water management), in order to apply the work to a broad spectrum of policy sectors, including climate change, agriculture, biodiversity conservation, and water resource management. Similarly, the Welsh government includes ecosystem health and sustainability prominently in their recent Environment (Wales) Act 2016 and in their 2015 Payment for Ecosystems approach to natural resource management and decision-making.

Valuing ES benefits, or undertaking a 'natural capital' valuation, can help inform a wide range of decisions relating to management of environmental risks such as urban pollution (POSTnote 542, 2016). However, valuation of ES and natural capital is very challenging for many reasons (e.g. data availability, spatial and temporal considerations, assessment of dis-benefits, and trade-offs); these are beyond the scope of this chapter. An overview of ES in relation to strategic improvement in urban green infrastructure elements is provided by Wentworth (2017).

15.5 Enhancing Urban ES to Mitigate Urban Pollution

Air pollution is one of the main environmental risks for human health worldwide (WHO, 2016). In this context, abatement of pollution has become of major concern especially in areas with high pollutant concentrations, typically associated with cities. Maintaining and developing green urban areas can be part of an integrative strategy to help improve air quality in cities. Trees reduce temperatures in cities by evaporating water and remove air pollutants and particulate matter via their leaves through dry deposition. In general, green plants reduce air pollutants by taking up gaseous pollutants primarily through leaf stomata. Once inside the plant, these gases react with water to form acids and other chemicals (Baldocchi *et al.*, 1987). Green plants can also intercept particulate matter as wind currents blow them into contact with sticky plant surfaces (Bidwell and Fraser, 1972). Some of these particulates can be absorbed into the plant, while others simply adhere to the surface. Vegetation can be a temporary site for particulates as they can be re-suspended into the atmosphere by winds or washed off by rain water.

Urban street trees, green roofs and green walls, green areas, and forests surrounding cities have the capacity to remove significant amounts of pollutants, thereby increasing environmental quality and improving human health (Pugh *et al.*, 2012; Maes *et al.*, 2017). Some dis-benefits have been associated with certain species of street trees, in relation to their location and management, additionally, several studies have cautioned their use for air pollution mitigation without careful consideration.

Pugh *et al.* (2012) attempted to model the complex geometry of urban surfaces in dense urban areas to gauge the effect of street-level air pollution concentrations. People are primarily exposed to pollutants at the street level, they were particularly interested in the occurrence and concentrations of air pollution in street canyons. They present results which show that in-canyon vegetation offers a method to improve urban air quality substantially. The results of their analysis show that street-level reductions by as much as 40% for NO_2 and 60% for PM_{10} are achievable using green walls, which is much higher than has been previously estimated. This suggests that the potential benefits of green infrastructure for air quality improvements have been substantially undervalued. This effect is particularly important for pollutants with atmospheric lifetimes long enough to be transported long distances, such as PM_{10} and ozone. Hence, greened urban canyons may ultimately experience better air quality than in surrounding rural areas. The study by Pugh *et al.* (2012) also considered the use of street trees as well as green walls. They conclude that street trees must be considered on a case-by-case basis. In streets with low street-level emissions (i.e., light traffic), improvements in air quality are predicted (they term this 'the filtered avenue effect'). They go on to caution that if street-level emissions are high, however, tree planting must be used with the utmost caution. The specific combination of tree species, canopy volume, canyon geometry, and wind speed, and direction must be modelled on a case-by-case basis. By not considering the adverse effects of tree planting on canyon ventilation, urban greening initiatives that concentrate on increasing the number of urban trees, without consideration of location, risk actively worsening street-level air quality while missing a considerable opportunity for air quality amelioration.

Salmond *et al.* (2014) used an urban ES framework to evaluate the direct, and locally generated, ES and disservices provided by street trees. They focussed their study on the services of major importance to human health and well-being that included 'climate regulation', 'air quality regulation', and 'aesthetics and cultural services'. While recognising both benefits and dis-benefits, they conclude that street trees can be important tools for urban planners and designers in developing resilient and resourceful cities in an era of climatic change. Kiss *et al.* (2015) identified and evaluated two important regulating services of urban trees: carbon sequestration and air pollution removal in the city centre of Szeged (Hungary). The analyses revealed the main tendencies in differences between tree species considering the tree condition, which affects the service-providing capacity to a high degree. From their observations, clear cuts and complete tree alley changes are not advisable from an ES point of view.

Many other studies have been conducted on green roofs and walls as well as street trees and other green spaces. For example, Currie and Bass (2008) conducted an early study to quantify the contribution made by green walls and green roofs on air contaminant levels in an urban neighbourhood. Results of the study indicated that grass on roofs (extensive green roofs) could augment the effect of trees and shrubs in air pollution mitigation, whereas placing shrubs on a roof (intensive green roofs) would have a more significant impact. By extension, a 10%–20% increase in the surface area for green roofs on downtown buildings would contribute significantly to the social, financial, and environmental health of all citizens (Currie and Bass, 2008). In addition to the external benefits, green roofs provide insulation to buildings, resulting in reduced costs for heating and cooling. An experimental green roof on a municipal building in Benaguasil, Spain, resulted in a 20%–25% saving in electricity for the air-conditioning systems' power consumption of the building (Pérez-Navarro Gómez *et al.*, 2015). In urban areas, roofs, a critical part of the built environment, are highly susceptible to solar radiation and other environmental changes, thereby influencing indoor comfort and energy consumption. Figure 15.3 shows a green roof installed on a school building in Xativa, Valencia region, Spain.

Urban waterbodies are typically in receipt of diffuse pollution from a range of sources. While managing water in urban spaces is an essential infrastructure requirement, it is generally undertaken in isolation from other urban functions, and has not been considered in conjunction with other agendas, thereby missing opportunities for optimising multiple ES benefits. Because of the limits of space and the need to respond to key drivers (e.g. mitigation of diffuse pollution), more sustainable approaches to urban water management are being applied which can have multiple functions and benefits (Lundy and Wade, 2011). Many of these approaches to urban water management are driven by practical requirements, for example, water reuse and natural flood management or 'slowing the flow' generated by rainfall in impermeable urban environments to mitigate flooding/inundation from small-scale storm events. Depending on the location and pressures, the 'drivers' or incentives for sustainable water management will differ. For instance, drivers include the need to reduce the pressure on existing sewer infrastructure and wastewater treatment plants in terms of both capacity and treatment efficiency as well as the need to control urban diffuse pollution as part of the approach to achieving EU WFD objectives and to comply with the EU Floods Directive (EU, 2000; EU, 2007; Lundy and

Figure 15.3 Green roof on a school building in Xativa, Valencia region, Spain. The green roof provides insulation to the building, and additional ecosystem service benefits are realised through air pollution mitigation (augmenting the effect of the adjacent street trees), habitat provision, aesthetic values (neighbouring residents prefer the view of green, rather than a grey roof from their apartments). Photo credit: Rebecca Wade, 2013.

Wade, 2011) and the national Sustainable Flood Management policy (Scot Gov 2011).

Given that urban waterbodies receive urban surface water run-off, they are typically in receipt of diffuse pollution from a range of sources. Sustainable drainage systems (SuDS) are now commonly in use throughout the United Kingdom and parts of Europe for new developments and redevelopments, and they provide a level of protection to existing watercourses. The SuDS features are engineered, but some vegetated features can become ES providers, particularly if they are well designed, constructed, and maintained (Jose *et al.*, 2015). Figure 15.4 shows a stormwater management feature, a retention pond, in Ardler, Dundee, Scotland; the pond slows and retains the flow of surface run-off from the residential development. Sediments and pollutants are deposited in the pond before the water continues through the SuDS treatment train.

While poor-quality waterbodies may at first appear to exclude the generation of services such as the supply of genetic information, the presence of species able to tolerate and/or degrade elevated pollutant levels offers interesting opportunities in relation to, for example, the field of microbial bioremediation (Galvao *et al.*, 2005). The potential for plants, bacteria, and fungi which can degrade or immobilise organic and inorganic pollutants has received considerable attention (Desai *et al.*, 2010), with drivers such as the EU Environmental Quality Standards Directive (2008) providing a strong impetus for the development of cost-effective technologies to mitigate identified and emerging priority substances.

15.6 Conclusions

As urbanisation increases, urban pollution impacts on ES will also increase. While good

Figure 15.4 SuDS pond in Ardler, Dundee, Scotland. The pond provides regulating ES (flood reduction and water quality improvement) by slowing the flow of surface run-off from the surrounding residential development and by retaining the sediments and pollutants which are carried by the run-off. Additional ES are being provided by the pond; the water and vegetation are providing a habitat for nesting and visiting birds and for insects and pollinators. In addition, aesthetic and recreational ecosystem benefits are provided, as the local residents enjoy looking at and walking to the pond to view the birds (Jose *et al.*, 2015). Preferred views such as this can increase house prices adjacent to SuDS features. Photo credit: Rebecca Wade, 2016.

progress has clearly been made in evidence gathering, and in some areas of policy development and practice, more work effort is still required in order to achieve pollution reduction and ES provision. Urban planners, architects, and landscape architects, along with policymakers, need to continue their efforts to consider how natural resources and urban green infrastructure can be strategically developed and managed sustainably to meet the needs of urban populations. The valuation of ES (including the services associated with urban pollution reduction) can support urban decision-making and budget planning, and contribute to reduce costs and improve people's well-being.

Demuzere *et al.* (2014) are optimistic in their assessment of the impact of green infrastructure on climate change; they state that if green infrastructure continues to be applied in its current state, climate change adaptation benefits will be achieved, as long as it is based on a synergistic relationship with biodiversity conservation and ES efforts. This contribution marks progress towards a more climate-proofed future, but also marks an opportunity unfulfilled. The multi-functionality of green infrastructure characterises environmental policies that will make the most significant contribution in the future by being both politically popular and scientifically robust. It is also this multifunctional ability that currently underpins the majority of the potential weaknesses of green infrastructure. While achieving multiple objectives simultaneously may not be an unprecedented approach, it is clearly one that stakeholders at each level have yet to master. It requires effective interdisciplinary working and cooperation across policy sectors, and this in itself poses a perpetual challenge for environmental governance generally, not just for green infrastructure alone.

Reduction of pollution and emissions is surely the first and best approach to tackling the impacts of urban pollution on ES. In combination with a reduction in emissions, the ecosystem benefits that urban greening offers must also be recognised and enhanced. For instance, unlike exhaust-pipe-based emission reduction strategies, greening also offers multiple benefits, including reduced surface temperature and noise pollution, and increased biodiversity and amenity value (Pugh *et al.*, 2012). But it also offers challenges in ensuring vegetation health and minimising damage to non-green infrastructure (e.g. tree roots damaging underground water infrastructure pipes), and in recognising and minimising the dis-benefits as well as the benefits.

References

Baldocchi, D.D., Hicks, B.B., and Camera, P. (1987). A canopy stomatal resistance model for gaseous deposition to vegetated surfaces. *Atmospheric Environment*, 21, 91–101.

Bidwell, R.G.S. and Fraser, D.E. (1972) Carbon monoxide uptake and metabolism by leaves. *Canadian Journal of Botany*, Vol. 50, No. 7: pp. 1435–1439. 10.1139/b72-174

Currie, B.A. and Bass, B. (2008) Estimates of air pollution mitigation with green plants and green roofs using the UFORE model. *Urban Ecosystems*, 11, 409–422. doi 10.1007/s11252-008-0054-y. https://link.springer.com/article/10.1007/s11252-008-0054-y

Davies, L., Kwiatkowski, L., Gaston, K.J., Beck, H., Brett, H., Batty, M., Scholes, L., Wade, R., Sheate, W.R., Sadler, J., Perino, G., Andrews, B., Kontoleon, A., Bateman, I., and Harris, J.A. (2011) Chapter 10 Urban. In The UK National Ecosystem Assessment Technical Report. UNEP-WCMC, Cambridge.

DEFRA (2013) Ecosystems Services Guidance. Published online by Department for Environment, Food and Rural Affairs: Guidance documents: Environmental management (Assessing environmental impact). https://www.gov.uk/guidance/ecosystems-services

Demuzere, M., Orru, K., Heidrich, O., Olazabal, E., Geneletti, D., Orru, H., Bhave, A.G., Mittal, N., Feliu, E., and Faehnle, M. (2014) Mitigating and adapting to climate change: Multi-functional and multi-scale assessment of green urban infrastructure. *Journal of Environmental Management*, 146, 107–115. http://ac.els-cdn.com/S0301479714003740/1-s2.0-S0301479714003740-main.pdf?_tid=2c4388e8-2e80-11e7-a8c7-00000aab0f6b&acdnat=1493651549_2fc72a0dca2fdb7f8f17dc9a5be237fc

Desai, C., Pathak, H., and Madamwar, D. (2010) Advances in molecular and '-omics' technologies to gauge microbial communities and bioremediation at xenobiotic/anthropogen contaminated sites. *Bioresource Technology*, 101 (6), 1558–1569.

European Union (EU) Environmental Quality Standards Directive (2008) Directive 2008/105/EC of the European Parliament and of the Council of 16 December 2008 on Environmental Quality Standards in the Field of Water Policy. Available at: http://ec.europa.eu/environment/water/water-dangersub/index.htm

European Union (EU) Floods Directive (2007) Directive 2007/60/EC of the European Parliament and of the Council of 23 October 2007 on the Assessment and Management of Flood Risks. Available at: http://eur-lex.europa.eu/LexUriServ/LexUriServ.do?uri=OJ:L:2007:288:0027:01:EN:HTML

European Union (EU) Water Framework Directive (WFD) (2000) Directive 2000/60/EC of the European Parliament and of the Council of 23 October 2000 Establishing a Framework for Community action in the Field of Water Policy Water Framework Directive. Available at: http://eur-lex.europa.eu/LexUriServ/LexUriServ.do?uri=CELEX:32000L0060:EN:HTML

Galvao, T.C., Mohn, W.W., and Lorenzo, V. (2005) Exploring the microbial biodegradation and biotransformation gene pool. *Trends in Biotechnology*, 23, 497–506.

GCV Green Network Partnership (no date) Integrating Green Infrastructure. http://www.gcvgreennetwork.gov.uk/igi/introduction

Jose, R., Wade, R., and Jefferies, C. (2015) Recognising the multiple-benefit potential of SUDS. *Water Science & Technology*, 71.2.

Kallis, G. and Butler, D. (2001) The EU water framework directive: Measures and implications. *Water Policy*, 3, 125–142.

Kiss, M., Takács, A., Pogácsás, R., and Gulyás, A. (2015). The role of ES in climate and air quality in urban areas: Evaluating carbon sequestration and air pollution removal by street and park trees in Szeged (Hungary). *Moravian Geographical Reports*, 23, 3/2015. https://www.itreetools.org/eco/resources/int_research/2015Mar_Kiss_Szeged_Hungary_Eco_Journal.pdf

Kong, F., Yin, H., Nakagoshi, N., and Zong, Y. (2010) Urban green space network development for biodiversity conservation: Identification based on graph theory and gravity modeling. *Landscape and Urban Planning*, 95, 16–27.

Lundy, L. and Wade, R. (2011) Integrating sciences to sustain urban ecosystem services. *Progress in Physical Geography*, 35, 653. doi: 10.1177/0309133311422464

MA (2005) Millennium Ecosystem Assessment, Ecosystems and Human Well-Being: Synthesis, 2005, Washington, DC: Island Press.

Maes, J., Polce, C., Zulian, G., Vandecasteele, I., Perpina, C., Rivero, I., Guerra, C., Vallecillo, S., Vizcaino, P., and Hiederer, R. (2017) Ecosystem services mapping. Chapter 5 in Benjamin Burkhard and Joachim Maes (eds.), *Mapping Ecosystem Services*. Pensoft Publishers, Bulgaria.

Pérez-Navarro Gómez, Á., Morales Torres, A., Peñalvo López, E., Andrés Doménech, I., Alfonso Solar, D., Perales Momparler, S., and Pablo Peris García, P. (2015) Report on the Green Roof Monitoring in Benaguasil. Part of the EU MED E^{2}STORMED PROJECT (Improvement of energy efficiency in the water cycle by the use of innovative storm water management in smart Mediterranean cities) www.e2stormed.eu

POSTnote 281. (2007) Ecosystem Services. March 2007.

POSTnote 376. (2011) Natural Capital Accounting. May 2011.

POSTnote 542. (2016) Natural Capital: An Overview. December 2016.

Pugh, T.A.M., MacKenzie, A.R., Whyatt, J.D., and Hewitt, C.N. (2012) Effectiveness of green infrastructure for improvement of air quality in urban street canyons. *Environmental Science & Technology*, 46, 7692–7699. http://citiesalive.greenroofs.org/resources/GreenInfrastructurePaper.pdf

Rogers, C.A., Wayne, P.M., Macklin, E.A., Muilenberg, M.L., Wagner, C.J. *et al.* (2006) Interaction of the onset of spring and elevated atmospheric CO_2 on ragweed (Ambrosia artemisiifolia) pollen production. *Environmental Health Perspectives*, 114, 865–869.

Salmond, J.A., Tadaki, M., Vardoulakis, S., Arbuthnott, K., Coutts, A., Demuzere, M., Dirks, K.M., Heaviside, C., Lim, S., Macintyre, H., McInnes, R.N., and Wheeler, B.W. (2014) Health and climate related ecosystem services provided by street trees in the urban environment. From The 11th International Conference on Urban Health Manchester, UK. 6 March 2014. *Environmental Health*, 2016, 15 (Suppl 1):36. doi 10.1186/s12940-016-0103-6. https://ehjournal.biomedcentral.com/articles/10.1186/s12940-016-0103-6

Säumel, I., Weber, F., and Kowarik, I. (2016) Toward livable and healthy urban streets: Roadside vegetation provides ecosystem services where people live and move. *Environmental Science & Policy*, 62, 24–33.

Scot Gov (2011) The Flood Risk Management (Scotland) Act 2009 – Sustainable Flood Risk Management – Principles of Appraisal: A Policy Statement. http://www.gov.scot/Publications/2011/07/20125533/0

URBES (2013) Valuing Ecosystem Services in Urban Areas. URban Biodiversity and Ecosystems Services (URBES) Project. Factsheet 3. www.urbesproject.org

Wentworth, J. (2017) Urban Green Infrastructure and Ecosystem Services. POST brief Number 26, July 2017.

WHO (2016) Ambient (Outdoor) Air Quality and Health. Fact sheet. http://www.who.int/mediacentre/factsheets/fs313/en/

Wise, S. (2008) GI Rising – Best Practices in Stormwater Management. American Association of Planners. August/September 2008.

16

Greywater Recycling and Reuse

Katherine Hyde and Matthew Smith

School of Construction Management and Engineering, University of Reading, United Kingdom

16.1 Introduction

The generation of greywater from human bathing and washing is as ubiquitous as the generation of sewage. Untreated or 'raw' greywater is a wastewater that is relatively lightly loaded in organic pollutants, and commonly comprises approximately one third of all human wastewater arising in many developed countries. In traditional and existing drainage systems in the built environment, either inside or outside of properties, all untreated greywater is combined and mixed with more polluted wastewaters flowing into the sewer. More polluted wastewaters can include sewage and faecal matter from toilets, dishwasher discharges, food waste, garage drainage, oils, and other effluents. Due to this mixing of wastewaters, the sewage causes degradation in the quality of the less polluted greywater.

The opportunity to separate out the less polluted greywater is thus foregone. This standard practice means that any original volumes of untreated greywater subsequently require more intensive treatment at the wastewater treatment plant (WTP) than they would have done had they not been discharged into the mixed sewer environment. In many respects, it may be significantly more environmentally prudent, CO_2 emission reducing, and resilience enhancing to maintain the separation between polluted sewage and less polluted greywater flows.

Furthermore, as more greywater is treated, recycled, and reused, the volume of pristine mains water required for basic functions such as toilet flushing and garden watering is reduced (see Hatfield *et al.*, 2014). This leads to better overall urban water cycle efficiency, and in certain circumstances, may also lead to carbon savings and system cost reductions. Unfortunately, influencing changes in wastewater system configurations is currently not predicated in strategy and policy in the water industry at local or national levels.

There is an option for utilising greywater for irrigation, and this has potential for sustaining green infrastructure during times of drought. While this principle achieves wide, cursory acceptance and would be unlikely to be rejected during an intensely dry period, significant barriers exist to regular irrigation using greywater.

16.2 The Ubiquitous Nature of Pollutants in Wastewater from Baths, Showers, and Handbasins

So far, it is assumed that all wastewater from bathrooms remains in the piped urban environment, whether being discharged to the

Urban Pollution: Science and Management, First Edition. Edited by Susanne M. Charlesworth and Colin A. Booth.

foul sewer or being recycled in a building's plumbing system.

In fact, it is not necessarily an environmental offence to discharge small volumes of treated bathroom greywater directly to the environment, provided they are not polluting. For example, treated effluent from a septic tank or a small sewage treatment plant can be discharged into the environment as long as the discharge point does not lie within a Source Protection Zone 1, and is less than 15,000 litres per day (EA, 2016).

This approach allows treated greywater to be applied to kitchen gardens, orchards, allotments, and so on. Clearly, this must be carefully controlled because any exceedances of limits and quality standards would attract a legal warning and possibly a penalty or fine from the Environment Agency. Legality under pollution prevention legislation and regulations depends on the greywater source, nature, and volume/ flow characteristics (see Section 16.9). Furthermore, the concept of discharging untreated greywater directly to the environment for the purpose of irrigating green infrastructure might sometimes be seen as unpalatable in environmental quality terms, even if the pollutant concentrations in the discharge are low. Visual and sensory observations, such as smell, colour, and clarity may occasionally be useful as an 'early warning indicators', but are rarely accurate or scientific.

One of the practical problems is that if, for example, a basic in-line filter is installed to treat discharges of shower and bath water, it may not be possible to monitor the quality of the treated greywater. In order to reach a reliable view about treated water quality and compliance with standards and regulations, appropriate monitoring and recording equipment is essential. There is a range of equipment available from basic monitors for pH, to more competent monitoring and recording instruments. Even the lowly pH measurement, when taken on a regular basis, can assist the householder in becoming familiarised with the performance of a basic filter. In some cases, this may be sufficient to alert the homeowner about the start of deterioration in the normal performance of the filter, thus triggering maintenance checks and remediation.

One of the reasons for raising this important question is that, in cases where chemical and microbiological analysis has been undertaken, the analysis demonstrates the degree of safety in discharging the treated greywater into the environment. Finley *et al.* (2009) found no significant difference in contamination levels between crops irrigated with tap water, untreated greywater, and treated greywater. They stated that faecal coliform levels and faecal streptococcus levels were highest on carrots and lettuce leaves, respectively, but that contamination for all the crops they tested were low, and constituted an insignificant health risk.

16.3 The Quality of Untreated Greywater and Its Water Resource Value

The integration of physical and environmental sciences combined with socio-economic factors and socio-technological behaviours and practices is fundamentally influential when considering anthropogenic engagement in greywater recycling and reuse.

Human engagement may often exhibit polarised reactions, in the form of enthusiastic alignment with the principles and with the environmental benefits gained through the recycling of water. In contrast, disengagement may be observed, for example, in the inherent right of access to pristine water for all needs, or because water is "free" (rain arrives without any payment), and therefore, wastage of water appears to some degree, to be inconsequential. Possibly more complex reactions present unwillingness to use recycled greywater based upon religious reasons, or upon health or safety concerns (Hyde *et al.*, 2016). These views are largely founded upon perception, though often scientifically quantifiable.

For this reason, it is becoming more pressing to engage with customers about the sharing of information on water quality. In future, as customers become more familiar with water quality parameters such as acidity and alkalinity, nitrate (NO_3), biochemical oxygen demand (BOD), and more, there is likely to be increased dialogue between customers concerning water quality results and suppliers.

16.4 Greywater Terminologies Used in this Chapter

One important point of terminology relates to definitions of different greywater flows. Most importantly, all existing flows from handbasins, baths, and showers comprise untreated greywater at their point of discharge to the collecting waste pipes. In this chapter, the use of the term 'treated greywater' is used as consistently as possible to distinguish it from greywater that has not been treated, filtered, or chlorinated.

16.5 Pollutants in Untreated Greywater

The quality of greywater relates directly to the source from which the greywater originated and the uses to which it has been put. In some cases, relatively clean handwash water may often contain few other constituents than the supply water from tap(s), the surface dirt and bacteria from hands, the dispensed hand soap or steriliser, and other common constituents from the use of handbasins. Most of the domestic products used for cleaning utilities and floors, plus the wastewater from the cleaning regimes in bathrooms will be disposed of either via the waste pipe(s) from the handbasins, which could supply a greywater system; or alternatively via the toilets to sewer.

Circumstantial evidence regarding bathroom behaviour that influences pollutant loading in greywater is sometimes available. This often has to be excluded from rigorous data collection and analysis. Commonly, bathroom behaviours are private and good data may be difficult to elicit. Greywater polluting events could be more commonplace than reported. To provide one example, babies 'might be bathed' in handbasins, an activity that is likely to give rise to higher levels of bacteria in the basin wastewater. This leads to one important observation about raw greywater, which is that, while assumptions about the average pollutant levels of a typical raw greywater from one specific location in one specific building may be reasonable, the actual distribution and frequency of variation in the pollutant concentrations can rarely be assumed. They can be predicted only in generalities, or after data sampling and analysis at specific sites.

On the other hand, since any waste stream may demonstrate consistently low pollutant concentrations after being subjected to a robust treatment process, so may be the case with greywater. Indeed, this is what is required to make greywater resources increasingly useful, by producing a known, reliable, and acceptable water quality. This is further explored in Section 16.6.

Shower and bath wastewater give rise to pollutants of the types already mentioned and also exhibit a range of constituents and concentrations from bath oils, shampoos, skin beauty treatments, hair and hair dyes, to pharmaceutical preparations, medical and wound treatments, and small solid objects. In some buildings, muddy clothes and outer garments are washed in bathrooms. Also, dust and substances from various trades such as brick and plaster dust, paint, agricultural dusts, liquid treatments used in any trade including pesticides and repellents, oils, grease, and fuels, and many other substances may be encountered. Many workers in trades do not return from site to the office at the end of their working day, and consequently, washing and showering at home introduces these typical occupational contaminants into greywater. Without further evidence, it might be assumed that such pollutant concentrations arising could be relatively low.

A number of basic tests can be conducted to check the variability of certain key quality characteristics of any greywater. Although it is preferable to know the concentrations of a full range of quality parameters of the greywater being discharged, on a routine basis this very often has to be reduced to a few parameters that can be monitored easily by the use of hand-held probes and indicator solutions. Most domestic greywaters are not 'regulated' (Environmental Permitting Regulations, 2010), in the sense that they are not required by UK regulations to be tested daily or weekly according to a frequent and specific timetable, and there is no requirement to keep daily quality records. Where discharges are regulated, there is likely to be a more rigorous regime in place.

In experiments using greywater, some authors such as Sawadogo *et al.* (2014) have used untreated rather than treated greywater, while others such as Pinto *et al.* (2010) and Christova-Boal *et al.* (1996) have tested various categories of greywater, which can lead to misunderstanding of what greywater actually is. Wiel-Shafran *et al.* (2006) tested greywater that included a significant proportion of laundry detergent solution, typical of washing machine wash cycles. Washing powders and solutions often contain aggressive ingredients, and as a consequence, some of the greywater currently recycled in the United Kingdom excludes laundry detergents. Most recently published studies have focused on greywater collected only from handbasins, baths, and showers, thus excluding greywater obtained from washing machines. In summary, this is a critical differentiator since studies using only bathroom and handbasin greywater will derive their results from greywaters containing lower concentrations of surfactants and salts than are used in washing machines. Avoiding the use of wastewater from kitchen sinks ('sullage') avoids problems associated with, for example, the accumulation of fats, oils, and greases, known in the water industry as F.O.Gs.

At a local scale, it is theoretically possible for households to control their own greywater quality by controlling the volumes of water used for showers, baths, and handbasins, and by controlling the amounts of soap and other personal hygiene products that are used. In practice, this requires procedures currently too onerous for an average household to control on a regular basis.

16.6 Standardising Greywater Treatment Systems: Removing and Minimising Pollutant Concentrations

When designing systems for treating and recycling greywater, the quality of the untreated greywater should be assumed to be contaminated, in the absence of definitive data. The terminology used in this chapter follows, as far as possible, that set out in BS: 8525 Parts 1 and 2 (2010 and 2011, respectively).

In order to facilitate discussions about wider reuse, greywater treatment will be discussed, reviewed, and evaluated where it is relevant to the development of integrated urban systems. Underlying the scientific approach is the general baseline assumption that greywater contains more pollutants than 'clean' water. In the United Kingdom, clean water is usually taken to be piped mains water delivered to consumers at potable and drinking water standards for human consumption.

There are a number of attributes of greywater quality and greywater supply that require scrutiny before greywater can be recycled, treated, and reused. Some of the pollutants found in greywater are also present in mains water, although usually at lower concentrations in the latter. For this reason, the reference term 'constituents' is preferable to the use of the term 'pollutants' within controlled environments such as in-building water systems. This will be explained according to the principles of constituent concentration profiles, whereby a range of concentrations is measured during longitudinal studies, for example, while being compared to the variability of a number of other factors. Once outside the built environment, greywater starts to mix with other sources of water, and here, greywater constituents are generally more accurately described as pollutants.

The Environment Agency's guidance and requirements (EA, 2016) must be followed with respect to the use and applications of greywater. The guidance requires an environmental risk assessment to be undertaken as a critical precursor prior to any greywater discharge. While the specific pathways for particular discharges of treated greywater must be subjected to individual assessment, as described in the EA guidance (2016), the following points have been noted in relation to the discharge of domestic greywaters. First, sensitive receptors must be identified and a risk assessment undertaken. Receptors include any places used to grow food, or to farm animals or fish; fields and allotments used to grow food; drain and sewer systems; local groundwater protection zones or Nitrate Protection Zones; homes or groups of homes; private drinking water supplies; schools, hospitals, or other public buildings; playing fields and playgrounds; conservation and habitat protection areas, including Sites of Special Scientific Interest (SSSIs); ponds, streams, rivers, lakes, and the sea. Second, although a treated greywater is likely to be of good quality, there should be no hesitation in comparing its constituents with those specified for the minimum requirements for treated sewage or trade effluents when the discharge compliance arrangements are being considered. In these cases, risks to surface water from hazardous pollutants, risks from "sanitary" and other pollutants, and risks to groundwater must all be considered (EA, 2010).

Third, once the potential receptors have been identified, the applicant has to state what the potentially consequential risks are; these include uncontrolled, unexpected, and unintended emissions as well as regular and frequent discharges. Where no such risks are relevant to the circumstances, the applicant must state that they are not relevant and may have to provide substantiating evidence, which may include an exercise to screen out potential risks (further details given in Section 16.9) by testing whether the discharges are likely to be of acceptable environmental standards or discharge limits. Constituents arising from washing and bathing include personal care products as sold over the counter. These might broadly present a similar toxicological risk to those encountered during product use itself, although the concentrations of exposure and incidental mixtures with other pollutants will be different. Personal care products sold in the United Kingdom provide details of the ranked order of constituents according to proportion, from the highest to the lowest, but rarely provide specific concentrations. The allowable discharge concentration for each substance or determinand to achieve 'no deterioration' in the environment must be calculated by the applicant. This condition for achieving no deterioration is not optional. Each applicant must calculate or determine appropriate limits for the discharge that are environmentally protective, technically feasible, and whose cost is proportional to potential benefits. For new discharges, the overall polluting load must be managed so that no individual concentration of ammonia, phosphorus, or BOD increases. Where this is not feasible or cost-effective, the Environment Agency may allow a within-class deterioration of up to 10%. Applicants for new discharges must be very careful to check that recent surface water quality concentrations have not improved since figures were last published, leading to potentially unexpected water quality impacts.

Personal care products must not contain substances that are pathogenic or capable of producing disease. Thus, they are manufactured by including either bacteriocidal or bacteriostatic constituents so that the product is not capable of supporting the proliferation or transference of bacteria. However, one pollutant of note due to increasing evidence of its environmental persistence in wastewaters is Triclosan, a bacteriocide commonly included as an ingredient in cleaning and personal care products (Ricart, 2010).

16.7 Managing the Environmental Characteristics, Applications, and Urban Uses of Treated Greywater

Sawadogo *et al.* (2014) conducted tests and evaluations on plant growth during irrigation with untreated greywater. They focused on irrigation using laundry water at low, medium, and high concentrations of detergent, whereby the highest concentration was a worst-case scenario in comparison with concentrations normally found in greywater from baths and showers. The high concentrations of detergent produced adverse effects, causing death in both lettuce and okra, whereas medium to low concentrations produced between low to no-discernible effects. Wiel-Shafran *et al.* (2006) reported that irrigation using insufficiently treated greywater is 'mistakenly considered safe'. They suggested that known surfactant concentrations in greywater range from 0.7 to 70 mgL^{-1} and that accumulation in soils can cause hydrophobia in soils, thus affecting soil productivity.

However, both Wiel-Shafran *et al.* (2006) and Lado and Ben-Hur (2009) state that lightly loaded greywaters are more reliable in terms of avoiding potentially damaging, unwanted polluting effects for crops and soils. Thus, if the inorganic and organic loading of greywater can be reduced by treatment, then that is likely to lead to less significant effects on crops. Further research is needed to determine the appropriate extent of treatment required for greywater from different sources that results in a quality which reduces adverse impacts and encourage crop growth.

Sawadogo *et al.* (2014) reported that surfactants in irrigation waters containing detergents have been recognised as a major contributor in the reduction of hydraulic conductivity of soils. In the cases of irrigation water with higher detergent concentrations, more advanced soil degradation can occur, leading to water-repellent soils. Lado and Ben-Hur (2009) showed that these have negative impacts on agricultural productivity and environmental sustainability. The question of the application of dilute greywaters for sustaining green infrastructure appears less widely understood.

Pinto *et al.* (2010) reported the key chemical characteristics of potable and greywater quality used in their irrigation and growth trials of silverbeet, over a period of 60 days (see Table 16.1).

Christova-Boal *et al.* (1996) applied greywater with a pH in excess of 8.0 during some of their growth tests. The authors observed that greywater has the potential to increase the soil alkalinity if applied to gardens for a long period. Pinto *et al.* (2010) reported pH results in excess of 10.0 and described the phytotoxicity arising from

Table 16.1 Greywater and tap water quality used in irrigation and growth trials (Pinto *et al.*, 2010).

Sample	pH	Electrical conductivity μS/cm	Total nitrogen, mg/L	Total phosphate, mg/L
Greywater	10.5	1358.0	0.2	4.4
Potable	7.0	277.0	0.16	0.0

greywater reuse as being principally due to anionic surfactants altering rhizosphere microbial communities. Such phytotoxicity effects arising from greywater reuse demonstrate significant variability of impact according to plant species.

In Table 16.2, the authors Sawadogo *et al.* (2014) provided electrical conductivity (EC) results in comparison to low, normal, and high concentrations of surfactant, expressed in terms of mgL^{-1} of linear alkylbenzene sulphonate (LAS). Of those three surfactant compositions, the greywater applied in Pinto *et al.* (2010) aligns approximately with the EC of the normal concentration surfactant used in Sawadogo *et al.* (2014). In the latter case, the pH, 9.9, of the normal concentration surfactant also lies in a reasonably comparable range to the pH 10.5 greywater applied by Pinto *et al.* (2010). The increasing concentrations of detergent LAS gave rise to increasing pH and EC values, as well as increasing the values of total N, C, and P.

16.8 University of Reading's 2016 Experimental Irrigation of Sedum using Treated Greywater

The University of Reading monitored growth trials of sedum grown in floor-standing boxes irrigated using bathroom greywater for a 6-month period during 2015–2016. The statistical mean values calculated from six-months of data are shown in Table 16.3. Irrigation was conducted using untreated greywater, treated greywater, and a mains tap water control. This enabled the assessment of analytical variables that were likely to produce any potential growth effects to be specifically attributed to irrigation. The chemical characteristics of the greywater were based on the British Standard 'recipe' for synthetic test greywater, which constituted a 'lightly loaded' greywater quality according to the EC and TP results, although

Table 16.2 Constituents in greywater tests (Sawadogo *et al.*, 2014).

Watering solutions	Distilled water	Low concentration surfactant, 0.1 g L^{-1}	Normal concentration surfactant, 1.0 g L^{-1}	High concentration surfactant, 5.0 g L^{-1}
pH	6.9	9.1	9.9	10.2
EC (mS cm^{-1})	28	159	1082	4870
Detergent as LAS (mg L^{-1})	ND	13.5	135.6	678
N_{total} (mg L^{-1})	ND	0.01	0.12	6.6
C_{total} (mg L^{-1})	ND	15.3	153	765.1
P_{total} (mg L^{-1})	ND	13.2	132.3	661.6

LAS = linear alkylbenzene sulphonate; ND = not detected; EC = electrical conductivity.

Table 16.3 Treated greywater and tap water quality monitored from 02 September 2015 to 02 March 2016.

Sample	pH	Electrical conductivity, μS/cm	Total nitrogen, mg/L	Total phosphate, mg/L
Treated greywater	7.6	669.9	4.5	0.02
Potable water	7.8	520.2	0.5	0.01

the TN is significantly higher than the TN used in Pinto *et al.* (2010).

The potable (mains) water comprised the largest component of the synthetic greywater and so, unsurprisingly for a lightly loaded greywater, the pH of the tap water and the pH of the greywater were very similar, as were the phosphate results in both cases. The EC was higher than that of the potable water, which was mainly due to the organic and inorganic loading associated with the presence of surfactants such as detergents or shampoo, as well as the constituents of the treated sewage final effluent (FE). The total nitrogen was also higher in the greywater than in the potable water due to the nitrogen content of the treated sewage effluent included in the synthetic greywater recipe, as well as the surfactants.

Treated sewage effluent was not added because:

1) The synthetic mix required a source of pollutants containing similar components to shower greywater, which follows the UK standard for Bacteriological Examination test procedures that would be difficult or impossible to replicate.
2) The test procedure needed to challenge the greywater treatment system to demonstrate that higher bacterial loads were safely eradicated from the treated greywater by a satisfactory process.

16.9 Soil Results Evaluated during Irrigation using Greywater Constituents

In Sawadogo *et al.* (2014), the plant growth parameters in okra and lettuce were measured every week, including the dry weights of stems and leaves of all plants. Soil pH and soil EC tended to increase as the detergent concentrations increased in the irrigation solutions. There were no significant differences in okra fruit growth (fresh and dry weight) using distilled water, low concentration (LC), and normal concentration (NC) treatments. However, plants in HC treatments died 20 days after planting (DAP). No significant difference was noticed in lettuce shoots (dry weight) between LC, NC, and DW treatments, but lettuce in HC treatments died 12 DAP. EC significantly increased in all the treatment regimes. Laundry detergent can inhibit plant growth, and the application of greywater containing high concentrations of detergent can increase soil salinity.

In the University of Reading's study, the sedum plants showed a gradual decrease in size and plant health during the winter months which was found regardless of the test category of irrigation water, whether greywater, treated greywater, or mains (potable) water. With the return of average ambient temperatures above approximately 10°C, the plants in all the irrigation categories and soil types exhibited more leaf growth. In test sedum boxes kept in the greenhouse during the first 1.5 months of the trials, a significant increase in plant development and growth volume was found in comparison with plants grown outside. Subsequently, the greenhouse-grown plants were moved outside in order to compare their performance during UK winter weather conditions with those plants that were already outside. The health of the latter group of plants deteriorated more quickly than that of the plants that had been outside from the beginning of the trials.

Sodium is an environmental pollutant when it is present at concentrations above background levels (and in some cases, at concentrations at background levels). Figure 16.1 shows that all plants had similar leaf sodium concentrations at the start of the trials, which reduced during the first two months and did not subsequently recover.

16.10 Applying the Principles of Controlled Waters to Greywater Discharges for Sustaining Green Infrastructure

Evidence from studies reviewed in this chapter has established that greywater can affect

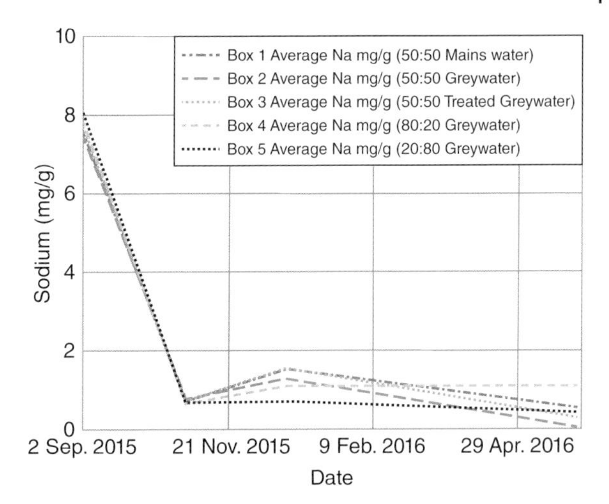

Figure 16.1 Longitudinal changes in plant leaf sodium concentrations for differing watering regimes; '50:50' refers to a plant substrate of 50 parts growing medium to 50 parts compost; '80:20' refers to a plant substrate of 80 parts growing medium to 20 parts compost; and '20:80' to a plant substrate of 20 parts growing medium to 80 parts compost. (*See colour plate section for the colour representation of this figure*)

crops and other plants when higher concentrations of detergent are applied, at the highest limit even causing plant death. The evidence from the Reading trials demonstrated that impacts from watering sedum using dilute greywater gave rise to only limited impacts during a winter period when the plants were stressed due to low ambient temperatures.

If significantly less impacts arising from low concentrations of detergent could be demonstrated, consideration would then be given to the concept of allowing treated greywaters to be used for watering green infrastructure. The Environment Agency (EA) (2016) states that risk assessments are not required for greywater discharges from domestic properties unless:

- a trade discharge is included in the effluent, or
- there is a discharge to ground or surface water of >15 m^3 per day, or
- the discharge to ground is more than 2 m^3 per day, and the location is in a groundwater Source Protection Zone, SPZ1 (an area of highest risk to groundwater quality).

The EA (2016) also does not require screening tests if the water is discharged to the same river or groundwater that the water was originally taken from or if no hazardous pollutants are added to the water. This is potentially of assistance to small greywater discharges that contain no hazardous pollutants, since the greywater discharge could potentially support green infrastructure in the same area of the catchment.

The stages required for screening for a particular discharge include:

- Identification of the pollutants released from the greywater treatment plant or source
- Gathering data on the environmental impacts of the pollutants prior to screening
- Undertaking the screening itself

This is likely to apply to discharges of greywater containing surfactants. Pollutants must be measured if they are hazardous and released to freshwaters, estuaries, and coastal waters, or to sewers. EA (2016) specifies that pollutants could be present in the discharge if:

1) They have been detected by chemical analysis.
2) They are allowed to be added to the discharge (e.g. water company trade effluent consent or discharges from installations).
3) They have been added to the discharge by means of a treatment process (e.g. using iron or aluminium to remove phosphorus).

Membrane greywater treatment systems (Hatfield *et al.*, 2014) do not introduce chemicals into the greywater as part of the treatment process; thus, its quality, once treated could therefore comply with EA requirements. However, bathing and personal hygiene products in greywater may require routine testing to demonstrate that the greywater is of a suitable quality for reuse.

The average flow rates, times, and duration of discharge, for example, 12 hours per day, will be required to be stated for assessing and monitoring compliance. The risks arising at the site are assessed, including the risks of the discharge creating environmental pollution and the sources of the risks. The assessment steps required by EA (2016) include:

1) Identification of the receptors at risk from the site (people, animals, property, etc.)
2) Identification of the possible pathways from the sources of the risks to the receptors
3) Assessment of the relevant risks to the specific activity; check that they are acceptable
4) Statement of the plans for controlling the risks if they are unacceptably high

The risk assessment must be submitted as part of a permit application, and a copy must be included in the management system. The criteria for unsatisfactory overflows include operating a breach of permit conditions and/or causing a breach of water quality standards and other regulatory standards EA (2016).

16.11 Concluding Comments and Review

Separating greywater before it combines with sewage and laundry effluents can be used to sustain urban environments and their green infrastructure by irrigation using treated greywater. An analytical approach to the quality and reuse opportunities for greywater leads to the determination of water reuse and irrigation potential, feasibility, and water stewardship. The evidence from various studies shows that lightly loaded greywaters can probably be used with confidence for watering certain types of green infrastructure. This is an important potential means for sustaining vegetation during times of water stress or drought.

For larger volume discharges, affordable system configurations for achieving licensable discharges to support plants need to be identified and approved. For smaller, domestic-scale discharges, it appears that there is often little legislative barrier to direct irrigation using treated greywater.

Rationalisation of the technology configuration, and thus the potential costs to be incurred, will provide an opportunity to carefully design and implement landscaping specifically to work in conjunction with irrigation systems. This will reduce levels of pollutants entering the urban environment while establishing and developing the opportunity for more greywater sources to be utilised in a sustainable manner.

Acknowledgements

1) European funding from Climate-KIC is acknowledged. The study sponsors had no involvement in the study design, collection, analysis, and interpretation of data or in the writing of the manuscript.
2) The EA, United Kingdom, for extracts from www.environment-agency.gov.uk.

References

Christova-Boal, D., Eden, R.E., and Mcfarlane, S. (1996) An investigation into greywater reuse for urban residential properties. *Desalination*, 106, 391–397.

Environment Agency. (2016) Risk Assessments for Your Environmental Permit; Environmental Management-Guidance. First published 1 February 2016. https://www.gov.uk/guidance/risk-assessments-for-your-environmental-permit.

Environmental Permitting Regulations 2010 relate to the regulation of discharges to controlled waters; The Environmental Permitting (England and Wales) Regulations 2010 No. 675. Hansard.

Finley, S., Barrington, S., and Lyew, D. (2009) Reuse of domestic greywater for the irrigation of food crops. Water, Air and Soil Pollution, 199(1), 235–245. doi:10.1007/s11270-008-9874-x

Hatfield, E., Booth, C.A., and Charlesworth, S.M. (2014) Greywater harvesting – Reusing, recycling and saving household water. In: Booth, C. and Charlesworth, S.M. (eds.), *Water Resources in the Built Environment – Management Issues and Solutions.* Wiley Blackwell.

Hyde, K., Smith, M., and Adeyeye, K. (2016) Developments in the quality of treated greywater supplies for buildings, and associated user perception and acceptance. *International Journal of Low-Carbon Technologies*, 0, 1–5.

Lado, M. and Ben-Hur, M. (2009) Treated domestic sewage irrigation effects on soil hydraulic properties in arid and semi-arid zones: A review. *Soil and Tillage Research*, 106 (1), 152–163.

Pinto, U., Maheshwari, B.L., and Grewal, H.S. (2010) Effects of greywater irrigation on plant growth, water use and soil properties. *Resources, Conservation and Recycling*, 54, 429–435.

Ricart, M. *et al.* (2010) Triclosan persistence through wastewater treatment plants and its potential toxic effects on river biofilms. *Aquatic Toxicology*, 100 (4), 346–353. 10.1016/j.aquatox.2010.08.010

Sawadogo, B., Sou, M., Hijikata, N., Sangare, D., Maiga, A.H., and Funamizu, N. (2014) Effect of detergents from greywater on irrigated plants: Case of Okra (*Abelmoschus esculentus*) and Lettuce (*Lactuca sativa*). *Journal of Arid Land Studies*, 24 (1), 117–120.

Wiel-Shafran, A., Ronen, Z., Weisbrod, N., Adar, E., and Gross, A. (2006) Potential changes in soil properties following irrigation with surfactant-rich greywater. *Ecological Engineering*, 26, 348–354.

17

Containment of Pollution from Urban Waste Disposal Sites

Isaac I. Akinwumi[1], *Colin A. Booth*[2], *Oluwapelumi O. Ojuri*[3], *Adebanji S. Ogbiye*[4], *and Akinwale O. Coker*[5]

[1] *Department of Civil Engineering, Covenant University, Nigeria*
[2] *Architecture and the Built Environment, University of the West of England, United Kingdom*
[3] *Department of Civil & Environmental Engineering, Federal University of Technology, Akure, Nigeria*
[4] *Department of Civil Engineering, Covenant University, Nigeria*
[5] *Department of Civil Engineering, Faculty of Technology, University of Ibadan, Nigeria*

17.1 Introduction

The last century has witnessed a tenfold rise in global waste generation (Hoornweg *et al.*, 2013). There is now an estimated 1.3 billion tonnes of municipal solid waste (MSW) generated annually worldwide, and this is expected to rise to 2.2 billion tonnes by the year 2025 (Hoornweg and Bhada-Tata, 2012). So what happens to wastes when they are disposed?

Waste management is needed to protect both the environment and public health (Bartkowiak *et al.*, 2016). A waste hierarchy can be used to rank the available management options (Department for Environment, Food and Rural Affairs [DEFRA], 2011). The hierarchy attempts to prioritise the prevention and minimisation of waste. Waste reduction, reuse, recycle, and recovery can be classified as waste diversion techniques. Waste disposal is the last and least-favoured option in the waste hierarchy. In order of increasing preference or decreasing environmental impact, it involves dumping, incinerating (with energy recovery), or landfilling of the waste.

Waste reduction, reuse, recycle, value recovery, and deposition per capita are greater in high-income countries than in low-income countries; whereas dumping of wastes predominates in most low- and lower-middle-income countries (Hoornweg and Bhada-Tata, 2012). Collected waste in Africa is almost entirely disposed of without any consideration of the waste hierarchy, while <40% of waste collected in member countries of the Organisation for Economic Co-operation and Development (OECD) gets eventually disposed of (Hoornweg and Bhada-Tata, 2012). This is partly because waste sorting at the sources of their generation is not strictly implemented and enforced in Africa. Many developing countries have limited capacity to manage waste, and this contributes to diseases and deaths (UNEP, 2011). For instance, the 2016 Lassa fever outbreak in some cities in Nigeria resulted in over 130 fatalities, according to the Nigeria Centre for Disease Control (NCDC). If wastes had been well managed (especially well disposed of) and urban sanitation had been taken seriously, the disease vector (rats) would have had little or no contact with humans, and these casualties would have been fewer. Given the seriousness of the problem in developing countries (Ogundipe

Urban Pollution: Science and Management, First Edition. Edited by Susanne M. Charlesworth and Colin A. Booth.

and Jimoh, 2015), it is therefore, essential that the planning, design, construction, and management of waste disposal facilities are effective in protecting society and the environment. The environmental impact of improper waste disposal does not only affect public health but also its socio-economic activities (Dagiliūtė and Juozapaitienė, 2015). Groundwater, a valuable source of drinking water, can become contaminated from improper disposal of wastes. According to the United Nations Development Programme (2006), about half of the hospital beds in the world are occupied by people suffering from water-related illnesses. Therefore, productive man-hours are being lost to sick leave taken by working or business class people.

The focus of this chapter is to highlight the challenges of managing urban waste, with particular emphasis on the contribution of improper waste disposal to pollution and how landfill liners are essential in protecting the environment and public health. It presents a brief overview of worldwide solid waste generation, together with the implications of improper waste disposal and how the lining of landfills can help to minimise the environmental impacts.

17.2 Generation of Waste Worldwide

It has been estimated that the worldwide annual waste generated is about 2.12 billion tonnes (The World Counts, 2016). "If all this waste was put on trucks, they would go around the world 24 times" (The World Counts, 2016). Using the world population estimate of 7.4 billion people, this means each person (on the average) generates 0.78 kg of waste per day.

According to UNEP and ISWA (2015), 50% of the total waste generated worldwide in 2012 is from developed countries. Member countries of the OECD generate the largest quantity of wastes – together, generating around 1.75 million tonnes per day (Hoornweg *et al.*, 2013), while Africa and South Asia generate the least quantities of waste (Hoornweg and Bhada-Tata, 2012). Figure 17.1 presents the mass of MSW generated per person per day for the regions of the world, in accordance with values provided by Hoornweg and Bhada-Tata (2012).

Figure 17.1 shows that the countries of the OECD generate the highest MSW, while South Asia and Sub-Saharan Africa generate a relatively lower MSW. According to World Economic Forum (2015), the countries generating the largest masses of MSW per capita per day include the following: Trinidad and Tobago (14.40 kg), Kuwait (5.72 kg), Antigua and Barbuda (5.5 kg) and St. Kitts and Nevis (5.45 kg), Guyana (5.33 kg), Sri Lanka (5.10 kg), Barbados (4.75 kg), St. Lucia (4.35 kg), and the Solomon Islands (4.30 kg). The top five leading MSW generators among the developed countries include New Zealand (3.68 kg/capita/day), Ireland (3.58 kg/capita/day), Norway (2.80 kg/capita/day), Switzerland (2.61 kg/capita/day), and the United States (2.58 kg/capita/day), while the countries generating the least MSW were Ghana (0.09 kg/capita/day) and Uruguay (0.11 kg/capita/day) (World Economic Forum, 2015).

Though we advocate the minimisation of waste generation, what is more important is how waste generated is managed in order to protect public health and the environment. Some of the countries that generate a large quantity of waste manage them better than those that generate less waste.

17.3 Waste Management Issues

17.3.1 Sorting, Collection, Reuse, and Recycling

After every attempt by waste generators to minimise or prevent waste, at-source sorting of wastes is an important step that facilitates waste collection, reuse, recycle, and disposal. In some countries, laws and policies have been in place to ensure that wastes are sorted

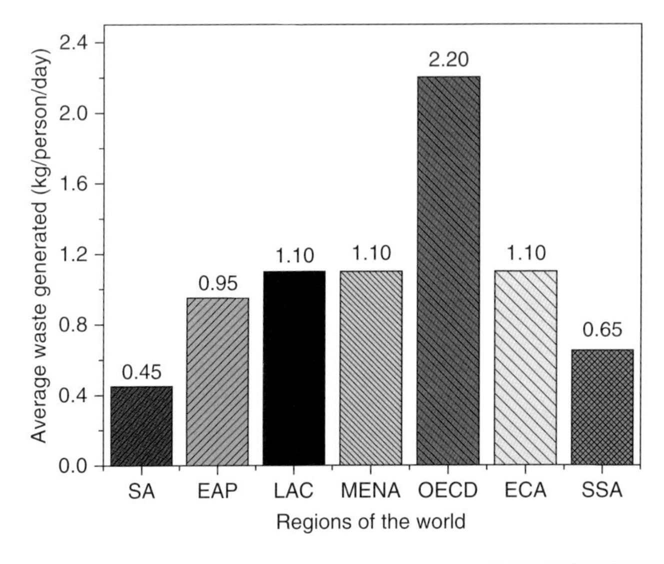

Figure 17.1 Regions of the world and their kilogram per capita per day of waste generated (*Source*: Adapted from Hoornweg and Bhada-Tata (2012)).

at their points of generation (Philippine Ecological Solid Waste Management Act, 2000; Sri Jayawardenapura Kotte Municipal Council, 2005; DEFRA, 2011). However, this is not strictly enforced in some countries, thereby making it more difficult for waste reuse and recycle.

Local councils or local governments of many nations are usually responsible for collecting municipal wastes (DEFRA, 2011), but this is not so in cities of some developing countries. Figure 17.2 presents street wastes disposed on a dual carriageway at Ota in Nigeria.

Figure 17.2 shows that the carriageway was reduced by the waste disposed on it. During the collection of these wastes disposed on roads, waste collection trucks usually obstruct traffic.

It is not unusual to see fly-tipping notices or signposts around some vacant land areas placed by their owners – who have discovered that their land awaiting development has been turned to a dumpsite – warning against continued disposal of wastes on it (see Chapter 12 for more on fly-tipping). In some municipalities in Lagos, Nigeria, for instance, residents engage unregulated private contractors or individuals to collect their wastes daily, bi-weekly, weekly, or bi-monthly, depending on the quantity and rate of wastes they generate. Many of these residents are unaware or careless about where the collected waste is eventually disposed. In contrast, in some parts of the United Kingdom, householders have been required by law to check that anyone (other than weekly council contracted collectors) removing waste from their property is authorised (with a Waste Carrier licence issued by the Environment Agency) to do so.

Reuse and recycling of wastes are important for their effective management. However, these need to be done within acceptable frameworks and should be regulated. We observed that this is not the case in

Figure 17.2 Waste disposed beside the median crash barrier of a dual carriageway. Photo: Isaac I. Akinwumi.

some developing countries, where waste reuse and recycling is mostly done at a small scale. In some cities of these countries, it is common to see waste pickers going through dumpsites and searching for items that still have value. However, some of these items, especially spare parts of vehicles, are re-painted or rebranded and sold to the unsuspecting public as new. This fraudulent engagement poses a risk to the health and safety of the public.

17.3.2 Disposal

In centuries past, waste disposal entailed the deposition of waste into a hole in the ground that had been created naturally or excavated by man, with minimal or no thought about what happens afterwards. This meant that no regulation or control was exercised. According to Sarsby (2012), this practice assumed any leachate contaminants from dumped waste would get dispersed and diluted in groundwater, thereby reducing their concentrations to acceptable levels. This assumption seemed right, then, and the real environmental hazards it posed were not evident. However, it is now known that dispersion and dilution do not eliminate the risk of the toxicity from the contaminants. Even at low concentrations, some contaminants pose significant threats to public health and the environment, either in the short term or in the long term. Sadly, open dumping is still a common practice in some developing countries, especially in Africa and Asia (UNEP, 2013).

Waste disposal facilities available include open dump, semi-controlled dump, controlled dump, engineered landfill, and sanitary landfill. Table 17.1 summarises each of the available waste tipping methods and their characteristic components.

Open dumps are usually poorly sited, with minimum or no site preparation prior to waste disposal. Normally, their capacities are unknown, and there are no records of the quantities and types of wastes that have been disposed. Open dumps are typically characterised by no operational and engineered management measures, including leachate and gas management. By comparison, semi-controlled dumps tend to also be poorly sited but may be fenced to control access;

Table 17.1 Available options for waste disposal on land.

	Operation on measures							Engineered measures			Leachate management		Landfill gas management		
Waste tipping method	Fenced	Placement of waste	Some siting considerations made	Waste record-keeping	Compaction of waste	Prohibition of waste picking	Application of daily soil cover	Proper siting	Engineered liner system	Infrastructure, such as good access road	Leachate collection	Leachate treatment	Gas capture	Passive ventilation or flaring	Flaring
Open dump	X	X	X	X	X	X	X	X	X	X	X	X	X	X	X
Semi-controlled dump	✓	✓	X	X	X	X	X	X	X	X	X	X	X	X	X
Controlled dump	✓	✓	✓	✓	✓	X	X	X	X	X	X	X	X	X	X
Engineered land fill	✓	✓	✓	✓	✓	✓	✓	✓	✓	✓	✓	✓	✓	✓	X
Sanitary land fill	✓	✓	✓	✓	✓	✓	✓	✓	✓	✓	✓	✓	✓	✓	✓

Source: Adapted from Hoornweg and Bhada-Tata (2012).

although they have no waste record-keeping, directions are sometimes given to where waste should be placed to allow informal waste picking. However, they do not have any engineering measures for leachate containment, collection, and treatment, nor gas management (Johannessen and Boyer, 1999; Hoornweg and Bhada-Tata, 2012). Controlled dumps, on the other hand, are characterised by site selection that takes cognisance of hydrogeological suitability, site preparation that involves grading and provision for drainage, access control, basic waste record-keeping or registration, controlled waste picking and trading, compaction of waste, regular (even though not daily) application of soil cover, lack of any engineering measure for leachate containment, plus collection and treatment, and gas management (Uriarte Jr., 2008). The Olusosun dumpsite in Lagos (Figure 17.3), which is one of the largest dumpsites in the world, can at best be described as a controlled dump.

The Olusosun dumpsite is fenced to control access and has roads, drainage channels, weighbridge, and facilities for waste compaction. The activities of waste pickers at the dumpsite are controlled by the management of the dump and the association of waste pickers. This dumpsite poses a great risk to the public, especially to the waste pickers, people living on and around the dumpsite.

Compared to the types of dump sites, engineered landfills have additional provisions, such as liner systems for leachate containment; leachate collection and some level of treatment; facilities for complete record-keeping of waste volumes, types, and sources; application of daily soil cover; and passive ventilation or flaring of gas generated (Hoornweg and Bhada-Tata, 2012). Sanitary landfill involves appropriate attention to all technical aspects of landfill development: siting, design, operation, and long-term environmental impacts (Johannessen and Boyer, 1999). The siting and capacity of a sanitary landfill are properly planned, with full consideration given to the environment, community, and cost. Waste disposal at landfills is typically done using cell systems, with application of daily, intermediate, and final soil cover on compacted waste. Despite being more environmentally friendly than dumpsites, the sad reality is that contaminants

Figure 17.3 Heap of waste within an area at the Olusosun dumpsite in Lagos, Nigeria. Photo: Isaac I. Akinwumi.

from landfills cannot be totally or indefinitely contained (Sarsby, 2000). Attention should be focused on effective (temporary) containment and contaminant collection (leachate collection and gas capture) to minimise their migration from landfills and the controlled release of contained contaminants. An adsorption process can be used to minimise the migration of contaminants, such as heavy metals, from landfills. According to Hoornweg and Bhada-Tata (2012), biological and physic-chemical treatment of collected leachate is often employed, while captured gas is usually flared (though energy may be recovered). No waste picking is permitted. Furthermore, plans are usually made for full closure and post-closure management of the landfill.

Landfills can also be classified according to the wastes they receive, namely, those receiving hazardous, non-hazardous, or inert wastes (European Union Council Directive, 1999). The type of waste a proposed landfill receives affects some of its design parameters and the risk it poses to the environment.

A major public nuisance associated with most waste disposal sites in developing countries is odour. Though there are no criteria for judging the level of acceptable odour from a landfill, measures such as odour control by design; operational management; physical prevention and minimisation of odour release; and odour-counteracting measures can be taken to control odorous emissions (Sniffer, 2013). Some of these measures entail considerations for odour control at the planning stage, waste delivery routing, operational cell sizes, and waste (daily) cover; accepting only permitted wastes, rejecting odorous wastes, disposing odorous wastes in cells further away from sensitive land use, immediate application of cover to the waste, and rapid capping of cells; and the use of products, such as masking agents, chemical counteractants, digestive deodorants, or chemical scavengers to treat foul air from landfilled wastes (Sniffer, 2013). The application of a daily cover to the landfilled wastes also serves to minimise the presence of flies and scavenging birds.

As part of the management of a landfill site, it is necessary to have a monitoring programme in place for leachate, gas, groundwater, and surface water. According to SEPA (2003), a waste management licence or Pollution Prevention Control (PPC) permit necessitates a monitoring programme to ensure that operation at a landfill site, and even post-closure, does not pose a threat to public health or the environment. Provision of leachate level monitoring wells at landfills serve to ensure that leachate accumulation is measured and monitored to avoid leachate breakouts, which have the potential of rendering groundwater unsuitable for drinking or causing death of aquatic wildlife in local surface water courses. Leachate collection and treatment systems also play vital roles in protecting the environment. In monitoring landfill gas, there are three major components to be considered: source monitoring, monitoring emissions, and monitoring air quality (SEPA and EA, 2004). According to SEPA and EA (2004), a gas source can be monitored using gas collection and monitoring wells, or gas-measuring instruments; surface emissions can be monitored using devices such as a flame ionisation detector (FID); lateral gas emissions can be monitoring using wells installed outside the perimeter of a landfill; combustion monitoring entails monitoring landfill gas flares and engines; and the air quality within and around a landfill is monitored to determine the level of odour and particulate matter. Routine monitoring of the characteristics of groundwater and surface water around a landfill for comparison with their initial characteristics prior to landfill development is very important. In the case of pollution, its spread must be prevented and prompt clean-up actions taken. Monitoring is one of the most important phases in the management of a landfill site. Therefore, prior and ongoing assessment of the risk posed by a landfill helps to determine the level of monitoring to put in place.

The location of a landfill must take into consideration requirements relating to the distances from the boundary of the site to residential areas, waterways, waterbodies, and other agricultural or urban sites (European Union Council Directive, 1999). According to the UK Environment Agency (EA) (2009), when a permit is issued for the development of a landfill, a list of permitted wastes is also included. For landfill sites where non-hazardous waste is to be disposed, 500 m is the minimum distance of the closest residential area, waterway, or waterbody from the landfill boundary, while 2000 m is specified for landfill sites receiving hazardous waste (European Parliament, 2007). However, this is not the case in many developing countries, as unregulated communities choose to build makeshift homes on or near to dumpsites, sometimes because the sites support a livelihood of scavenging.

The cost of improper waste management to the society is 5–10 times greater than the financial costs per capita of managing waste properly (UNEP and ISWA, 2015). With the intention of protecting both public health and the environment, plans for landfill construction should incorporate engineered lining systems and leachate and gas collection and treatment systems.

17.4 Landfill Liners

Landfill containment systems (lining and capping systems) serve as barriers to contain wastes and their derivatives (leachates and gases) within waste disposal facilities, so that they do not pollute groundwater resources, soil, air, and the environment, in general. Consequently, it can be said that how a landfill affects or impacts its environment is mainly dependent on its lining and capping systems (Chakradhar and Katoch, 2016). As stated in Section 17.3.2, total or permanent containment of contaminants in landfills cannot be practically achieved. However, migration of contaminant can be minimised to near zero (Sarsby, 2000). Lining and capping of landfills can be provided using either natural (mineral or clay liners) or artificial (synthetic) material, or their combination. According to Sarsby (2000), there are three main types of landfill liners: simple liners, composite liners, and double liners. A simple liner is a containment system having only one hydraulic barrier material, while a composite liner has two different hydraulic barrier materials, with one material placed directly on the other. An example of a composite liner is the placement of a geomembrane on a compacted clay layer. A double liner differs from a composite liner in that its two hydraulic barrier materials are separated by an intermediate drainage layer that ensures the uniform dispersion of leaked leachate from the top barrier material. Sarsby (2000) disapproves its use. The differences between mineral liners and synthetic liners are presented in Table 17.2.

From Table 17.2, it can be seen that mineral liners are less effective at containment of leachate and gas in the landfill. However, mineral liners are still preferred in some developing countries (Widomski, 2016) because they are readily available.

For lining the base, sides, and engineered capping of a landfill designed to receive non-hazardous and hazardous wastes, it is required that the mineral layer should be prepared such that its permeability does not exceed 1×10^{-9} m/s, while the minimum thickness of the mineral layer should be 1 m and 5 m for non-hazardous and hazardous wastes, respectively (European Union Council Directive, 1999). An artificial sealing layer is also necessary for both the lining and capping systems, and in combination with the mineral layer, they function to prevent the migration of leachates from the landfill and to prevent groundwater from gaining access to the landfill. A leachate drainage layer not <0.5 m thick is normally required (Burnley *et al.*, 2005). Of compacted clay and synthetic liners, compacted clay liners are cheaper and most readily available worldwide. However, the lining system used in countries like Austria, France, Germany, Italy, Portugal, and the United States usually comprises both a compacted clay layer and geomembrane materials (Viswanadham, 2017).

Sarsby (2000) identified two mechanisms by which flow through a liner can be described: advective flow and diffusion. When the permeability of the liner material is less than 1×10^{-9} m/s, diffusion becomes an important transport mechanism; otherwise, it does not predominate. Advective flow is dependent on the hydraulic gradient. If a higher hydraulic head is in the landfill compared to that of the water table beneath or around the landfill, then the advective flow will be out of the landfill; otherwise, the advective flow will be into the landfill (Sarsby, 2000). However, diffusion moves contaminants from landfilled wastes towards the landfill liner in their journey out of the landfill to groundwater.

The suitability of using locally available clays of various cities/region of the world for lining landfills has been investigated or the properties of compacted clay liners used have been reported: Ankara, Turkey (Met and Akgun, 2015); Argentine North Patagonia

Table 17.2 Mineral liners versus synthetic liners.

Factors or considerations	Mineral liner	Synthetic liner
Migration of leachate out of landfill	Relatively permits more	Relatively permits less
Susceptibility to damage	Not easily damaged	Easily damaged
Gas containment	Relatively less effective at restricting gas flow	Relatively more effective at restricting gas flow
Production or construction cost	Relatively lesser	Relatively higher
Sophisticated sealing material or technique	Not required	Required
Satisfying principles of sustainable development	Yes	No in some developing countries
Ease of installation and maintenance	Relatively easier	Relatively difficult
Indigenous materials, technology, and expertise	Available and usually used	Unavailable in some developing countries

Source: Adapted from Sarsby (2000) and Widomski (2016).

(Musso *et al.*, 2014); North America (Benson *et al.*, 1994; Benson *et al.*, 1999); Edo, Nigeria (Ojuri *et al.*, 2017), to mention only a few. It is important that materials used for lining landfills possess a hydraulic barrier and reactive properties, in order to prevent the migration of toxic substances from landfills. In some situations, it is necessary to amend soils so as to meet standard requirements for compacted soil liners or to improve on their hydraulic barrier and adsorption properties. Researchers have investigated the application of several reactive materials to clay in order to improve its capacities to adsorb toxic substances. Some of these materials include fly ash (Chakradhar and Katoch, 2016), sawdust (Akinwumi *et al.* 2016a; 2016b; 2017), blast furnace slag (Osinubi *et al.*, 2009), rice husk ash (Eberemu *et al.*, 2013), and bagasse ash (Eberemu, 2013).

Any clay or clay mineral to be used as a compacted clay liner should have geotechnical properties, such that its permeability $\leq 1 \times 10^{-9}$ m/s (for landfills to receive hazardous and non-hazardous wastes) or $\leq 1 \times 10^{-7}$ m/s (for landfills to receive inert wastes) (European Union Council Directive, 1999; Burnley *et al.*, 2005; Osinubi and Nwaiwu, 2006) and unconfined compressive strength >200 kN/m^2 (Guney *et al.*, 2014); and liquid limit <90%, plasticity index <65%, and clay content >10% (NRA, 1989; Environment Agency, 2014).

17.5 Conclusions

Improper waste disposal poses risk of air, water, and soil pollution. Many cities are finding it difficult to cope with the increasing quantity of wastes they generate. In many developing countries, the proportion of the generated waste that gets disposed of is incredibly high per capita. The quantity of waste disposed by a city is influenced by the effectiveness of waste prevention, source separation, collection, reuse, recycling, and value recovery techniques adopted by that municipality. Consequently, the challenges of managing wastes in many countries of the world are initiated by lack of policies to encourage or strictly enforce at-source sorting of wastes by waste generators. This in turn affects how generated wastes are collected and the possibilities of reusing, recycling, or recovering value from them.

Wastes that eventually get discarded are disposed in open dumps, semi-controlled dumps, controlled dumps, engineered landfills, or sanitary landfills. These wastes could be hazardous, non-hazardous, or inert. A landfill liner is a component of the landfill that helps to isolate disposed wastes and their derivatives from the environment. Compacted soil liners, synthetic membrane liners, and amended soil liners are common classes of landfill liners. However, compacted clays are the most widely used materials in the lining system of landfills. Clay that is suitable for use as landfill liner should have low permeability and be stable enough to withstand the load imposed by the wastes. The environmental impact of the disposal of wastes greatly relies on how effective wastes and their derivatives are contained within the waste disposal facility. The use of landfill liners ensures that public health is protected by preventing the contamination of water resources, air, and soil.

References

Akinwumi, I.I., Ojuri, O.O., Edem, D., and Ogbiye, A.S. (2016a) Sawdust stabilization of lateritic clay as a landfill liner to retain heavy metals. *Proceedings of Geo–Chicago 2016 International Conference on Sustainability, Energy and the Geoenvironment*, Chicago, Illinois, American Society of Civil Engineers (ASCE), 478–487.

Akinwumi, I.I., Ojuri, O.O., Ogbiye, A.S., and Booth, C.A. (2016b) Engineering of clay lining system for landfills using bio-material to improve its adsorption of Pb^{2+}. Book of Abstract, Engineering and Life Conference, p. 42, 19–21 October 2016, Hanover.

Akinwumi, I.I., Ojuri, O.O., Ogbiye, A.S., and Booth, C.A. (2017) Engineering properties of tropical clay and bentonite modified with sawdust. *Acta Geotechnica Slovenica,* 14 (2), 47–56.

Bartkowiak, A., Lemanowicz, J., and Siwik–Ziomak, A. (2016) Assessment of selected heavy metals and enzymes in soil within the range of impact of illegal dumping sites. *International Journal of Environmental Research*, 10 (2), 245–254.

Benson, C.H., Daniel, D.E., and Boutwell, G.P. (1999) Field performance of compacted clay liners. *Journal of Geotechnical and Geoenvironmental Engineering,* 125 (5), 390–403.

Benson, C.H., Zhai, H., and Wang, X. (1994) Estimating hydraulic conductivity of compacted clay liners. *Journal of Geotechnical Engineering,* 120 (2), 366–387.

Burnley, S., Cooke, D., and Gladding, T. (2005) T308 Environmental Monitoring, Modelling and Control: Block 4 Solid Wastes Management. The Open University, Milton Keynes.

Chakradhar, V. and Katoch, S.S. (2016) *Study of Fly Ash in Hydraulic Barriers in Landfills – A Review*. Proceedings of IRF International Conference, Goa, India.

Dagiliūtė, R. and Juozapaitienė, G. (2015) Socio-economic assessment in environmental impact assessment: Experience and challenges in Lithuania. *Journal of Environmental Engineering and Landscape Management,* 23 (3), 211–220. 10.3846/16486897.2015.1002842

Department for Environment, Food and Rural Affairs (2011) Government Review of Waste Policy in England 2011. Retrieved from https://assets.publishing.service.gov.uk/government/uploads/system/uploads/attachment_data/file/69401/pb13540-waste-policy-review110614.pdf

Department for Environment, Food and Rural Affairs (2011) Guidance on Applying the Waste Hierarchy. Department for Environment, Food and Rural Affairs (DEFRA), London.

Eberemu, A.O. (2013) Evaluation of bagasse ash treated lateritic soil as a potential barrier material in waste containment application. *Acta Geotechnica*, 8 (4), 407–421.

Eberemu, A.O., Amadi, A.A., and Osinubi, K.J. (2013) The use of compacted tropical clay treated with rice husk ash as a suitable hydraulic barrier material in waste containment applications. *Waste and Biomass Valorization*, 4 (2), 309–323.

Environment Agency (2009) How to Comply with Your Environmental Permit, Additional Guidance for: Landfill (5.02), Environment Agency, Bristol.

Environment Agency (2014) LFE4 – Earthworks in Landfill Engineering: Design, Construction and Quality Assurance of Earthworks in Landfill Engineering. United Kingdom Environment Agency, Bristol.

European Parliament (2007) Parliamentary Question: Answer Given by Mr Dimas on Behalf of the Commission. Retrieved from http://www.europarl.europa.eu/sides/getAllAnswers.do?reference=E-2007-2568andlanguage=DE#def1 [Accessed: 6 March 2016].

European Union Council Directive (1999) Council Directive 1999/31/EC of 26 April 1999 on the landfill of waste. *Official Journal of the European Communities*, 16 (7), L182/1–L182/19.

Guney, Y., Cetin, B., Aydilek, A.H., Tanyu, B.F., and Koparal, S. (2014) Utilization of sepiolite materials as a bottom liner material in solid waste landfills. *Waste Management*, 34 (1), 112–124.

Hoornweg, D. and Bhada-Tata, P. (2012) What a Waste: A Global Review of Solid Waste Management. World Bank, Washington, DC.

Hoornweg, D., Bhada-Tata, P., and Kennedy, C. (2013) Waste production must peak this century. *Nature*, 502, 615–617. 10.1038/502615a

Johannessen, L.M. and Boyer, G. (1999) Observation of Solid Waste Landfills in Developing Countries: Africa, Asia, and Latin America, The International Bank for Reconstruction and Development/The World Bank, Washington, DC.

Met, I. and Akgun, H. (2015) Geotechnical evaluation of Ankara clay as a compacted clay liner. *Environmental Earth Sciences*, 74 (4), 2991–3006.

Musso, T.B., Parolo, M.E., Pettinari, G., and Francisca, F.M. (2014) Cu(II) and Zn(II) adsorption capacity of three different clay liner materials. *Journal of Environmental Management*, 146, 50–58.

NRA (1989) Earthworks to landfill sites. North-West Region: National Rivers Authority (NRA).

Ogundipe, F.O. and Jimoh, O.D. (2015) Life cycle assessment of municipal solid waste management in Minna, Niger State, Nigeria. *International Journal of Environmental Research*, 9 (4), 1305–1314.

Ojuri, O.O., Akinwumi, I.I., and Oluwatuyi, O.E. (2017) Nigerian lateritic clay soils as hydraulic barriers to adsorb metals: Geotechnical characterization and chemical compatibility. *Environment Protection Engineering*, 43 (4), 209–222.

Osinubi, K.J. and Nwaiwu, C.M. (2006) Design of compacted lateritic soil liners and covers. *Journal of Geotechnical and Geoenvironmental Engineering*, 132 (2), 203–213.

Osinubi, K.J., Eberemu, A.O., and Amadi, A.A. (2009) Compacted lateritic soil treated with blast furnace slag as hydraulic barriers in waste containment systems. *International Journal of Risk Assessment and Management*, 13 (2), 171–189.

Philippine Ecological Solid Waste Management Act (2000) Implementing Rules and Regulations of Republic Act 9003. Retrieved from http://faolex.fao.org/docs/texts/phi68080.doc

Sarsby, R. (2000) *Environmental Geotechnics*. Thomas Telford Publishing, London.

Sarsby, R.W. (2012) Climate Change and the Geotechnical Stability of 'Engineered' landfill Sites. In: Booth, C.A., Hammond, F., Lamond, J., and Proverbs, D. (eds.), *Solutions to Climate Change Challenges in the Built Environment*. Wiley-Blackwell, Oxford.

SEPA (2003) Guidance on Monitoring of Landfill Leachate, Groundwater and Surface Water v 2, July 2003. Scottish Environment Protection Agency (SEPA).

SEPA and EA (2004) Landfill Directive: Guidance on the Management of Landfill Gas, Environment Agency.

Sniffer (2013) Odour Monitoring and Control on Landfill Sites: ER31 Final Report, SLR Consulting.

Sri Jayawardenapura Kotte Municipal Council (2005) Solid Waste Management Strategy Guiding Principles and Strategic Options. Retrieved from http://www.fukuoka.unhabitat.org/programmes/scp/sri_lanka/pdf/CCA_2-2_Summary_Kotte_SWM_strategy.pdf

The World Counts (2016) World Waste Facts. Retrieved from http://www.theworldcounts.com/counters/shocking_environmental_facts_and_statistics/world_waste_facts [Accessed: 3 December 2016].

United Nations Development Programme (2006) Human Development Report 2006: Beyond Scarcity: Power, Poverty and the Global Water Crisis. Retrieved 17 November 2015, from http://www.hdr.undp.org/en/reports/global/hdr2006

UNEP (2011) Waste: Investing in Energy and Resource Efficiency. United Nations Environment Programme (UNEP).

UNEP (2013) Municipal Solid Waste Open Dumpsite Juba, South Sudan: Preliminary Environmental Assessment. United Nations Environment Programme, South Sudan.

UNEP and ISWA (2015) Global Waste Management Outlook: Summary for Decision-Makers. United Nations Environment Programme (UNEP) and International Solid Waste Association (ISWA). http://web.unep.org/ietc/what–we–do/global–waste–management–outlook–gwmo

Uriarte, Jr. F.A. (2008) Solid Waste Management Principles and Practices: An

Introduction to the Basic Functional Elements of Solid Waste Management, with Special Emphasis on the Needs of Developing Countries. The University of the Philippines Press, Quezon City.
Viswanadham, B.V.S. (2017) Geotechnical Aspects of Landfill Design. Retrieved from https://www.civil.iitb.ac.in/~dhingra/ce152_files/ce152_BVSV.pdf
Widomski, M.K. (2016) Sustainability of Compacted Clay Liners and Selected Properties of Clays. Monografie Komitetu Inżynierii Środowiska PAN vol. 127, Komitet Inzynierii Srodowiska Polska Akademia Nauk.
World Economic Forum (2015) *Which Countries Produce the Most Waste?* https://www.weforum.org/agenda/2015/08/which–countries–produce–the–most–waste/ [Accessed: 3 December 2016]

18

Mitigating Urban Pollution through Innovative Use of Construction Materials

Jamal M. Khatib[1,2], *A. A. Elkordi*[2], *and Z. Abou Saleh*[2]

[1] *Faculty of Science and Engineering, School of Architecture and Built Environment, University of Wolverhampton, United Kingdom*
[2] *Faculty of Engineering, Beirut Arab University, Lebanon*

18.1 Introduction

With increases in world population and a growing trend of people moving to urban areas, there has been an associated increase in urban pollution. This is due to many factors, including the provision of new infrastructure (including housing) and maintenance of the existing infrastructure. Construction activities contribute to a substantial part of urban pollution (e.g. gas emission and noise). Generally, materials for construction are brought from outside the urban environment into construction sites inside and around cities, thus creating more pressure on the transportation network and increasing pollution. Furthermore, construction materials require excavation and processing, which can also create more pollution.

Urban pollution reduction can be achieved by minimising the construction materials required, by efficient design and the use of recycled and waste materials. Using recycled and waste materials in construction avoids extraction of virgin materials and their associated processing and transporting. Less extraction of materials and less transport can contribute towards a cleaner and more sustainable urban environment.

Numerous materials are used in construction. These can include aggregate, concrete, steel, timber, glass, water, soil, plastic, and bitumen. These materials are used in the construction of building, factories, tunnels, sewer and water networks, roads, and factories. The appropriate choice of these materials can play a vital role in reducing urban pollution. This chapter will explore the various ways that urban pollution may be reduced through the use of waste and recycled materials in construction projects and the use of innovation in construction.

18.2 Recycled Materials

In cities across the world, many existing buildings and infrastructure are in need of renovation, expansion, demolition, or renewal. Increases in population in urban areas require efficient utilisation of available spaces. In some cases, for example, old buildings have to be demolished and replaced with new, and perhaps taller ones, to accommodate more people. Infrastructure, such as sewers and water mains, has to be maintained or may be renewed due to an expected increase in flow. Road and rail network capacity may have to be enlarged. This may

Urban Pollution: Science and Management, First Edition. Edited by Susanne M. Charlesworth and Colin A. Booth.

be achieved through either widening roads, extending tracks, building bridges, or creating tunnels.

Waste materials can be generated due to urban activities, such as demolition, tunnelling, and other industrial activities. In the case of demolition of buildings and old structures, large quantities of crushed brick and concrete are generated. These wastes can be screened into various sizes and the material reused as aggregates (coarse or fine) in the production of concrete and/or in bituminous mixtures (Khatib, 2005; Widyatmoko, 2016). They can also be used as hard core materials for construction of bottom layers of roads and rail tracks. There are different grades of concretes where these crushed aggregates can be used. These include plain concrete, where high strength is not normally required, and also structural concrete, which is a high-grade application. There are challenges and restrictions to using recycled materials, such as crushed concrete or brick, in structural applications due to the current legislation that limits the amount of this type of aggregate. However, construction professionals should have sufficient knowledge to convince the regulatory bodies of where and when larger amounts can be used.

Other waste materials generated from construction activities include glass, timber, plastic, polystyrene, foundry sand, rubber tyres, and plasterboard. Waste glass can also be crushed and used as aggregate in concrete and bituminous applications (Achintha, 2016) or ground and used as partial substitution of cement (Khatib *et al.*, 2012b). Waste timber can be either reused or processed in order to produce, for example, new engineered timber products that can be used as structural/construction elements (Milner and Woodard, 2016). Examples of structural elements include cross-laminated timber, structural I-beams, and oriented strand boards. Other construction elements include plywood, chipboard, and fibre board. Less imported timber would certainly reduce the long distances involved in transporting materials and any concomitant pollution.

Waste rubber tyres from vehicles are an environmental burden. Shredded and crumb rubber tyres can be used in concrete mixtures to partially substitute for the aggregate. While the compressive strength of concrete may be reduced in the presence of waste rubber, the deformability properties and crack propagations are improved, and this can be advantageous in particular applications (Topçu and Uygunoglu, 2016). Topçu and Demir (2007) showed that concrete containing waste rubber is able to provide better resistance against freeze-thaw compared with concrete without waste rubber. Furthermore, waste rubber used in concrete can improve the thermal insulation of building materials and noise reduction, and this makes it attractive to use in populated areas adjacent to roads (Topçu and Uygunoglu, 2016). For instance, use of shredded tyres in the construction of concrete pavement or concrete barriers is possible. Moreover, the density of concrete containing waste tyres is reduced, and this makes it attractive for structures, as the dead load is reduced if other properties such as strength requirements are met. Although the compressive strength is reduced with the addition of waste rubber, adequate strength can be achieved if the rubber particles are treated. Treatment includes washing with water or chemicals. This will provide a rough texture of the rubber particles and provide better bond if used in concrete (Topçu and Uygunoglu, 2016).

Waste materials, such as waste foundry sand (Khatib and Ellis, 2001; Khatib *et al.*, 2013c), waste-expanded polystyrene (Herki and Khatib, 2013; Herki and Khatib, 2016), expanded clay (Khatib *et al.*, 2015a), foamed glass (Khatib *et al.*, 2012a), and vegetable fibres (Savastano *et al.*, 2016), can be used as partial replacement of fine and coarse aggregates or as fibres. While the overall strength may be reduced, concrete mixes can be designed so acceptable strengths can be achieved for many applications.

18.3 Cement Replacement and Geopolymer-Based Materials

The cement industry is a major contributor to environmental issues. For instance, it is estimated that it contributes ~10% towards global carbon dioxide emissions. This is partly due to the high temperatures used (~1500°C) in its manufacture and partly due to the decomposition of limestone. For each tonne of cement used, up to 1 tonne of carbon dioxide emissions are released (Rubenstein, 2012). Therefore, reducing the amount of cement used in construction can have a positive impact on the environment. This can be achieved by replacing the cement with appropriate waste, such as pozzolanic and other materials. An example of these materials is coal fly ash, ground-granulated blast-furnace slag, silica fumes, volcanic ash, natural pozzolan, limestone fines, and cement kiln dust (Kurdi *et al.*, 1996). Other pozzolanic processed materials include metakaolin and calcined clay.

Coal fly ash is a by-product from coal power stations, and it is generated in large quantities in many parts of the world, particularly in China, India, Russia, the United States, Indonesia, Poland, Germany, and Turkey. Sadly, more than half of the fly ash generated ends up in landfill. Dust-sized fly ash can cause air pollution due to the fineness of the particles and can leach into the ground contaminating groundwater. While fly ash has successfully been used to replace cement, however, there is still greater potential to use more fly ash in various construction activities using novel methods and techniques. Research has shown that the mechanical and durability performance of concrete using coal fly ash can be improved, and these include the long-term compressive strength, resistance to sulphate attack, and alkali-aggregate reaction (Mangat and El-Khatib, 1992). This is mainly due to the consumption of calcium hydroxide produced during cement hydration.

Ground-granulated blast-furnace slag (GGBS) is a by-product from the steel manufacturing industry, and currently large amounts of this material are used in the concrete industry to replace cement. In fact, more than 70% of cement can be replaced with GGBS, and this can improve the mechanical and durability performance of concrete materials if cured properly (Hadjsadok *et al.*, 2012; Khatib and Hibbert, 2005; Khatib, 2014).

Silica fume is the by-product of the ferrosilicon industry, and it too can be used to partially replace cement to produce a concrete material with enhanced performance (Khatib and Mangat, 1995). However, despite being a waste material, it is expensive to purchase (around 4 or 5 times the price of cement) due to the enhancement caused in concrete properties when silica fume is included (Mangat and Khatib, 1995).

Metakaolin is an amorphous dehydration product of Kaolinite, which, depending on the purity, exhibits strong pozzolanic activity. It is processed from kaolin clay and calcined at a temperature of between 600°C and 800°C. At higher temperatures, metakaolin undergoes further reactions to form crystalline compounds. Metakaolin contains silica and alumina in an active form that will react with the calcium hydroxide ($CaOH_2$) resulting from cement hydration, to produce more calcium silicate hydrate (C–S–H). Depending on its purity, the mechanical and durability performance of concrete can be improved when calcined clay or metakaolin is present (Khatib, 2008; Khatib, 2009; Wild *et al.*, 1997). For instance, the compressive strength can be up to 40% higher than cement-only concrete (Wild *et al.*, 1996). Also, if cement is replaced with 10%–30% metakaolin, the sulphate resistance is drastically improved (Khatib and Wild, 1998). Furthermore, metakaolin can be used in conjunction with fly ash of GGBS to replace cement, and adequate performance can be achieved (Khatib and Hibbert, 2005; Khatib *et al.*, 2009).

Other waste products that can be used as a partial replacement of cement are limestone fines (Bai, 2016; Black, 2016; Sonebi *et al.*, 2016). These fine-size wastes are generated from the crushing of limestone into aggregate. Menadi *et al.* (2009) and (2014) showed that concrete with an adequate performance can be achieved when small amounts of limestone fines, typically 5%–15% cement replacement, are included in concrete mixes. This is somewhat in agreement with results reported elsewhere (El-Kurdi *et al.*, 2014a, 2014b). This has been permitted by the Canadian cement standard since the early 1980s and by the European Standard EN 197–1 (CEN, 2000). However, the percentage addition depends upon the type of cement used. For instance, a 5% addition is permitted when CEM I is used. This can go up to 20% and 35% when CEM II/A–L and CEN II/B–L are used, respectively. About 20% of all cement sold in Europe contains between 6% and 35% limestone (CEMBUREAU, 2001). Limestone fines can be added to Portland cement as an addition, either at the factory to create blended cement or at the batch plant to form combination cement as a replacement or additive by mass. The use of limestone fine as additive by 15% of cement content reduces the cracks of concrete when heated at 400°C and 600°C, which may be due to the filling effect of unreacted limestone fine and due to formation of new hydrated compounds, such as the carboaluminates, that filled the capillary pores of the cementitious matrix (El-Kurdi *et al.*, 2014b).

To achieve the desired concrete properties, adequate curing should be employed in order to retain the water for the hydration to continue. This is particularly the case when pozzolanic materials, such as fly ash and slag, are included in the mix. If concrete is subjected to drying in the early stages of hydration, the compressive strength is reduced, particularly so when pozzolanic materials (such as fly ash or slag) are used (Khatib, 2014). Khatib and Mangat (1995, 1999) showed that the initial curing method has a noticeable effect on porosity and the pore size distribution of cement paste. This undoubtedly affects the durability properties and the ingress of harmful ions, such as chloride, into concrete (Bennacer *et al.*, 2016; Khatib and Mangat, 2003). However, some properties that negatively affect strength may not affect the durability in the same way. Mangat and Khatib (1995) showed that concrete subjected to some initial air curing exhibits greater resistance to a sulphate-bearing environment compared with concrete subjected to moist curing in the initial periods of hydration. This was attributed to the formation of a carbonation layer under air curing, which hinders the ingress of sulphate ions into concrete and the associated expansive chemical reaction products.

In recent years, there has been an increasing interest in using large quantities of fly ash or GGBS without the use of cement (Khatib *et al.*, 2014a; Khatib *et al.*, 2015a; Mangat and Lambert, 2016). This is possible using alkali activators, such as sodium hydroxide, sodium silicate, and potassium hydroxide. The construction materials produced are referred to as alkali-activated or geopolymer materials. With careful selection and proportioning of the materials, a compressive strength comparable to that of cement-based materials can be achieved (Mangat and Lambert, 2016). However, a high temperature is normally required during the early stages of hydration. Therefore, assessment of the material's performance, its carbon footprint, and cost are required.

18.4 Innovative Ways of Using Waste Clay

Waste clay is generated in urban areas in various places such as the tunnelling operation and other excavation activities. Normally these wastes are transported outside the city, and virgin materials are brought in to use them as ingredients in the production of construction materials, such as concrete. Recent research suggests that waste clay can be used in the production of lightweight aggregates through calcination (Owens *et al.*, 2016). There are different types of lightweight aggregate products available from waste clay; however, the proposed technique is expected to produce lightweight aggregates from waste clay that will have relatively low water absorption. Therefore, the new lightweight aggregate can be used in the production of concrete materials, including structural concrete. Due to the presence of lightweight aggregates, the concrete produced will have a lower density compared to traditional concrete, thereby reducing the quantity of constituents used in the making of concrete. Employing this innovative approach for manufacturing suitable materials out of waste in urban settings is greatly advantageous. This will avoid the need for long-distance transportation and the associated emission of harmful gases, in addition to reducing the noise level and its frequency, particularly in the urban environment. The research conducted thus far is encouraging as it shows that acceptable properties such as in situ strength can be achieved. Furthermore, the adiabatic temperature generated by concrete using lightweight aggregates produces more heat than that with normal aggregates.

18.5 Treatment and Stabilisation of Contaminated Sites

Industry goes through various cycles, and some may even dwindle or disappear. Recently, in the United Kingdom, for example, the steel industry has faced major challenges and closures, which may leave large areas of brownfield land unoccupied and derelict (Bowler, 2016). In order to use this land for other developments, such as housing and commercial development, brownfield sites may have to be treated, depending upon the level of contamination left behind in the ground. To be safe and sustainable for future users, treatment of contaminated land has to be efficient, environmental, and economical. That said, it is possible to use waste materials (such as fly ash) to treat contaminated soil, and Hassan *et al.* (2012, 2013) showed that using lime can reduce leaching in lead-contaminated soil. Since the soil properties are improved, it allows construction of buildings at lower cost, due to the type of foundations that would otherwise be required. For instance, improving the properties of the soil can allow pad foundations to be used as opposed to pile foundations.

18.6 Incineration of Municipal Solid Waste

In some urban areas, in order to deal with the large amounts of municipal solid waste sent to landfill, it is incinerated. For instance, in the United Kingdom, some cities such as Sheffield, Nottingham, and Wolverhampton have opted for incinerating their municipal solid waste, and the incinerators are located very near to the city centres in order to avoid long-distance transport of the large quantities of waste. After incineration, the volume of the waste is reduced by more than 80% (Morales-Hernandez *et al.*, 2005). However, ashes are generated out of this process. The ashes are predominantly bottom ash, and there are some fly ashes too. At present, much of the ashes produced are sent to landfill. However, it is possible to use incineration ashes in the production of concrete materials to replace fine aggregates. Moreover, in the Netherlands, incineration bottom ash is used in road construction. Morales-Hernandez *et al.* (2005) showed that adequate mortar strength can be achieved where mortar containing municipal solid waste incineration bottom ash either as partial substitution of sand or cement. However, the strength tends to be lower than that of mortar without ash. The volume and utilisation of incineration bottom ash as aggregate replacement in road construction or concrete application varies between countries. This is more crucial in the case of incineration fly ash, where the concentration of heavy metals tends to be higher. There is a potential to use these types of fly ash in cement-based material. However, more research is required in order to assess the environmental impact of these wastes, as leaching of heavy metals from mortar or concrete may contaminate the ground or groundwater.

18.7 Flue Gas Desulphurisation (FGD) Wastes

Coal-powered industry contributes immensely to the pollution of the environment. Pollution does not only include the emission of CO_2 but also SO_x and NO_x. Many governments throughout the world have opted for a process referred to as clean coal technology in order to reduce the emission of SO_x. The process involves the injection of alkali sorbent into the flue gases, causing SO_x to react, resulting in solid residues referred to as FGD waste. There are different processes of desulphurisation; wet, dry, and semi-dry. For each process, different types of FGD are obtained. The wet process is the most efficient in terms of waste residue produced, which is calcium sulphate (i.e., gypsum), and this waste can be fully utilised in the production of plaster board for buildings. However, this process is expensive to install. With regards to the other processes (i.e., dry and semi-dry) that many developing countries are using, the waste produced is a combination of coal fly ash and gypsum. The FGD produced from these processes are referred to as SiO_2–Al_2O_3–$CaSO_4$-based FGD waste. The content of these chemicals varies from one installation to the next and also within the same installation. This varying chemical composition makes it difficult to use this waste in construction applications (Wright and Khatib, 2016). However, provided thorough research is done, there is potential to use this type of waste in cement-based materials, including concrete. Cement is the most expensive and polluting ingredient in the making of concrete. Research has shown that there is a potential to use FGD to partially replace the cement in the making of concrete (Wright and Khatib, 2016). It was also shown that using simulated FGD waste in the making of mortar or concrete yields similar results to that made with actual FGD (Mangat *et al.*, 2006). In plain concrete, FGD can be used and produces satisfactory properties (Khatib *et al.*, 2016a). Khatib *et al.* (2008) showed that the sulphate resistance of simulated FGD is better than normal concrete, indicating that it can be used in concrete exposed to sulphate environments. Khatib *et al.* (2013a, 2014a) concluded that there is a difference in the total pore volume and pore size distribution of pastes containing large amounts of simulated FGD, depending on the SO_3 content. However, the porosity of pastes containing small amounts of FGD can be similar to the control paste. Also, the type and length of the curing regime influence the properties of the materials produced (Khatib *et al.*, 2013b). Further research is required if FDG is to be used in reinforced concrete as the effect on steel corrosion is not yet completely understood.

18.8 Paper Industry Waste

Waste paper sludge is the by-product of the paper manufacturing industry. In the recent past, this sludge was normally sent to landfill, thus exerting pressure on the built environment in terms of pollution and landspace. On combustion, the cellulose left in the waste paper sludge has calorific value and can be used to generate power or heating. Therefore, many parts of the world are opting for the combustion of the sludge, partly to reduce its quantity and partly to generate power. After combustion, waste paper ash is generated, and this can be used as a cementitious material to produce building materials (Kinuthia, 2016). Kinuthia *et al.* (2001b) and O'Farrell *et al.* (2001) found after a series of trials that combining waste paper ash with GGBS can produce a binder that is equivalent in properties to Portland cement. These properties include setting time and compressive strength. This makes the waste paper ash suitable for inclusion in cement-based products such as masonry and compressed earth. Kinuthia *et al.* (2001b) showed that resistance to sulphate attack of mortar containing waste paper ash and GGBS as a cementitious binder was superior to that of mortars based on Portland cement only. However, there are some problems in using paper sludge ash in cementitious systems as the water demand is high. This would necessitate the use of chemical admixtures, such as superplasticiser, which would increase the cost of the material produced.

Waste paper is also generated in large amounts in urban areas. Recent research has shown that waste paper can be shredded, moistened, and split into aggregate sizes without using burning or incineration. These aggregates can be used as a component in building materials, such as the production of light blocks, which can be suitable for inner non-loading walls in buildings (Okeyinka *et al.*, 2015b). The blocks can also contain a binder based on food waste (Okeyinka *et al.*, 2015a). The cost of these blocks is expected to be low compared with the alternatives as it contains large quantities of wastes (e.g. waste paper and waste food).

18.9 Shelled Compressed Earth

Use of local materials without too much processing, particularly in urban environments, to produce building materials is advantageous in that it reduces the urban pollution owing to their low embodied energy. Rammed earth is a potential material that can be used in the urban environment. However, rammed earth is susceptible to erosion due to rain, so its use in houses and multi-storied buildings is limited. If rammed earth is going to be widely used in urban environment buildings, then the erosion resistance has to be increased. In addition, the rammed earth blocks produced are still relatively heavy, and attempts must be made to reduce the weight. Recent research has focused on using innovative techniques for producing rammed earth blocks with adequate resistance to water (Egenti and Khatib, 2016). Shelled compressed earth blocks comprise an outer shell that is stabilised with up to 10% cement. The rest of the block consists of rammed earth only or rammed earth with a small proportion of cement. The compressive strength of the blocks is higher than the sand cement block (Egenti *et al.*, 2014). Also, the water penetration rate is lower than that of standard sand cement blocks. As the compaction pressure during block manufacturing is increased, the compressive strength increases. There is an inverse correlation between strength and water absorption at different compaction pressures. In other words, as the compressive strength increases, the water penetration rate decreases. The shelled blocks have higher resistance to surface abrasion compared to sand cement blocks for compaction pressures above 2.5 N/mm^2.

Despite the improved performance of rammed earth blocks using the above innovative technique, there are still some issues with the density of the blocks, which makes it heavier for handling. In order to reduce the weight of the blocks, the technique has been modified to include holes in the middle. Even then, the compressive strength of the new block is acceptable and is still better than that of conventional sand cement blocks. It is expected that rammed earth offers other advantages, including better thermal resistance. This makes its usage more attractive in buildings located in hot regions in the world where the need for air conditioning would be reduced.

18.10 Innovative Green Construction Materials

Recent developments in the field of construction materials suggests that carbon dioxide emitted from cement manufacturing or other industries can be used and mixed with other materials to produce a concrete-like material. This may be achieved through adding carbon dioxide to an optimum ratio of calcium to silica, the main raw materials in cement manufacture, to produce a new cementitious material (Pyper, 2014). It is expected that the new concrete would be stronger than normal concrete, in that it has higher compressive strength, higher fracture resistance, and low permeability. This can advantageous, not only in the urban environment, but also in the oil and gas industry around oil casings to prevent leakage. However, the research is still in the early stages, and further development is still required. The potential benefits of this innovation are huge in that carbon dioxide emission is reduced by 60% compared to that with normal concrete. This will reduce pollution, particularly in urban areas, where many industries such as power and steel are located.

A similar recent development in the area of new building materials and carbon reduction/capture is the use of carbon dioxide emitted from industrial activities: it is mixed with a material such as lime to produce a cement-like material. At this stage, this technology uses 3D printing to produce solid products on a small scale. The research is still in the early stages, but the initial results are encouraging (Molitch-Hou, 2016). Other indirect measures to reduce carbon dioxide is the production of solid materials using bacteria that can be used for filling the cracks in damaged concrete structures such as bridges (Lofgren, 2012). Often, traffic would have to be diverted in order to repair a bridge. This will cause congestion elsewhere and increase urban pollution. Diversion may take vehicles via longer routes, which would further increase pollution (e.g. noise and exhaust emissions). Therefore, the new technique for repairing concrete can be beneficial in reducing pollution.

18.11 Innovative Chemical Admixtures for Construction Materials

Large amounts of food and drink wastes are generated in urban areas, and normally these wastes are transported for disposal or processing. A recent European project, completed in 2015, suggests that brewery waste can be chemically modified to produce new admixtures to improve the workability of concrete. A wide range of admixtures were produced, which is predominantly based on brewery waste for different concrete applications (El-Sayed *et al.*, 2014; Zhansaya *et al.*, 2014). One of these applications is self-compacting concrete (SCC), where the novel chemical admixture can be used. SCC eliminates the requirement of compaction after casting and the associated noise. However, SCC is expensive to produce compared to traditional concrete, so its use has been limited.

New chemical admixtures can be used in the production of other construction materials which are non-cement based. It is possible to use these admixtures in making building blocks containing waste paper (Section 18.8) in order to improve their engineering and durability performance (Okeyinka *et al.*, 2015b).

18.12 Conclusions

Reducing urban pollution is mandatory for the health and well-being of inhabitants, flora, and fauna of towns and cities. Any industrial activity should always be examined and measures taken to reduce pollution into the air, water, and land. This chapter has focused mainly on construction activities and how pollution may be minimised by utilising waste. There are various ways in which urban pollution due to construction activities can be reduced and how waste generated from other industries (e.g. coal and steel) can be used in the production of construction materials. This avoids transporting waste to landfill, as often happens, and postpones bringing yet more virgin materials into urban areas. Transportation is costly and generates noise and vehicle pollution in the environment. Reducing the use of virgin materials that require quarrying and processing, in addition to transport, will undoubtedly have a positive impact on the environment.

Continued innovation in the way society uses construction materials is required. However, current regulations need to be altered in order to allow for more utilisation of waste in construction materials so that the new products can be created and pollution reduced.

References

Achintha, M.L. (2016) Sustainability of glass in construction. In: Khatib, J.M. (ed.), *Sustainability of Construction Materials.* 2nd edition. Elsevier, pp. 79–104; ISBN 978–0–08–100995–6

Bai, J. (2016) Durability of sustainable construction materials. In: Khatib, J.M. (ed.), *Sustainability of Construction Materials.* 2nd edition. Elsevier, pp. 399–416, ISBN 978–0–08–100995–6

Black, L. (2016), Low clinker cement as a sustainable construction material, In: Khatib, J.M. (ed.), *Sustainability of Construction Materials.* 2nd edition. Elsevier, pp. 417–460, ISBN 978–0–08–100995–6

Bennacer, R., Abahri, K., and Belarbi, R. (2016). Intrinsic properties affecting the sustainability of construction materials. In: Khatib, J.M. (ed.), *Sustainability of Construction Materials.* 2nd edition. Elsevier, pp. 33–54, ISBN 978–0–08–100995–6

Bowler, T. (2016), Britain's Steel Industry: What's Going Wrong? *BBC News*, 30 March, http://www.bbc.co.uk/news/business–34581945

CEMBUREAU (2001) Annual Report. CEMBUREAU, Brussels, 2001. http://www.cembureau.be/Documents/Publications/Annual%20Reports/Annual%20Report%202001.pdf.

Egenti, C., Khatib, J.M., and Oloke, D. (2014) Conceptualisation and Pilot Study of Shelled Compressed Earth Block for Sustainable Housing in Nigeria. *International Journal of Sustainable Built Environment*, 3 (1), 72–86, June, ISSN: 2212–6090, DOI: 10.1016/j.ijsbe.2014.05.002, http://www.journals.elsevier.com/international–journal–of–sustainable–built–environment/

Egenti, C. and Khatib, J.M. (2016). Sustainability of Compressed Earth. In: Khatib, J.M. (ed.), *Sustainability of Construction Materials.* 2nd edition. Elsevier, pp. 311–344, ISBN 978–0–08–100995–6

El-Kurdi, A.A., Abdel-Hakam, A., and El-Gohary, M.M. (2014a) Study the effect of silica fume, polypropylene fiber, steel fiber, limestone powder and bentonite on the fire resistance of concrete. *International Journal for Research and Analysis in Allied Sciences and Engineering*, 1 (1), 13–29, April.

El-Kurdi, A.A., Abdel-Hakam, A., El-Gohary, M.M. (2014b) Impact of elevated temperature on properties of limestone concrete. *International Journal of Innovative Technology and Exploring Engineering (IJITEE)*, 4 (4), 1–9, ISSN: 2278–3075

Hadjsadok, A., Kenai, S., Courard, L., Michel, F., and Khatib, J.M. (2012) Durability of mortar and concretes containing slag with low hydraulic activity. *Journal of Cement and Concrete Composites*, 34 (5), 671–677, May 2012, ISSN 0958–9465. doi:10.1016/j.cemconcomp.2012.02.011

Hassan, M., Gardiner, P.H.E., Khatib, J.M., and Mangat, P.S. (2012) Stabilisation of Lead Contaminated Soil Using Portland Cement. International Workshop on Earthquake and Sustainable Materials, Sakarya University, Faculty of Technology, Civil Engineering Department, Turkey, 19 January 2012, pp 157–167, ISBN: 978–975–7988–92–2. http://www.iwesm.sau.edu.tr/?s=s9

Hassan, M., Khatib, J.M., Mangat, P.S., Naseef, A., and Gardiner, P.H.E. (2013) FTIR and XRD characterized lime stabilised lead contaminated soil. *2nd International Conference on Environment, Chemistry and Biology – ICECB 2013*, 13–14 December 2013, Stockholm, Sweden, co-sponsored by Asia–Pacific Chemical, Biological

& Environmental Engineering Society (APCBEES). www.cbees.orghttp://www.icecb.org/index.htm

Herki, B.A. and Khatib, J.M. (2016) Valorisation of waste expanded polystyrene in concrete using a novel recycling technique. *European Journal of Environmental and Civil Engineering*, Taylor & Francis pub, http://dx.doi.org/10.1080/19648189.2016.1170729

Herki, B.A. and Khatib, J.M. (2013) Lightweight concrete incorporating waste expanded polystyrene. *Advanced Materials Research Journal*, 787, 131–137, Trans Tech Publications, Switzerland, ISBN print: 978–3–03785–802–8, doi:10.4028/www.scientific.net/AMR.787.131

Khatib, J.M. and Mangat, P.S. (1995) Absorption characteristics of concrete as a function of location relative to the casting position. *Cement and Concrete Research Journal*, 25 (5), 999–1010, July 1995. ISSN: 0008–8846

Khatib, J.M. and Wild, S. (1998) Sulphate resistance of metakaolin mortar. *Cement and Concrete Research Journal*, 28 (1), 83–92. ISSN: 0008–8846

Khatib, J.M. and Ellis, D.J. (2001) Mechanical properties of concrete containing foundry sand. F*ifth CANMET/ACI International Conference on Recent Advances in Concrete Technology*, SP 200 (Ed. V M Malhotra), 29 July–1 August, Singapore, 2001, pp. 733–748, American Concrete Institute. ISBN 0–87031–029–1. http://www.normas.com/ACI/pages/SP–200.html

Khatib, J.M. and Mangat, P.S. (2003) Porosity of cement paste cured at 45°C as a function of location relative to casting position. *Journal of Cement and Concrete Composites*, 25 (1), 97–108.

Khatib, J.M. (2005) Properties of concrete containing fine recycled aggregates. *Cement and Concrete Research Journal*, 35 (4), 763–769, April 2005. ISSN: 0008–8846 doi: 10.1016/j.cemconres.2004.06.017

Khatib, J.M. and Hibbert, J.J. (2005) Selected engineering properties of concrete incorporating slag and metakaolin. *Construction and Building Materials Journal*, 19 (6), 460–472, July 2005. ISSN 0950–0618 doi: 10.1016/j.conbuildmat.2004.07.017

Khatib, J.M., Mangat, P.S., and Wright, L. (2008) Sulphate resistance of blended binders containing FGD waste. *Construction Materials Journal – Proceedings of the Institution of Civil Engineers (ICE)*, Vol. 161, Issue CM3, August 2008, pp. 119–128. ISNN 1747–650X doi: 10.1680/coma.2008.161.3.119

Khatib, J.M. (2008) Metakaolin concrete at a low water to binder ratio. *Construction and Building Materials Journal*, 22 (8), 1691–1700, August 2008. ISSN 0950–0618 doi: 10.1016/j.conbuildmat.2007.06.003

Khatib, J.M. (2009) Low curing Temperature of Metakaolin Concrete. *American Society of Civil Engineers (ASCE) – Materials in Civil Engineering Journal*, 21 (8), 362–367, August, ISSN 0899–1561/2009/8–362–367 DOI: 10.1061/(ASCE)0899–1561(2009)21:8(362)

Khatib, J.M., Kayali, O., and Siddique, R. (2009) Strength and Dimensional Stability of Cement–Fly Ash–Metakaolin Mortar. *American Society of Civil Engineers (ASCE) – Materials in Civil Engineering Journal*, 21 (9), 523–528, September, ISSN 0899–1561/2009/9–523–528, DOI: 10.1061/(ASCE)0899–1561(2009)21:9(523), CODEN: JMCEE7

Khatib, J.M., Shariff, S., and Negim, E.S.M. (2012a) Effect of incorporating foamed glass on the flexural behaviour of reinforced concrete beams. *World Applied Sciences Journal*, 19 (1), 47–51, ISSN 1818–4952,DOI: 10.5829/idosi.wasj.2012.19.01.2763

Khatib, J.M., Negim, E.M., Sohl, H.S., and Chileshe, N. (2012b) Glass powder utilisation in concrete production. *European Journal of Applied Sciences*, 4 (4), 173–176, ISSN 2079–2077. doi; 10.5829/idosi.ejas.2012.4.4.1102

Khatib, J.M., Wright, L., and Mangat, P.S. (2013a) "Effect of fly ash–gypsum blend on porosity and pore size distribution of cement pastes. *Journal of Advances in Applied Ceramics, Structural, Functional and Bioceramics*, 112 (4), 197–201, online, ISSN 17436753, DOI: 10.1179/1743676112Y.0000000032

Khatib, J.M., Mangat, P.S., and Wright, L. (2013b) Early Age Porosity and Pore Size Distribution of Cement paste with Flue Gas Desulphurisation (FGD) Waste. *Journal of Civil Engineering and Management*, 19 (5), 622–627, Taylor & Francis, ISSN 1392–3730 print/ISSN 1822–3605 online, doi: 10.3846/13923730.2013.793609

Khatib, J.M., Herki, A.B., and Kenai, S. (2013c) Capillarity of concrete containing foundry sand. *Journal of Construction and Building Materials*, 47, 867–871, October, Elsevier 10.1016/j.conbuildmat.2013.05.013

Khatib, J.M. (2014) Effect of initial curing on absorption and pore size distribution of paste and concrete containing slag. *Korean Society of Civil Engineers (KSCE) Journal*, 18 (1), 264–272, January. pISSN 1226–7988, eISSN 1976–3808 DOI 10.1007/s12205–014–0449–7

Khatib, J.M., Mangat, P.S., and Wright, L. (2014a) Pore Size Distribution of Cement pastes Containing Fly Ash–Gypsum Blends Cured for 7 Days. *Korean Society of Civil Engineering (KSCE) Journal*, 18 (4), 1091–1096. doi: 10.1007/s12205–014–0136–8 Online ISSN: 1976–3808; Print ISSN: 1226–7988

Khatib, J.M., Halliday, C., Negim, E.S.M., and Khatib, S. (2014b) Activation of fly ash paste with lime in the presence of small amount of metakaolin. *Proceeding of International Forum on "Engineering Education and Science in the XXI Century: Challenges and Perspectives" – Devoted to the 80th Anniversary of Satpaev Kazntu*, Vol. 1, 22–24 October 2014, Kazakh National University under the auspices of the Ministry of Education and Science, Almaty City Akimat, Kazakhstan , pp504–506, ISBN 978–601–228–666–3

Khatib, J.M., Onaidhe, E., Sonebi, M., and Abdelgader, H. (2015a) Lime activation of fly ash in mortar in the presence of metakaolin. *Proceedings of the 1st International Conference on Bio-based Building Materials (ICBBM 2015)*, Amziane and Sonebi (eds.), 21–24 June 2015, Claremont–Ferrand, France. RILEM publication, pp 107–110, ISBN PRO 99: 978–2–35158–154–4 https://sites.google.com/site/icbbm2015/home

Khatib, J.M., Jefimiuk, A., and Khatib, S. (2015b) Flexural behaviour of reinforced concrete beams containing expanded glass as lightweight aggregates. *Slovak Journal of Civil Engineering*, 23 (4), 1–7, De Gruyter Publishing. doi: 10.1515/sjce–2015–007, http://www.svf.stuba.sk/generate_page.php?page_id=2075ISSN: 1210–3896

Khatib, J.M., Wright, L., and Mangat, P.S. (2016a) Mechanical and Physical Properties of Concrete Containing FGD Waste. Magazine of Concrete Research, Article number: MACR–D–15–00092. doi: 10.1680/macr.15.00092 (In Press).

Khatib, J.M., Wright, L., and Mangat, P.S. (2016b) Effect of desulphurised waste on long–term porosity and pore structure of blended cement pastes. *Sustainable Environment Research Journal*, Manuscript ID SER2014–082; http://ser.cienve.org.tw (In Press).

Kinuthia, J.M., O'Farrell, M., Sabir, B.B., and Wild, S. (2001b) A preliminary study of the cementitious properties of wastepaper sludge ash (WSA) – ground granulated blast–furnace slag (GGBS) blends. Ravindra K Dhir, Mukesh C Limbachiya, and Moray D Newlands (eds.), *Proceedings of the International Symposium on Recovery and Recycling of Paper*, Dundee University, 19th March 2001, pp. 93–104, Thomas Telford ISBN: 0 7277 2993.

Kinuthia, J.M. (2016). Sustainability of waste paper in construction. In: Khatib, J.M. (ed.), *Sustainability of Construction Materials*. 2nd edition. Elsevier, pp. 569–598, ISBN 978–0–08–100995–6

Kurdi, A., Awad, A., Kassim, M., and El-Zaafraney, M. (1996) Properties and usage of cement kiln dust in concrete. *Alexandria Engineering Journal*, 35 (5), 209–222.

Lofgren, K. (2012) Dutch Scientists Create Concrete that Heals Itself With Built–in Bacteria, Inhabitat, Green Materials, Green Technology, Innovation, News, 11 May, http://inhabitat.com/dutch–scientists–create–concrete–that–heals–itself–with–built–in–bacteria/

Mangat, P.S. and El-Khatib, J.M. (1992) Influence of initial curing on sulphate resistance of blended cement concrete. *Cement and Concrete Research Journal*, 22, (6), 1089–1100, November 1992. ISSN: 0008–8846

Mangat, P.S. and Khatib, J.M. (1995) Influence of fly ash, silica fume and slag on sulphate resistance of concrete. *American Concrete Institute–Materials Journal*, 92 (5), 542–552.

Mangat, P.S., Khatib, J.M., and Wright, L. (2006) Optimum Utilisation of Flue Gas Desulphurisation (FGD) Waste in Blended Binder for Concrete. *Construction Materials Journal – Proceedings of the Institution of Civil Engineers*, 1 (2), pp 60–68, August 2006. ISNN 1747–650X

Mangat, P.S. and Lambert, P. (2016) Sustainability of alkali-activated cementitious materials and geoplymers. In: Khatib, J.M. (ed.), *Sustainability of Construction Materials.* 2nd edition. Elsevier, pp. 461–478, ISBN 978–0–08–100995–6

Menadi, B., Kenai, S., Khatib, J.M., and Ait-Mokhtar, A. (2009) Strength and durability of concrete incorporating crushed limestone sand. *Construction and Building Materials Journal*, 23 (2), 625–633, February 2009. ISSN 0950–0618, doi:10.1016/j.conbuildmat.2008.02.005

Menadi, B., Kenai, S., and Khatib, J.M. (2014) Fracture behaviour of concrete containing limestone fines. *Proceedings of the ICE; Construction Materials Journal* ,167 (3), 162–170, June 2014, Paper 1200041 DOI: 10.1680/coma.12.00041.

Milner, S.J. and Woodard, L. (2016) Sustainability of engineered wood products. In: Khatib, J.M. (ed.), *Sustainability of Construction Materials.* 2nd edition. Elsevier, pp. 159–180, ISBN 978–0–08–100995–6, ISBN 978–0–08–100995–6

Molitch-Hou, M. (2016) From carbon to concrete: Pollution converted into 3D printing feedstock for construction, 3D Printing Industry, Featured, News, Research, 15 March, http://3dprintingindustry.com/2016/03/15/from–carbon–to–concrete–pollution–converted–into–3d–printing–stock/

Morales-Hernandez, B., Khatib, J.M., and Gardiner, P. (2005) Use of Municipal Solid Waste Incineration Bottom Ash (MSWI–BA) in Cement Mortar. *Proceedings of the 1st Global Slag Conference – From Problem to Opportunity*, Dusseldorf, Germany, 14–15 November 2005, pp. 210–218. http://www.propubs.com/gsc/gs05reviewed.html

O'Farrell, M., Chaipanich, A., Kinuthia, J.M., Sabir B.B., and Wild, S. (2001) A new concrete incorporating wastepaper sludge ash (WSA). Innovations and Developments in Concrete Materials and Construction. 149 – 158. Thomas Telford Publishers, 2001, edited by Ravindra K. Dhir, P. C. Hewlett, Laszlo J. Csetenyi; ISBN: 0 7277 3179 3

Okeyinka, O.M., Oloke, D.A., and Khatib, J.M. (2015a) Self-compacting grout to produce two-stage concrete. *2nd International Sustainable Buildings Symposium (ISBS 2015)*, 28–30 May 2015, Gazi University, Ankara, Turkey, pp. 406–410 http://www.isbs2015.gazi.edu.tr ISBN: 978–975–507–278–4

Okeyinka, O.M., Oloke, D.A., and Khatib, J.M. (2015b) A review of recycle use of post–consumer waste paper in construction. *Proceedings of the 1st International Conference on Bio–based Building Materials (ICBBM 2015)*, Eds. Amziane and Sonebi, 21–24 June 2015, Claremont–Ferrand, France. RILEM publication, pp. 711–717, ISBN PRO 99: 978–2–35158–154–4 https://sites.google.com/site/icbbm2015/home

Owens, P.L., Boarder, R., and Khatib, J.M. (2016). Sustainability of lightweight aggregates manufactured from waste clay for reducing the carbon footprint of concrete. In: Khatib, J.M. (ed.),

Sustainability of Construction Materials. 2nd edition. Elsevier, pp. 209–246, ISBN 978-0-08-100995-6

Pyper, J. (2014) New Formula Could Cut Pollution from Concrete. *Scientific American*, ClimateWire, Sustainability, 26 September; http://www.scientificamerican.com/article/new-formula-could-cut-pollution-from-concrete/

Rubenstein, M. (2012) Emission from the Cement Industry, State of the Planet, Climate, Earth Institute. Columbia University, http://blogs.ei.columbia.edu/2012/05/09/emissions-from-the-cement-industry/

Savastano, H., Jr., Santos, S.F., Fiorelli, J., and Agopyan, V. (2016) Sustainable use of vegetable fibres and particles in civil construction, In: Khatib, J.M. (ed.), *Sustainability of Construction Materials.* 2nd edition. Elsevier, pp. 479–522, ISBN 978-0-08-100995-6

Sonebi, M., Ammar, P., and Diederich, P. (2016) Sustainability of cement, concrete and cement replacement materials in construction. In: Khatib, J.M. (ed.), *Sustainability of Construction Materials.* 2nd edition. Elsevier, pp 417–460, ISBN 978-0-08-100995-6

Topçu, İ.B. and Demir, A. (2007) Durability of rubberized mortar and concrete. *Journal of Materials in Civil Engineering*, 19 (2), 173–178.

Topçu, I.B. and Uygunoglu, T. (2016). Sustainability of waste rubber in construction. In: Khatib, J.M. (ed.), *Sustainability of Construction Materials.* 2nd edition. Elsevier, pp. 599–626, ISBN 978-0-08-100995-6

Widyatmoko, I. (2016). Sustainability of bituminous materials in construction. In: Khatib, J.M. (ed.), *Sustainability of Construction Materials.* 2nd edition. Elsevier, pp. 345–372, ISBN 978-0-08-100995-6

Wright, L. and Khatib, J.M. (2016) Sustainability of desulphurised waste materials in construction. In: Khatib, J.M. (ed.), *Sustainability of Construction Materials*, 2nd edition. Elsevier, pp. 685–720, ISBN 978-0-08-100995-6

Wild, S., Khatib, J.M., and O'Farrell, M. (1997) Sulphate resistance of mortar containing ground brick clay calcined at different temperatures. *Cement and Concrete Research Journal*, 27 (5), 697–709. ISSN: 0008-8846

Wild, S., Khatib, J.M., and Jones, A. (1996) Relative strength, pozzolanic activity and cement hydration in superplasticised metakaolin concrete. *Cement and Concrete Research Journal*, 16 (10), 1537–1544. ISSN: 0008-8846

Zhansaya, N., Negim, E.S.M., Rauash, A.M., Khatib, J.M., Mun, G.A., and Williams, C. (2014) Effect of pH on the physico-mechanical properties and miscibility of MC/PAA blends. *Carbohydrate Polymers Journal*, 101, 415–422, Elsevier, http://dx.doi.org/10.1016/j.carbpol.2013.09.047

19

Application of Zeolites to Environmental Remediation

Craig D. Williams

Faculty of Science and Engineering, University of Wolverhampton, United Kingdom

19.1 Introduction

The use of zeolites for environmental remediation will be examined in this chapter. Both natural and synthetic zeolites have been used for remediation purposes; however, due to their cost, the synthetic ones are normally only used in special circumstances.

A surfactant-modified zeolite has been used to selectively remove the perchlorate anion from contaminated waters containing other anions such as hydroxide, carbonate, chloride, and sulphate. However, when waters containing nitrate were treated, the selectivity for perchlorate dropped, and the final selectivity sequence was perchlorate>> nitrate> sulphate> chloride> carbonate/hydroxide. Under batch conditions, the zeolite sorption capacity for perchlorate was 40–47 mmol kg^{-1}, while in flow systems, this reduced to 34 mmol kg^{-1}. When the perchlorate-loaded zeolite was exposed to ultra-pure water, only 1% leached out; however, when exposed to a 0.5 M nitrate solution, about 40% of the perchlorate leached out (Zhang *et al.*, 2007).

19.2 Heavy Metal Removal

Naturally occurring arsenic contaminates groundwater in many countries at levels higher than the WHO guidelines (10 mg L^{-1}). Zeolites have been widely studied for the removal of arsenic (both As^{3+} and As^{5+}) from both contaminated wastewaters and groundwaters. Most work has been performed using natural clinoptilolite, chabazite, or mordenite, usually in the form of tuffs. Several studies have assessed increasing zeolite selectivity by modifying it; for example, Dousova *et al.* 2006 modified a clinoptilolite tuff with an iron (II) solution, which deposited iron (III) oxi-hydroxide on the surface of the zeolite. When compared with untreated zeolite, there was a marked increase in the sorption of arsenate from 0.5 to >20 mg/g. Payne and Abel-Fattah (2005) also assessed the efficiency of iron-modified clinoptilolite, chabazite, and synthetic zeolites X and A, finding that the iron-modified chabazite performed the best, removing approximately 50% of the arsenate and 30% of the arsenite. Campos and Buchler (2007) modified a natural stilbite/laumontite tuff using the surfactant hexadecyltrimethylammonium; this greatly improved the sorption of arsenate from aqueous solution. Qui and Zheng (2007) synthesised a cancrinite zeolite from fly ash and studied its sorption capacity for arsenate compared to other adsorbents such as activated carbon, silica gel, and synthetic zeolites Y and A. The study found that cancrinite had a capacity of 5.1mg L^{-1} compared to 4.0 mg L^{-1} for activated carbon, 0.46 mg L^{-1} for silica gel, 1.4 mg L^{-1} for zeolite Y, and

Urban Pollution: Science and Management, First Edition. Edited by Susanne M. Charlesworth and Colin A. Booth.

4.1 mg L^{-1} for zeolite A. Shevade and Ford (2004) examined the use of synthetic zeolites for arsenate removal from wastewaters. They found that ammonium-loaded zeolite Y showed significant arsenate removal, as the buffering effect of the ammonium kept the solution within the pH range 3.5–7.0. As it is known that extremes of pH adversely affect the sorption of arsenate into zeolites, this buffering effect should lead to savings in cost and processing time for effluent treatment due to the elimination of a pH pre-conditioning step prior to arsenate removal. Xu *et al.* (2002) showed that an aluminium-loaded natural zeolite P could effectively remove low concentrations of arsenate from solutions containing chloride, nitrate, sulphate, chromate, and acetate anions; however, the presence of phosphate greatly interfered with arsenate adsorption.

19.3 Pesticide Removal

With the increasing use of pesticides in agriculture, it was inevitable that they would eventually be found in surface and ground waters. The EU imposed a 0.1 ppm limit for total pesticide concentration in drinking water in the 1990s. This has encouraged a number of water treatment companies to investigate the use of zeolites as potential adsorbents for a variety of pesticides. Clinoptilolite has been used to removed lindane and aldrin from aqueous solution. Both kinetic and equilibrium tests have been conducted by Sprynskyy *et al.* (2008). They found that during kinetic tests, most of the aldrin and half of the lindane was adsorbed during the first hour. After 48 hours, the equilibrium values were 95% aldrin and 68% lindane removal. Modified zeolites have also been studied; Jovanovic *et al.* (2006) have modified the following zeolites: synthetic zeolites A, X, Y and natural clinoptilolite with a range of quaternary ammonium-type surfactants. These surfactants changed the zeolite surface from hydrophilic to hydrophobic and greatly increased the efficiency of pesticide removal. Lemic *et al.* (2006) has studied the removal of atrazine, lindane, and diazinone from water by organo-zeolites. They found that, like Jovanovic *et al.* (2006), the hydrophobic character of the organo-zeolite surface enhanced pesticide removal with adsorption capacities of 2.0 mmol/g for atrazine, 4.4 mmol/g for diazinone, and 3.4 mmol/g for lindane. The pesticide 2,4–D was shown to be photo degraded by titanium dioxide supported on the zeolite synthetic HZSM-5. The high activity of the supported titanium dioxide is believed to be due to the synergistic effects of improved adsorption of 2,4-D and efficient delocalisation of photo-generated electrons by the zeolite support (Shankar *et al.*, 2006). The zeolite HZSM-5 has also been shown to be a very efficient adsorbent for paraquat and looks like a promising primary treatment for acute paraquat poisoning (Walcarius and Mouchotte, 2004).

19.4 Zeolites Used in Transport

The desulphurisation of transportation fuels is becoming increasingly mandatory for many governments; however, the present method of high pressure and temperature catalytic reactors is quite costly. Several studies have looked at the possibility of using zeolites for desulphurisation. Faghihian *et al.* (2007) showed that mercury- or silver-loaded clinoptilolite could adsorb sulphur compounds from commercial fuels selectively and with a high capacity. An 83.6% reduction in the sulphur content of fuel was observed with a liquid-to-solid ratio of 501,100 gm L^{-1} at 353K and a flow rate of 3 mL min^{-1}. De Lasa *et al.* (2006) showed that the zeolite ZSM-5 could catalytically perform dehydrosulphidation of petrol. Zeolites have also been shown to remove nitrogen compounds from transportation fuels. Hernandez-Maldonado *et al.* (2006) used copper-loaded zeolite Y to desulphurise and denitrogenate diesel fuel. The zeolite adsorbent could be readily regenerated by heating it in a stream of air at 350°C, followed by reduction of the copper (II) to copper (I). Xue *et al.* (2006) showed the efficiency of cerium-loaded zeolite Y for the selective removal of trace organic sulphur compounds from diesel. They found the uptake of sulphur compounds was strongly dependent on the valency of the cerium, with cerium (IV) having the highest adsorptive capacity. With the addition of copper (II) oxide to the cerium-loaded zeolite, the initial 5 ppm sulphur was reduced to 0.01 ppm, making this a promising desulphurisation process to prepare clean fuels for fuel cells.

19.5 Zeolites Used in Wastewater Treatment

Zeolites have been used since the 1930s to remove heavy metals from water streams; they are particularly good at the removal of Cd^{2+}, Pb^{2+}, Zn^{2+}, Cu^{2+}, and Cr^{3+}. Both natural and synthetic zeolites have been used, but due to the lower cost of natural zeolites, these have been studied to a greater extent. Widiastuti *et al.* (2008) provides a very good review of the use of natural zeolites for greywater treatment. The transformation of fly ash into a variety of zeolites, which were subsequently used for the removal of mercury and lead ions from wastewater, was studied by Somerset *et al.* (2008). They prepared the following zeolites – Faujasite, sodalite, and zeolite A – and then examined their ability to remove both mercury and lead at 5–10 gL^{-1} concentrations. Their study showed that the zeolites effected a 95% reduction for lead and a 30% reduction for mercury at a pH of 4.5. Kocasoy and Sahin (2007) looked at the ability of clinoptilolite to remove copper, iron, and zinc from industrial wastewaters and found the clinoptilolite had 100% efficiency in removing these ions. Zeolites have also been modified for wastewater treatment. Roque-Malherbe *et al.* (2007) supported a zeolite on expanded clay aggregates; these aggregates help to provide microporosity, which increased the diffusion of heavy metal ions into the zeolite framework, thereby increasing the efficiency of the ion removal process. Rios *et al.* (2008) looked at treating acid mine drainage with zeolites. They compared commercial zeolite Y, synthetic phillipsite, and faujasite synthesised from natural clinker. They found faujasite to be the most efficient; it increased the pH from 1.69 to 8.00 and a metal adsorption selectivity as follows Fe>As>Pb>Zn>Cu>Ni>Cr. Their findings indicate that the use of zeolites has the potential to provide improved methods of treatment for acid mine drainage.

19.6 Zeolites Used in Nuclear Clean-Up

Nuclear power plants use the fission of uranium to produce power; however, this fission process releases a variety of radioactive elements, which need to be absorbed selectively. Once a nuclear fuel rod has reached the end of its lifespan, it is stored under water in 'ponds'. During this storage, there is a build-up of fission products such a $^{137}Cs^{+}$ and $^{90}Sr^{2+}$, the 'pond' water is very alkaline (pH ~ 11) and contains both sodium hydroxide and carbonate. Zeolites, in particular, clinoptilolite and mordenite, can selectively remove picograms of Cs^{+} and Sr^{2+} from solutions containing 100 mgL^{-1} Na^{+}, 1.5 mgL^{-1} Ca^{2+}, and 0.7 mgL^{-1} Mg^{2+} (Dyer, 1988). Zeolites have also been considered as an inner liner for spent fuel canisters being housed in a geological repository. A study by Puig *et al.* (2008) found that dehydrated zeolites were a promising material. Kaygun and Akyil (2007) examined the adsorption of thorium into zeolites but found that a composite adsorbent performed best. Studies have also investigated adsorbing the radioactive ions into a zeolite and then vitrifying the zeolite. Lima *et al.* (2007) looked at the adsorption of ^{60}Co into zeolite X and then vitrifying the zeolite. It was shown that no cobalt leached from the vitrified sample. Watanabe *et al.* (2006) studied a zeolite/apatite composite for the long-term immobilisation of radioactive ions. They used pulse electric current sintering to fabricate the composite, which showed increased micro-hardness compared to zeolite or apatite alone.

19.7 Zeolites in Organic Clean-Up

Zeolites have also found use as adsorbents for organic molecules including phenols, aniline, and nicotine. Kamble *et al.* (2008) looked at the adsorption of phenol and o-chlorophenol on zeolite Y formed from fly ash, surface-modified fly ash, and commercial zeolite. The study found that surface-modified fly ash zeolite Y adsorbed 4.05 and 3.24 times more phenol than fly ash zeolite Y and commercial zeolite Y, respectively. For o-chlorophenol, the surface-modified zeolite Y adsorbed 2.29 and 1.8 times more than fly ash zeolite Y and commercial zeolite Y, respectively. The performance of the surface-modified zeolite was attributed to the increased hydrophobicity imparted by the surface modification. Eriksson (2008) looked at the release of m-cresol from dealuminated zeolite Y. He found that the zeolite provides a slow release of the m-cresol and used this as a slow release agent for preventing microbial growth. One study by O'Brien *et al.* (2007) looked at the adsorption of aniline on the zeolite ZSM-5 loaded with copper. They found an uptake level of 40 mg g^{-1}; the zeolite was then heated, and the catalytic decomposition of the aniline resulted in carbon dioxide, water, and nitrogen, with minor amounts of nitrogen oxide being formed. Stosic *et al.* (2008) examined the adsorption of nicotine from aqueous solutions. The study looked at clinoptilolite, ZSM-5, zeolite P, and zeolite beta and found that zeolite P performed the best.

19.8 Zeolites used in Agriculture

Zeolites have been used in agriculture and animal husbandry since the 1960s, mainly in Southeast Asia, which has widespread natural deposits. There are three natural zeolites that are widely used – clinoptilolite, chabazite, and phillipsite – and one synthetic zeolite A. The use of zeolites as soil amendments has been studied by Rafiee and Saad (2008), who examined the use of natural Iranian clinoptilolite for growing lettuce. The study found that compared to soil with no zeolite, the yield of lettuce was significantly increased: 1507 g compared to 275 g. Zeolites (clinoptilolite) have also been shown to enhance the utilisation of urea nitrogen in maize; it was shown that the zeolite reduced ammonia loss in acid soils, thereby increasing crop yields (Ahmed *et al.*, 2008). Clinoptilolite has also been used to reduce soil erosion; Andry *et al.* (2007) showed that the addition of 10% artificial zeolite to an acid soil greatly promoted plant growth, which reduced soil erosion significantly. The addition of zeolites to loamy soil helps reduce nitrate and ammonia leaching. Sepaskhah and Yousefi (2007) found that the addition of 60 kg ha^{-1} reduced nitrogen leaching by 40% and at levels of 8 tonnes ha^{-1} this increased to 65%. Natural mordenite has been shown to have a beneficial effect on the growth of oats (Spinola *et al.*, 2007). Clinoptilolite has also shown beneficial effects for wheat growth. Tsadilas and Argyropoulos (2006) studied the effect of adding clinoptilolite rich tuff and ammonium sulphate fertiliser. The study found a 52% increase in wheat yield due to the reduction in fertiliser leaching. Kavoosi (2007) studied the addition of clinoptilolite on rice fields and found that the addition of 8 tonnes ha^{-1} significantly improved the nitrogen utilisation of urea-based fertiliser by 65%. The zeolite was acting as a slow release fertiliser, and crop yields were significantly improved. Jakkula *et al.* (2006) examined the use of Linde F zeolite as a soil amendment for growing maize. This study showed that the high affinity for ammonium exchanged into the zeolite; this subsequently led to a slow release of ammonium during the experimental period and an increase in plant growth. Leggo and Ledesert (2008) have studied a combination of zeolite and chicken litter. This study found that the zeolite entrapped the nutrients from the chicken litter and then released them slowly over a prolonged period, leading to significantly higher growth rates for a number of plants. The study also showed how this organo-zeolite mixture can be used to ameliorate contaminated mine spoil. Studies on the use of zeolites in animal husbandry are numerous due to their ability to improve health and protect against mycotoxin intoxication. Mohri *et al.* (2008) looked at the effects of short-term supplementation of clinoptilolite on dairy calves. The study showed that a 2% zeolite addition had significant effects on concentrations of calcium, phosphorous sodium, and iron. The increased levels of calcium and iron should promote faster healthier growth, while the reduction in phosphorous should be considered during zeolite supplementation. Thilsing *et al.* (2007) also found that the addition of synthetic zeolite A had a marked effect on calcium and phosphorus levels in cattle; however, they also noted that there was an increase in serum aluminium levels due to the partial breakdown of the zeolite A in the intestinal tract. The effect of zeolite supplements in pigs has also been studied. Prvulovic *et al.* (2007) examined the effects of a clinoptilolite supplement in pigs' diet on performance and serum parameters. The zeolite was added at 5 g per kg of diet, and individual live weights were recorded at 45, 90, and 135 days. The zeolite-supplemented group had higher body weight gain compared to the control group (at 90 days +7%). It was also noted that the zeolite-fed pigs had higher triglyceride and lower cholesterol concentrations. The effects of natural zeolite addition to sheep rumen fermentation has also been studied. Ruiz *et al.* (2007) looked at the addition of 1.5%, 3.0%, and 4.5% clinoptilolite addition to a sheep feed of 70% alfalfa and 30% concentrate. They found that 3.0% clinoptilolite significantly increased the concentration of propionic acid, while the concentrations of acetic and butyric acids were not affected. Zeolites have also been studied as potential silicon supplements for horses. O'Connor *et al.* (2008) examined the use of zeolite A and orthosilicic acid as silicon supplements. They found that the zeolite increased calcium retention, but the zeolite promoted urinary silicon excretion. The largest use of zeolites in animal husbandry is in the poultry field. However, the effects of zeolite on weight gain, growth rate, food intake, egg production, and egg properties are not universally proved. Some studies show beneficial effects due to zeolite inclusion in bird feed, while others show no effects or even detrimental effects. Several reasons have been put forward, such as the type of zeolite or whether it is natural or synthetic. It should be noted, however, that the majority of studies show a beneficial effect on water consumption, feed ratio efficiency, nutrient utilisation, aflatoxicosis, and litter condition. Sehu *et al.* (2007) reported on the use of 0.5% zeolite A in bird feed to reduce aflatoxin in quails. They found that the zeolite promoted body weight increase and significantly protected the birds from aflatoxin. Strakova *et al.* (2008) investigated the effects of clinoptilolite feed supplements at 0.5%, 1.5%, and 2.5%. The study found that at the end of the experiment the birds fed zeolite showed significantly higher live weight: 3% for female birds and 5% for male birds. The effect of zeolite addition on broiler feeding schemes has been investigated by Acosta *et al.* (2005). They found that that addition of clinoptilolite to a bird's diet at the age of 21 days had a beneficial effect, yielding better-quality meat and leaner carcasses. The addition of zeolite did not improve live weight, but the feed conversion with lower feed and protein intake was improved. Due to the very high capacity for clinoptilolite to adsorb ammonia, it has been widely used in poultry houses to remove or entrap the ammonia from chicken litter. The zeolite is normally added to the bedding in a finely ground form at about 5% wt/wt. Several studies have shown the efficiency of ammonia entrapment by zeolite. Cai *et al.* (2007) showed that the addition of zeolite in the range 2.5% to 10% significantly reduced odour. A 67% reduction was noted for a topical application on fresh hen manure. Eleroglu and Yalcin (2005) also found that clinoptilolite addition to poultry house litter had a positive effect on broiler performance, poultry house conditions, and litter moisture content. For litter consisting of wood shavings, they found that the optimum addition rate was 25% zeolite.

19.9 Zeolites as Slow Release Agents

Recently zeolites have been examined as potential slow release agents for a number of drugs. The majority of zeolites studies have been the synthetic ones, where the purity of the zeolite can be guaranteed. Rimoli *et al.* (2008) studied zeolite X and a co-crystallised zeolite X and A for their ability to release the drug ketoprofen. Both zeolites were activated with 800 mg of drug per 2 g of zeolite. Then release studies were carried out at different pH conditions to mimic gastrointestinal fluids. The absence of release under acidic conditions and the double-phased release at two different pH values (5 and 6.8) show that these materials offer a good, potential, slow release delivery system for ketoprofen. Horcajada *et al.* (2006) investigated the use of dealuminated faujasite as a slow release agent for ibuprofen. They found 15 weight per cent ibuprofen was adsorbed into the zeolite cavities. The release profile was a two-stage process with the initial release being governed by a diffusion process which was independent of the aluminium present in the zeolite. The second stage, however, was dependant on the aluminium content, the release being slower when the Si/Al ratio is increased. Zeolite composites have also been shown to be potential slow release agents; Arruebo *et al.* (2006) examined the sustained release of doxorubicin from a zeolite–magnetite nanocomposite. They used high-energy ball milling at room temperature to produce the nanocomposites, which were able to store and subsequently releases substantial amounts of doxorubicin. The advantage of the nanocomposite was its ability to not agglomerate, unlike zeolite without the magnetite. Zeolites have also been shown to be potential slow release agents for veterinary drugs. Dyer *et al.* (2000) studied the use of commercial zeolite Y for a number of anthelminitic drugs. They found that zeolite Y was a very efficient slow release agent for the drug dichlorovos; however, the release of the drugs pyrantel and fenbendazole was too fast. This was attributed to the large size of these two drugs, because of which they could not diffuse into the zeolite structure but simply stuck to the surface.

Zeolites have also been investigated as slow release agents for antibacterial silver release. The silver has been incorporated either by ion exchange or isomorphous substitution of silver for aluminium in the zeolite framework. Bingshe *et al.* (2008) examined the effectiveness of silver-loaded zeolite A against *Escherichia coli*. They found that by heat-treating the silver-loaded zeolite at 400°C, 450°C, and 500°C, the release of silver was slowed and prolonged antibacterial activity was provided. Kwakye-Awuah *et al.* (2008) studied the efficiency of silver-loaded zeolite X against *Escherichia coli*, *Pseudomonas aeruginosa*, and *Staphylococcus aureus*. They found that the silver-loaded zeolite X completely killed the bacteria with 1 hour of exposure. The zeolite was then retrieved and reused a total of three times, and in each case the zeolite killed fresh bacteria with 1 hour of exposure. The influence of light on the antimicrobial efficiency of silver-loaded zeolite was investigated by Inoue *et al.* (2008). Their study showed that the onset of bactericidal activity can be controlled by light irradiation.

19.10 Zeolite Safety

Zeolites are generally thought to be benign when either digested or inhaled; this is due to their intrinsic chemical stability. Table 19.1 shows the data for synthetic zeolite A. The only recorded danger is from the natural zeolite erionite, which has been linked to occurrences of lung cancer in Turkey. This is due to the physical form of the erionite crystals, which occur as long thin needle-like crystals, which become lodged in the lung lining.

Table 19.1 Tests on zeolite A for human safety purposes.

Test	Result
Acute toxicity	
LD50	Essentially non-toxic
Percutaneous toxicity	Essentially non-toxic
Irritation of the eye	None
Irritation of the skin	Some; transient
Sensitisation of the skin	None
Inhalation	No effects
Subchronic toxicity	
Oral	No effects
Percutaneous	No systematic effects
Inhalation	No effects
Metabolism	Hydrolysed and excreted
Teratogy	No effects
Chronic toxicity	
Oral	No effects, non-carcinogenic at 0.1% in diet
Inhalation	No effects at 50 mg m^{-3} for 12 months or at 20 mg m^{-3} for 22 months

Source: Synthesis, characterisation and use of zeolitic microporous materials, J.B. Nagy, P. Bodart, I. Hannus, and I. Kirisci, DecaGen Ltd., 1998.

19.11 Conclusions

Zeolites have found numerous uses around the world for remediation of both inorganic and organic pollutants. They have found to be useful as slow release agents for both drugs and fertilisers. Due to the cost implications, natural zeolites have found favour in remediation, while synthetic zeolites have found favour as slow release agents.

References

Ahmed, O.H., Hussin, A., Ahmad, H.M.H., Rahim, A.A., and Majid, N.M.A. (2008) Enhancing the urea-N use efficiency in maize cultivation on acid soils amended with zeolite and TSP. *The Scientific World Journal*, 8, 394–399.

Andry, H., Yamamoto, T., and Inoue, M. (2007) Effectiveness of hydrated lime and artificial zeolite amendments and sedum plant cover in controlling soil erosion from an acid soil. *Australian Journal of Soil Research*, 45, 266–279.

Arruebo, M., Fernandez-Pacheco, R., Irusta, S., Arbiol, J., Ibarra, M.R., and Santamaria, J. (2008) Sustained release of doxorubicin from zeolite–magnetite nanocomposites prepared by mechanical activation. *Nanotechnology*, 17, 4057–4064.

Bingshe, X., Wensheng, H., Shuhua, W., Liqiao, W., Husheng, J., and Xuguang, L. (2008) Study on the heat resistant property of Ag/4A antibacterial agent. *Journal of Biomedical Materials Research B Applied Biomaterials*, 84, 394–399.

Cai, L.S., Koziel, J.A., Liang, Y., Nguyen, A.T., and Xin, H.W. (2007) Evaluation of zeolite for control of odorants emissions from simulated poultry manure storage. *Journal of Environmental Quality*, 36, 184–193.

Campos, V. and Buchler, P.M. (2007) Anionic sorption onto modified natural zeolites using chemical activation. *Environmental Geology*, 52, 1187–1192.

De Lasa, H., Enriquez, R.H., and Tonetto, G. (2006) Catalytic desulphurisation of gasoline via dehydrosulphurdation. *Industrial and Engineering Chemistry Research*, 45, 1291–1299.

Dousova, B., Grygar, T., Martaus, A., Fuitova, L., Kolousek, D., and Machovic, V. (2006) Sorption of As-V on alumino silicates treated with Fe-II nanoparticles. *Journal of Colloid and Interface Science*, 302, 424–431.

Dyer, A. (1988) *An Introduction to Zeolite Molecular Sieves*. Wiley, Chichester, ISBN 0471 91981 0

Dyer, A., Morgan, S., Wells, P., and Williams, C. (2000) The use of zeolites as slow release anthelmintic carriers. *Journal of Helminthology*, 74, 137–141.

Eleroglu, H. and Yalcin, H. (2005) Use of natural zeolite-supplemented litter increased broiler production. *South African Journal of Animal Science*, 35, 90–97.

Eriksson, H. (2008) Controlled release of preservatives using dealuminated zeolite Y. *Journal of Biochemical and Biophysical Methods*, 70, 1139–1144.

Faghihian, H., Vafadar, M., and Tavokil, T. (2007) Desulphurization of gas oil by modified clinoptilolite. *Iranian Journal of Chemistry & Chemical Engineering – International English edition*, 26, 19–25.

Hernandez-Maldonado, A.J., Yang, F.H., Qi, G.S., and Yang, R.T. (2006) Sulphur and nitrogen removal from transportation fuels by pi-complexation. *Journal of the Chinese Institute of Chemical Engineers*, 37, 9–16.

Horcajada, P., Marquez-Alvarez, C., Ramila, A., Perez-Pariente, J. and Vallet-Regi, M. (2006) Controlled release of ibuprofen from dealuminated faujasites. *Solid State Science*, 8, 1459–1465.

Inoue, Y., Kogure, M., Matsumoto, K., Hamashima, H., Tsukada, M., Endo, K., and Tanaka, T. (2008) Light irradiation is a factor in the bacterial activity of silver loaded zeolite. *Chemical and Pharmaceutical Bulletin*, 56, 692–694.

Jovanovic, V., Dondur, V., Damjanovic, L., Zakrzewska, J., and Tomasevic-Canovic, M. (2006) Improved materials for environmental applications: Surfactant modified zeolites. In Uskovic, D.P., Milonjic, S.K., and Rakovic, D.I. (eds.), 7th Conference of the Yugoslav Materials Research Society. *Recent Developments in Advanced Materials and Processes*, 518, 223–228.

Kamble, S.P., Mangrulkar, P.A., Ansiwal, A.K.B., and Rayalu, S.S. (2008) Adsorption of phenol and o-chlorophenol on surface altered fly ash based molecular sieves. *Chemical Engineering Journal*, 138, 73–83.

Kaygun, A.K. and Akyil, S. (2007) Study of the behaviour of thorium adsorption on PAN/ zeolite composite adsorbent. *Journal of Hazardous Materials*, 147, 357–362.

Kocasoy, G. and Sahin, V. (2007) Heavy metal removal from industrial wastewater by clinoptilolite. *Journal of Environmental Science and Health Part A Toxic / Hazardous Substances & Environmental Engineering*, 42, 2139–2146.

Kwakye-Awuah, B., Williams, C., Kenward, M., and Radecka, I. (2008) Antimicrobial action and efficiency of silver loaded zeolite X. *Journal of Applied Microbiology*, 104, 1516–1524.

Leggo, P.J. and Ledesert, B. (2008) Organic zeolitic soil systems: A new approach to plant nutrition. Chapter 9 in Elsworth, L.R. (ed.), *Fertilizers: Properties, Application and Effects*. Nova Scientific Publishers, Canada, pp 223–239. ISBN 1604564830

Lemic, J., Kovacevic, D., Tomasevic-Canovic, M., Stanic, T., and Pfend, R. (2006) *Water Research*, 40, 1079–1085.

Lima, E., Bosch, P., and Bulbulian, S. (2007) Immobilization of cobalt in collapsed non-irradiated and gamma-irradiated X zeolites.

Applied Radiation and Isotopes, 65, 259–265.

Mohri, M., Seifi, H.A., and Maleki, M. (2008) Effects of short term supplementation of clinoptilolite in colostrums and milk on the concentration of some serum minerals in neonatal dairy calves. *Biological Trace Element Research*, 123, 116–123.

O'Brien, J., Curtin, T., and O'Dwyer, T.F. (2007) Removal of organic compounds from waste streams: A combined approach. In Brebbia, C.A. and Kungolos, A. (eds.), 4th International conference on water resources management. *Water Resources Management IV*, 103, 447–456.

O'Connor, C.I., Nielsen, B.D., Woodward, A.D., Spooner, H.S., Ventura, B.A., and Turner, K.K. (2008) Mineral balance in horses fed two supplemental silicon sources. *Journal of Animal Physiology and Animal Nutrition*, 92, 173–181.

Payne, K.B. and Abel-Fattah, A. (2005) Adsorption of arsenate and arsenite by iron–treated activated carbon and zeolites. *Journal of Environmental Science and Health Part A Toxic/ Hazardous Substances and Environmental Engineering*, 40, 723–749.

Prvulovic, D., Jovanovi-Galovic, A., Stanitic, B., Popovic, M., and Grubor-Lajsic, G. (2007) Effects of a clinoptilolite supplement in pig diets on performance and serum parameters. *Czech Journal of Animal Science*, 52, 159–164.

Puig, F., Dies, J., de Pablo, J., and Martinez-Esparza, A. (2008) Spent fuel canister for geological repository: Inner material requirements and candidates evaluation. *Journal of Nuclear Materials*, 376, 181–191.

Qui, W. and Zheng, Y. (2007) Arsenate removal from water by alumina-modified zeolite recovered from fly ash. *Journal of Hazardous Materials*, 148, 721–726.

Rafiee, G.R. and Saad, C.R. (2008) Roles of natural zeolite as a bed medium on growth and body composition of red tilapia and lettuce seedlings in a pisciponic system. *Iranian Journal of Fisheries Science*, 7, 47–58.

Rimoli, M.G., Rabaioli, M.R., Melisi, D., Curcio, A., Mondello, S., Mirabelli, R., and Abignente, E. (2008) Synthetic zeolites as a new tool for drug delivery. *Journal of Biomedical Materials Research A*, 87, 156–164.

Rios, C.A., Williams, C.D., and Roberts, C.L. (2008) Removal of heavy metals from acid mine drainage (AMD) using coal fly ash, natural clinker and synthetic zeolites. *Journal of Hazardous Materials*, 156, 23–35.

Roque-Malherbe, R.M., Hernandez, J.J.D., Del Valle, W., and Toledo, E. (2007) Lead, copper, cobalt and nickel removal from water solutions by dynamic ionic exchange in LECA zeolite beds. *International Journal of Environment and Pollution*, 31, 292–303.

Ruiz, O., Castillo, Y., Miranda, M.T., Elias, A., Arzola, C., Rodriguez, C., and La, O.O. (2007) Levels of zeolite and their effects on the rumen fermentation of sheep fed alfalfa hay and concentrate. *Cuban Journal of Agricultural Science*, 41, 241–245.

Sehu, A., Ergun, L., Cakir, S., Ergun, E., Cantekin, Z., Sahin, T., Essiz, D., Sareyyupoglu, B., Gurel, Y., and Yigit, Y. (2007) Hydrated sodium calcium aluminosilicate for reduction of aflatoxin in quails. *Deutsche Tierarztliche Wochenschrift*, 114, 252–259.

Sepaskhah, A.R. and Yousefi, F. (2007) Effects of zeolite application on nitrate and ammonium retention of a loamy soil under saturated conditions. *Australian Journal of Soil Research*, 45, 368–373.

Shankar, M.V., Anandan, S., Venkatachalam, N., Arabindoo, B., and Murugesav, V. (2006) Fine route for an efficient removal of 2, 4-dichlorophenoxyacetic acid (2, 4-D) by zeolite-supported TiO2. *Chemosphere*, 63, 1014–1021.

Shevade, S. and Ford, R.G. (2004) Use of synthetic zeolites for arsenate removal from pollutant water. *Water Research*, 38, 3197–3204.

Somerset, V., Petrik, L., and Lwuoha, E. (2008) Alkaline hydrothermal conversion of fly ash precipitates into zeolites: The removal of mercury and lead ions from wastewater.

Journal of Environmental Management, 87, 125–131.

Spinola, A.G., Mendoza, T.M.H., Gonzalez, F.D., and Zelaya, F.P. (2007) Effect of zeolite amended andosols on soil chemical environment and growth of oat. *Interciencia*, 32, 692–696.

Sprynskyy, M., Ligor, T., and Buszewski, B. (2008) Clinoptilolite in study of lindane and aldrin sorption processes from water solution. *Journal of Hazardous Materials*, 151, 570–577.

Stosic, D.K., Dondur, V.T., Rac, V.A., Rakic, V.M., and Zakrzewski, J.S. (2008) Adsorption of nicotine on different zeolite types from aqueous solutions. *Hemijska Industrija*, 61, 123–128.

Strakova, E., Pospisil, R., Suchy, P., Steinhauser, L., and Herzig, I. (2008) Administration of clinoptilolite to broiler chickens during growth and its effect on growth rate and bone metabolism indicators. *Acta Veterinaria Brno*, 77, 199–207.

Thilsing, T., Larsen, T., Jorgensen, R.J., and House, H. (2007) The effect of dietary calcium and phosphorus supplementation in zeolite A treated dry cows on periparturient calcium and phosphorus homeostasis. *Journal of Veterinary Medicine Series A – Physiology Pathology Clinical Medicine*, 54, 82–91.

Tsadilas, C.D. and Argyropoulos, G. (2006) Effect of clinoptilolite addition to soil on wheat yield and nitrogen uptake. *Communications in Soil Science and Plant Analysis*, 37, 2691–2699.

Walcarius, A. and Mouchotte, R. (2004) Efficient in vitro paraquat removal via irreversible immobilization into zeolite pellets. *Archives of Environmental Contamination and Toxicology*, 46, 135–140.

Watanabe, Y., Ikoma, T., Suetsugu, Y., Yamada, H., Tamura, K., Komatsu, Y., Tanaka, J., and Moriyoshi, Y. (2006) The densification of zeolite/ apatite composites using a pulse electric current sintering method. A long term assurance material for the disposal of radioactive waste. *Journal of the European Ceramic Society*, 26, 481–486.

Xu, Y.H., Nakajima, T., and Ohki, A. (2002) Adsorption and removal of arsenic (V) from drinking water by aluminium loaded Shirasu-zeolite. *Journal of Hazardous Materials*, 92, 275–287.

Xue, M., Chitraker, R., Sakane, K., Hirotsu, T., Ooi, K., Yoshimura, Y., Toba, M., and Feng, Q. (2006) Preparation of cerium loaded zeolite Y for removal of organic sulphur compounds from hydrodesulphurized gasoline and diesel oil. *Journal of Colloid and Interface Science*, 298, 535–542.

Zhang, P.F., Avudzega, D.M., and Bowman, R.S. (2007) Removal of perchlorate from contaminated waters using surfactant-modified zeolite. *Journal of Environmental Quality*, 36, 1069–1075.

20

Bioremediation in Urban Pollution Mitigation: Theoretical Background and Applications to Groundwaters

Alan P. Newman[1]*, Andrew B. Shuttleworth*[2]*, and Ernest O. Nnadi*[3]

[1] *Centre for the Built and Natural Environment, Coventry University, United Kingdom*
[2] *SEL Environmental Limited, Lancashire, United Kingdom*
[3] *GITECO-UC, University of Cantabria, Santander, Spain*

20.1 Introduction

In urban areas, groundwater pollution has resulted from numerous causes, but the most common are acute losses from ruptured tanks and long-term chronic effects from infiltration of rainwater through contaminated ground. Many cities have aquifers that are badly contaminated, but even smaller communities and many rural areas have examples of both localised and extensively polluted groundwater. Bioremediation is a useful approach to both the clean-up of contaminated groundwater and the prevention of its ongoing pollution through the remediation of contaminated land. This chapter will briefly introduce the theory of bioremediation, which applies to both solid and aqueous media and then discuss the application of this important technology to the in situ biological treatment of groundwater. The treatment of polluted *solid* media is covered separately in Chapter 21. Biosorption (e.g. Gupta *et al.*, 2015) is addressed, but only where it plays a supplementary role, for example, in biologically mediated permeable reactive barriers where sorption of the contaminant in the biofilm is a precursor to degradation.

The great majority of organic environmental pollutants are highly reduced; indeed, it is often this property which make them attractive as both fuels and chemical feedstocks. Because the pollutants are highly reduced, their complete oxidation to carbon dioxide and water is often thermodynamically feasible. There is also a more limited range of compounds able to undergo reduction, of which the best examples are halogenated organic compounds or inorganic oxy-anions such as perchlorate. This chapter will thus examine both oxidative and reductive mechanisms relating to both organic and inorganic contaminants.

20.2 Essentials for Bioremediation

All microorganisms have basic requirements for survival and growth. One of the most fundamental requirements is water. Microorganisms consist of between 60% and 80% water and require a film of water around them to carry out metabolic processes. If water is limited, the organisms do not always die, but they will always show reduced activity.

A carbon source is obviously essential in a carbon-based life system. The organism gains its energy by transferring electrons through a series of energetically favourable

Urban Pollution: Science and Management, First Edition. Edited by Susanne M. Charlesworth and Colin A. Booth.

steps, which release energy by promoting various 'energy currency' compounds from a low-energy form to a high-energy form. In addition to the carbon, the organism needs an electron donor to provide the energy it needs to grow. The vast majority of organisms, and certainly many of those relevant to bioremediation, use organic matter as the electron donor, and those capable of generating the most energy in this way tend to dominate.

If these energy-releasing electron transfer steps are to continue, the final step in the chain requires a terminal electron acceptor. The higher the redox potential of the terminal electron acceptor, the more energy will be extracted from a given electron donor. Because of the high redox potential of oxygen, species which can utilise oxygen as a terminal electron acceptor dominate wherever oxygen is present.

Phosphorus, nitrogen, and sulphur are always required as nutrients to form components of both structural and metabolic molecules in the organism. Providing an ongoing supply of these elements is often one of the major requirements in any bioremediation process, but nitrogen and sulphur compounds can also play the role of either electron donor or electron acceptor (e.g. Häggblom *et al.,* 1993;Cuthbertson and Schumacher 2010; Tang *et al.*, 2005). Nitrate is the usual alternative option for electron acceptor after oxygen has been exhausted, and many aerobic bacteria also possess enzyme systems enabling them to use nitrate in this way. Organisms which can switch from aerobic to anaerobic metabolism are termed facultative anaerobes (in contrast to obligate aerobes or anaerobes). Reduction of nitrate generates various by-products consisting of nitrite ions and the gases nitric oxide, dinitrogen oxide, and finally nitrogen. This process is termed *denitrification* and can be used as an end in itself in some key parts of bioremediation processes.

Sulphate is also used as an electron acceptor by a wide range of organisms, and the potential of this pathway to assist in bioremediation has been highlighted by Miao *et al.* (2012). Sulphide, generated by this process, plays an important role in precipitation of heavy metals as sulphide.

When all the preferred inorganic terminal electron acceptors have been exhausted, many organisms can use organic molecules as both electron acceptors and donors in a metabolic pathway called fermentation. This generates much less energy and results in the incomplete breakdown of organic compounds producing much smaller ones, typically, acids and alcohols. The least energetically favourable option for microbial communities is the formation of methane by the process of methanogenesis. Two of the commonest processes are the hydrogenation of carbon dioxide and the conversion of short-chain carboxylic acids to methane. During the degradation process, electron acceptors become depleted (unless replenished), while hydrogen and carbon dioxide accumulate along with light organics produced by fermentation. Without methanogenesis, a great deal of carbon (in the form of fermentation products) would accumulate in anaerobic environments.

In addition to the macronutrients, some organisms also require a range of micronutrients, often metal ions which form parts of active sites of enzymes, but also organic micronutrients such as vitamins (e.g. Radwan and Al-Muteirie, 2001).

In summary, the main requirements for bioremediation to be successful include:

1) Microorganisms (suitably adapted for both the target material and the environment).
2) Suitable water content. Microorganisms will not operate, and some will not survive in an environment which is too dry (often an important consideration in the clean-up of solid media).
3) A source of carbon and energy (which may be the target itself or some additional material).
4) Inorganic nutrients (and sometimes organic micronutrients).

5) A suitable environment with respect to electron acceptors, that is, oxygen or other electron acceptors in the case of oxidative processes, but a suitably anoxic environment when the microbiological process is reductive.

However, while fulfilling these requirements, it is not necessarily all that is needed for successful bioremediation. This is because it is rare that organisms can simply carry out bioremediation without needing considerable intervention. There are often time constraints whereby target pollutant concentrations need to be achieved in order to fit in with project management aims and/or to drive the process along one particular metabolic pathway rather than another. To the list above therefore should be added the infrastructure required for the microbes to be supplied with oxygen (or in some cases deprived of it), ensuring the reactions produce intended products. In some circumstances, temperature may need to be controlled to ensure that the degradation achieves an acceptably fast rate.

Direct metabolism (see Singh *et al.*, 2009), where either a single species or several species working in a mixed population (e.g. Grostern and Edwards., 2006) transform the target compounds and in doing so derive all or at least the majority of their energy from the process, is a commonly used approach, but in some situations, this process switches to an in-series process involving a sequence of organisms (sequential metabolism). Sometimes, the assemblage of organisms is not capable of deriving energy (or sufficient energy) from the degradation of the target compound, but the enzyme systems it produces are able to cause the pollutant compound to degrade. In the presence of supplementary compounds to provide the energy and biomass, pollutants can often be broken down in useful parallel reactions. In these examples of co-metabolism, the supplementary source of energy/carbon is either already available in the polluted environment or it needs to be added as part of the process of managing the bioremediation (Hazen, 2009). The co-metabolic approach has the advantage that it can address even trace levels of the contaminant, as long as the substrate the organisms require for growth is maintained at acceptable concentrations (because the bacteria do not rely on the contaminant for energy), and thus, substrates which are particularly toxic can continue to be oxidised even when their concentration is insufficient to support the organism being exploited.

In another scenario, an organism can derive its energy from degradation of some material other than the target compound and, coincidentally, partially degrades the target compound, but at some point the products derived from the target material become resistant to further degradation and it stops (gratuitous metabolism). In some cases, remediation may still be achieved if the product is either much less hazardous due to changes in either its mobility or toxicity, or is more easily treatable by chemical or physical means. However, care must be taken that the partially degraded product is not more harmful than the target chemical being degraded (e.g. Parsons, 2004). In anaerobic situations, the process can get even more complex as the contaminant itself may take the role of either the electron donor or the electron acceptor, and in some cases, one contaminant serves as donor while another as acceptor.

Prior, subsequent, and parallel non-biotic transformations can be helpful since the conditions encouraging bioremediation will also encourage abiotic chemical transformation of the contaminants. These reactions can also play both positive and negative roles. For example, under anaerobic conditions, sulphide produced from anaerobic sulphate reduction can help by precipitating dissolved metal contaminants, such as lead, cadmium, zinc, and copper (Miao *et al.*, 2012) but, if present in large concentrations, sulphides can also reduce the permeability of media either directly or following subsequent oxidation to elemental sulphur if conditions change.

20.3 Bioremediation of Groundwater

The technique of 'monitored natural attenuation' is beyond the scope of this chapter since, while it almost always involves some microbiological degradation of pollutants, it does not require human intervention. However, it is covered in Carey and Fletcher (2005) and US EPA (1999), and a basic introduction is given in US EPA (2012). 'Enhanced Monitored Natural Attenuation' (sometimes just 'Enhanced Attenuation') is not natural at all, but involves adding nutrients or electron acceptors to speed up the natural processes (EUGRIS, n.d.; Early *et al.*, 2006). Bioremediation processes used in Pump and Treat groundwater remediation technology (Mercer *et al.*, 1990; Eastern research group, 1996) is also not covered since this biologically mediated treatment (e.g. Voudrias, 2001) is similar to sewage and industrial effluent treatment. In particular, in situ groundwater bioremediation will be considered, including the use of those permeable reactive barriers which have a biological element to their technology. In traditional in situ groundwater remediation, the process usually aims at remediating both the dissolved phase plume and the adsorbed or physically trapped pollutants in the saturated aquifer matrix, which can be seen as a source-centred approach. This must almost always be accompanied by, or follow on from, a source removal or remediation effort in the unsaturated zone (see Chapter 21). With in situ groundwater treatment, it is often hard to distinguish between attempts to clean up just the groundwater plume from those aiming to remediate both the plume and the contaminated phreatic zone. Indeed, if there is a contaminant present in the dissolved phase, thermodynamics dictates that there must also be adsorbed contaminant on the grains of the strata. Any attempt to remediate one will remediate the other. However, both near or at the source or in the extent of influence of a body of separate phase pollutant, there may also be either microscopic or macroscopic bodies of physically entrapped separate phase material, and in situ clean-up attempts are often designed to achieve both aims.

Permeable reactive barriers (PRBs), on the other hand, are considered to be a receptor-centred approach as they are placed down gradient of the contaminated strata and up gradient of receptors (such as surface waters or abstraction points). As long as the hydrology of the aquifer is well understood and the plume is well defined, it is not even necessary to know where the contamination originates. However, if source remediation is not possible, ongoing monitoring and maintenance must be undertaken until the inputs reduce to levels that make the process no longer necessary or sometimes until the abstraction point is replaced by another which is unaffected.

An understanding of the ways pollutants move as plumes in groundwater is essential to the successful understanding of groundwater bioremediation, whether by traditional means or using a PRB. Masters and Ela (2007) give a basic introduction, but much other material has been published including both mathematical and experimental approaches (e.g. Sudicky *et al.*, 1983; Mackay *et al.*, 1986; Klenk and Grathwohl, 2002). One of the best available publications on in situ groundwater remediation is the publication by the US EPA (2013) which was prepared by the office of solid waste and emergency response, and readers who desire a very detailed approach to this particular topic are directed there.

20.4 In Situ Plume Treatment

The commonest sources of pollutant plumes are probably from hydrocarbon spills arising either from the production, transportation, storage, or use of petroleum or the destructive distillation of coal (in either manufactured gas plants or coking plants) and the utilisation of the liquid products which arise. Both petroleum and benzole-based hydrocarbons are immiscible in water (and, usually, less dense) and are capable of forming macroscopic bodies of Light Non-Aqueous Phase Liquid (LNAPL) which sit above the water-saturated zone. This separate phase will continue to feed the dissolved phase plume, and whatever is used to clean up the plume, it can never be successful until the free product is removed. Thus, because free product recovery is an essential adjunct to bioremediation of groundwater – and also because some associated techniques have biological elements – it will be briefly covered before examining the process of treating the plume itself.

The distribution of free product is highly variable, depending on the nature of the rock strata and the degree of fracture. Even if it was a simple fine gravel unconfined aquifer, the way in which the free product distributes itself is highly complex, and because of the tendency of water to show a greater capillary rise than the LNAPL, it is rare for the free product to show 100% hydrocarbon saturation (for a report on LNAPL behaviour, see Newell *et al.*, n.d.). Another issue is the smearing of product into the intermittently saturated zone as the water table is raised and lowered. This can tie up free product in isolated, difficult-to-recover pockets in the strata which should guide the response to fresh spillages. Attempts to enhance recovery by lowering the water table can be counterproductive. In such circumstances, it is best to respond by skimming free product from the surface of the recovery wells or trenches.

This works well provided the recovery characteristics of the well are reasonable, and it will avoid the smearing and entrapment that will be inevitable (Giadom and Tse, 2015) if the water table is lowered. Thus, following, for example, a tanker accident, the free product should be recovered as quickly as possible in order to make the bioremediation step easier. This can be achieved by installing numerous recovery wells or extensive networks of recovery trenches and removing only the product itself, taking care not to lower the water table. In the past, this has been a difficult task, but nowadays a variety of devices are available for use when the free product thickness in the well starts to reduce (Abanaki, n.d.; Kassab, 2010).

If a spillage is relatively old, and particularly if there has been repeated seasonal raising and lowering of the water table, it is likely that a good proportion of the LNAPL has been temporarily trapped below the water table, some permanently so. If all sites are occupied due to repeated exposure to smearing of free product, a lot of damage will be done because the entrapped material will be hard to recover, and remediation will be difficult by whatever method is chosen. If the LNAPL is trapped in pockets below the water table, even if the main body of free product is removed, it will inevitably continue to provide a feed for the dissolved phase plume for a very long time. The common way to access this material is to remove as much as possible by lowering the water table, exposing the trapped pockets in the hope that it is sufficiently mobile to move towards the recovery well; this will create thicker layers in recovery wells and enhance recovery rates. If the strata are badly contaminated with pockets that do not move under gravity, then the usual response is vacuum-enhanced recovery, where induced airflow in the strata causes the hydrocarbon to move. There are two traditional approaches:

1) A single pump draws the water table down, recovering oil from the wells by pumping at the oil water interface while the vacuum is applied at the well head using a separate system. This recovers both oil and water together.
2) Two pumps are used in addition to the vacuum system. One submersible pump draws the water down, while the other, usually pumping more slowly, recovers oil from above the oil–water interface while the oil is thick enough for the upper pump to distinguish it. In both cases, the

Figure 20.1 Typical plant used for vacuum-enhanced recovery. Image: Geo2 Ltd.

vacuum system causes rapid air movement in the strata, encouraging the trapped LNAPL to move towards the well.

An alternative approach to these traditional methods, called *slurping*, is increasingly finding favour. When airflow in the unsaturated ground is used to achieve biodegradation, this is often called *bioslurping* (see Miller, 1996). Such a system is made up of a well into which an adjustable length of 'slurp tube' is connected to a vacuum pump and lowered into the LNAPL layer, removing free product with some groundwater. This air movement through the unsaturated zone increases oxygen content and enhances aerobic bioremediation (termed *bioventing*; see Chapter 21), as long as the conditions are otherwise favourable (moisture, nutrients, etc.). When the applied vacuum causes a rise in the water table, the slurp cycles back, removing both liquid phases (LNAPL and groundwater) together; these are sent to an oil/water separator. The airflow is usually treated before release, first to remove liquid droplets and then to remove hydrocarbon vapour. Typically, the off-gas treatment is by adsorption (US EPA, 1991; US EPA, 2006), a bio-filter (Cox and Russell, 2003; US EPA, 2006), or by a combustion system (US EPA, 2006). Once the separated phase is no longer significantly contributing to the plume, the dissolved phase can be addressed.

20.5 Electron Acceptor Management in Groundwater Bioremediation

One of the earliest approaches to enhancing aerobic biodegradation of organic pollutants in groundwater was simply to inject water saturated with air or pure oxygen similarly to enhanced biodegradation systems used in the unsaturated zone (Johnson *et al.*, 1993, Wilson *et al.*, 2002). However, air-saturated water contains only 8–10 mg/L dissolved oxygen, and even water saturated with pure O_2 can only carry up to 40 mg/L. If the technique was only aimed at plume remediation, this might provide sufficient oxygen, but about 3 g of O_2 is required for complete mineralisation of 1 g of hydrocarbon, and if the saturated zone is badly contaminated, large amounts of saturated water would be required if a significant improvement was to be achieved. Johnson *et al.* (1993) calculate this to be as much as thousands of pore volumes of water that would need to be exchanged. Some improvement may be achieved by injecting hydrogen peroxide solutions, which can effectively contain 500 mg/L of available oxygen (Johnson *et al.*, 1993), but even that amount may not be enough.

Since the early 1990s, in situ air sparging has been an alternative means of supplying the electron acceptor for in situ groundwater remediation, but it must be recognised that the active mechanism of this process is not limited to biodegradation. It is a combination of physical, chemical, and biological processes that even after several decades of use remains relatively poorly understood. This is a technique in which air is injected into water-saturated zones to reduce contaminant concentrations by a combination of aerobic biodegradation and, for the more volatile fraction, volatilisation. Volatilisation is an inevitable consequence of the airflow, and unless rates are kept sufficiently small to allow the volatilised compounds to be aerobically biodegraded as they are carried through the vadose zone, account needs to be taken of the emissions generated. In most cases, this means applying soil vapour extraction (USACE, 2002) to the area under treatment to both protect nearby receptors and prevent unacceptable release of volatile organic compounds into the atmosphere. Off-gas treatment by sorption (US EPA, 1991; US EPA, 2006) or bio-filter (Cox and Russell, 2003; US EPA 2006) or combustion (US EPA, 2006) is often required.

Energy demands for biosparging can be significant. Obviously, when the injection well is installed, the water level will equilibrate with the water table, and for air to be injected, it must first displace the water from the well, requiring a minimum pressure to be applied. This 'air entry pressure' can be as much as a few metres of water head for fine-grained strata (Johnson *et al.*, 1993). The air movement is a continuous stream following preferential channels to the vadose zone; if the saturated zone contains a layer of finer-grained material above the injection depth, it can form a barrier to the air which can become trapped under the fine-grained material until it has enough pressure to penetrate through it. Lateral movement can lead to bubbling in monitoring wells, which is often misinterpreted as being favourable air distribution in the strata (Johnson *et al.*, 1993).

Inevitably the design of most air sparging systems will be based on incomplete knowledge of the site, indicating that that the system should ideally be designed to allow for flexible operation (including expansion) where necessary. One of the difficulties is following progress, because of the potential heterogeneity of the waterbody and the likelihood of more active bioremediation close to injection wells. To confirm successful remediation, installing sufficient additional monitoring wells will give confidence, and needs to be factored into the budget and time scale. It is also important that suitable efforts be made to distinguish between pollutant removal by volatilisation to the atmosphere and removal by biodegradation, either in the waterbody or as the pollutants pass through the unsaturated zone in the gas phase.

20.6 Oxygen Releasing Compounds

Oxygen releasing compounds (ORCs) can provide electron acceptors via substances introduced into the subsurface layer which decompose to release O_2. Historically, the earliest use of ORCs utilised H_2O_2 (e.g. Hinchee *et al.*, 1991, Zappi *et al.*, 2000) or ozone injection. Hydrogen peroxide quickly decomposes to oxygen in the subsurface, usually within 4 hours (US EPA, 2016), significantly increasing O_2 in the saturated zone (Zappi *et al.*, 2000). However, H_2O_2 is toxic to microorganisms at concentrations greater than 100–200 ppm, and this must be taken into account. An alternative is ozone, which is 10 times more soluble in water than pure oxygen, and about half of dissolved ozone introduced into the subsurface decomposes to oxygen within 20 minutes (EPA, 2004).

Due to its oxidation potential, ozone can initiate chemically mediated reactions which can often be useful pre-treatments for recalcitrant compounds or even facilitate remediation without biological involvement (e.g. Schwartz *et al.*, n.d.; Siegrist *et al.*, 2000). Ozone can also be toxic to microbes (ozone solutions, n.d.), actually suppressing subsurface biological activity, but a sufficient population can usually survive and resume biodegradation after ozone application (EPA, 2004). Ozone may be injected into the subsurface in a dissolved or gaseous phase. Injection or sparging of gaseous ozone, typically at a 5% concentration, into a contaminated area in the subsurface is typical (Plummer *et al.*, n.d.). A soil vapour extraction and treatment system may need to be used, as ozone injection rates can sometimes be high enough to emit excess ozone to the unsaturated zone. Calcium and magnesium peroxide can also be injected into the saturated zone as solids or slurries (EPA, 2004). Magnesium peroxide is more commonly used because it dissolves more slowly, prolonging the release of oxygen. The results from ORC use appear to be mixed, and they are probably somewhat dependent on the site conditions. For example, Kunuku (2007) found that the implementation of an in situ bioremediation scheme to address petroleum in groundwater at the site of a former gasoline station was not fully successful. Site investigations had indicated that groundwater beneath the site was contaminated with up to 34,300 μg/L benzene, toluene, ethylbenzene, and xylenes (BTEX). The results indicated that over time, the levels of BTEX had significantly decreased, and a kinetic study showed that its removal fitted a zero-order kinetic, but at the end of the study, remediation was not complete. Other studies do report success (e.g. Mackay *et al.*, 2002), and the most well-known manufacturer of a range of ORCs presents a number of case studies claiming significant success; however, many of these are combined treatments with sorption or chemical oxidation incorporated.

20.7 Anaerobic Bioremediation of Groundwater

While aerobic conditions allow for a higher rate of biodegradation of reduced contaminants than anaerobic conditions, overall oxygen demand from dissolved metals such as iron and manganese in groundwater is often overlooked and underestimated. Even if oxygen demand is accounted for, it can readily react with dissolved iron(II) to form an insoluble iron(III) precipitate, which can clog the aquifer or may foul injection tools and well screens. Therefore, when oxygen levels are already depleted in groundwater, it can be better to facilitate oxidative anaerobic bioremediation using an alternative electron acceptor if an appropriate degradation pathway exists for the target contaminants (Alvares and Ilman, 2005).

The most commonly used electron acceptor in the absence of oxygen is nitrate; for example, Bewley and Webb (2001) targeted phenols, BTEX, and PAHs. Nitrate is highly soluble in water and mobile in aquifers; after oxygen, it provides the most free energy for microbial action. However, above 10 mg/L nitrate concentrations in groundwater have negative toxicological effects on humans and animals; therefore, care must be taken if groundwater is to be amended with nitrate, and regulatory interest will be intense. Nitrate is reduced to nitrite, nitric oxide (reactive intermediate), dinitrogen oxide (reactive intermediate), nitrogen gas, or a combination of these bi-products, depending on the microbes present. The latter three are soluble gaseous bi-products and but will generally escape into the vadose zone; however, they may become trapped in the pore spaces, displacing water and reducing the hydraulic conductivity of the saturated matrix. If the process is not completed, dinitrogen oxide is a greenhouse gas with a 100 year global warming potential nearly 300 times greater than CO_2; it is therefore very important that the process is as complete as possible.

Sulphate is also very soluble in water, will not sorb appreciably, and is generally unreactive. Sulphide, the end product of sulphate reduction, precipitates with iron(II) and is effectively immobilised. However, in acidic environments, sulphide can produce hydrogen sulphide, with obvious health and aesthetic consequences. Anderson and Lovley (2001) carried out the first field study showing that it is possible to use sulphate to stimulate anaerobic benzene degradation in a petroleum-contaminated aquifer. They introduced sulphate using an infiltration gallery and used isotope tracers to demonstrate that sulphate was the dominant terminal electron acceptor. They concluded that, in their particular aquifer, the presence of significant quantities of iron(II) was preventing free sulphide generation.

Sometimes the electron acceptors are the pollutants of interest, and in such circumstances remediation of groundwater containing nitrate and other electron acceptors such as sulphate and perchlorate (Xu *et al.*, 2003) can be achieved by injecting electron donors, usually organic compounds, in sufficient quantities that anaerobic conditions are established.

20.8 Reductive Anaerobic Degradation

The removal of nitrate and perchlorate are examples of reductive anaerobic degradation (RAD), which is far less common than oxidative bioremediation. Partially oxidised organic compounds such as halogenated solvents, where a halogen atom (most commonly chlorine) has replaced hydrogen in a short-chain hydrocarbon, are also amenable to RAD (e.g. McCue *et al.*, 2003). The most common chlorinated solvents released to the environment include tetrachloroethene, trichloroethene, trichloroethane, and carbon tetrachloride (terachloromethane). These chlorinated solvents are hazardous to health, deplete ozone, and resist natural degradation. They exist in an oxidised state, and thus they are generally not susceptible to aerobic oxidation processes (with the possible exception of co-metabolic degradation). However, oxidised compounds are susceptible to reduction under anaerobic conditions by either biotic or abiotic processes (Parsons Corporation, 2004). Anaerobic bioremediation technology primarily exploits biotic anaerobic processes in degrading mainly, but not exclusively, chlorinated solvents in groundwater. Other chlorinated organic contaminants that can also potentially be treated with this approach include chlorobenzenes, chlorinated pesticides, and polychlorinated biphenyls (PCBs). Explosive and ordnance compounds have also been targeted (Parsons Corporation, 2004). The technique is commonly referred to as 'anaerobic reductive dechlorination', which includes both direct and co-metabolic biotic processes and non-biotic reactions produced under the same conditions. In practice, it may not be possible to distinguish the three different reactions at the field scale since all three reactions may be occurring (Parsons Corporation, 2004).

In direct anaerobic reductive dechlorination, which is the fastest approach if the compound is amenable, bacteria gain energy and grow when one or more chlorine atoms in a molecule are replaced with hydrogen. In this reaction, the chlorinated compound serves as the electron acceptor, often with hydrogen serving as the electron donor.

Many different substrates can be used to stimulate anaerobic biodegradation depending on the biogeochemistry and hydrology of the site. Common substrate types include soluble ones including lactic acid (Kennedy *et al.*, 2006), sugar-based materials such as molasses (Suthersan *et al.*, 2002), and sucrose esters of fatty acids (Borden and Rodrigues, 2006). Slow-release substrates can also be used including both non-emulsified and emulsified vegetable oil (EVO) (Borden and Lee, 2006; AFCEE, 2007), as well as commercial products referred to as hydrogen release compounds (HRCs), which incorporate engineered controlled release of the electron

Figure 20.2 Whey powder and lactate injected into aquifer for TCE biodegradation. Image: https://www.flickr.com/photos/departmentofenergy/7609886626/, public domain.

donor (e.g. Koenigsberg, 1999; Fiore and Zanetti, 2011). This technology is commercially well exploited, and numerous useful case studies are available online (e.g. ETEC, n.d.; Regenesis, n.d.).

20.9 PRBs and Bioremediation

PRBs work by directing groundwater under the influence of the hydraulic gradient through a zone where remedial processes are optimised (Powell *et al.*, 1998; FRTR, 2002). As mentioned previously, they are generally considered to be devices aimed at protecting *receptors* rather than for providing a permanent solution to a particular *source*, and if the contaminant plume can be well defined, there is no specific need to know the location of the source. While they can be constructed as a continuous permeable wall across the contaminant plume in an aquifer, such systems are less common than the combination of hydraulic control of groundwater using impermeable barriers and shorter lengths of permeable barriers designed to transform the contaminants (see Figure 20.3). Of the PRBs shown in Figure 20.3, the most common geometry is the funnel and gate, a term originally proposed by Starr and Cherry, 1994 (see also McGovern *et al.*, 2002; Courcelles, 2014), but the trench and gate (e.g. Bowles, 1997; Bowles *et al.*, 2000) is also common.

In the trench and gate system, a preferential pathway is created against the impermeable barrier to assist flow towards the active gate. The impermeable barrier is often a sheet piled system, while for the funnel and gate (where the impermeable barriers are often more extensive) cement–bentonite walls are more commonly used.

PRBs use a variety of mechanisms and media to achieve their aim (e.g. Grajales-Mesa and Malina, 2016), and microbiologically mediated transformations (Scherer *et al.*, 2000) are not often used. The commonest chemical approach is the use of the zero-valent iron barrier to reduce chlorinated solvents (e.g. Vogan *et al.*, 1999) and chromate (Kjeldsen and Locht, 2002). Physical removal by adsorption of the contaminants onto a solid medium is also commonly used. However, the adsorption capacity of sorbents will eventually be exhausted, and where the chemical reactive mechanism is via a solid reagent deposited in the barrier, these can be exhausted too. Replaceable cassettes have been used without much success. The beauty of biologically mediated mechanisms is that as long as the nutrients can be replenished and appropriate conditions with respect to oxygen status maintained, the organisms will continue to carry out the required transformations.

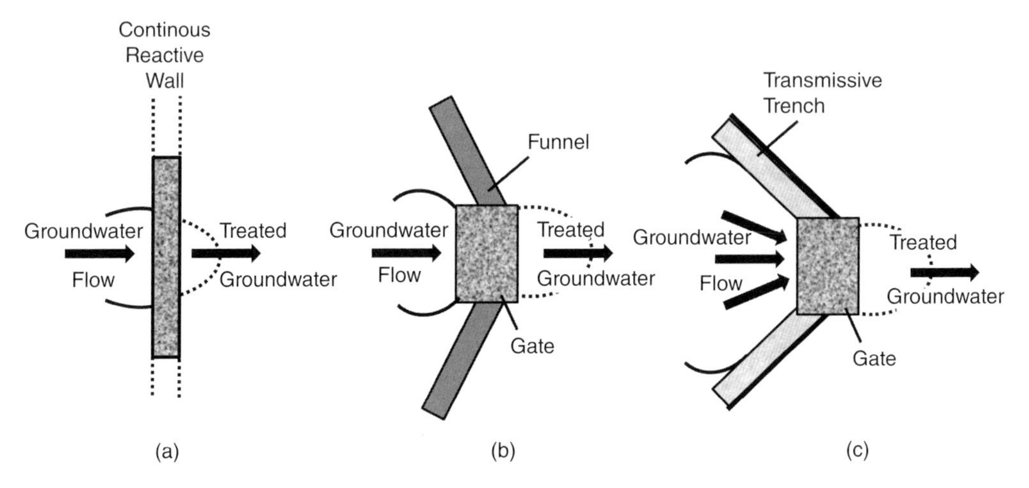

Figure 20.3 PRB arrangements: (a) continuous wall, (b) funnel and gate, and (c) trench and gate

Reactive barriers with biological treatment are described in great detail by Powell *et al.* (1998) and ITRC (2005).

If microorganisms are to be used in a PRB system, they will need to be maintained on a solid support on which they can form an active biofilm. These materials may be either totally inert materials such as granulated stone, ceramics, or plastics, but sometimes a degradable organic material is used which provides support, a supplementary source of carbon, and adsorptive capacity. Sometimes the barriers can be quite complex; for example, Kao and Borden (1997) proposed a two-layer PRB for BTEX removal which had an up-gradient layer of nutrient-laden concrete briquettes followed by a layer of peat. Doherty *et al.* (2006) used an upstream bed of modified fly ash in front of the biobarrier, and in a lab-scale experiment, Choi *et al.* (2006) proposed a complex combination of a palladium-coated zero-valent iron barrier for chemical treatment of PCP up gradient of a biowall. However, the medium is often as simple as a bed of sand or fine gravel; for example, a case study reported by CL:AIRE (2005) targeted a wide range of organic compounds from a former gasworks site and utilised aerobic bioremediation, the oxygen supplied via an air sparging unit, similar to biosparging previously described except the contaminants are constrained to an area where all the conditions for bioremediation have been optimised. Where air sparging is used, there is, as with in situ biosparging, often a combined mechanism taking place. The flow of air through or just before the gate will encourage volatilisation of lighter contaminants, with biological oxidation acting on the heavier fraction. Furthermore, the increased biomass in the growing biofilm will increase the sorption of organics from solution, holding them there long enough for degradation to occur: in effect, a combined physical-chemical-biological mechanism.

Supplying the final electron acceptor in bio-oxidative PRBs is highly varied, and oxidative anaerobic processes are also used. In the approach of Kao and Borden (1992), the nitrate released from specially manufactured concrete briquettes served as the electron acceptor, and biodegradation proceeded via denitrification. Conventional means of supplying nitrate is also commonly applied. In these devices, the most common means of maintaining the anaerobic conditions necessary to encourage denitrification is to incorporate solid degradable organic matter into the gate, although other options are available (Parsons, 2008).

20.10 Reductive Biodegradation in Permeable Reactive Barriers (PRBs)

The biological degradation of organic pollutants is not the sole, or even the dominant, application of microorganisms in PRB technology. Treatment strategies employing biologically mediated reactions have also been proposed for direct treatment of ions (e.g. Blowes *et al.,* 1998) including nitrate (Robertson and Cherry, 1995). The processes involved are identical to those discussed above, but theoretically, they occur in a managed subsoil environment rather than ambient in situ biodegradation schemes. Biological involvement has also been used for indirect removal of other ions by precipitation as sulphides, where biologically mediated reduction of sulphate to sulphide, accompanied by the formation of metal sulphides, occurs. An example is of a field-scale reactive barrier for the treatment of acid mine drainage and removal of dissolved Ni, which was installed in 1995 at the Nickel Rim mine site near Sudbury, Ontario. It comprised municipal compost, leaf compost, and wood chips. Monitoring of the reactive barrier in 1998 indicated continual removal of the acid generating capacity of the groundwater flowing through the PRB and a decrease in dissolved Ni from 10 mg/L to <0.01mg/L (Powell *et al.,* 1998). An example of an anaerobic process using direct reduction is the use of bacteria which have been isolated and capable of reducing U(VI) to U(IV), and sustaining metabolism based on this reaction (Lovley *et al.,* 1993).

20.11 Conclusion

Microorganisms provide an effective means of degrading the wide range of pollutants prevalent in the urban environment. Given that the microorganisms used might well be operating outside their optimal range of conditions, the key to any bioremediation strategy is to provide the organisms with an environment that maximises their activity, but does so in the desired metabolic direction and at a speed that makes these efforts worthwhile. It is usually attempts to direct the organisms to achieve a particular, human-centred task that presents the challenges rather than any inherent shortcomings in the organism. Often, physical treatments must be used in combination with biological treatments if remediation goals are to be achieved in an acceptable time frame.

References

Abanaki (n.d.) Oil Skimmer Belt Selection Guide. http://www.abanaki.com/brochures/belt-selection-guide.pdf

AFCEE (2007) Protocol For *In Situ* Bioremediation of Chlorinated Solvents using Edible Oil. United States Air Force Center for Engineering and the Environment http://www.eosremediation.com/download/Anaerobic%20Bioremediation/EVO/AFCEE%20Oils%20Protocol.pdf

Alvares, P.J. and Illman, W.A. (2005) *Bioremediation and Natural Attenuation: Process Fundamentals and Mathematical Models.* Volume 27 of Environmental Science and Technology: A Wiley-Interscience Series of Texts and Monographs. John Wiley & Sons ISBN 0471738611, 9780471738619

Anderson, R.T. and Lovley, D.R. (2001) Anaerobic remediation of benzene under sulphate reducing conditions in a petroleum contaminated aquifer. *Environmental Science & Technology*, 34, 2261–2266. http://www.geobacter.org/publication-files/Environ_Sci_Technol_2000_Nov.pdf.

Bewley, R.J.F. and Webb, G. (2001) In situ bioremediation of groundwater contaminated with phenols, BTEX and PAHs using nitrate as electron acceptor. *Land Contamination & Reclamation*, 9 (4), 335–347.

Blowes, D.W., Ptacek, C.J., Bain, J.G., Waybrant, K.R., and Robertson, W.D. (1995) Treatment of mine drainage water using in situ permeable reactive walls. In *Proc. Sudbury 95, Mining and the Environment*, CANMET (Ottawa, Ontario), 3, 979–987.

Blowes, D.W., Ptacek, C.J., Benner, S.G., McRae, C.W.T., and Puls, R.W. (1998) Treatment of dissolved metals using permeable reactive barriers. *Groundwater Quality: Remediation and Protection* (Proceedings of the GQ'98 Conference held at 4g 3 Tubingen, Germany, September 1998). IAHS Publ. no. 250, http://hydrologie.org/redbooks/a250/iahs_250_0483.pdf

Borden, R.C., Goin, R.T., and Kao, C.-M. (1997) Control of BTEX migration using a biologically enhanced permeable barrier. *Groundwater Monitoring and Remediation*, pp. 70–80. https://info.ngwa.org/GWOL/pdf/971462618.PDF

Borden, R.C. and Lee, M.D. (2006) Method for Remediation of Aquifers. Patent, US RE40734 E1 https://www.google.com/patents/USRE40734

Borden, R.C. and Ximena Rodriguez, B. (2006) Evaluation of slow release substrates for anaerobic bioremediation. *Bioremediation Journal*, 10 (1–2), 59–69.

Bowles, M.W. (1997) The Trench and Gate Groundwater Remediation System. MSc Thesis, University of Calgary. https://clu-in.org/download/rtdf/prb/thesismb.pdf

Bowles, M.W., Bentley, L.R., Hoyne, B., and Thomas, D.A. (2000) In situ ground water remediation using the trench and gate system. *Groundwater* 38 (2), 172–181.

Carey, M. and Fletcher, I. (2005) Management of a Petroleum Plume by Monitored Natural Attenuation. CL: AIRE Site Bulletin SB1. http://www.claire.co.uk/component/phocadownload/category/14-site-bulletins?download=57:sitebulletin01

Choi, J.-H., Kim, Y.-H., and Choi, S.J. (2006) Reductive dechlorination and biodegradation of 2,4,6-trichlorophenol using sequential permeable reactive barriers: Laboratory studies. *Chemosphere*, 67 (8), 1551–1597.

CL: AIRE (2005) Laboratory and Field Evaluation of a Biological Permeable Reactive Barrier for Remediation of Organic Contaminants in Soil and Groundwater. Case Study Bulletin (CSB) 03. CL:AIRE, London. http://www.claire.co.uk/component/phocadownload/category/3-case-study-bulletins?download=5:casestudybulletin03

CLU-IN (n.d.) Bioremediation, Anaerobic Bioremediation (Direct). CLU-IN Website. https://clu-in.org/techfocus/default.focus/sec/Bioremediation/cat/Anaerobic_Bioremediation_(Direct)/

Cox, L. and Russell, D. (2003) Using bioreactors to control air pollution, The Clean Air Technology Centre, U.S. Environmental Protection Agency Research Triangle Park, North Carolina 27711 https://www3.epa.gov/ttncatc1/dir1/fbiorect.pdf

Courcelles, B. (2014) Proceedings of the 4th International Conference on Environmental Pollution and Remediation Prague. Czech Republic, August 11–13, 2014 Paper No. 81 81-1 Guidelines for Preliminary Design of Funnel-and-gate Reactive Barriers http://avestia.com/ICEPR2014_Proceedings/papers/81.pdf

Cuthbertson, J. and Schumacher, M. (2010) Full Scale Implementation of Sulfate Enhanced Biodegradation to Remediate Petroleum Impacted Groundwater. *Proceedings of the Annual International Conference on Soils, Sediments, Water and Energy*, 14, Article 15, 1–26.

Doherty, R., Phillips, D., McGeough, K., Walsh, K.P., and Kalin, B. (2006) Development of a modified flyash as a permeable reactive barrier media for a former manufactured gas plant site Northern Ireland. *Environmental Geology*, 50 (1), 37–46.

Early, T., Borden, R., Heitkamp, M., Looney, B.B., Major, D., Waugh, J., Stoller, S.M., Wein, G., Wiedemeier, T., Vangelas, K.M., Adams, K.M., and Sink, C.H. (2006) Enhanced Attenuation: Approaches to Increase the Natural Treatment Capacity of a System, U.S. Department of Energy Contract Number DEAC09-96-SR18500. https://clu-in.org/download/contaminantfocus/tce/DOE_EA_doc.pdf

Eastern Research Group (1996) Pump-and-Treat Ground-Water Remediation: A Guide for Decision Makers and Practitioners. US EPA Document EPA/625/R-95/005 https://nepis.epa.gov/Exe/ZyPDF.cgi/30004PC8.PDF?Dockey=30004PC8.PDF

ETEC (n.d.) Aggressive Chlorinated Solvent Remediation using the 40-Gpm In Situ Delivery (Isd™) System Industrial Facility. Eugene, OR. http://www.etecllc.com/docs/ISD_Full-scale_Eugene_OR.pdf

EUGRIS (n.d.) Portal for Soil and Water Management in Europe. http://www.eugris.info/FurtherDescription.asp?eugrisid=27&Category=Content_Digests&Title=Monitored+Natural+Attenuation&GlossaryID=111&DocTitle=&showform=&ContentID=3&CountryID=0&ResourceTypes=&DocID=&Tools=Selected+list+of+abbreviations

FRTR (2002) Evaluation of Permeable Reactive Barrier Performance, Federal Remediation Technologies Roundtable. https://clu-in.org/download/rtdf/2-PRBperformance_web.pdf

Giadom, F.D. and Tse, A.C. (2015) Groundwater contamination and environmental risk assessment of a hydrocarbon contaminated site in eastern Niger Delta, Nigeria. *Journal of Environment and Earth Science*, 5 (14), 166–176.

Grajales-Mesa, S.J. and Malina, G. (2016) *Environmental Earth Sciences*, 75, 772.

Grostern, A. and Edwards, E.A. (2006) A 1,1,1-Trichloroethane-degrading anaerobic mixed microbial culture enhances biotransformation of mixtures of chlorinated ethenes and ethanes. *Applied and Environmental Microbiology*, 72 (12), 7849–7856.

Gupta, V.K., Nayak, A., and Agarwal, S. (2015) Bioadsorbents for remediation of heavy metals: Current status and their future prospects. *Environmental Engineering Research*, 20 (1), 1–18.

Häggblom, M.M., Rivera, M.D., and Young, L.Y. (1993) Influence of alternative electron acceptors on the anaerobic biodegradability of chlorinated phenols and benzoic acids. *Applied and Environmental Microbiology*, 59 (4), 1162–1167, 1993 April.

Hinchee, R.E., Downey, D.C., and Aggarwal, P.K. (1991) Use of hydrogen peroxide as an oxygen source for in situ biodegradation: Part I. Field studies. *Journal of Hazardous Materials*,, 28, 287–299.

ITRC (2005) Permeable Reactive Barriers: Lessons Learned/New Directions, PRB-4. Washington, D.C., Interstate Technology & Regulatory Council.

Johnson, R.L., Johnson, P.C., McWhorter, D.B., Hinchee, R.E., and Goodman, I. (1993) An overview of in situ air sparging. *Ground Water Monitoring & Remediation*, 13, 127–135. doi:10.1111/j.1745-6592.1993.tb00456.x

Kao, C.M. and Borden, R.C. (1997) Enhanced biodegradation of BTEX in a nutrient Briquet-Peat barrier system. *Journal of Environmental Engineering*, ASCE, 123 (1), 18-24, 1997.

Kassab, S.Z. (2010) Empirical correlations for the performance of belt skimmer operating under environmental dynamic conditions. *International Journal of Water Resources and Environmental Engineering*, 2 (5), 121–129.

Kennedy, L.G., Everett, J.W., Becvar, E., and DeFeo, D. (2006) Field-scale demonstration of induced biogeochemical reductive dechlorination at Dover Air Force Base, Dover, Delaware. *Journal of Contaminant Hydrology*, 88, 119–136.

Kjeldsen, P. and Locht, T. (2002) Removal of chromate in a permeable reactive barrier using zero-valent iron. *Groundwater Quality: Natural and Enhanced Restoration of Groundwater Pollution. Proceedings of the Groundwater Quality 2001 Conference*, Sheffield. UK. June 2001. IAHS Publ. 275. 409. http://hydrologie.org/redbooks/a275/iahs_275_409.pdf

Klenk, I.D. and Grathwohl, P. (2002) Transverse vertical dispersion in groundwater and the capillary fringe. *Journal of Contaminant Hydrology*, 58 (1–2), 111–128.

Koenigsberg, S.S. (1999) Hydrogen release compound (HRC): A novel technology for the bioremediation of chlorinated hydrocarbons. *Proceedings of the 1999 Conference on Hazardous Waste Research.* St Louis Missouri, May 24–27, 144–157. https://www.engg.ksu.edu/HSRC/99Proceed/koenigs.pdf

Kunuku, Y.C. (2007) In situ bioremediation of groundwater contaminated with petroleum constituents using oxygen release compounds (ORCs). *Journal of Environmental Science and Health Part A: Toxic /Hazardous Substances and Environmental Engineering*, 42 (7), 839–845. http://www.tandfonline.com/doi/abs/10.1080/10934520701373174

Lovley, D.R., Roden, E.E., Phillips, E.J.P., and Woodward, J.C. (1993) Enzymatic iron and uranium reduction by sulfate-reducing bacteria. *Marine Geology*, 113, 41–53.

Mackay, D.M., Freyberg, D.L., Roberts, P.V., and Cherry, J.A. (1986) A natural gradient experiment on solute transport in a sand aquifer 1. Approach and overview of plume movement. *Water Resources Research*, 22 (13), 2017–2029.

Masters, G.M. and Ela, P. (2007) *Introduction to Environmental Engineering and Science.* 3rd edition. Pearson Pub, ISBN-13: 978-0131481930 ISBN-10: 0131481932

McCue, T., Hoxworth, S., and Randall, A.A. (2003) Degradation of halogenated aliphatic compounds utilizing sequential anaerobic/aerobic treatments. *Water Science and Technology*,. 47 (10), 79–84.

McGovern, T., Guerine, T.F., Horner, S., and Davey, B. (2002) Design, construction and operation of a funnel and gate in-situ permeable reactive barrier for remediation of petroleum hydrocarbons in groundwater. *Water, Air, and Soil Pollution*, 136 (1), 11–31.

Mercer, J.W., Skipp, D.C., and Giffin, D. (1990) Basics of Pump-and-Treat Ground-Water Remediation Technology. US EPA Publication EPA/600/8-90/003. https://nepis.epa.gov/Exe/ZyPDF.cgi/30001IV1.PDF?Dockey=30001IV1.PDF

Miao, Z., Brusseau, M.L., Carroll, K.C., Carreón-Diazconti, C., and Johnson, B. (2012) Sulfate Reduction in groundwater: Characterization and applications for remediation. *Environmental Geochemistry and Health*, Aug; 34 (4), 539–550.

Miller, R.R. (1996) Biosluping. Technology Overview Report TO-96-05. Ground-Water Remediation Technologies Analysis Center. https://clu-in.org/download/toolkit/slurp_o.pdf

Newell, C.J., Acree, S.D., Ross, R.R., and Huling, S.G. (n.d.) Light Nonaqueous Phase Liquids, Groundwater Issue EPA/540/S-95/500. US EPA Office of Solid Waste and Emergency Response. https://www.epa.gov/sites/production/files/2015-06/documents/lnapl.pdf

Ozone Solutions (n.d.) Ozone Effects on Specific Bacteria, Viruses and Molds. http://www.ozoneapplications.com/info/ozone_bacteria_mold_viruses.htm

Park, B.H., J.S., Namkoong, W., Hwang, E.-Y., and Kim, J.-D. (2008) Effect of co-substrate on anaerobic slurry phase bioremediation of TNT-contaminated soil. *Korean Journal of Chemical Engineering*, 25, 102.

Parsons Corporation (2004) Principles and Practices of Enhanced Anaerobic Bioremediation of Chlorinated Solvents. https://frtr.gov/costperformance/pdf/remediation/principles_and_practices_bioremediation.pdf

Parsons Infrastructure & Technology Group (2008) Technical Protocol for Enhanced Anaerobic Bioremediation using Permeable Mulch Biowalls and Bioreactors, Pub. Air Force Center for Engineering and the Environment Technical Directorate. https://clu-in.org/download/techfocus/prb/Final-Biowall-Protocol-05-08.pdf

Plummer, C.R., Lucket, M.D., Porter, S., and Moncrief, R. (n.d.) Ozone Sparge Technology for Groundwater Remediation Pub. H_2O Engineering Inc. http://www.ecosafeusa.

com/documents/Ozone%20Documentation/Soil%20Remediation/OZONE%20SPARGE%20TECHNOLOGY%20FOR%20GROUNDWATER%20REMEDIATION.pdf

Powell, R.M., Blowes, D.W., Vogan, J.L., Guelph, R.W.G., Powell, P.D., and Sivavec, T. (1998) Permeable Reactive Barrier Technologies for Contaminant Remediation. US EPA Report EPA/600/R-98/125. https://clu-in.org/download/rtdf/prb/reactbar.pdf

Radwan, S.S. and Al-Muteirie, A.S. (2001) Vitamin requirements of hydrocarbon-utilizing soil bacteria. *Microbiological Research*, 155 (4), 301–307. https://www.ncbi.nlm.nih.gov/pubmed/11297361

Regenesis (n.d.) Large-Scale *In Situ* Enhanced Reductive Dechlorination of a 6 Hectare Site Treatment of Chlorinated Solvents at a Complex Site with Sensitive Receptors. https://regenesis.com/project/large-scale-in-situ-enhanced-reductive-dechlorination-of-a-6-hectare-site/

Robertson, R.D. and Cherry, J.A. (1995) In situ denitrification of septic system nitrate using reactive porous medium barriers: Field trials. *Ground Water*, 33, 99–111.

Scherer, M.M., Richter, S., Valentine, R.L., and Alvarez, P.J.J. (2000) Chemistry and microbiology of permeable reactive barriers for *in situ* groundwater clean-up. *Critical Reviews in Environmental Science and Technology*, 30 (3), 363–411.

Schwartz, O.R., James, A., Berndt, L.A., and Mundell, J.A. (2005) The Use of Ozone Sparging to Remove MTBE from Groundwater in a Uniform Sand Aquifer. http://mundellassociates.com/media/2012/01/Ozone-Sparging-to-Remove-MTBE-from-Groundwater-Uniform-Sand-Aquifer.pdf

Siegrist, R.L., Urynowicz, M.A., and West, O.R. (2000) In situ chemical oxidation for remediation of contaminated soil and ground water. *Groundwater Currents*, Issue 37. US EPA Document EPA 542-N-00-006. http://www.ozonesolutions.com/files/research/groundwater_remediation.pdf

Singh, A., Kuhada, R.C., and Ward, O.P. (2009) *Advances in Applied Bioremediation*. Springer Science & Business Media, ISBN 978-3-540-89621-0

Starr, R.C. and Cherry, J.C. (1994) In-situ remediation of contaminated ground water: The funnel-and-gate system. *Ground Water*, 32 (3), 465–476.

Sudicky, E.A., Cherry, J.A., and Frind, E.O. (1983) Migration of contaminants in groundwater at a landfill: A case study: 4. A natural-gradient dispersion test. *Journal of Hydrology*, 63 (1–2), 81–108.

Suthersan, S.S., Lutes, C.C., Palmer, P.L., Lenzo, F., Payne, F.C., Liles, D.S., and Burdick, J. (2002) Technical Protocol for Using Soluble Carbohydrates to Enhance Reductive Dechlorination of Chlorinated Aliphatic Hydrocarbons. ARCADIS G&M, Inc., Online: https://clu-in.org/download/contaminantfocus/tce/BioTechProtocol.pdf

Tang, Y.J., Carpenter, S., Deming, J., and Krieger-Brockett, B. (2005) Controlled release of nitrate and sulfate to enhance anaerobic bioremediation of phenanthrene in marine sediments. *Environmental Science & Technology*, 39 (9), 3368–3373.

USACE (2002) Engineering and Design Manual No. 1110-1-4001, Soil Vapor Extraction and Bioventing. US Army Corps of Engineers. http://www.publications.usace.army.mil/Portals/76/Publications/EngineerManuals/EM_1110-1-4001.pdf

US EPA (1991) Granular Activated Carbon Treatment, Engineering Bulletin Oct 1991. US EPA Document EPA/540/ 2-91/024 https://nepis.epa.gov/Exe/ZyPDF.cgi/10001KAJ.PDF?Dockey=10001KAJ.PDF

US EPA (1999) Use of Monitored Natural Attenuation at Superfund, RCRA Corrective Action, and Underground Storage Tank Sites. U.S. Environmental Protection Agency Office of Solid Waste and Emergency Response Directive 9200.4-17P. https://www.epa.gov/sites/production/files/2014-02/documents/d9200.4-17.pdf

US EPA (2006) Off-Gas Treatment Technologies for Soil Vapor Extraction

Systems: State of the Practice. US EPA Document EPA-542-R-05-028. https://clu-in.org/download/remed/EPA542R05028.pdf

US EPA (2012) A Citizen's Guide to Monitored Natural Attenuation. https://clu-in.org/download/Citizens/a_citizens_guide_to_monitored_natural_attenuation.pdf

US EPA (2013) Introduction to in Situ Bioremediation of Groundwater. Office of Solid Waste and Emergency Response Report 542-R-13-018. https://clu-in.org/download/remed/introductiontoinsitubioremediationofgroundwater_dec2013.pdf

US EPA (2016) How to Evaluate Alternative Cleanup Technologies for Underground Storage Tank Sites. ES EPA Document, EPA 510-B-16-005, https://www.epa.gov/sites/production/files/2014-03/documents/tum_ch12.pdf

Vogan, J.L., Focht, R.M., Clark, D.K., and Graham, S.L. (1999) Performance evaluation of a permeable reactive barrier for remediation of dissolved chlorinated solvents in groundwater. *Journal of Hazardous Materials*, 68 (1–2), 97–108.

Voudrias, E.A. (2001) Pump-and-treat remediation of groundwater contaminated by hazardous waste: Can it really be achieved? *Global Nest: The International Journal*, 3 (1), 1–10.

Wilson, R.D., Mackay, D.M., and Scow, K.M. (2002) In situ MTBE biodegradation supported by diffusive oxygen release. *Environmental Science & Technology*, 36 (2), 190–199.

Xu, J., Song, Y., Min, B., Steinberg, L., and Bruce, E., Logan, B.E. (2003) Microbial degradation of perchlorate: Principles and applications. *Environmental Engineering Science*, 20 (5) [Online] http://online.liebertpub.com/doi/pdfplus/10.1089/109287503768335904

Zappi, M., White, K., Hwang, H.-M., Bajpai, R., and Qasim, M. (2000) The fate of hydrogen peroxide as an oxygen source for bioremediation activities within saturated aquifer systems. *Journal of the Air & Waste Management Association*, 50 (10), 1818-1830. doi: 10.1080/10473289.2000.10464207

21

Bioremediation in Urban Pollution Mitigation: Applications to Solid Media

Andrew B. Shuttleworth[1], *Alan P. Newman*[2], *and Ernest O. Nnadi*[3]

[1] *SEL Environmental Limited, Lancashire, United Kingdom*
[2] *Centre for the Built and Natural Environment, Coventry University, United Kingdom*
[3] *GITECO-UC, University of Cantabria, Santander, Spain*

21.1 Introduction

Contaminated land remains both a barrier to the redevelopment of urban areas and a threat to the wider environment. However, in densely occupied areas, land which has suffered a degree of contamination from its previous use is often all that is available for reuse; thus, strategies for the remediation of these sites have received a major economic boost in recent decades. The most widely used method of remediation is bioremediation (Singh et al., 2009), which was introduced in Chapter 20. This chapter concentrates on the applications of bioremediation to the clean-up of solid substrates, both as a means of allowing redevelopment and also to prevent ongoing groundwater pollution. This chapter will examine both oxidative and reductive mechanisms; however, it will not include such methods as vermiculture (which make use of macroscopic animals) or phytoremediation. In order to study the performance of remediation efforts, whether for research purposes or simply for evaluating success, a wide range of monitoring approaches will be required. This is particularly true for solid substrates, where the concentrations determined are often defined by the extraction method. This will include monitoring of target compounds and their degradation products in the environment in which the organisms operate and the potential off-site effects of that remediation effort (e.g. off-gas and perimeter monitoring). Apart from a brief coverage of automated online monitoring, space does not permit a detailed coverage of this area, but it is important to maintain good information during any bioremediation effort.

Bioremediation is now one of the commonest approaches to the remediation of contaminated soil and subsoil, but in the United Kingdom, the use of remedial methods other than “dig and dump” were held back for many years by poor legislation and the relatively low cost of landfill (now a thing of the past). Numerous sites were cleaned up during the 1980s and 1990s by landfilling soil into co-disposal landfills, many of which were totally unsuitable hydraulic containment or, worse still, “dilute and disperse” or “dilute and pollute” landfills. Since the start of the 1990s, the United Kingdom’s contaminated land community has been putting considerable efforts into research and development, with some success. The United States has led in this regard, and this is reflected in the body of literature available (e.g. Sims *et al.*, 1990; Bajpai *et al.*, 1996; Powell *et al.*, 1998; USACE 2002). However,

Urban Pollution: Science and Management, First Edition. Edited by Susanne M. Charlesworth and Colin A. Booth.

more recently, the United Kingdom has made contributions to the area with documents from the Environment Agency, Defra, and in particular, CL:AIRE (e.g. Thomas *et al.*, 2004; Pearl, 2007).

There are two basic approaches to cleaning up the strata from a contaminated site:

1) Dig the material up and treat it at the surface: the ex situ approach.
2) Leave the material where it is and treat it in situ. A basic coverage of these two distinct approaches is provided by the US EPA (Rawe and Hodge, 2006).

Furthermore, it is essential that the process itself should not do environmental harm, and the maxim *Primum non nocere* should be as much applied by environmental professionals as by physicians. Thus, the pollutants (and sometimes the degradation products) must be contained, controlled, and safely dispersed. Therefore, leachate drainage collection and treatment systems and off-gas treatment systems are often essential to bioremediation, if not to the microorganisms. In many cases, the physical state of the medium in which the pollutant is contained is the limiting factor. Particle size and homogeneity of solid media are examples of important factors that often need addressing.

21.2 In Situ Treatment above the Water Table

In situ approaches tend to be undertaken when the contamination is at great depth, with little space available, and in some cases, where the site must be kept in use during the process. One example of an approach allowing continued use of the site is bioventing (FRTR, n.d.; Hinchee and Leeson, 1996), which stimulates naturally occurring soil microorganisms to break down contaminants in soil and subsoil, by providing oxygen. This is done when the rate of natural biodegradation is effectively limited by the lack of electron acceptors rather than by the lack of nutrients (see Chapter 20). Oxygen is most commonly supplied through direct air injection into contaminated strata (active bioventing), but in certain circumstances, passive bioventing systems (e.g. Foor *et al.*, 1995) using natural air circulation to deliver oxygen to the subsurface via bioventing wells have been shown to work. This approach requires the strata to be permeable to gas, which typically limits it to sandy or gravelly soils. In one of these systems, vent wells were provided with a one-way valve to prevent air movement from strata to surface (Foor *et al.*, 1995), relying on changes in atmospheric pressure to produce air exchange in the strata. Another example of a passive venting technique is the commercially available "Virtual Curtain" system, which produces air flow requirements for passive bioventing systems (Wilson and Shuttleworth, 2002) and depends on wind-induced ventilation rather than simple changes in atmospheric pressure. The extraction side of the system uses highly efficient geocomposite vent nodes, which are driven into the ground without excavation and are connected to a distribution duct, situated just beneath the surface. Ventilation of the duct can be achieved using a combination of vent stacks, bollards, or ground-level boxes, depending on the topography and wind conditions at a particular site (see Figure 21.1). This can encourage oxygen to be introduced into the soils and subsoils from the atmosphere, sometimes without the need for air injection wells but, where ground conditions demand, either additional rows of vents can be added, or traditional infiltration wells can be drilled. The advantage of the no-dig system is that it minimises spoil, contact by workers with contaminated soil, and can be installed in restricted spaces (see Chapter 6).

The conditions in which passive bioventing systems work are not universal, and the active option is often the only choice (see Figure 21.2). In active bioventing systems, the technology used is often similar to that used in soil vapour extraction systems (see Wilson, 1995), but the air flow rates are much lower. It is important that the flow of air induced by injection and extraction devices (e.g. blowers

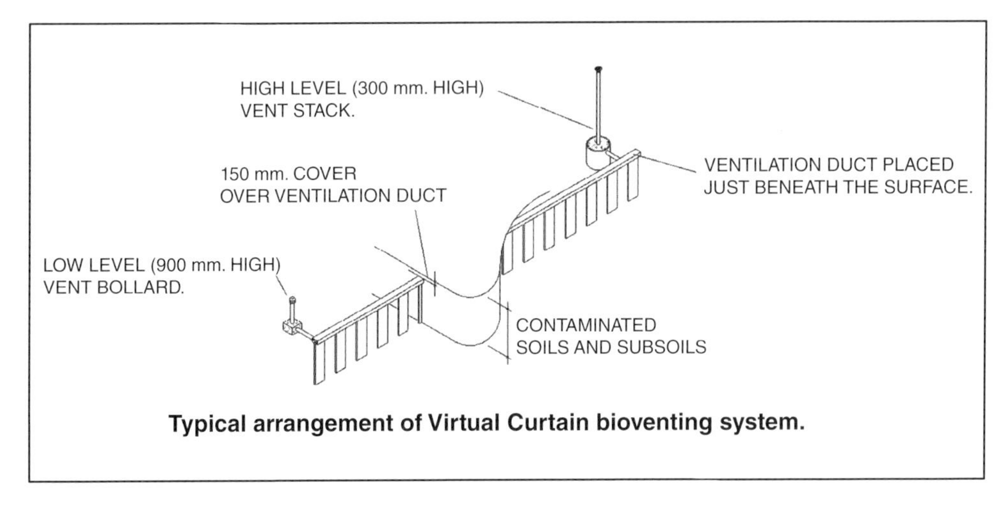

Figure 21.1 Schematic of Virtual Curtain system showing driven nodes, ventilation duct, and wind-powered extraction bollards.

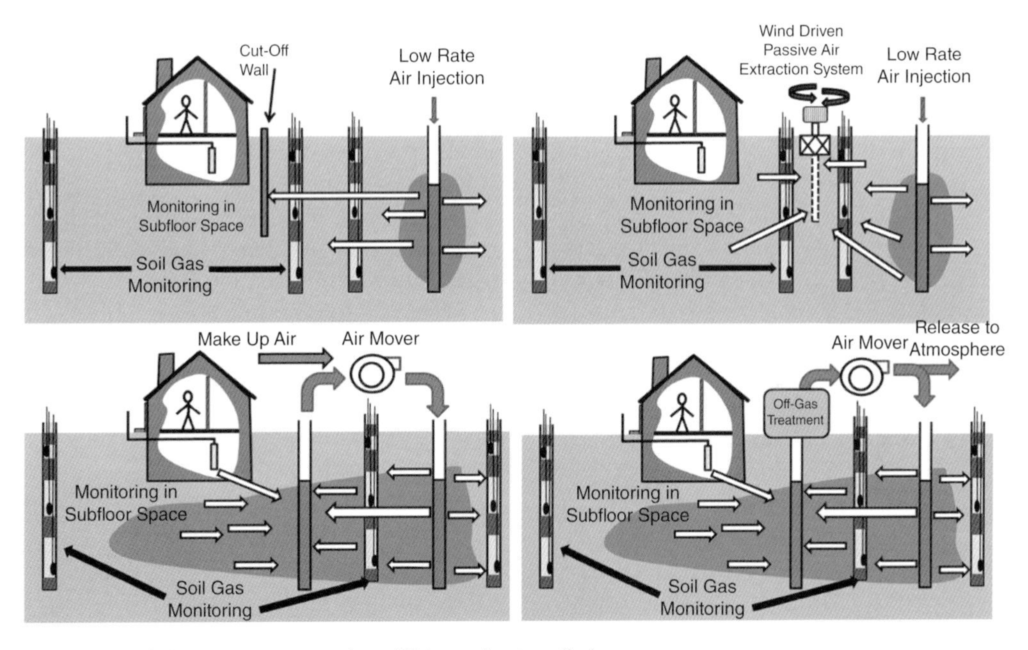

Figure 21.2 Various arrangements for soil bioventing installations.

and vacuum pumps) does not dry out the strata too much as this will inhibit the microorganisms. Even at low air flows, volatile organic compounds (VOCs) can be transported in the flow of moving air, which can be both advantageous, in that they will be subject to biodegradation as they pass through the actively degrading strata, and disadvantageous. In some cases, displacement of VOCs by positive air injection may drive them into nearby properties, necessitating either barrier systems or supplementary passive gas extraction devices (such as the Virtual Curtain system described above and illustrated in Figure 21.2). In the worst case, a positive extraction system may be needed. Furthermore, negative pressure or extraction systems may necessitate off-gas monitoring at the point where the extracted air is discharged. All aerobically biodegradable contaminants are potentially treatable by this method; it is the often the nature of the strata which limits applicability (low-permeability soils and very high or very low water content can limit performance) (FRTR, n.d.). Additionally, if the local concentrations of contaminants are too high, they can be toxic to microorganisms.

If the target compounds are volatile enough to be transported in the gas phase, then monitoring of the extracted air stream can be used to follow remediation progress, but the alternatives are to either carry out periodic sampling with, for example, a direct push boring device or a window sampler (the site variability can require greater efforts in such cases) or, if there is an ongoing release of contaminants (often the incentive for carrying out the remediation), the groundwater located just down gradient of the contaminated area can be continuously monitored by systems linked to on-site data loggers. However, telemetry and, more recently, Internet-based monitoring and control systems are now being used for various applications, and this technology can be readily applied to both gas and water monitoring (e.g. Glasgow *et al.*, 2004; Lorentz, n.d.). In a gas phase monitoring application, a typical system consists of on-site monitoring sensors that detect the volumes of gas extracted from a site in real time along with pollutant concentrations. The information from the sensors is transmitted via a wireless signal to a cloud-based portal, where the sensor information is gathered and processed. The portal subsequently emits operational instructions to on-site control hardware to adjust parameters like air flows for management of a range of environmental applications. Management of such systems has been significantly facilitated by cloud-coupled technology as engineers can log into the portal on any Web-enabled device to control systems from remote destinations, view real-time data, and download historical data for data interpretation and system performance reviews.

Where VOCs are being remediated and the intention is to protect a receptor building, there is often the option of alternating low-flow bioventing with high-flow soil vapour extraction, when conditions are such that the receptor is affected by the VOCs (falling atmospheric pressure). This reverts back to flow velocities suitable for bioventing when conditions allow. Cloud control can also be usefully applied to this application. Anaerobic bioventing, using inert gas to displace oxygen, has also been used (e.g. Shah *et al.*, 2001), but it is relatively rare.

When a site is to be remediated, the aim is to break the existing or potential pollution linkages. Hence, during a bioventing process, it is possible to continue to use the site provided there are no human health implications. This can be achieved by placing (or retaining) a barrier between the pollutants and site users as long as this does not interfere with the bioventing process. Even if there are long-term reuse aims that require the site to be cleaned up at depth before development proceeds, because the plant footprint requirements are relatively small, it is often a financially worthwhile to use the site during the clean-up programme, for example, car parking and self-storage containers.

21.3 Enhanced In Situ Bioremediation

If nutrient injection and application of exogenous microorganisms are required to sustain biodegradation, the bioventing system begins to take on the characteristics of more advanced in situ techniques, such as Biostimulation and Bioaugmentation (e.g. Mrozik and Piotrowska-Seget, 2010; Kalantary *et al.*, 2014), often grouped together as Enhanced Biodegradation. These techniques involve the addition of microorganisms to the subsurface environment (e.g. fungi, bacteria, and other microbes) or nutrients or both, possibly also electron acceptors, in order to accelerate the natural biodegradation process. Enhanced bioremediation of soil typically involves the percolation or injection (or re-injection) of water mixed with nutrients, saturated with dissolved oxygen, or, as outlined in Chapter 20, provided with another electron acceptor.

While injection wells are used for deeper contaminated soils, spray irrigation or an infiltration gallery can both be used for those that are shallow, but spray irrigation often interferes with the current use of the site on account of the space-demanding infrastructure. Figure 21.3 shows one example of how water and nutrients may be supplied while allowing continued use of the land. Water dosed with the necessary additives can be introduced into the contaminated subsoils via polypropylene void-forming conduits known as Permavoid boxes (avoiding the need for above-ground tanks on site). The Permavoid boxes are enclosed along the base and sides with a waterproof polypropylene liner forming a series of gutters through which the dosed water, supplied from pipework connections above, can flow, forming a series of dispersed underground reservoirs. The gutters are laid in parallel lines typically at 3–5 m centres across the whole of the area to be treated and can then be covered in block paving or asphalt. The boxes contain special wicking cones, located in their structural columns and providing a vertical hydraulic bridge between the bottom and top of the box. The cones use capillary action to distribute the dosed water at a controlled rate into the surrounding area with the assistance of a wicking geotextile fabric that is laid horizontally over the whole area, before the final temporary hard standing is applied in contact with the top of the box to maintain hydraulic continuity.

Enhanced bioremediation techniques have been successfully applied to remediate soils contaminated with petroleum hydrocarbons, VOCs, semi-VOCs, and pesticides, but it is particularly effective for remediation of low-level residual contamination. If there is continual input from a part of the site highly contaminated with a mobile contaminant, then its use in conjunction with source removal and/or localised treatment by another method is generally essential.

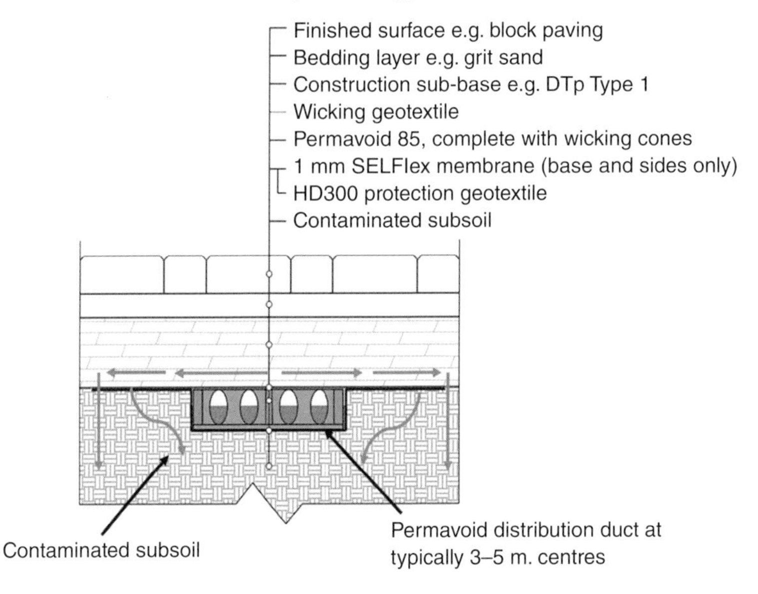

Figure 21.3 Use of load bearing void formers and capillary liquid fed to create an infiltration gallery for nutrient provision (SEL Environmental Ltd image used with permission).

The techniques can also be applied to compounds susceptible to anaerobic biodegradation, for example, the anaerobic microbial degradation of nitro toluenes in contaminated soils (Uche and Dadrasnia, 2017). These techniques can be cost competitive, and, as with bioventing, the scheme frequently allows the site to be used during the clean-up.

21.3.1 Landfarming

In situ bioremediation does not usually involve the homogenisation of materials on site; there will possibly be hotspots of very high contaminant concentrations locally toxic to microorganisms. An in situ soil remediation technique which attempts, in a limited way, to mitigate against this heterogeneity, known as landfarming, has been used in the petroleum industry for many years. However, care must be used, as the term is also applied to an extensive form of ex situ biodegradation in which very thin layers of excavated soils are spread out on a large, drainage-system-equipped liner (New South Wales, 2014). Even when such a liner is not used, the idea that landfarming is genuinely an in situ technique is arguable since the activity of turning and mixing will require soil to be removed from its original position even if it is immediately replaced.

In the in situ variant of the process, contaminated soils are simply mixed with soil amendments such as soil bulking agents and nutrients and then mixed with earth at their original location. The soil is periodically turned to improve aeration and increase homogeneity, ensuring all parts of the tilled soil are treated equally, but no attempt is made to control drainage to protect the underlying groundwater. In many systems, turning can be achieved with a deep plough, but other devices have also been used, for example, front-end loaders (Fingas, 2015). Soil conditions are controlled to optimise the rate of contaminant degradation. Enhanced microbial activity due to better oxygen supplies results in increased rates of degradation, particularly of some of the more intransigent compounds. This in situ approach is only suitable if contaminated soils are shallow.

21.4 Ex Situ Bioremediation in Unsaturated Strata

21.4.1 Ex Situ Landfarming

If contaminated soils are deeper than about 1 m then, for landfarming to work, the soils must be excavated and reapplied on the ground surface over a wider area (New South Wales, 2014). It is at this point that the technique becomes ex situ, representing the simplest form of a "prepared bed" ex situ bioremediation technique (Figure 21.4). In this case, contaminated soil is excavated and spread out in layers approximately 0.3 m deep, and usually, this is on a lined treatment area to prevent leaching (or surface run-off) of contaminants. As with the in situ process, bioremediation is enhanced by regular turning of the bed and addition of nutrients. Due to the limiting thickness of soil layers, such ex situ landfarming techniques often require areas of land larger than that of the contamination. They are therefore rarely suitable for the small sites that usually arise in urban areas.

Landfarming has been used most successfully in treating petroleum hydrocarbons, but can also be applied to certain halogenated volatile and semi-volatile compounds, non-halogenated-semi-volatiles, pesticides, and wood-preserving wastes such as creosote and polycyclic aromatic hydrocarbons (PAHs). However, the method has significant limitations since a large amount of space is required, and it is rarely as quick as turned windrows and aerated biopiles, which are less space-dependent alternatives. In urban environments, a quick turnaround time on a project is normally demanded for both financial and project management reasons, and public/political pressure will always require restoration of the site to as aesthetically pleasing a state as soon as possible. Furthermore,

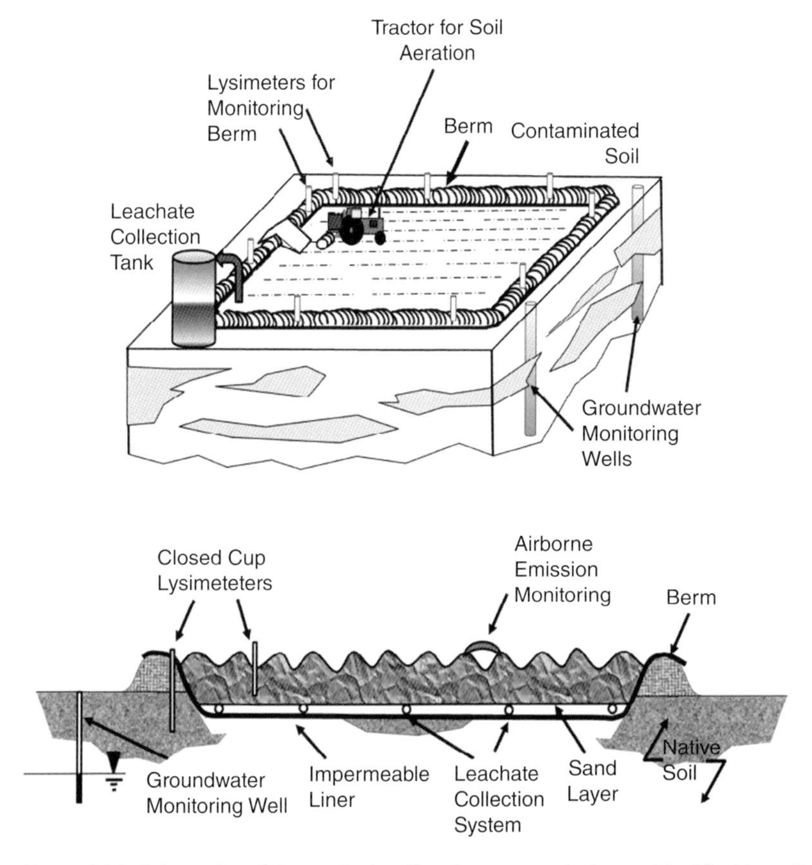

Figure 21.4 Schematics of the ex situ landfarming approach. *Source*: Modified from Battelle (1996).

odours from landfarming can be difficult to control, and even where a populace had lived with the smells of a refinery or similar plant for many years, they are far less accepting of odour when the plant is closed and no longer providing jobs for the community. An example of this in the United Kingdom is the Coalite site at Bolsover, which was tolerated for decades but, since closure, has generated vociferous complaints (Derbyshire Times, 2017).

The process will inevitably combine contaminated soil with material that is uncontaminated, including on-site soils, often from below the contaminated layer, and imported bulking agents. This creates a larger volume of contaminated material in the first instance.

Another problem with landfarming is that conditions affecting biological degradation of contaminants such as temperature and excessive soil moisture (which is largely rainfall dependent) are, for the most part, uncontrolled, and variations outside those expected occurring during the project may increase the time to achieve remedial targets. While landfarming is considered a mature technology, with numerous full-scale operations having been completed, particularly for sludges produced by the petroleum industry, for more confined urban sites it is rarely the preferred option.

21.4.2 Turned Windrows

For smaller sites with a large depth of contaminated soil, another approach is the use of turned windrows (Coker, 2006). Here, the excavated soil is placed into a shaped pile up to about 2 m height and 6 m wide in a lined area provided with appropriate drainage. Aeration is achieved by periodical turning of the windrows, sometimes using traditional earth moving plant but these days more commonly by specialised machinery. Soil amendments, including nutrients and organic amendments such as bulking agents (e.g. straw, sawdust, or sometimes green waste), can be easily added to the windrows during turning (Fingas, 2015). This is often the most cost-effective method of ex situ biological treatment. An option is to provide both extra control on the internal environment of the bioremediation system and protection of the outside environment, providing an opportunity for odour and fugitive control by carrying out the process in an enclosed area, such as a polytunnel or mobile lightweight roof systems (Figure 21.5).

In open air windrow systems, waterproof coverings (Figure 21.6) are usually used to both control water input from rain and reduce evaporation. For a basic overview of windrow applications to contaminated soil, see Coker (2006).

If airborne emissions are not to be contained with a structure, monitoring at appropriate boundaries is essential to satisfy both regulatory and public concerns. An example of this is the effort expended during the remediation of coal-tar-contaminated sludges from Grassmoor Lagoons near Chesterfield (RSK, 2012). At this semi-urban site, air quality monitoring for VOC releases was undertaken using near-real-time GC instruments along the two most sensitive boundaries. They measured naphthalene, benzene, toluene, ethylbenzene, and xylene

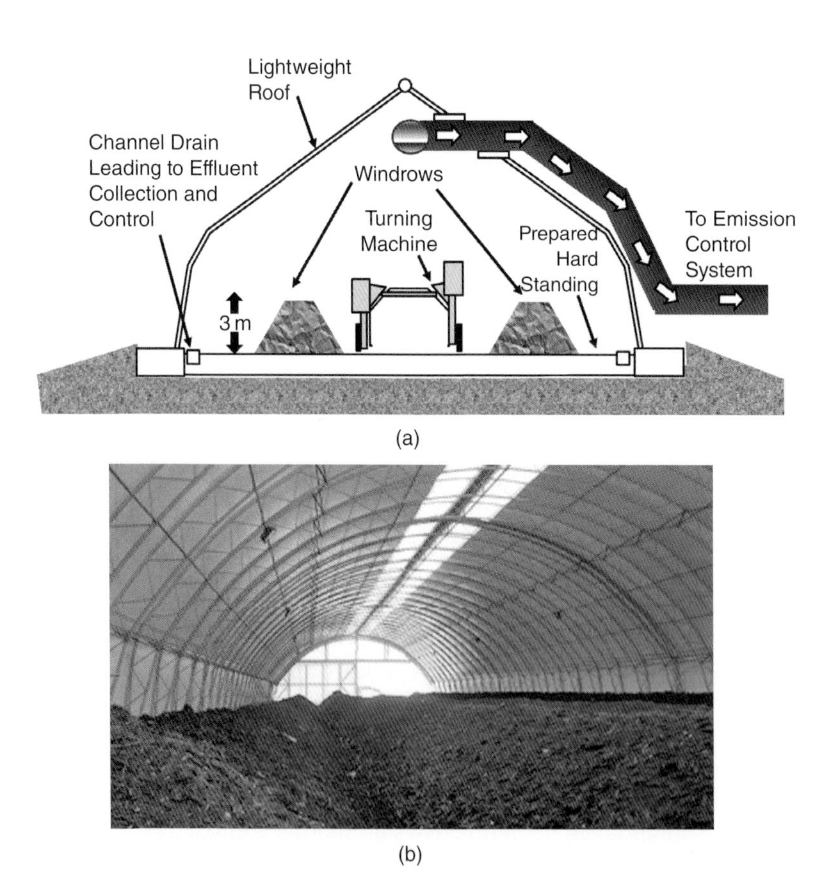

Figure 21.5 Turned windrows in an enclosed situation. *Source*: (a) Adapted from Coker (2006), (b) Reproduced with permission from Compost Systems GmbH, Wels, Austria.

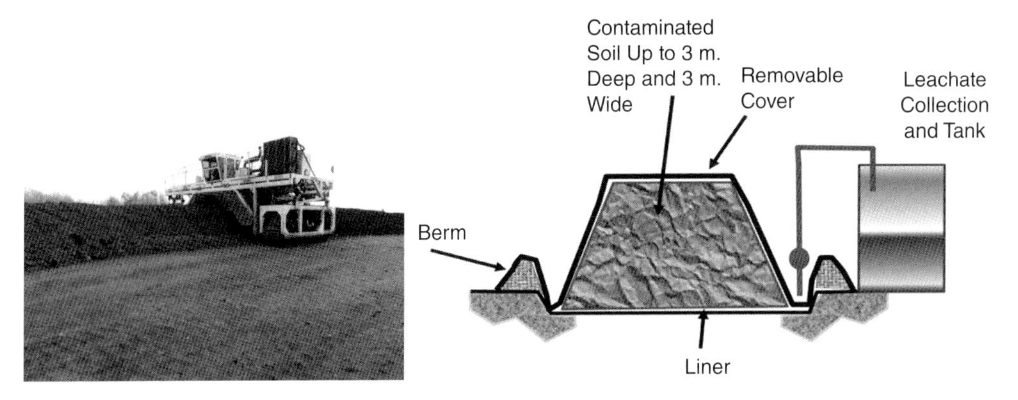

Figure 21.6 Use of turned windrows in an un-enclosed situation requires flexible covers to maintain required humidity levels. Photograph used with permission REMDEX Ltd., Chesterfield.

with the results sent by telemetry link to the site office at 30 minute intervals. They were backed up using Tenax tubes, later analysed by thermal desorption/GC-MS for the listed compounds as well as some others. Particulates (PM_{10}) were also measured every hour using near-real-time instruments, including a beta attenuation monitor located along the downwind boundary of the site. In addition, pairs of low-volume samples were co-located with the GC instruments to provide a reference measurement of PM_{10} along with heavy metal and PAH concentrations. Nuisance dusts were determined using dust deposition gauges (see Vallack, 1995)

21.4.3 Ventilated Biopiles

An alternative to turning windrows, which both requires less space and is easier to provide atmospheric controls for, is the use of actively vented biopiles. In this technology, the excavated soil is placed into a somewhat higher and wider pile which is not turned, but aerated by introducing air into the pile by pressure or vacuum (e.g. Michaud *et al.*, 1999; Hazen *et al.*, 2003) or both (Figure 21.7). Again, they are constructed in a lined area equipped with an adequate drainage system. Treatment times in actively vented biopiles typically range from 3 to 6 months depending on the contaminant type, concentration, soil properties, and time of year (Battelle ERD, 1996), but times as short as 6 weeks have been reported (Vertase FLI, n.d.). However, this is highly dependent on starting and target concentrations.

The ambient temperature has a significant effect on bioremediation, with the best degradation rates occurring during the warmer summer months; cold winter weather can significantly slow and even halt biodegradation, but the larger cross section, compared to a windrow, often improves heat retention. Vapours from the biopile often need to be collected and treated, typically by granular activated carbon (GAC) (US EPA, 1991; US EPA, 2006), a bio-filter (Cox and Russell, 2003; US EPA, 2006), or an afterburner system (US EPA, 2006). If an afterburner is used, waste heat recovery can be used to increase the bed temperature during winter periods, which is particularly useful in temperate areas if overwintering is required. Even if not mandatory for health reasons, where the process is taking place near residential areas, such treatment is often advisable if only to satisfy the concerns of the local populace or prevent odour issues. Leachate and run-off control will also be required. Landfarming, turned windrows, and aerated biopile options are normally grouped together under the term 'prepared bed systems'.

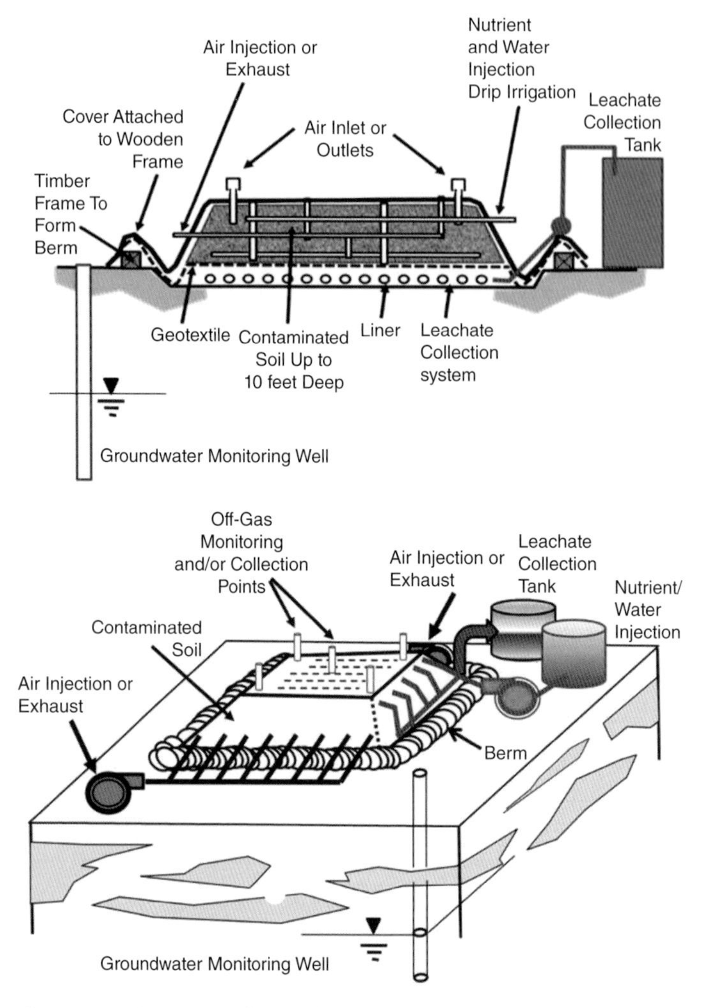

Figure 21.7 Operation of aerated biopiles. *Source*: Modified from Battelle (1996).

21.4.4 Infrastructure Requirements

When carrying out ex situ bioremediation in a prepared bed, the protection of both the underlying groundwater and clean unsaturated strata is paramount. This means that the underlying waterproof liner or geomembrane used for leachate control must be of the appropriate quality. It is not uncommon to integrate a leak detection layer into the system for added security. These take the form of a sandwich comprising two impermeable geomembrane layers separated by a relatively thin (25 mm, typically) voided layer in between. Adjacent leak detection chambers connected to the leak detection layer can be monitored.

The waterproof liner or geomembrane selected must be able to withstand the environment to which it will be exposed. This also means that in addition to being able to withstand chemical and biological attack if present, it will need to be resilient to other physical factors including UV radiation when exposed to light in outdoor applications and resistant to fatigue and stress cracking due to continued expansion and contraction during temperature changes. Jointing of geomembranes is a very important aspect when considering the

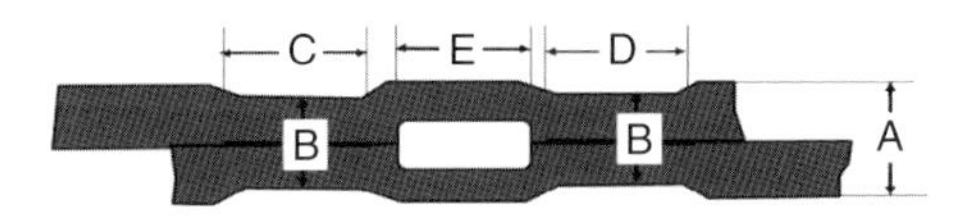

Figure 21.8 Cross-sectional diagram of an overlap weld showing test channel created in the weld for pressure testing.

overall integrity of the completed installation. Even with an exceptionally well-selected and functioning geomembrane, the whole system can fail if the method of jointing is inappropriate.

If a tape is selected for jointing, then the selected material must also be able to withstand attack from the same environmental influences encountered by the geomembrane. If the system is intended to retain leachate or other fluids, then either gaseous or hydrostatic pressures will likely develop. It is advisable to avoid taped joints for these types of applications since prolonged exposure to geomembranes subjected to pressure can cause jointing tapes to creep and eventually cause rupture. Welded joints are preferred, which must be given as much attention as if they were permanently installed landfill liners (Figure 21.8); the fact they are temporary does not reduce the need for assured quality. If sheets of membrane are welded together, they need to be tested for integrity, usually by double-welding the joints and pressure-testing the enclosed volume. Details of how this is achieved are available online (SEL Environmental, n.d.).

Where the membranes are to be penetrated with pipes, and so on, the integrity of the joints is also very important. Furthermore, if an aerated pile is to be used, the integrity of the aeration pipework is fundamental to the economics of the system (leaks are expensive) and achieving desired aims in the target time. High-quality control mechanisms are also important; thus, only experienced contractors and subcontractors should be used for the installation of infrastructure. Electrical installations also need to meet all requirements for the jurisdiction in which they are installed; in particular, if volatile and flammable materials are being handled, intrinsically safe electrical components must be used.

21.4.5 Slurry Phase Biodegradation

Prepared bed systems represent the vast majority of bioremediation projects that are actually put into use, but there is an alternative, less commonly used approach which offers a potentially faster achievement of targets: the use of slurry phase reactors (e.g. Thomas *et al.*, 2004; Thomas *et al.*, 2006). Such bioreactors are more suitable for heterogeneous soils, those with low permeability, and project plans demanding relatively short treatment times. Biological treatment in slurry phase reactors is usually a batch process, and has been successfully used to remediate soils, sludges, and groundwater contaminated by hydrocarbons, petrochemicals, solvents, pesticides, wood preservatives, and other organic chemicals. They can be utilised either to treat a slurry created by suspending the entire soil matrix (or more usually the finer fractions separated by a dry or semi-dry screening process) or the fine particle products of a soil washing process (see Pearl, 2007).

The systems are usually designed such that slurry contains between 5% and 40% solids by weight, but this is highly variable and depends on the soil type, the contaminant, and its concentration. The soil is suspended in a reactor vessel (see Figure 21.9) and mixed with nutrients, with O_2 usually supplied as air although pure O_2 can reduce aeration rates and hence reduce foaming problems

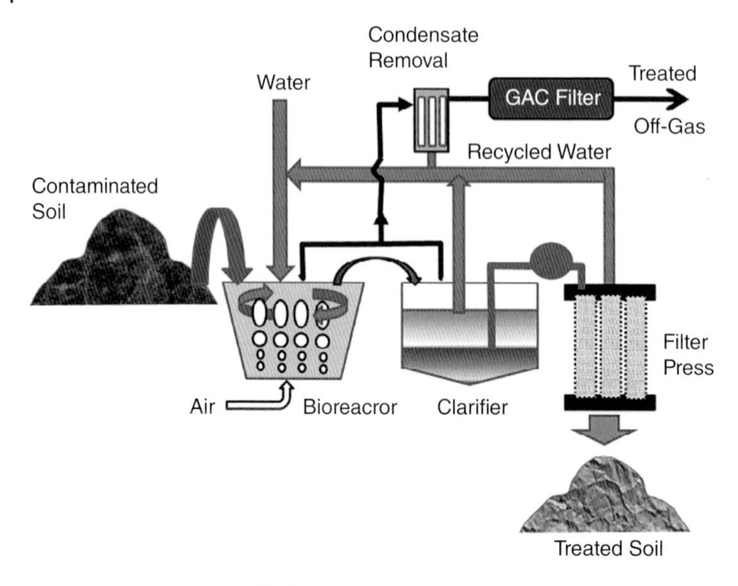

Figure 21.9 Schematic of a slurry phase bioreactor system.

(Bajpai *et al.*, 1996). Microorganisms and reagents such as acids, alkalis, or surfactants (which may already be part of the soil washing process and may later need degrading themselves) may be added to the slurry depending on the requirements of the system. Stirring may be achieved by mechanical paddles, induced by the flowing air stream or by continually withdrawing and re-injecting a proportion of the slurry using pumps outside the vessel. The relatively homogeneous nature of the system means monitoring is much easier and more reliable than for prepared bed and in situ systems. Slurry-based systems are capable of breaking down contaminants that are not amenable to prepared bed systems such as 5- and 6-ring PAHs (Thomas *et al.*, 2004).

When biodegradation is complete, the soil slurry must be dewatered, usually by a clarifier followed by a filter press or a centrifuge. This part of the process is often the most expensive, but is based on easily available technology since the equipment is widely used in the mining industry as well as in sewage treatment. As previously mentioned, Thomas *et al.* (2004) gives good information on dewatering problems and possible solutions; a summary of general dewatering systems (aimed at sediment dewatering but very relevant) is available online (Englis and Hunter, n.d.).

21.5 Conclusion

Microorganisms provide an effective means of degrading a wide range of pollutants in both shallow and deep deposits and in excavated materials. Whenever biodegradation is used, however, care needs to be taken that the selected process does not cause unexpected (or, at least, unidentified) harm to other environmental media. Hence, methods of containment, monitoring, and control demand as much attention as the microbiological system itself. Furthermore, for ex situ methods it is often the physics which dominates the success of process rather than the biology or even the chemistry. For ex situ processes, particle size reduction and other forms of pretreatment, as well as maintaining the balance between gas and liquid flows are often the keys to success. For in situ techniques, the key could quite probably be mathematics, or at least statistics, as the inherent inhomogeneity provides a real challenge in ensuring quality control.

References

Bajpai, R.K., Banerji, S.K., Puri, R.P., and Zappi, M.E. (1996) Remediation of Soils Contaminated with Wood-Treatment Chemicals (PCP and Creosote). Final Report: US EPA:EPA Grant Number: R825549C056 https://cfpub.epa.gov/ncer_abstracts/index.cfm/fuseaction/display.highlight/abstract/5241/report/F

Battelle (1996) Biopile Design and Construction Manual. Technical Memorandum TM-2189-ENV, Naval Facilities Engineering Service Center, Port Hueneme, California. https://clu-in.org/download/techfocus/bio/Biopile-design-and-construction-1996-tm-2189.pdf

Coker (2006) Environmental remediation by composting. *Biocycle*, 47 (12), 18. https://www.biocycle.net/2006/12/14/environmental-remediation-by-composting/

Cox, L. and Russell, D. (2003) Using Bioreactors to Control Air Pollution. The Clean Air Technology Centre, U.S. Environmental Protection Agency Research Triangle Park, North Carolina 27711. https://www3.epa.gov/ttncatc1/dir1/fbiorect.pdf

Derbyshire Times (2017) http://www.derbyshiretimes.co.uk/news/environment-agency-investigating-nasty-smells-from-coalite-site-1-8338479

Englis, M. and Hunter, D.W. (n.d.) *A* Description of Sediment Dewatering Techniques. The Mcilvaine Company Repository. http://www.mcilvainecompany.com/Decision_Tree/subscriber/Tree/DescriptionTextLinks/DescripSedDewater.pdf .

Fingas, M. (2015) *Handbook of Oil Spill Science and Technology*. John Wiley & Sons, Print ISBN: 9780470455517 Online ISBN: 9781118989982 doi: 10.1002/9781118989982

Foor, D.C., Zwick, T.C., Hinchee, R.E., Hoeppel, R.E., Kyburg, C., and Bowling, L. (1995). Passive bioventing driven by natural air exchange: In: Hinchee, L., Miller, R.E., Miller, R.N., and Johnson, P.C. (eds.), *Bioremediation*, 3 (2), 630 p. United States: Battelle Press. ISBN 1-57477-003-9; https://info.ngwa.org/GWOL/pdf/940160820.PDF

FRTR (n.d.) Remediation Technologies Screening Matrix and Reference Guide, 4.1 Bioventing. https://frtr.gov/matrix2/section4/4_1.html

Glasgow, H.B., Burkhlder, J.M., Reed, R.E., Lewitus, A.J., and Kleinman, J.E. (2004) Real-time remote monitoring of water quality: A review of current applications, and advancements in sensor, telemetry, and computing technologies. *Journal of Experimental Marine Biology and Ecology*, 300 (1–2), 409–448.

Hazen, T.C., Tien, A.J., Worsztynowicz, A., Altman, D.J., Ulfig, K., and T. Manko, T. (2003) Biopile for remediation of petroleum contaminated soils: A Polish case study. In: *The Utilization of Bioremediation to Reduce Soil Contamination: Problems and Solutions*, Vol. 19, NATO Science Series, pp. 229–246. ISBN: 978-1-4020-1142-9

Hinchee R.E. and Leeson A. (1996) *Soil Bioventing: Principles and Practice*. CRC Press. ISBN 9781566701266

Hinchee, R.E., Downey, D.C., and Aggarwal, P.K. (1991) Use of hydrogen peroxide as an oxygen source for in situ biodegradation: Part I. Field studies. *Journal of Hazardous Materials*, 28, 287–299.

Kalantary, R.R., Mohseni-Bandpi, A., Esrafili, A., Nasseri, S., Ashmagh, F.R., Jorfi, S., and Ja'fari, M. (2014) Effectiveness of biostimulation through nutrient content on the bioremediation of phenanthrene contaminated soil. *Journal of Environmental Health Science and Engineering*, 12, 143.

Lorentz (n.d.) Leachate Management and Monitoring Solution. at a UK Landfill Site. Bernt Lorentz GmbH & Co. KG, Krögerskoppel 724558 Henstedt-Ulzburg Germany.http://enitial.co.uk/wp-content/uploads/2016/04/lorentz_uk_leachate_pump_case_study.pdf

Michaud, J.-R., Dutil, F.,Viel, G.,Ouellette, Y., and Gourdeau, F. (1999) Rotamix Process for the Biotreatment of Soil Contaminated with Pentachlorophenol and Petroleum Hydrocarbons F 1999 Data Sheet Pub. Environment Canada Eco-Technology Innovation Section Montreal, Quebec. ISSN: 1188-8903 ISBN: 0-662-27542-X http://publications.gc.ca/collections/Collection/En1-17-43-1999E.pdf

Morris, R.S. and Jones, S.D. (1988) Coal Tar Bioremediation in Land Farms. Internal Report GRTC R2621, BG Technology (cited in *Thomas et al.*, 2004).

Mrozik, A. and Piotrowska-Seget, Z. (2010) Bioaugmentation as a strategy for cleaning up of soils contaminated with aromatic compounds. *Microbiological Research*, 165 (5), 363–37.

New South Wales (2014) Best Practice Note: Landfarming Pub. Environment Protection Authority 59 Goulburn Street, Sydney NSW 2000. http://www.epa.nsw.gov.au/resources/clm/140323landfarmbpn.pdf

Pearl, M. (2007) Understanding Soil Washing, CL:AIRE Technical Bulletin 13. http://www.claire.co.uk/component/phocadownload/category/17-technical-bulletins?download=54:technicalbulletin13

Powell, R.M., Blowes, D.W., Vogan, J.L., Guelph, R.W.G., Powell, P.D., and Sivavec, T. (1998) Permeable Reactive Barrier Technologies for Contaminant Remediation. US EPA Report EPA/600/R-98/125. https://clu-in.org/download/rtdf/prb/reactbar.pdf

RSK (2012) Grassmoor Tar Lagoons Remediation Scheme. http://www.rsk.co.uk/images/technical-library/casestudies/CS0186.pdf

SEL Environmental Ltd (n.d.) *On Line*: http://selel.co.uk/Welding-Methods.php

Shah, J.K., Sayles, G.D., Suidan, M.T., and Kaskassian, S.R. (2001) Anaerobic bioventing of unsaturated zone contaminated with DDT and DNT. *Water Science & Technology*, 43 (2), 35–42. US EPA Document EPA 542-N-00-006. http://www.ozonesolutions.com/files/research/groundwater_remediation.pdf

Sims, J.L., Sims, R.C., and Matthews, J.E. (1990) Approach to Bioremediation of Contaminated Surface Soils, Biological Engineering Faculty Publications. Paper 43. http://digitalcommons.usu.edu/bioeng_facpub/43

Singh, A., Kuhada, R.C., and Ward, O.P. (2009) *Advances in Applied Bioremediation.* Springer Science & Business Media. ISBN 978-3-540-89621-0

Thomas, R., Hughes, D.E., and Daly, P. (2006) Proceedings of the International Symposium and Exhibition on the Redevelopment of Manufactured Gas Plant Sites 4–6 April 2006, Reading, UK. *Land Contamination & Reclamation*, 14 (2). The use of slurry phase bioreactor technology for the remediation of coal tars.

Thomas, R., Hughes, D., Harries, N., Sweeney, R., and Wallace, S. (2004) Technology Demonstration Report TDP4. Slurry Phase Bioreactor Trial. CL:AIRE. https://www.google.co.uk/url?sa=t&rct=j&q=&esrc=s&source=web&cd=1&cad=rja&uact=8&ved=0ahUKEwig4r2uuMvRAhVHmBoKHWGwAMUQFggfMAA&url=http%3A%2F%2Fwww.claire.co.uk%2Findex.php%3Foption%3Dcom_phocadownload%26view%3Dcategory%26download%3D105%3Aslurry-phase-bioreactor-trial%26id%3D19%3Atechnology-demonstration-project-reports&usg=AFQjCNHgB5X_XDgxQtwCBn232p7W4kjerg

Thomas, R.A.P., Gustavsen, E.L., Daly, P.J., and Jones, S.D. (2000). The Bioremediation of Soils Contaminated with Coal Tar At Pilot and Commercial Scales. March, internal report GRTC3605, BG Technology (cited in Thomas *et al.*, 2004).

Uche, E.C. and Dadrasnia, A. (2017) Biodegradation of hydrocarbons. In: Heimann, K., Karthikeyan, O.P., and Muthu, S.S. (eds.), *Biodegradation and Bioconversion of Hydrocarbons. Springer*, Chap. 3, pp. 105–135. https://www.researchgate.net/publication/309895073_HC-0B-06_Biodegradation_of_Hydrocarbons

USACE (2002) Engineering and Design Manual No. 1110-1-4001. Soil Vapor Extraction and Bioventing. US Army Corps of Engineers. http://www.publications.usace.army.mil/Portals/76/Publications/EngineerManuals/EM_1110-1-4001.pdf

US EPA (1991) Granular Activated Carbon Treatment. Engineering Bulletin Oct 1991. US EPA Document EPA/540/ 2-91/024 https://nepis.epa.gov/Exe/ZyPDF.cgi/10001KAJ.PDF?Dockey=10001KAJ.PDF

US EPA (2006) Off-Gas Treatment Technologies for Soil Vapor Extraction Systems: State of the Practice. US EPA Document EPA-542-R-05-028. https://clu-in.org/download/remed/EPA542R05028.pdf

Wilson, S. and Shuttleworth, A.J. (2002). Design and Performance of a Passive Dilution Gas migration Barrier, Ground Engineering January 2002. https://www.geplus.co.uk/technical-papers/technical-paper-design-and-performance-of-a-passive-dilution-gas-migration-barrier/8680635.article

Vallack, H.W. (1995) A field evaluation of Frisbee-type dust deposit gauges. *Atmospheric Environment*, 29 (12), 1465–1469. doi: 10.1016/1352-2310(95)00079-E

Vertase FLI (n.d.) Ex Situ Bioremediation. http://www.vertasefli.co.uk/fr/our-solutions/expertise/ex-situ-bioremediation

Wilson, D.J. (1995) *Modelling of In Situ Techniques for Treatment of Contaminated Soils: Soil Vapour Extraction, Sparging, and Bioventing*. Technomic Publishing Company, Inc. Lancaster USA, ISBN: 1566762340

22

Use of Environmental Management Systems to Mitigate Urban Pollution

Rosemary Horry[1], *and Colin A. Booth*[2]

[1] *College of Life & Natural Sciences, University of Derby, United Kingdom*
[2] *Architecture and the Built Environment, University of the West of England, United Kingdom*

22.1 Introduction

An environmental management system (EMS) is an instrument that can help organisations to manage and positively improve their level of impact on the environment (Christini *et al.*, 2004; Oke, 2004). An EMS helps a company measure its environmental performance and provides the framework for the integration of sustainable development goals within the organisation's corporate plan (Ilinitch *et al.*, 1998; Jolevski, 2013; Owolana and Booth, 2016). Within organisations, environmental management should be treated as an integral part of any operation, alongside the option for corporate social responsibility (CSR), to ensure they not only acknowledge the impacts of their operations but also that they have plans and strategies in place to minimise their harm to the environment and society. This is relevant to both their daily operations and also any resulting risk from an emergency situation (e.g. pollution events).

Typically, an EMS encompasses policies, goals, plans, and regulatory requirements and is usually reflected in the company's annual reports. Christini *et al.* (2004) give the following basic characteristics for an organisation's EMS: (1) goals, methods, and a timeline for achieving environmental criteria; (2) procedures for maintaining a paper trail in relation to those goals; (3) a defined structure and a matrix of responsibilities, as well as allocated resources; (4) corrective actions and emergency procedures; (5) an employee training plan; and (6) a plan for monitoring and auditing the organisation's performance in achieving the EMS goals (Owolana and Booth, 2016).

This chapter provides an overview of the importance for organisations to have an established EMS in place. It then employs a series of infamous case studies to highlight where an EMS could have served as a useful means of mitigating pollution events.

22.2 Why Is Environmental Management Important?

When thinking about environmental management, it is easy to immediately consider our own needs as being paramount and that we manage the environment for our own benefit. To some degree, that probably is the case. However, an increasing number of companies are not just preserving the environment for the resources they can then use, but also to promote a green company image, to foster positive public relations, to attract better employees through their ethical performance/operations or, in some instances, because morally it is the right thing to do. After all, society (hereinafter referred to as 'we') lives on one planet and, as such, has

Urban Pollution: Science and Management, First Edition. Edited by Susanne M. Charlesworth and Colin A. Booth.

finite, limited resources. Thankfully, humans are immensely creative and, as technology improves, we seem to find ways to use resources more wisely (e.g. reduce, reuse, recycle). However, it is essential for us to manage not just the resources we use but also the pollution we create and leave behind too.

There will, of course, always be those organisations that just want to make a profit, and fail to appreciate the adverse legacy they leave behind for others or future generations to clear up. Thankfully, this trend is declining, as both people and organisations come to realise that the environment where they live and work does matter. Without doubt, we, or our future generations, will have to deal with the environmental and pollution issues created in the past or those still to be created.

Until recent times, humans chose to concentrate their attention on progress and industrial achievement with minimal thought to the environment. Fortunately, there were a group of individuals who had the foresight to recognise potential problems caused by pollution, and they fought for the introduction of environmental legislation (e.g. the Alkali Act). In that respect, environmental management has a short history in terms of the human occupation of the planet, whereas most environmental issues are historic in creation. Many nations have long industrial heritages but, sadly, our predecessors were not always careful in respect to their environmental and pollution impacts. Nowadays, we appear to have greater awareness of our impacts, and while our knowledge is always improving, and with appropriate environmental management we seem to have acquired a better understanding and regulation of the perils associated with our modern industrial and domestic processes.

22.3 Organisational Benefits and Barriers of Implementing an Environmental Management System (EMS)

As the number of organisations adopting a third-party certifiable EMS linked to international standards (such as Eco–Management and Audit Scheme (EMAS) and ISO 14001) increases, the question that many organisations want answered is, what are the benefits and barriers for them? An unpublished UK study has identified 'reduction of environmental risks' and 'contribution to environmental protection' as the main benefits; while a similar Nigerian study has recognised 'reduction of environment-related sickness and injuries' and 'contribution to environmental protection' as the main benefits (Owolana and Booth, 2016); and 'cost saving due to the reduction of fines associated with convictions' and 'improving staff work environment, thus increasing their morale' are acknowledged as the main benefits from a study in Hong Kong (Shen and Tam, 2002). The unpublished UK study also identifies 'lack of technological support within organisation' and 'lack of government legal enforcement' as the main barriers; the Nigerian study is similar and highlights 'lack of government legal enforcement' and 'lack of technological support within organisation' as the main barriers; whereas the Hong Kong study finds 'change of existing practice of company structure and policy' and 'lack of government legal enforcement' as the main barriers against the adoption of EMS. Stakeholder perceptions of the main benefits of and the main barriers to adopting an EMS are clearly different for different nations, but the assortment of advantages and challenges is diverse (Tables 22.1 and 22.2).

22.4 What can Companies do in Relation to their Environmental Impacts?

All businesses face a challenge in terms of their environmental impacts; environmental work is not just a concern for multinational organisations, which is sometimes how it is viewed. Small- and medium-sized enterprises (SMEs) form an important part of all economies, accounting for 99% of businesses in the United Kingdom (Revel and Blackburn, 2007). They are a major part of society,

Table 22.1 Identified organisational benefits of implementing an EMS.

EMS Benefits
Reduced liability through compliance
Enhanced relations with stakeholders
Integration with other management systems
Efficient resource management
Amplify financial operations and performance
Strengthened access to markets
Augmented business competitiveness
Boost environmental, ethical, and sustainable business operations
Reinforce business reputation
Encourage innovation to reduce business impacts
Facilitate agile business
Improve business operations
Be responsive to external pressures
Augment risk management and mitigation
Empower staff performance
Strengthen business strategy
Opportunities for collaboration within the supply chain

Table 22.2 Identified organisational barriers of implementing an EMS.

EMS Barriers
Keeping pace with requirements
Lack of perceived value to the business, supply chain, and wider stakeholders
Necessity for environmental expertise for implementation and training
Not a legal requirement
Inability to compare across business sectors
Environment not an issue acknowledged by the management and absent from the business strategy
Viewed as superficial
Deemed irrelevant within the sector
Concern over accountability or liability through the required documents
Disincentive due to cost of implementation

providing jobs and contributing financially to the economy. But there is an issue here, in that in any environmental work there are always constraints of limited resources, including cash flows versus customer demands, and this can be a particular problem for the typical SME (Hudson *et al.*, 2001), where staff capacity and cash are not always readily available. So, in the case of many organisations, the focus can be on current performance rather than on taking a strategic approach; this in conjunction with their size contributes to their having what is termed a flat organisational structure (Hudson *et al.*, 2001). It can also result in their being unsure of the tangible business benefits of environmental management (Hillary, 2004; Revel and Blackburn, 2007). As with any organisation, whether they be multinationals or SMEs, they will have both positive and negative impacts within society. However, while we always think of the multinationals being the pollution kings of the world, SMEs have been linked to approximately 60% of carbon dioxide emissions (Marshall Report, 1998; Revell and Blackburn, 2009), and on the global scale it has been suggested they are responsible for 70% of all pollution (Revell *et al.*, 2009). However, this is a problem with a solution in the form of an EMS, as high adopters of such systems are more likely to report increased recycling and reductions in air emissions, solid waste, and energy usage (Florida and Davidson, 2001). Organisations just need to be able to see the benefits of a system and have the confidence to work towards being more environmentally sound. The ISO 14001 system is not about being perfect, it is about making continual improvements in your environmental performance.

Implementation of an EMS not only provides improved environmental performance but also economic benefits. Research into the performance of some Italian firms operating an EMS identified economically quantifiable benefits, including raw material conservation, improved productivity, energy conservation, smoother production processes, waste reduction, improved access to government regulatory incentives, and reduced insurance costs (Alberti *et al.*, 2000). Benefits which are less easily quantifiable are those such as risk reduction and improved company image (Alberti *et al.*, 2000). However, we are currently witnessing an increasing importance in these in terms of the opportunity for improved customer perception and even the recruitment of better, more productive staff.

An ISO 14001 enables companies to focus strategically on their environmental performance, and it has been suggested that companies which adopt such a system improved their environmental performance more significantly than non-adopters (Potoski and Prakash, 2005). However, it must be remembered that organisations need to have the correct management support in place to deliver an EMS, and also the financial capacity necessary to implement and certify such systems (Carraro and Leveque, 1999). There needs to be a commitment and support to make the potential a reality; the system will not fix the problem, it can only ever provide a mechanism to work towards a solution.

We are living in a changing world where companies face the immediacy of social media in terms of their operations; so, if they are linked with an environmental incident, it is global in a matter of hours. Panapanaan *et al.* (2003) suggested that CSR and environmental management had become increasingly important as a result of the operation of all organisations impacting on the wider communities in which they exist. This is increasingly the case; companies need to be seen as responsible and sustainable. We have moved to a situation where the major companies now aim to implement their environmental systems through the supply chain to ensure that they, the major players, are seen to be acting in a socially responsible manner (Roberts, 2003; Enderle, 2004). These larger companies are in a position which enables them to encourage such socially responsible behaviour within their supply chain (Lepoutre and Heene, 2006). They can accomplish this via compliance with mandatory requirements or capacity building. The compliance approach sets standards for suppliers and tries to prevent non-compliance by a strict monitoring programme, which is time intensive (Lepoutre and Heene, 2006). If a non-compliance is discovered, then the customer can terminate the contract or cease trading with the SME until corrective action is taken. Capacity building is more supportive in helping build the supplier's own capacity to handle CSR issues (Lepoutre and Heene, 2006) but tends to be more resource intensive for the customer. If this process involves SMEs in developing countries, the larger companies are also obligated to take responsibility for the well-being and performance of upstream producers that work in those countries (Wolters, 2003). Therefore, it is clear that any supply chain work potentially has significant cost implications for the larger organisations that sit at the top of the supply chain.

Larger organisations in the past 30 years have increasingly adopted environmental management, but opportunities still exist for some of them and for many SMEs to boost their environmental performance through the provision of practical, easy-to-understand knowledge (Shearlock *et al.*, 2000). Another issue for some SMEs, in particular, is that they are '*vulnerably compliant*', being unaware of existing, particularly environmental legislation (CEC, 2002). This highlights a need within organisations, particularly SMEs, to obtain assistance in relation to their environmental impacts and legislative compliance.

Currently a situation exists where organisations at the top of the supply chain are trying to manage their suppliers, many of whom are disparate. The easiest method would be to insist that all suppliers must have or be working towards an EMS, such as ISO 14001, which is a voluntary standard. However, research suggests that voluntary standards such as ISO 14001 and Acorn, which are self-management and industry-driven approaches, are not effective in promoting environmental management in SMEs due to lack of take-up (e.g. Masurel, 2007; Revell and Blackburn, 2007). So, what are the reasons for and against organisations engaging with ISO 14001 (Figure 22.1)?

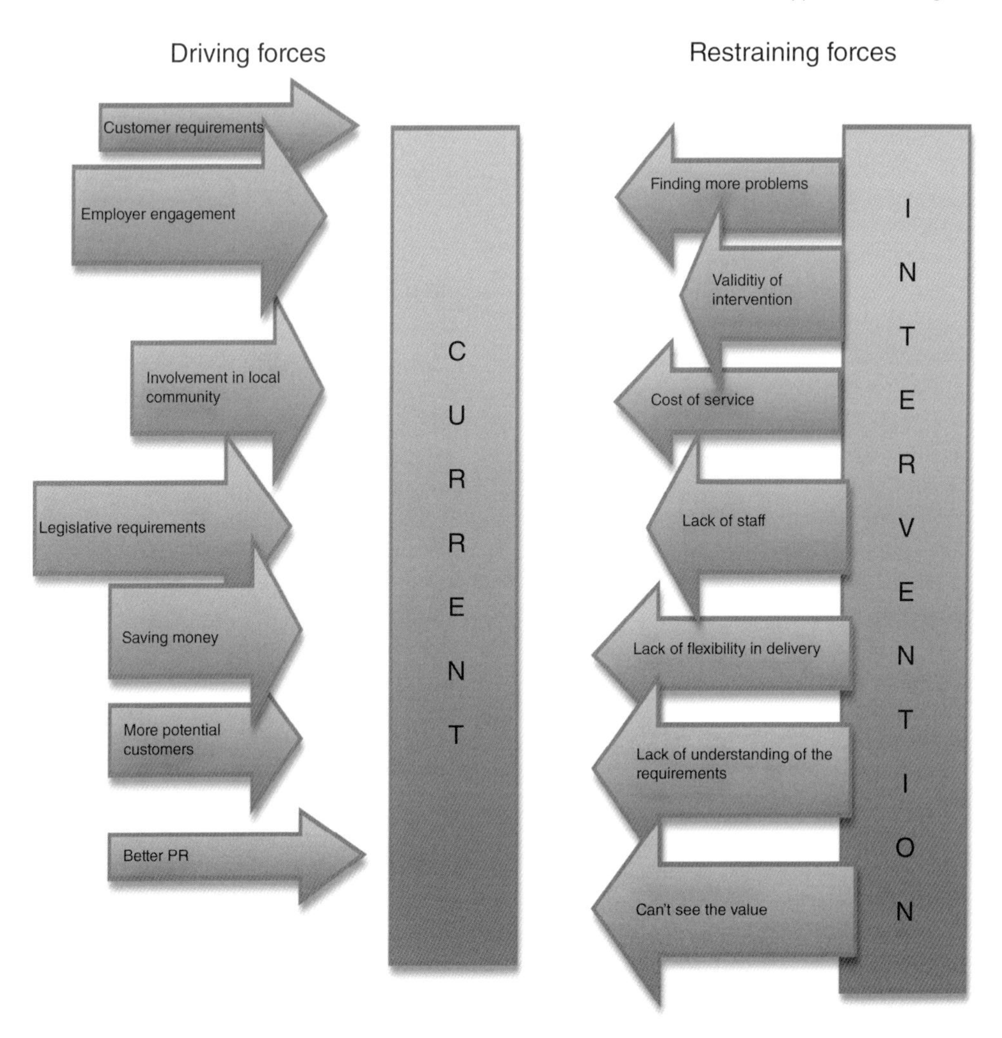

Figure 22.1 Lewin's force field analysis applied to environmental management.

If an EMS were to be implemented, the resulting actions could bring about a stronger, more dedicated and productive workforce, which would be more outwardly focused. This could result in greater attention to the customers' requirements, in turn leading to a better working relationship within the supply chain.

It is not just about environmental matters; there has been considerable discussion over the last decade around the topic of CSR (Polonsky and Jevons, 2009), which again is implemented through the supply chain to create the idea of the major companies being seen as acting in a socially responsible manner. All organisations will impact their local environment; therefore, they should consider their CSR, as no organisation can operate within a vacuum (Roberts, 2003; Enderle, 2004). Traditionally, some of these impacts were mitigated through the use of legislation, but what if companies governed themselves? Or is this just not possible due to human nature?

22.5 What Happens when Things Go Wrong?

This section details some of the well known pollution incidents and how they were managed. While reading about these, consider if we have actually learnt from these events, or if the same situation could still occur again today somewhere in the world.

Love Canal – This has to be one of the worst environmental pollution incidents in the history of the United States, not just in relation to the incident but also the consequences. William T. Love had a dream of creating a short canal between the upper and lower Niagara rivers to produce cheap hydroelectric power for industry and homes. Unfortunately, Nikola Tesla had just made the momentous discovery of how to transmit electricity over vast distances at a rate which made it economically viable, and there was an economic downturn. So, in 1910, construction of the canal ceased. The canal in the 1920s started to be used as a municipal and industrial chemical dump. Finally, in 1953 the dump was sealed and sold to the city for one dollar. A school and housing was subsequently built on the site. No one really understood the issues or that residents were living on a toxic waste dump thought to contain 82 different compounds (11 being suspected carcinogens). In 1978, unprecedented rainfall resulted in the leaching of chemicals from the site. Children and pets were getting chemical burns, and there was an unusually high rate of miscarriages in the area. It had also been noted that patients were experiencing high white blood cell counts, which is potentially linked to leukaemia. Residents had to be evacuated from their homes, and the state had to purchase the land and property. Now we are more aware of the need to manage landfill sites. In the United Kingdom, we have regulations that prevent the mixing of different types of hazardous waste and have processes in place to ensure that the required geology is present to contain the materials. However, in 1920 there was no such knowledge, and it is in fact through events such as Love Canal that we have come to realise the necessity of environmental management to ensure that landfills are safe.

But how many other sites exist that we are just not aware of? How many companies have buried chemicals or materials in the past, with the records of that location now being lost? We cannot know; however, what we can do is to stop this happening in the future by the effective management of our impacts enabling us to reduce the potential damage.

We now try to ensure that we manage chemicals safely, with legislation such as REACH. We also ensure that at the time of their disposal they go to places where they can be managed safely by ensuring that landfill sites are appropriately licensed for all the wastes taken, in the United Kingdom. However, what about other countries of the world?

Of course, licensing and legislation do not prevent accidents, and there is still that historic pollution that can cause problems for us or future generations. This is where environmental management is so helpful in ensuring that we are not only aware of the issues, but that there are operational procedures in place to contain any incidents and that the organisations consider their operations and reduce the risks where possible and mitigate against others.

In the case of historical sites, there is a problem. If, for example, a foundry was on a site and the owners sold it to another foundry, how could it be established who caused the pollution? If one of the companies ceases to exist or both, someone else is potentially left with the bill for clean-up. In the case of water pollution, that someone is the Environment Agency, but that means that in effect the tax payer is paying. So there are still issues to be considered and worked through, but we are now at least thinking about them.

Bhopal – Another notable environmental disaster was that of Bhopal where on 3 December 1984 a cloud of 40 tonnes of methyl isocyanate (MIC) gas was released from the Union Carbide Factory, which had become surrounded by a shanty town with a large population (Broughton, 2005). Four months after the incident, the Indian government reported that 1,430 people had died but in 1991 said the figure was more than 3,800 dead and many thousands with resulting disabilities (Broughton, 2005). There is still debate on the cause of the disaster, as some say that the plant had numerous safety issues, while the company stated that it was an act of sabotage. Whatever the cause, there was a business risk, whether the safety systems were faulty or someone had the opportunity to cause sabotage which resulted in the deaths of nearly 4,000 people. So, were safety issues met? If not, then an effective EMS with procedures in place would have helped to ensure that safety systems were functioning and operating procedures followed. If it was sabotage, then procedures stating that on the MIC system two members of staff had to be in place during any work and that the system was secure at all other times would have reduced the risk and saved those lives lost and the suffering of those who are still subject to the health consequences of such exposure.

Camelford – This is a small town in Cornwall that became the centre of a very public water pollution incident in July 1988, when the drinking water supply was accidently contaminated with 20 tonnes of aluminium sulphate, which raised the concentration of aluminium to 3,000 times the permissible limits. The Lower Moor Works was originally a fully manned water treatment station. Later, it was converted to operate automatically and was, therefore, unsupervised. Due to the geology of the area, the incoming water had a tendency to be acidic, so it had to go through a process where lime was added to neutralise it, and then to remove the brown colouring and any remaining bacteria it was necessary to add aluminium sulphate, which was then itself removed by filtration. It may then be necessary to add more lime to ensure that the final pH is normal before the water reached the mains. On the fateful day, a different driver delivered the aluminium sulphate; the key he had fitted both the aluminium storage tank and the water tank, but the tanks were not labelled, and the driver put the aluminium in the wrong one. As the aluminium sulphate passed through the supply system, its chemical structure broke down to produce several tonnes of sulphuric acid which stripped the pipe network, causing leaching of the metals such as lead and copper. Over the years, there have been numerous enquiries and court cases. However, while compensation has been paid out, we still do not know the real implications of this incident. Some people who were resident at the time have died of dementia. In 2012, at an inquest into the death of one of the victims, it was stated by the coroner that South West Water had been 'gambling with as many as 20,000 lives' due to their slowness in acknowledging the incident and by stating that there was no health risk when they had no certainty of this being the case. In 2013, a report was produced by the Lowermoor subgroup of the Committee on Toxicity of Chemicals in Food, Consumer Products and the Environment which concluded that exposure to the chemicals was unlikely to cause '*delayed or persistent harm*' and was also unlikely to cause future ill health. But if high levels of aluminium are in the body, can we really be certain that no harm will be caused? Do we know enough about how our brains function to be sure that a huge dose of a chemical such as aluminium which has been found to have passed through the blood/brain barrier will not have severe implications for the people concerned? Also, what of the environment, fish killed in the river, and farm animals that suffered and had to be humanely destroyed? Aluminium is still used in many water purification processes, and the debate over the health implications continues.

Camelford is an example of another incident which occurred as a result of a series of errors: the same lock on two very different unlabeled storage tanks, a new driver who did not know the site, and the fact that it was an unmanned site. Hopefully we are more aware today and more likely to avoid such incidents. But are we? Do all the companies that deal with dangerous chemicals have safety procedures that are followed to the letter? Are there any companies that have no new staff, or if new staff arrive, their induction happens on day one and not several months later?

When we think of major environmental disasters, probably one of the most frightening is a nuclear incident. It may be the severity of the potential impact, or the fact that you cannot see the pollution and therefore avoid it. So when **Chernobyl** became news on 26 April 1986, suddenly the world was looking towards what is now Ukraine but formerly the USSR. Those affected included people involved in the emergency operations on the day of the incident (~600), those sent to the area in the period 1986–1989 to do decontamination work, and of course the residents who lived within 30 km of the site; around 116,000 people were evacuated from the area (Hatch *et al.*, 2005).

There was an explosion and fire at the site which released radioactive particles into the atmosphere which spread over much of what was then the USSR and Europe. This was the worst nuclear disaster due to not only the costs but also the casualties involved. It, along with Fukushima, is one of the only two level 7 classifications for nuclear incidents. The cost implications to the Ukraine government as a result is estimated at USD 148 billion for the period 1986–2015 (Oughton *et al.*, 2009).

The accident caused 31 deaths, but this does not include the long-term issues with the increase in cancers rates still being studied. Questions remain: what are the long-term health implications? Was there significant long-term environmental damage? Is the area really safe? Again, we think we are invincible sometimes, that nothing can possibly go wrong; but how many times do we see incidents that could have been prevented? An EMS is not immediately perfect, but with such a system corrective action is required and, therefore, any incidents would be carefully reviewed to ensure that they are not repeated. Potentially if organisations were to incorporate information from other disasters, then their systems would be improved through the increased knowledge and risk awareness.

Sea Empress – On 15 February 1996, a single-hull oil tanker entered the Cleddau Estuary on route to Milford Haven heading for the Texaco Oil Refinery. Due to the nature of the port, there was a requirement for two experienced pilots from the port to bring it safely in. The cause of the incident was pilot error and miscalculation of the effects of the tides. Other contributory factors were found to include poor weather and tugs with insufficient pulling power. Due to strong currents, the ship was pushed off course and hit the rocks, resulting in the perforation of the single hull and creation of the third largest oil spill to enter British waters, killing many birds and polluting a large stretch of the Pembrokeshire coastline. If an EMS had been in place, would it have been different? Potentially yes, if procedures had existed to ensure that pilots were sufficiently trained and accurate tidal information was utilised along with the allocation of tugs of sufficient capacity for the tasks required.

Have we learnt from these events? It appears not, as on the evening of Saturday, 10 December 2005, there was a major incident at a tank at the Hertfordshire Oil Storage Limited (HOSL) part of the **Buncefield Oil Storage Facility**. At the site, there were two forms of fill level control in place: a gauge that enabled the employees to monitor the filling operation and an independent high-level switch (IHLS) which was designed to close down automatically if the tank was in danger of being overfilled. However, it appears that the first gauge got stuck and the high-level switch was not working, resulting in the control room being unaware that the tank was being overfilled. As the tank overflowed, the resulting vapour ignited, and the resulting fire lasted for five days. It appears that the issues were a gauge which although serviced had got stuck afterwards, and no action had been taken to fix it. There was a requirement with the high-level switch for a padlock to secure it in the working position, but the installer, maintenance, and the site contractor were not aware of this; so no lock was fitted. As the petrol was overflowing, the bund (secondary containment) should have been effective in containing the liquid. If not, there was a system of drains (tertiary containment) to stop the fuel escaping into the environment. However, both failed due to inadequate design and maintenance. Is there a lesson to be learnt here? Buncefield had a management system in place, and despite independent audits the system failed. So even with a system, things can still go wrong. Here, we see a lack of regulation available to the staff in the control room, and it is noted that the throughput had increased, so the aspect of the operation had actually changed. It is easy to see how a change in events or operations can cause severe problems, but if these changes had been reviewed and discussed would the situation have happened?

22.6 Conclusions

It is always easy to be wise after the event, and all those involved with the incidents highlighted were doing what they thought to be the best at the time. There was no deliberate intent to damage or injure. However, when you consider these incidents, it becomes clear that they all have issues around human error, lack of management control, lack of planning, or increase in demand. We cannot control earthquakes, but we can build facilities which can withstand them. We cannot avoid using relief or new staff, but we can make sure they have all the skills and information they need to do the job. We have to expand processes to meet demand, but we can think of the risks involved and mitigate the potential for failures. Can we expect people to always get things right? Probably not! Can we put systems in place which can flag up issues and ensure that they are correct? Most definitely yes! Human error and attempts to save money were at the root of these incidents. We cannot stop companies taking shortcuts except with the threat of the legal implications, of course; however, we are more acutely aware of the issues of business and reputational risk now, and this may encourage more companies to take the route of environmental management to ensure that they are prepared in relation to environmental incidents.

So, we have all these problems which can occur, be it through sabotage, human error, or failure of equipment. But what can we do to prevent these events happening; one obvious route is through the use of an EMS. It may be that you want to use a recognised system, such as EMAS or ISO 14001, or you may just want to follow the route and apply the principles. Whatever you decide, at least you will be looking at the risks, and if you wish your organisation to be responsible, you will be able to address those risks.

References

Alberti, M., Caini, L., Calabrese, A., and Rossi, D. (2000) Evaluation of the costs and benefits of an environmental management system. *International Journal of Production Research*, 38 (17), 4455–4466.

Broughton, E. (2005) The Bhopal disaster and its aftermath: A review. *Environmental Health*, 4, 6.

Carraro, C. and Leveque, F.E. (1999) *Voluntary Approaches in Environmental Policy*. Dordrecht: Kluwer Academic.

Christini, G., Fetsko, M., and Hendrickson, C. (2004) Environmental management systems and ISO 14001 certification for construction firms. *ASCE Journal of Construction Engineering and Management*, 130 (3), 330–336.

Commission of the European Communities (2002) European SMEs and Social and Environmental Responsibility. Observatory of European SMEs 4.

Enderle, G. (2004) Global competition and corporate responsibilities of small and medium-sized enterprises. *Business Ethics Euro*, 13 (1), 51–63.

Florida, R. and Davidson, D. (2001) Gaining from green management: Environmental management systems inside and outside the factory. *California Management Review*, 43 (3), 64–84.

Hatch, M., Ron, E., Bouvill, A., Zablotska, L., and Howe, G. (2005) The Chernobyl disaster: Cancer following the accident at the Chernobyl Nuclear Power Plant. *Epidemiological Reviews*, 27, 56–66.

Hillary, R. (2004) Environmental management systems and the smaller enterprise. *Journal of Cleaner Production*, 12 (6), 561–569.

Hudson, M., Lean, J., and Smart, P.A. (2001) Improving control through effective performance measurement in SMEs. *Production Planning Control*, 12 (8), 804–813.

Lepoutre, J. and Heene, A. (2006) Investigating the impact of firm size on small business social responsibility: A critical review. *Journal of Business Ethics*, 67 (3), 257–273.

Marshall Report (1998) Economic Instruments and the Business Use of Energy. Task Force Report, Stationary Office, London.

Masurel, E. (2007) Why SMEs invest in environmental measures: sustainability evidence from small and medium-sized printing firms. *Business Strategy and the Environment*, 16 (3), 190–201.

Panapanaan, V.M., Linnanen, L., Karvonen, M.-M., and Phan, V.P. (2003) Road mapping corporate social responsibility in Finnish Companies, 44 (2–3), 133–148.

Polonsky, M. and Jevons, C. (2009) Global branding and strategic CSR: An overview of three types of complexity. *International Marketing Review*, 26(3), 327–347.

Potoski, M. and Prakash, A. (2005) Covenants and weak swords: ISO14001 and facilities' environmental performance. *Journal of Policy Analysis and Management*, 24 (4), 745–769.

Revell, A. and Blackburn, R. (2007) The business case for sustainability? An examination of small firms in the UK's construction and restaurant sectors. *Business Strategy and the Environment*, 16, 404–420.

Revell, A., Stokes, D., and Chen, H. (2009) Small businesses and the environment: Turning over a new leaf? *Business Strategy and the Environment*, 19, 273–288.

Roberts, S. (2003) Supply chain specific? Understanding the patchy success of ethical sourcing initiatives. *Journal of Business Ethics*, 44 (2), 159–170.

Shearlock, C., Hooper, P., and Millington, S. (2000) Environmental improvement in small and medium-sized enterprises: A role for the business-support network, *Greener Management International*, 30, 50–60.

Shen, L. and Tam, V.W. (2002) Implementation of environmental management in the Hong Kong construction industry. *International Journal of Project Management*, 20, 535–543.

Wolters T. (2003) Transforming international product chains into channels of sustainable production. The imperative of sustainable chain management. *Greener Management International*, 43, 6–13.

23

Role of Citizen Science in Air Quality Monitoring

Natasha Constant

Sustainable Places Research Institute, Cardiff University, United Kingdom

23.1 Introduction

Cities and towns are sources of innovation and wealth creation but are subject to crime, disease, and pollution (Bettencourt *et al.*, 2007). In Europe, North America, and other parts of the world, anthropogenic pressures within urban areas such as the growth of road transport, demands for housing, and the consumption of natural resources exert pressure on the environment and human health (Royal Commission on Environmental Pollution, 2007). Urban pollution originates from different sources including transportation, construction building, misconnections, run-off activities, and discharges from contaminated land (Department for Environment Food and Rural Affairs, 2015). Exposure to air pollution, temperature, and noise have adverse health effects, while, green space has been associated with both positive and negative health effects in urban settings and the ability to reduce personal exposure to air pollution (Royal Commission on Environmental Pollution, 2007). Considering the diverse sources and impacts of urban pollution, more innovative methods are required to monitor and mitigate pollution effects on the environment and human health.

Citizen science provides a step forward in the engagement of local populations in the monitoring and management of urban pollution matters. Citizen science has been described as a 'time-honoured, evolving practice that engages nonprofessional scientists in the practice of research' (Crain *et al.*, 2014, p. 1), and as a process whereby 'citizen scientists work with professional counterparts on projects that have been or adapted to give amateurs a role, either for the educational benefit of the volunteers themselves or for the benefits of the project' (Silvertown 2009, p. 467). The nomenclature surrounding citizen science is often uncertain and contradictory because wide-ranging definitions are being applied among academic and practitioner communities. Citizen science has also been used synonymously with other monitoring activities such as 'volunteer biological monitoring' (Lawrence, 2006), 'community-based monitoring' (Danielsen *et al.*, 2009), and 'participatory monitoring' (Bell *et al.*, 2008). Citizen science has also been associated with a movement to democratise science by bringing the public and science closer together through greater dialogue and decision-making processes on a range of environmental issues (Bonney *et al.*, 2015). Citizen science can also involve empowered individuals who monitor their local environment and share their knowledge locally, out of 'care for place' without necessarily having any explicit relationship with academic research activity (Nye *et al.*, 2011).

Urban Pollution: Science and Management, First Edition. Edited by Susanne M. Charlesworth and Colin A. Booth.

Citizen science has the potential to contribute to the monitoring and mitigation of urban pollutants, particularly in regions lacking resources and data, by improving the spatial and temporal scales of environmental research in urban areas. There is an increasing realisation that involving local citizens in monitoring campaigns may also foster learning opportunities and increase awareness of environmental issues (Sîrbu *et al.*, 2015). In this chapter, we explore new innovations and technologies in the field of citizen science to engage civil society in the monitoring and management of air pollution in urban contexts.

23.2 Air Pollution in Urban Environments

Clean air is fundamental for human life; however, in urban areas, air pollution has widespread implications for health, well-being, and the urban environment. Long-term exposure to particulate air matter, nitrogen dioxide (NO_2) and ozone (O_3) from traffic and domestic combustion processes is associated with ischaemic heart diseases, cardiovascular disease, and lung and bladder cancer mortalities (Anderson *et al.*, 2012, Cesaroni *et al.*, 2013, International Agency for Research on Cancer World Health Organisation, 2015). Furthermore, research has also indicated links between air pollution, low birth rates (Brauer *et al.*, 2015) and other diseases such as diabetes (Rao *et al.*, 2015).

Traditionally, personal exposure to air pollution was assessed by estimating population-wide exposure derived from static air quality monitoring networks (Steinle *et al.*, 2013). Static sensors measured annual ambient average concentrations interpolated over spatial maps for a variety of pollutants (Steinle *et al.*, 2013). Air quality monitoring networks established in many European countries fulfil regulatory requirements for monitoring air quality to serve the reporting standards embedded in national and international laws (Reis *et al.*, 2013). However, monitoring sites have to comply with strict requirements; the instruments used are expensive, necessitating specific expertise for their maintenance; and few monitoring sites are located in areas that give representative data for urban settings (Reis *et al.*, 2013). In cities, many people live in close proximity to traffic, where they may experience greater exposure to particulate matter and NO_2 levels (Reis *et al.*, 2013). While monitoring networks contribute to assessing trend and observations of the effect of emission control measures, their limited extent means they cannot provide high-resolution indications of air quality at finite scales (Reis *et al*, 2013). In urban environments, air quality is dependent on local conditions; for example, NO_2 levels vary in relation to anthropogenic impacts such as traffic density, distance to major roads, and the presence of major industries (Jerrett *et al.*, 2007; Luginaah *et al.*, 2006, Rijnders *et al.*, 2001). Many studies have also found that personal exposure of schoolchildren to traffic air pollutants is related to the proximity of schools and local residences to busy roadways (Singer *et al.*, 2004, Van Roosbroeck *et al.*, 2006). The link between air pollution and human health effects is well documented; however, the ability to identify attributing sources is problematic because of the temporal and spatial variability of pollutants and the difficulty in estimating the time spent in different micro-environments (Snyder *et al.*, 2013). Citizen science has the potential overcome these problems.

23.3 Citizen Science and New Advances in Air Quality Monitoring

Citizen science can lead to protection of public health and the environment by providing communities with better data on pollution in their neighbourhoods, reducing the costs of air pollution monitoring and better management of facilities to support regulatory and public agencies and researchers (Snyder *et al.*, 2013). Since the late 1970s, wearable and potable sensors formed the basis for measuring personal exposure, time-budget, and health impacts (Steinle *et al.*, 2013). Recently, the development of low-cost, easy-to-use portable air pollution sensors has enabled high-resolution data to be obtained in near real time (Reis *et al.*, 2013; Snyder *et al.*, 2013).

The MESSAGE Project was one of the first citizen science projects in Europe to test and evaluate the performance of a network of low-cost sensors for urban air quality monitoring in the United Kingdom; using static and mobile sensors carried by people (Polak, 2007). Other projects such as CommonSense make use of 'mobile participatory sensing' methods, where a large number of sensors are carried by communities with Global Positioning System (GPS) devices to monitor individual movements to derive personal exposure measurements in urban settings (Dutta *et al.*, 2009, Willett *et al.*, 2010). Citi-Sense-Mob applies electrochemical sensors to measure an array of pollutants to test air quality by involving citizens who mount sensors on bicycles, cars, and buses (Castell *et al.*, 2015). The bus sensors connect with computer systems to collect data on bus speed, position, braking, and GPS information. Similarly, the bike sensors also record GPS data, which is transmitted to a central database. Sensors when interlinked with Information Communication Technologies (ICTs) improve the handling of large datasets and data availability, accessibility, and visualisation. The use of sensor technology has also expanded through wireless networks, allowing communication across different networks and Web services. The Citi-Sense-Mob project connects gathered data with mapped air pollution levels, on a website and mobile phone application presented in real-time data and in the future will be used to identify less polluted routes across cities (Castell *et al.*, 2015). The project will also pilot services to integrate the data presented on mobile phone applications with other applications such as air quality forecasts, UV indexes, and pollen indexes (Castell *et al.*, 2015).

Personal exposure to air pollution has usually been associated with time spent in outdoor environments; however, diffusion of outdoor air into buildings and the release of primary substances (tobacco smoke, cooking, and heating with natural gas or other fuels) can also affect personal indoor exposure levels (Braniš, 2010, Steinle *et al.*, 2013). Dons *et al.* (2013) determined personal exposure by considering the daytime mobility of people, concentration exposure at different locations, time spent in different microenvironments (indoors and outdoors), and exposure encountered on transportation using activity-based models. The models were validated using weeklong time–activity diaries and exposure data revealed from citizen monitoring campaigns. Air pollution sensors are also interlinked with physiological sensors and GPS data to understanding the relationship between personal exposure, microenvironments, and health indicators (Snyder *et al.*, 2013). Smartphones can also be used to track physical activity and geographical locations to use time–locations patterns and energy expenditure associated with physical activity when linked to space–time air pollution maps (De Nazelle *et al.*, 2013) and sensor-derived data (Nieuwenhuijsen *et al.*, 2015). A range of sensors can also be used in tandem to measure air pollution, noise, temperature, UV, physical activity, location, blood pressure, heart rate, and lung function, and emotional status/mood of a participant to obtain information on multiple environmental exposures and acute health effects in situ (Nieuwenhuijsen *et al.*, 2014).

The results of these studies have implications not only for urban planners but also for health care practitioners to track air quality, health-related impacts, and improve health diagnoses and treatment on an individual basis. Current advances in sensor technologies and the adoption of volunteers in air pollution monitoring have expanded the scope and scale of urban air pollution and health research. However, there are also numerous challenges associated with the development of new sensor technologies including the need to test new sensor technologies for data quality, develop strategies to integrate data from multiple sensors of differing quality, and improve data processing, storage, visualisation, and dissemination to different end users (Snyder *et al.*, 2013).

23.4 Citizen Science, Biomonitoring, and Plants

Citizen science has been applied to monitor the impacts of air pollution on plant communities and to measure the ecosystem service benefits of plants to mitigate air pollution in urban areas. Citizen scientists have engaged in biomonitoring to complement studies of direct air pollution monitoring by assessing the impacts of air pollution on the leaves of plants sensitive to O_3 effects (Pellegrini *et al.*, 2014). O_3 causes physiological damage and reduced yields for some common plant species (Fishman *et al.*, 2014). The oxidant properties of O_3 can also be toxic for plant species by interfering with the plants' ability to produce and store food; increasing susceptibility to disease, insect infestations, and reduce productive capabilities (Fishman *et al.*, 2014). Citizen scientists in Italy estimated the percentage area of cotyledon and leaf cover covered by lesions of tobacco plants affected by O_3 damage (Pellegrini *et al.*, 2014). The lesions were assessed visually by comparing spot standards on score sheets to assess the level of damage and the results compared with air quality measurements taken with air sensors (Pellegrini *et al.*, 2014). The project demonstrated that O_3 levels were above background levels, and plants exposed to ambient air pollution rapidly exhibited symptoms of ozone damage within a period of two days (Pellegrini *et al.*, 2014).

Seed *et al.* (2013) used citizen-science-generated data to show a significant association between the abundance of lichens and levels of nitrogen and sulphur containing pollutants on trunks and twigs. NO_2 from vehicle emissions is also predicted to influence urban lichen diversity and species composition close to roadways (Davies *et al.*, 2007; Gadsdon *et al.*, 2010). Tregidgo *et al.* (2013) carried out a citizen science survey where trends in lichen community composition were monitored along transects within selected sites where traffic volume was high and low. A high abundance of nitrophilic lichens were situated close to roadways, with a trend towards more nitrophoblic lichens further away from roadways in areas with low background nitrogen deposition and high NO_2. The study provides further evidence that vehicular pollution effects on lichens are largely attributable to the impacts of nitrogenous pollution. Gosling *et al.* (2016) asked citizen scientists to record the symptoms of tar spots created from NO_2 and other environmental factors in the atmosphere on sycamore trees. The results showed that: tar spots symptoms were reduced where there were fewer fallen leaves, which acted as a source of inoculum against the disease; elevated NO_2 concentrations reduced tar spots symptoms above a specific threshold; and tar spot symptoms were lower at sites with higher temperature and rainfall (Gosling *et al.*, 2016). Pellegrini *et al.* (2014) suggest that biomonitoring of airborne chemical pollutants may represent a key element of engaging citizens and communities in assessments of the state of the environment, where a "new discipline 'environmental horticulture' may be envisaged" (ibid, p.5).

Researchers and planners have also been interested in plants for their associated social and health benefits and the role they play in the removal of atmospheric pollutants (Setälä *et al.*, 2013). Plants provide a regulating ecosystem service; for example, trees and herb leaves remove particulates, NO_2, O_3, and other pollutants (Nowak, 2006, Yin *et al.*, 2011). The air filtering properties of vegetation, and the extent of air pollution immobilisation, is dependent on the species, different plant traits, location, and structure of vegetation (Bolund and Hunhammar, 1999; Weber et al., 2014). Citizen science has played a major role in urban research on trees which involves members of the public acting as 'urban forest inventory teams' or participating in other monitoring projects that have otherwise been undertaken by forestry practitioners and certified arborists. In the United States, volunteers have been engaged in 'volunteer-based urban forest inventory and monitoring programmes' since the 1990s to enhance public awareness of the benefits and value of urban forests (Tretheway *et al.*, 1999). The Philly Tree Map project is an urban tree mapping and monitoring project that has developed a website, and mobile phone application to allow citizen scientists to enter information about trees species, diameter, and height and to calculate and view the ecosystem services of trees, such as air quality improvements on visual maps (Urban Forest Map, 2016). The resulting maps have been used to educate residents about the benefits of urban forests in the areas they live and work (Urban Forest Map, 2016). The implementation of similar tree-monitoring projects across multiple cities and the standardisation of data collection methods may also expand research on the ecosystem services and mitigating effects of trees across large spatial scales.

The analytical software i-Tree was created by United States Forest Services to calculate the ecosystem services of trees and allows the public to calculate average annual benefits in energy savings, air quality improvements, storm water interception, aesthetics, and property value improvements for each tree (Hilde and Paterson, 2014). The inclusion of comprehensive management costs also allows for a cost-benefit analysis of street trees and their net benefits (Hilde and Paterson, 2014). The popularity of citizen science in biomonitoring and urban tree assessments provides information to inform regional planning and decision-making and may enhance understandings of the dynamics of air pollution and urban ecosystem services. Open-source software tools support methodological advances in planning practice, and enable the development of more user-friendly tools for different stakeholders.

23.5 Social Dimensions of Citizen Science Air Quality Monitoring

Research in the social sciences suggests that the public exhibit low levels of awareness on air pollution matters, highlighting a need to improve the amount of information made publicly available through different communication strategies (Bush *et al.*, 2001; Hedges, 1993). Citizen science may provide a mechanism to foster public engagement in air pollution issues. Recent studies highlight a wide range of social outcomes for individuals engaged in citizen science including enhanced knowledge, changes in attitudes or behaviours, attainment of social and personal benefits, conditions to decision-making, civic environmental actions, and environmental management policies (Ballard *et al.*, 2008; Cornwell and Campbell, 2012; Evans *et al.*, 2005; Fernandez-Gimenez *et al.*, 2008; Lawrence, 2006; O'Rourke and Macey, 2003).

Engagement in air quality monitoring may enhance local awareness, knowledge of the spatial trends in air pollutants among residents, educators, advocacy groups, and policymakers at local scales and facilitate actions towards improving air quality and health in local communities. A review of citizen participation in air quality monitoring in California and Louisiana where groups of residents living close to industrial zones monitored air pollution near soil refineries, chemical factories, and other plants exhibited a range of social benefits (O'rourke and Macey, 2003). Citizens enhanced their awareness, felt empowered by providing new sources of information on air emissions, highlighted gaps in existing monitoring and enforcement systems, and helped to increase regulatory and industry accountability (O'rourke and Macey, 2003).

A study in North Carolina in the United States worked with citizens in the monitoring of NO_2 and O_3 across five counties (Hauser *et al.*, 2015). The project developed collaborations with a local college to disseminate results to communities through newsletters, articles, public presentations, and provided educational material on personal exposure levels and the risks associated with poor air quality. Other projects such as the Citi-Sense-Mob project have developed a more bottom-up approach towards public engagement by allowing a broad range of stakeholders (local government, transport-related groups, and health interest groups) to contribute their views and needs to the development of the project (Castell *et al.*, 2015). The communication and dissemination strategies of Citi-Sense-Mob included public meetings, workshops, and events to exchange information; Web-based technologies and mobile phone applications to visualise and interact with air quality data; and social networking to effectively communicate and engage with a wider audience.

A large-scale citizen science project named the AirProbe International Challenge is one of the few published studies to evaluate the impact of air quality monitoring on the perceptions and behaviours of citizen scientists (Sîrbu *et al.*, 2015). Participants were asked to participate in a competition-based incentive scheme between four cities in Europe. Volunteers engaged in a Web-based game to estimate pollution levels in their cities by placing markers on a map with an estimated concentration and were then asked to measure actual air pollution levels using air-quality-sensing devices. The data was analysed to assess participation patterns and changes in the perception and behaviour of participants over time demonstrating that direct engagement in citizen science can enhance learning and environmental awareness outcomes.

Similarly, a series of Open Air Laboratories (OPAL) surveys conducted in the United Kingdom assessed the impacts of air pollution on plants. The programme developed a wide range of materials for different age groups and abilities to participate in the surveys, including schools, universities, and other education organisations (Davies *et al.*, 2016). An evaluation report of the programme reported that almost half of people interviewed (43%) changed the way they thought about the environment, 37% changed their behaviour towards the environment, 90% had learnt something new, and 83% had developed new skills (Davies, 2013). The programme provided an opportunity for schools to gain access to high-quality science, outdoor learning, and to contribute to research and work with scientists. The above examples provide evidence for the generation of wider social outcomes through participation in citizen science; however, published studies documenting social outcomes are limited to the context of air quality monitoring and largely biased towards documenting the positive benefits of participation.

23.6 Conclusions

Citizen science has expanded the scope of air quality research to move beyond static sensing techniques to engaging citizens in new sensor technologies and biomonitoring to understand the environmental and social factors influencing urban air pollution. When air pollution sensors are used in conjunction with GPS data, activity diaries, mobile phone applications, and other physiological sensors; connections can be made between a person's exposure to their environment and health indicators. New advances in ICTs, mobile phone applications, and social networks also allow for improved communication and dissemination of data to members of the public, regarding personal exposure levels and associated health implications. The future of citizen science will lead to engagement and empowerment in environmental governance through the establishment of new partnerships between academics, local agencies, transport and mobility engineers, urban planners, health groups, and communities to tackle environmental and social issues in cities. To overcome data quality, interpretation, and communication challenges, these partnerships will have to collaborate to test, evaluate, and provide guidance on the efficacy of new research tools and methodologies.

Citizen science has wide implications for air pollution monitoring and management, health literacy, and mobility, transport, and green space planning in urban regions. Although limited in scope, published data describes a variety of benefits for participating individuals and communities through increased awareness, changes in attitudes and behaviours, and community advocacy. Further work is needed for the development of more systematic and evidence-based approaches towards communications on air pollution. Furthermore, integrating social and behaviour research, and citizen scientist perspectives and experiences (both positive and negative) into the project lifecycle to enhance understandings of the value of citizen science as a public engagement tool. However, opportunities for knowledge exchange and dialogue will depend upon the aims, objectives, and assumptions of project designers to inform what type of knowledge counts and whose knowledge and decisions are relevant. A review of citizen scientist programme characteristics (e.g. see Conrad and Hilchey (2011) for a definitions of different forms of participation) and social outcomes is missing from the literature and would provide further information for project organisers to plan and implement participatory monitoring initiatives to achieve desired goals. Expanding the scope of social science research in citizen science can be achieved by collecting data on participants through questionnaires, interviews, and focus groups and mobile phone technologies to provide insights into the attitudes, motivations, and behaviours associated with air pollution and human health.

References

Anderson, J.O., Thundiyil, J.G., and Stolbach, A. (2012) Clearing the air: A review of the effects of particulate matter air pollution on human health. *Journal of Medical Toxicology*, 8, 166–175.

Ballard, H.L., Fernandez-Gimenez, M.E., and Sturtevant, V.E. (2008) Integration of local ecological knowledge and conventional science: A study of seven community-based forestry organizations in the USA. *Ecology and Society*, 13, 37.

Bell, S., Marzano, M., Cent, J., Kobierska, H., Podjed, D., Vandzinskaite, D., Reinert, H., Armaitiene, A., Grodzińska-Jurczak, M., and Muršič, R. (2008) What counts? Volunteers and their organisations in the recording and monitoring of biodiversity. *Biodiversity and Conservation*, 17, 3443–3454.

Bettencourt, L.M., Lobo, J., Helbing, D., Kühnert, C., and West, G.B. (2007) Growth, innovation, scaling, and the pace of life in cities. *Proceedings of the National Academy of Sciences*, 104, 7301–7306.

Bolund, P. and Hunhammar, S. (1999) Ecosystem services in urban areas. *Ecological Economics*, 29, 293–301.

Bonney, R., Phillips, T.B., Ballard, H.L., and Enck, J.W. (2015) Can citizen science enhance public understanding of science? *Public Understanding of Science*, 0963662515607406.

Braniš, M. (2010) Personal exposure measurements. *Human Exposure to Pollutants via Dermal Absorption and Inhalation.* Springer.

Brauer, M., Lencar, C., Tamburic, L., Koehoorn, M., Demers, P., and Karr, C. (2015) *A Cohort Study of Traffic-Related Air Pollution Impacts on Birth Outcomes.* University of British Columbia.

Bush, J., Moffatt, S., and Dunn, C.E. (2001) Keeping the public informed? Public negotiation of air quality information. *Public Understanding of Science*, 10, 213–229.

Castell, N., Kobernus, M., Liu, H.-Y., Schneider, P., Lahoz, W., Berre, A.J., and Noll, J. (2015) Mobile technologies and services for environmental monitoring: The Citi-Sense-MOB approach. *Urban climate*, 14, 370–382.

Cesaroni, G., Badaloni, C., Gariazzo, C., Stafoggia, M., Sozzi, R., Davoli, M., and Forastiere, F. (2013) Long-term exposure to urban air pollution and mortality in a cohort of more than a million adults in Rome. *Environmental Health Perspectives (Online)*, 121, 324.

Conrad, C.C. and Hilchey, K.G. (2011) A review of citizen science and community-based environmental monitoring: Issues and opportunities. *Environmental monitoring and assessment*, 176, 273–291.

Cornwell, M.L. and Campbell, L.M. (2012) Co-producing conservation and knowledge: Citizen-based sea turtle monitoring in North Carolina, USA. *Social Studies of Science*, 42, 101–120.

Crain, R., Cooper, C., and Dickinson, J.L. (2014) Citizen science: A tool for integrating studies of human and natural systems. *Annual Review of Environment and Resources*, 39, 641.

Danielsen, F., Burgess, N.D., Balmford, A., Donald, P.F., Funder, M., Jones, J.P., Alviola, P., Balete, D.S., Blomley, T., and Brashares, J. (2009) Local participation in natural resource monitoring: A characterization of approaches. *Conservation Biology*, 23, 31–42.

Davies, L. (2013) OPAL Community Environment Report: Exploring Nature Together. Open Air Laboratories (OPAL).

Davies, L., Bates, J., Bell, J., James, P., and Purvis, O. (2007) Diversity and sensitivity of epiphytes to oxides of nitrogen in London. *Environmental Pollution*, 146, 299–310.

Davies, L., Riesch, H., Fradera, R., and Lakeman Fraser, P. (2016) Surveying the citizen science landscape: An exploration of the design, delivery and impact of citizen science through the lens of the Open Air Laboratories (OPAL) programme.

De Nazelle, A., Seto, E., Donaire-Gonzalez, D., Mendez, M., Matamala, J., Nieuwenhuijsen, M.J., and Jerrett, M. (2013) Improving estimates of air pollution exposure through ubiquitous sensing technologies. *Environmental Pollution*, 176, 92–99.

Department for Environment Food and Rural Affairs (2015) Policy paper: 2010 to 2015 government policy: Environmental quality. In: Department for Environment, F., and Rural Affairs and the RT Hon Elizabeth Truss MP (ed.).

Dutta, P., Aoki, P.M., Kumar, N., Mainwaring, A., Myers, C., Willett, W., and Woodruff, A. (2009) Common sense: Participatory urban sensing using a network of handheld air quality monitors. *Proceedings of the 7th ACM Conference on Embedded Networked Sensor Systems*, ACM, 349–350.

Evans, C., Abrams, E., Reitsma, R., Roux, K., Salmonsen, L., and Marra, P.P. (2005) The Neighborhood Nestwatch Program: Participant outcomes of a citizen-science ecological research project. *Conservation Biology*, 19, 589–594.

Fernandez-Gimenez, M.E., Ballard, H.L., and Sturtevant, V.E. (2008) Adaptive management and social learning in collaborative and community-based monitoring: A study of five community-based forestry organizations in the western USA. *Ecology and Society*, 13, 4.

Fishman, J., Belina, K.M., and Encarnación, C.H. (2014) The St. Louis ozone gardens: Visualizing the impact of a changing atmosphere. Bulletin of the American Meteorological Society, 95, 1171–1176.

Gadsdon, S.R., Dagley, J.R., Wolseley, P.A., and Power, S.A. (2010) Relationships between lichen community composition and concentrations of NO_2 and NH_3. *Environmental Pollution*, 158, 2553–2560.

Gosling, L., Ashmore, M., Sparks, T., and Bell, N. (2016) Citizen science identifies the effects of nitrogen dioxide and other environmental drivers on tar spot of sycamore. *Environmental Pollution*, 214, 549–555.

Hauser, C.D., Buckley, A., and Porter, J. (2015) Passive samplers and community science in regional air quality measurement, education and communication. *Environmental Pollution*, 203, 243–249.

Hedges, A. (1993) *Air Quality Information, Report on Consultancy and Research*. Unpublished report prepared for the Department of the Environment, London, UK.

Hilde, T. and Paterson, R. (2014) Integrating ecosystem services analysis into scenario planning practice: Accounting for street tree benefits with i-Tree valuation in Central Texas. *Journal of Environmental Management*, 146, 524–534.

International Agency for Research on Cancer World Health Organisation (2015) Press Release No. 221 (17 October 2013). IARC: Outdoor Air Pollution a Leading Environmental Cause of Cancer Deaths. Retrieved: June.

Jerrett, M., Arain, M., Kanaroglou, P., Beckerman, B., Crouse, D., Gilbert, N., Brook, J., Finkelstein, N., and Finkelstein, M. (2007) Modeling the intraurban variability of ambient traffic pollution in Toronto, Canada. *Journal of Toxicology and Environmental Health, Part A*, 70, 200–212.

Lawrence, A. (2006) 'No personal motive?' Volunteers, biodiversity, and the false dichotomies of participation. *Ethics Place and Environment*, 9, 279–298.

Luginaah, I., Xu, X., Fung, K.Y., Grgicak-Mannion, A., Wintermute, J., Wheeler, A., and Brook, J. (2006) Establishing the spatial variability of ambient nitrogen dioxide in Windsor, Ontario. *International Journal of Environmental Studies*, 63, 487–500.

Nieuwenhuijsen, M.J., Donaire-Gonzalez, D., Foraster, M., Martinez, D., and Cisneros, A. (2014) Using personal sensors to assess the exposome and acute health effects. *International Journal of Environmental Research and Public Health*, 11, 7805–7819.

Nieuwenhuijsen, M.J., Donaire-Gonzalez, D., Rivas, I., De Castro, M., Cirach, M., Hoek, G., Seto, E., Jerrett, M., and Sunyer, J. (2015) Variability in and agreement between modeled and personal continuously measured black carbon levels using novel smartphone and sensor technologies. *Environmental Science & Technology*, 49, 2977–2982.

Nowak, D.J. (2006) Institutionalizing urban forestry as a "biotechnology" to improve environmental quality. *Urban Forestry & Urban Greening*, 5, 93–100.

Nye, M., Tapsell, S., and Twigger-RosS, C. (2011) New social directions in UK flood risk management: Moving towards flood risk citizenship? *Journal of flood Risk Management*, 4, 288–297.

O'Rourke, D. and Macey, G.P. (2003) Community environmental policing: Assessing new strategies of public participation in environmental regulation. *Journal of Policy Analysis and Management*, 22, 383–414.

Pellegrini, E., Campanella, A., Lorenzini, G., and Nali, C. (2014) Biomonitoring of ozone: A tool to initiate the young people into the scientific method and environmental issues. A case study in Central Italy. *Urban Forestry & Urban Greening*, 13, 800–805.

Polak, J. (2007) Mobile environmental sensor systems across a grid environment – The MESSAGE project. *ERCIM News*, 2007.

Rao, X., Montresor-Lopez, J., Puett, R., Rajagopalan, S., and Brook, R.D. (2015) Ambient air pollution: An emerging risk factor for diabetes mellitus. *Current Diabetes Reports*, 15, 1–11.

Reis, S., Cowie, H., Riddell, K., Semple, S., Steinle, S., Apsley, A., and Roy, H. (2013) Urban air quality citizen science Phase 1: Review of methods and projects. Report for the Scottish Environment Protection Agency (SEPA) on behalf of the Scotland's Environment Web Project, supported by the European LIFE+ programme. Available: https://www.environment.gov.scot/media/1129/urban-air-quality-citizen-science-phase-1.pdf [Accessed July 2016].

Rijnders, E., Janssen, N., Van Vliet, P., and Brunekreef, B. (2001) Personal and outdoor nitrogen dioxide concentrations in relation to degree of urbanization and traffic density. *Environmental Health Perspectives*, 109, 411.

Royal Commission on Environmental Pollution (2007) Twenty-Sixth Report: The Urban Environment. Available: https://assets.publishing.service.gov.uk/government/uploads/system/uploads/attachment_data/file/228911/7009.pdf [Accessed July 2016].

Setälä, H., Viippola, V., Rantalainen, A.-L., Pennanen, A., and Yli-Pelkonen, V. (2013) Does urban vegetation mitigate air pollution in northern conditions? *Environmental Pollution*, 183, 104–112.

Silvertown, J. (2009) A new dawn for citizen science. *Trends in Ecology & Evolution*, 24, 467–471.

Singer, B.C., Hodgson, A.T., Hotchi, T., and Kim, J.J. (2004) Passive measurement of nitrogen oxides to assess traffic-related pollutant exposure for the East Bay Children's Respiratory Health Study. *Atmospheric Environment*, 38, 393–403.

Sîrbu, A., Becker, M., Caminiti, S., De Baets, B., Elen, B., Francis, L., Gravino, P., Hotho, A., Ingarra, S., and Loreto, V. (2015) Participatory patterns in an international air quality monitoring initiative. *PLOS One*, 10, e0136763.

Snyder, E.G., Watkins, T.H., Solomon, P.A., Thoma, E.D., Williams, R.W., Hagler, G.S., Shelow, D., Hindin, D.A., Kilaru, V.J., and Preuss, P.W. (2013) The changing paradigm of air pollution monitoring. *Environmental Science & Technology*, 47, 11369–11377.

Steinle, S., Reis, S., and Sabel, C.E. (2013). Quantifying human exposure to air pollution – Moving from static monitoring to spatio-temporally resolved personal exposure assessment. *Science of the Total Environment*, 443, 184–193.

Tretheway, R., Simon, M., Mcpherson, E.G., and Mathis, S. (1999) Volunteer-Based Urban Forest Inventory and Monitoring Programs. Results from A Two-Day Workshop: February, 1999. 27.

Urban Forest Map. (2016) Available: http://www.phillytreemap.org. [Accessed July 2016].

Van Roosbroeck, S., Wichmann, J., Janssen, N.A., Hoek, G., Van Wijnen, J.H., Lebret, E., and Brunekreef, B. (2006) Long-term personal exposure to traffic-related air pollution among school children, a validation study. *Science of the Total Environment*, 368, 565–573.

Weber, F., Kowarik, I., and Säumel, I. (2014) Herbaceous plants as filters: Immobilization of particulates along urban street corridors. *Environmental Pollution*, 186, 234–240.

Willett, W., Aoki, P., Kumar, N., Subramanian, S., and Woodruff, A. (2010) Common sense community: Scaffolding mobile sensing and analysis for novice users. *International Conference on Pervasive Computing*, Springer, 301–318.

Yin, S., Shen, Z., Zhou, P., Zou, X., Che, S., and Wang, W. (2011) Quantifying air pollution attenuation within urban parks: An experimental approach in Shanghai, China. *Environmental Pollution*, 159, 2155–2163.

24

Unique Environmental Regulatory Framework Streamlines Clean-Up and Encourages Urban Redevelopment in Massachusetts, United States

Catherine M. Malagrida, Ileen Gladstone, and Ryan S. Hoffman

GEI Consultants, Inc., Massachusetts, United States

24.1 Introduction

The Commonwealth of Massachusetts in the United States originally enacted Massachusetts General Law, Chapter 21E, the state Superfund law, in 1983, to create the state's hazardous waste site clean-up programme (MGL, 1983). Soon after the programme started, it became clear that the Massachusetts Department of Environmental Protection (MassDEP), the state agency authorised to implement the law, could not oversee the clean-up of thousands of sites in a timely manner. In 1992, amendments to Chapter 21E effectively 'privatised' the environmental clean-up programme. The 1992 amendments and the 1993 Massachusetts Contingency Plan (MCP) (the implementing regulations) created a privatised and risk-based clean-up programme. The cornerstone of the privatised system was the licensing of environmental consultants known as Licensed Site Professionals (LSPs). Decision-making and authority for site investigation and remediation shifted from state workers to private sector LSPs (with limited state oversight) who worked directly for those responsible for cleaning up contamination. The amendment also streamlined the path for reaching a regulatory endpoint. Risk-based standards were developed that took into account current and future site activities and land uses and so allowed for more varied decisions regarding the appropriate clean-up remedy and the ultimate closure strategy. These changes removed regulatory obstacles, thereby speeding up the time needed to clean up and redevelop contaminated sites, particularly in urban areas.

The original environmental clean-up programme stifled site clean-ups and urban redevelopment. Any levels of contamination were reportable to MassDEP, resulting in thousands of sites entering into the clean-up system. Once in the system, sites lingered due to limited MassDEP resources and stringent clean-up standards that did not consider the location or current or future use of the contaminated site. For example, groundwater was required to be cleaned up to stringent drinking water standards even if the area was not in a drinking water aquifer.

The introduction of LSPs and a risk-based clean-up programme dramatically changed how contaminated sites, particularly in urban areas, were cleaned up in Massachusetts. Later amendments were introduced that further encouraged redevelopment of contaminated sites in urban areas, commonly referred to as brownfields (MGL, 1998). These amendments included providing certain liability protections to those willing to take on contaminated sites, and establishing loan and tax credit programmes to incentivise

Urban Pollution: Science and Management, First Edition. Edited by Susanne M. Charlesworth and Colin A. Booth.

the higher costs of addressing the historic contamination.

24.2 LSPs and the Privatised System

An LSP's responsibility is to arrive at evaluations regarding a hazardous waste site that protect people and the environment. LSPs and the privatised system streamlined assessment and clean-up across the state. Except for the most sensitive of hazardous waste sites, the MassDEP no longer needed to review and approve plans and reports. This not only removed the regulatory approval bottleneck, but it also allowed LSPs to use their own professional judgement on what was necessary to meet the performance standards in the MCP and to ensure compliance with the environmental regulations. MassDEP is required to audit a representative number of sites to confirm that LSPs are appropriately implementing the MCP requirements, and MassDEP has the authority to reopen cases that are not in compliance.

The authority of the LSP to effectively endorse their assessment and clean-up approaches made the process more efficient and provided a more predictable way to reach a regulatory endpoint. This certainty allowed property owners to better plan for development projects and encouraged redevelopment of brownfields in urban areas.

24.3 The Risk-Based Clean-Up Programme

The risk-based clean-up approach allows for the type of remedy and the ultimate closure strategy to be based on site activities and land use. The MCP requires contamination to be cleaned up to a level that protects people and the environment on the basis of how the site is being or will be used. Soil clean-up standards are dictated by whether the land use includes a residence, school, daycare, or recreational area (so-called unrestricted use) or whether the land use is limited to a commercial or industrial facility or soils encapsulated beneath a building or some other type of cap (so-called restricted use). These soil standards do not envision clean-up to pristine conditions, even for unrestricted use, and were developed taking into account not only a contaminant's toxicity, but also background conditions. Soil standards for unrestricted use are always more stringent than soil standards for restricted use. A deed restriction in the form of an Activity and Use Limitation (AUL) must be placed on a property if any standard other than the unrestricted use standard is used.

Groundwater clean-up standards are dictated by whether the site use includes a current or potential drinking water source, whether site conditions might result in a source of vapour intrusion into buildings, or whether there is the potential for contaminant discharge to a surface waterbody. Groundwater standards in drinking water areas are typically the most stringent, whereas standards for groundwater with the potential to discharge to surface water are typically less stringent. However, standards are based on toxicity to both people and the environment, so the most stringent standard may not always be in a drinking water area.

Not only are the clean-up standards to reach a regulatory endpoint based on risk, but initial notification thresholds that trigger entry into the clean-up system are also based on risk. This risk-based approach has reduced the number of sites entering the clean-up system since the notification thresholds also consider current land use and location of the property. Residential properties have lower notification thresholds for soil contamination than commercial or industrial properties.

The incorporation of risk-based standards into the Massachusetts clean-up programme has been instrumental in the clean-up and redevelopment of urban properties. Expensive and long-term groundwater treatment systems to meet drinking water standards have not been necessary in urban areas if they are not located within drinking water aquifers. Caps can be used to limit access to contaminated soils and allow levels of contamination that exceed residential clean-up standards to remain at the site. Caps and engineered barriers include asphalt paving, a clean soil layer, or a geotextile barrier, and warning layer.

Sites that meet the applicable clean-up standards can achieve regulatory closure using a Permanent Solution. A deed restriction is necessary for any site that does not meet unrestricted use. For example, controls are required for caps that eliminate exposure to contaminated soil or for vapour barriers and sub-slab depressurisation systems (SSDS) that mitigate vapours entering indoor air.

24.4 Brownfield Redevelopment Incentives

Within the context of the MCP and risk-based clean-up programme, Massachusetts has encouraged the redevelopment of contaminated properties, which are often in urban areas, through the use of Brownfields tax credits and other financial incentives. Brownfields tax credits offset the tax burden for a project proponent by an amount equal to 50% of eligible costs related to assessment and remediation (or 25% if the regulatory endpoint relies on a deed restriction). For non-profit organisations that do not pay some taxes, this tax credit can be sold on the open market to entities that can benefit from the tax credit, and the non-profit still realises a financial benefit. Other financial incentives offered by the state include low-cost loans for brownfields site assessment or remediation projects and low-cost environmental insurance to help mitigate risk associated with brownfields redevelopment.

24.5 Case Studies

Overall, the unique Massachusetts environmental regulatory and business climate encourages the development of urban brownfields sites. The following case studies highlight how the privatised and risk-based clean-up programme, sometimes with Brownfields financial incentives, have streamlined assessment and remediation and allowed for creative solutions for redeveloping contaminated properties in urban areas.

24.5.1 Case Study 1 – East End Veteran's Memorial Park

24.5.1.1 Peabody, Massachusetts

The City of Peabody, Massachusetts constructed a sanctuary downtown by developing an unused, contaminated parcel of land into a lovely park (Figure 24.1), where children can play, others can enjoy a peaceful place, and nature's waters can be accommodated. The 1.3 acre (approx. 5,260 m^2) East End Veteran's Memorial Park in the urban centre of Peabody, Massachusetts was historically a tannery, a coal and wood storage yard, and an automotive repair facility. Prior to being purchased by the City of Peabody, the abandoned brownfields property had become overgrown with vegetation, and regularly flooded during heavy rainfall by the adjacent North River. The property was originally placed on the MassDEP contaminated sites list due to petroleum hydrocarbons, arsenic, and lead that exceeded reportable thresholds for soil located near residential areas (GEI, 2013a).

To characterise environmental conditions at the property, subsurface investigations were conducted under the state environmental regulations. An initial risk assessment concluded that petroleum hydrocarbon and heavy metal contamination in soil across the property and in hotspots posed significant risks to the public.

With only limited available funding, the City was able to take advantage of the Massachusetts environmental regulations that allow for risk-based solutions tailored to site-specific conditions and considerations by constructing the park while leaving much of the contamination in place, beneath a

Figure 24.1 East End Veteran's Memorial Park in Peabody, Massachusetts, after brownfields redevelopment. Mesh grass cap shown in background, porous pavement cap shown in foreground.

suitable cap. To develop this urban downtown site into a public park, the design needed to cap and isolate the contaminants in soil to prevent exposure to future park users; facilitate adequate rainwater infiltration through the cap to minimise stormwater run-off; and accommodate the necessary flood storage capacity of the property required by the adjacent North River, which periodically overflowed its bank.

Under the state environmental regulations (MassDEP, 2014), contaminated soil may be left at a property if risk to current and future users is mitigated. Excavation and off-site disposal of all of the contaminated soil would have been cost prohibitive, and the brownfields site would have likely remained vacant and a community eyesore and hazard if a solution to leave soil on-site could not be developed. The original design concept was to place 3 ft (approx. 0.91 m) of clean soil across the site, providing a suitable barrier to the underlying contaminated soil. However, during design, the City was informed by the MassDEP Division of Water Resources that the property was critical in accommodating the North River flood waters and a 3 ft (approx. 0.91 m) cap across the site would not be an acceptable remedy as it would have reduced flood storage capacity.

The remedy was then re-designed to remove the most heavily contaminated soils, including a heavy metal hotspot to a depth of 6 ft (approx. 1.81 m) and contaminated soil from depths ranging from 12 to 36 in. (approx. 30 to 91 cm) across the entire property.

Three types of caps were installed at the property depending on the planned use of the area (GEI, 2013a). A porous pavement cap was used in hardscaped areas such as walking paths and seating areas, a mesh grass cap was used in areas for passive recreation, and a top soil cap was used in landscaped areas with plantings. Cross sections of each type of cap and where they were used are shown on Figure 24.2 and further described below.

24.5.1.2 Porous Pavement Cap

A porous pavement cap was placed in areas requiring hardscape. Use of porous pavement increased the site's capacity for stormwater infiltration, minimising run-off to the adjacent North River. The porous pavement cap is 2 in. (approx. 5 cm) thick underlain by 10 in. (approx. 25 cm) of subgrade. A geotextile layer and geogrid marker layer inscribed with the phrase 'Warning: Contaminated Soil Do Not Excavate Below This Tape' was placed directly over contaminated soil.

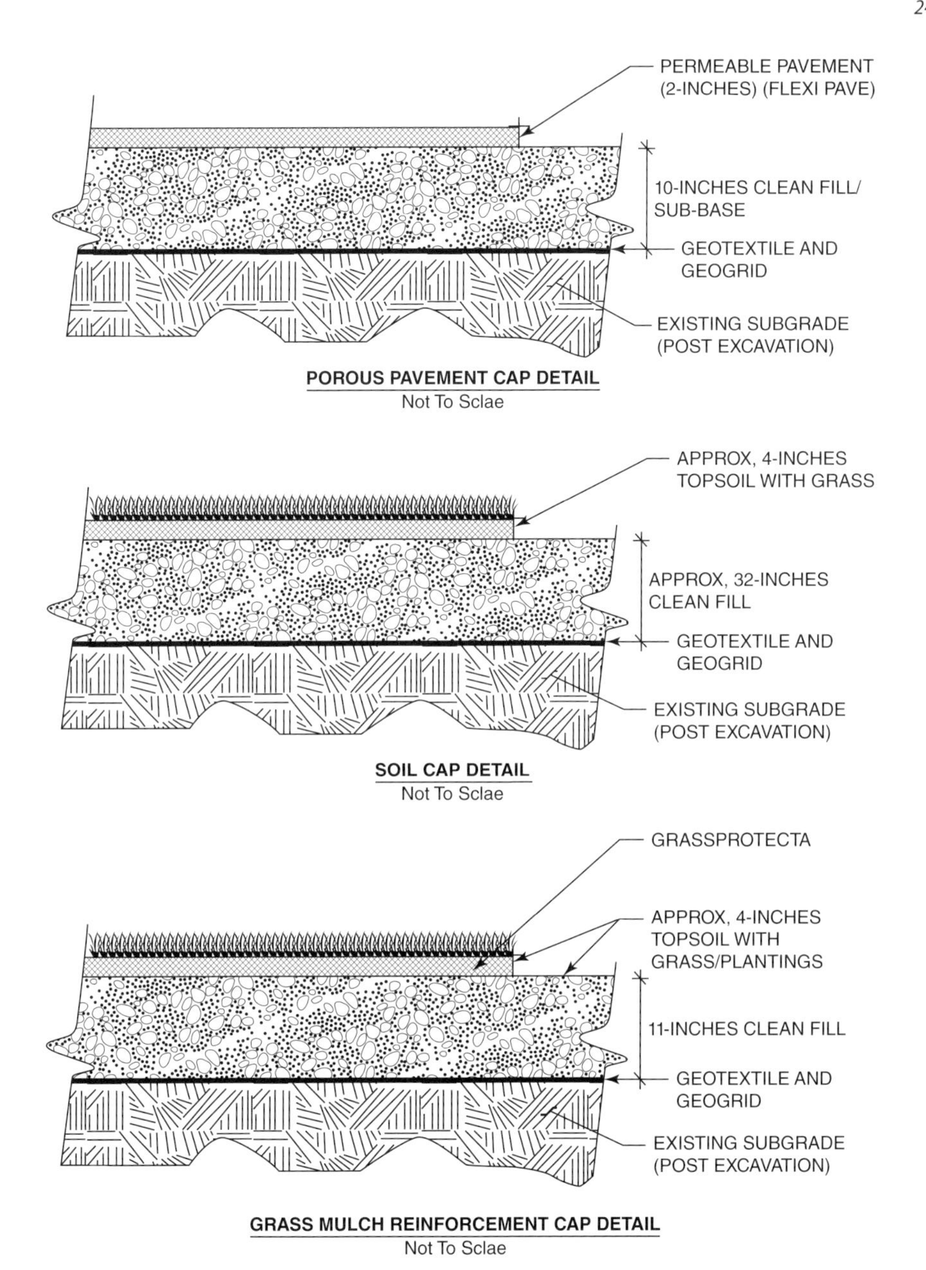

Figure 24.2 Cross sections of cap designs at East End Veteran's Memorial Park in Peabody, Massachusetts.

24.5.1.3 Top Soil Cap

In locations with landscaping and planting, which are not areas amenable to direct contact with soils, a 3 ft (approx. 0.91 m) cap was installed which comprised 32 in. (approx. 81 cm) of clean, imported fill, and 4 in. (approx. 10 cm) of imported top soil and hydroseed, placed over a geotextile layer, geogrid marker layer inscribed with the phrase 'Warning: Contaminated Soil Do Not Excavate Below This Tape'.

24.5.1.4 Mesh Grass Cap

To create an impenetrable barrier and to prevent direct contact with contaminated soil in grassed areas, a mesh grass reinforcement textile (Grassprotecta) was used. The Grassprotecta was placed over approximately

11 in. (approx. 28 cm) of clean fill, a geotextile layer, and geogrid marker layer inscribed with the phrase 'Warning: Contaminated Soil Do Not Excavate Below This Tape'. By using the Grassprotecta, the thickness of the cap was reduced from the required 3 ft (approx. 0.91 m) to approximately 1.5 ft (approx. 0.46 m).

The construction of a cap across this property was planned to eliminate the potential for soil exposure. Since underlying contaminated soil was to remain, the implementation of a deed restriction was necessary to define prohibited and permitted activities and uses at the property. Without the risk-based programme available in Massachusetts, it is unlikely this contaminated urban site would have been cleaned up and redeveloped as a vibrant park.

24.5.2 Case Study 2 – Affordable Housing, Boston, Massachusetts

24.5.2.1 The Community Builders

Charlesview Inc., an ecumenical non-profit that developed and owned the Charlesview Apartments in the wake of urban renewal in 1971, sought to preserve its affordable housing for new generations. Charlesview Inc. and The Community Builders Inc. (TCB), a non-profit development corporation headquartered in Boston, Massachusetts, in the United States, negotiated a land swap in Boston along with financial resources with Harvard University to relocate and reconstruct the ageing Charlesview Apartments to another nearby parcel. TCB redeveloped the former Harvard parcel as the new Charlesview Residences.

The original Charlesview Apartments was a 213-unit low and moderate-income housing development in Boston. The new Charlesview Residences is a 470,000 ft^2 (approx. 143,256 m^2) mixed use development consisting of 240 residential rental units and 100 new homeownership units; community, commercial, and open space uses; significant infrastructure; new streets and parks; and underground parking for 243 cars (Figure 24.3).

The Charlesview Residences was constructed on a 9 acre (approx. 36,421 m^2) contaminated brownfields site that Harvard made available in

Figure 24.3 The completed Charlesview Residences after transformation from vacant, underutilised brownfields parcel.

a mutually beneficial land exchange for the 4.5 acre (approx. 18,211 m^2) ageing Charlesview Apartments site. TCB worked with the United States Office of Housing and Urban Development (HUD) and Mass Housing on a large and innovative financing package that relied on the transfer of Charlesview's existing project-based rent subsidies to the new site and with private investment attracted by 4% Low Income Housing Tax Credits. The creative financing solutions enabled the $143 million rental phase to start and complete in the depth of an economic recession while utilising no competitive, public capital subsidies.

The land where the new Charlesview Residences was built was historically the site of various industrial, manufacturing, and commercial uses beginning in at least the 1880s. These uses included a rope-making factory, a silk factory, a laundry machinery company, and a metal-stamping factory. By the 1950s, the land was occupied by retail space including department and clothing stores. The buildings were abandoned in the early 2000s until redevelopment activities began in 2011 (GEI, 2015a).

Prior to the redevelopment activities, comprehensive environmental due diligence investigations were performed to evaluate impacts from over a century of different land uses. Investigations identified urban fill soil containing varying levels of metals and polycyclic aromatic hydrocarbons (PAHs); soil containing residual oil and polychlorinated biphenyls (PCBs); and groundwater containing chlorinated solvent compounds, primarily trichloroethylene (TCE) (GEI, 2015a).

On the basis of the contaminants identified during these investigations and the planned use of the land as residential, various mitigation strategies were used to address the environmental conditions, including:

- Bulk excavation and removal of soils to accommodate the underground parking garage and the building foundations. Approximately 65,000 yd^3 (approx. 49,696 m^3) of soil were displaced by the project and disposed of off-site at various landfills and other soil-receiving facilities. A comprehensive soil pre-characterisation study allowed the contractor to identify some off-site facilities that were more cost-effective than landfills, but still appropriate for receiving lightly contaminated soil from urban areas.
- Vapour mitigation systems to address potential vapour intrusion of volatile compounds (e.g. TCE) into the buildings. Mitigation for the townhomes with slabs on-grade consisted of sub-slab vent piping and a thicker vapour barrier with taped seams and pipe boots to seal utility and other penetrations. The venting systems were designed to be passive systems, but if necessary could be activated with the addition of blowers. Mitigation for the block buildings above the one-level underground parking garage consisted of the vapour barrier beneath the mechanically vented garage that de facto moves vapours out from beneath the building (Figure 24.4).
- A groundwater cut-off system for the underground parking garage, while intended to minimise groundwater beneath the garage, also acts as part of the vapour mitigation system since groundwater is kept away from the building. The cut-off system consists of a perimeter sheet pile wall and perimeter garage footings.
- Targeted removal of historical buried oil storage tanks encountered during building demolition. Excavation and removal of oily soils was performed as necessary.

The unique Massachusetts environmental regulatory and business climate encourages the development of urban brownfields sites, and in the case of Charlesview, allowed Charlesview to benefit from implementing the planned environmental mitigation. The environmental mitigation was integrated into the redevelopment activities to maximise cost-efficiencies and schedule, and the approaches were protective of human health and the environment. For example, the state recognises that especially in urban areas, decades and even centuries of use have

Figure 24.4 Typical configuration of vapour barrier with taped seams and pipe boots to seal utility and other penetrations; and venting system beneath a townhome at the Charlesview Residences (during construction).

potentially degraded groundwater resources in such a way that groundwater resources may never be restored to pristine conditions. Given this consideration, the most restrictive groundwater standards reserved for drinking water sources are not applicable and therefore do not need to be met. Instead, only groundwater standards protective of the indoor air and surface water discharge pathways are applicable.

The privatised system in Massachusetts allowed TCB and their LSP to dictate the timing and structuring of environmental compliance, as long as it met overall state requirements and deadlines. There were three portions of the project that had different regulatory closure approaches (Figure 24.5). During the project, the state environmental regulations were due to change, including decreasing the allowable groundwater standards (particularly TCE) that were protective of indoor air. Following completion of the underground parking garage (Area 1), this initial phase of the project was closed under the environmental regulations (a partial closure) to take advantage of the original standards prior to changes in the regulations (GEI, 2013b). The vapour mitigation system in place was protective of human health regardless of the clean-up standard, but under the new regulation closure would have required additional regulatory compliance and additional costs. In addition, early, partial closure avoided encumbering this garage portion of the project with a deed restriction that was ultimately necessary on a different portion of the project. Under state law, Brownfields tax credits are reduced by half if the site is encumbered with a deed restriction. Therefore, the strategy for an early, partial closure of this garage portion of the project was used to maximise the Brownfields tax credits on the

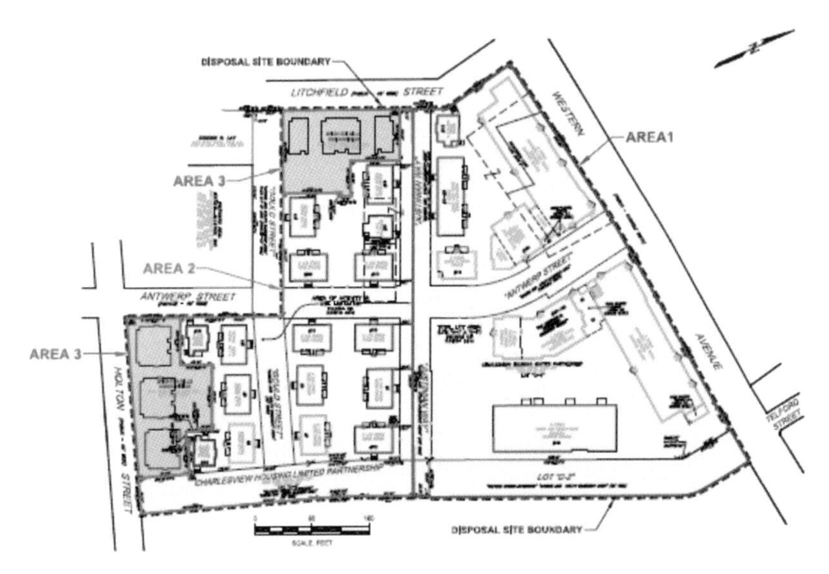

Figure 24.5 Site plan of the Charlesview Residences showing the three areas where three different regulatory approaches were taken to take advantage of regulations and maximise Brownfields tax credits.

portion of the project where most of the clean-up costs were incurred. The expedited arrival of the Brownfields tax credits enabled the rental phase to fit out and lease to locally owned preschool and restaurant tenants in the retail phase of the development. It also enabled the rental phase to pre-pay a portion of the second loan from MassHousing, which enabled MassHousing to re-deploy its valuable affordable housing subsidies in turn.

Another phase of the project where TCE concentrations in groundwater were highest (Area 2) required the installation of vapour mitigation systems and a deed restriction to ensure that the vapour mitigation systems remained in place and were protected from removal or damage (GEI, 2015b).

The final phase of the project (Area 3) was closed under the regulations before any construction of planned town homes even occurred (GEI, 2014). The early, partial closure of this portion of the site was originally intended to position it more quickly to be eligible for Brownfields tax credits. For these townhomes, vapour mitigation systems were installed as a prudent step given nearby groundwater conditions even though TCE concentrations were below the groundwater standard.

Several elements of the Massachusetts environmental regulations encouraged urban development of this contaminated site. These included the applicability of groundwater clean-up standards that are less stringent if a site is not located in a drinking water aquifer, the introduction of institutional controls, restrictions placed on the deeds of properties, to limit or restrict the use of properties for certain types of activities, and the introduction of Brownfields tax credits that encourage developers to take on legacy contaminated sites and put them back into productive use.

24.5.3 Case Study 3 – Thomas J. Butler Freight Corridor and Memorial Park

24.5.3.1 South Boston, Massachusetts

Conley Terminal, located on the Port of Boston, Massachusetts, in the United States and operated by the Massachusetts Port Authority, is the only full-service container terminal in New England (Figure 24.6). Conley Terminal is part of a much larger portion of Boston that sits on filled land, the result of hundreds of years of purposeful land forming activities undertaken by Boston to reclaim tidal flats and other low-lying areas.

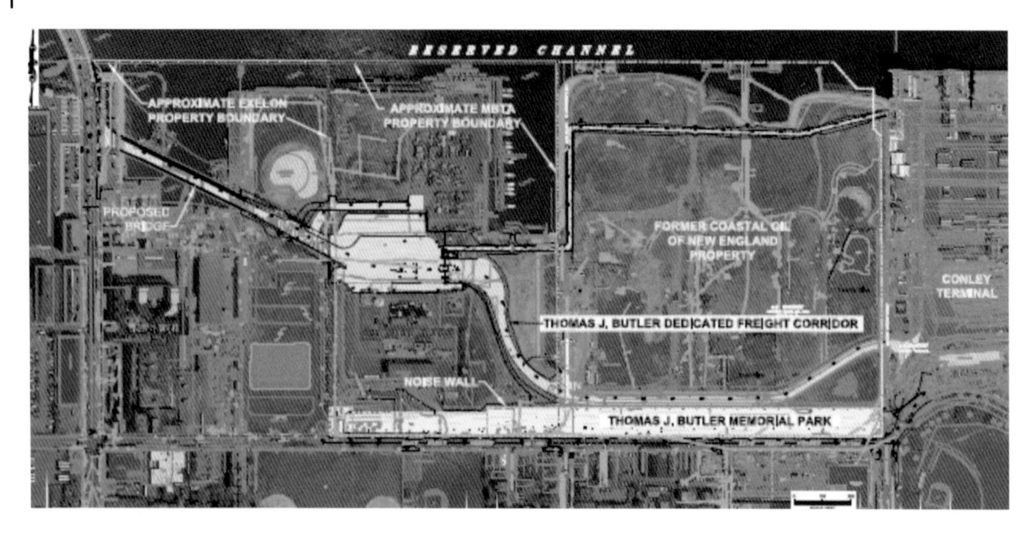

Figure 24.6 Site plan showing the Former Coastal Oil property and the layout of the Freight Corridor and Memorial Park.

The quality of fill used for these activities sometimes contained high levels of heavy metals and PAHs whose sources were commonly coal ash, wood ash, or other anthropogenic materials.

To accommodate the anticipated growth of container operations, Massport is undertaking a significant multi-project expansion of Conley Terminal including roadway, bridge, park, berth and container yard construction, all on contaminated filled land. An essential element of the expansion plans was the construction of the Thomas J. Butler Freight Corridor and Memorial Park. The project connects Conley Terminal to the main Summer Street thoroughfare, removes container truck traffic from a residential street (East First Street), and creates new community open space. The majority of the property along which the Freight Corridor and Memorial Park was built was contaminated, and its clean-up was integrated into the construction. The fast-paced construction schedule mandated by Massport was not impeded by the contamination because the Massachusetts environmental regulations and the privatised LSP programme allowed for more rapid assessment and remediation than a traditional state-led clean-up programme. The assessment and remediation elements of the project proceeded without the need for direct oversight or approval by the state environmental agency because of the responsibilities entrusted to the LSP. This encouraged faster development progress at the highly urban Port of Boston.

Key to expansion plans for Conley Terminal was the acquisition in 2008 and clean-up of the adjacent Former Coastal Oil of New England (CONE) property, a bulk oil tank terminal. Massport also acquired property along the proposed Freight Corridor right of way from the Massachusetts Bay Transit Authority (MBTA) and Exelon Corporation (GEI, 2015c). A new 425-ft-long (130 m) bridge was constructed from Summer Street across an inlet on Reserve Channel joining the Freight Corridor across the properties acquired from Exelon and the MBTA and then along the southern portion of the former CONE property. The 4.2 acre (approx. 16,997 m^2) Memorial Park, approximately 100 ft (approx. 30 m) wide and 1,000-ft-long (approx. 304.8 m), along with an 18-ft-tall (approx. 5.5 m) noise wall serves as a significant noise and visual buffer for the adjacent residents and creates a noteworthy new amenity for the neighbourhood.

The CONE property is a nearly square (approximately 1,100 by 1,200 ft [approx. 335

to 366 m]) parcel that was formerly a petroleum receiving and distribution terminal; it ceased operations in 2000. All tanks associated with the former CONE facility have been removed, and all associated piping has either been removed or cleaned and abandoned in place. Originally reported to the MassDEP in January 1987, petroleum-contaminated soil and light non-aqueous phase liquid (LNAPL) consisting of No. 2, No. 4, and No. 6 fuel oil were identified throughout the property. Since 1993, significant investigation and remedial activities have been conducted including installation of a multiphase extraction (MPE) system, construction of biopiles to treat petroleum-contaminated soils, automated and manual recovery of LNAPL following the shutdown of the MPE system when it became apparent that it was no longer effective, and a retrofit of the MPE system for use as a bioventing system.

Although these methods of LNAPL recovery were adequate for an inactive former oil terminal, a more aggressive and definitive method of treatment was required to address LNAPL that remained beneath the proposed Freight Corridor and Memorial Park. Efficient soil management and remediation by in situ solidification (ISS) were incorporated into the design of the Freight Corridor and Park.

Solidification reduces the mobility of hazardous substances and contaminants in the environment by immobilising contaminants within their 'host' medium (i.e. the soil, sand, and/or building materials that contain them) instead of removing them through chemical or physical treatment. Portland cement is a common additive used for solidification. ISS uses heavy construction equipment (e.g. auger or paddle rigs or excavators) to inject and mix the solidification additive into the soil matrix. The process is repeated in overlapping sections to encapsulate the contaminants in a solidified mass.

Approximately 28,700 ft^2 (approx. 2,666 m^2) of the property and 15,000 yd^3 (approx. 11,468 m^3) of LNAPL-impacted soils were remediated by ISS between 6 and 20 ft (approx. 1.8 to 6.1 m) below the original ground surface and incorporated into roadway construction to control potentially mobile LNAPL in the southeast portion of the site. Prior to performing ISS, the top 6 ft (approx. 1.8 m) of soil was excavated and stockpiled for eventual reuse as fill (Figure 24.7). ISS was then performed in the LNAPL-impacted zone between 6 and 20 ft (approx. 1.8 to 6.1 m) below the original ground surface, approximately 14 ft (approx. 4.3 m) of treatment. An aqueous slurry containing a mixture between approximately 10% and 10.7% Portland cement was mixed in situ into the LNAPL-saturated zone to create a uniform, stabilised soil mass. It was specified that the stabilised soil mass would have low permeability (10^{-5} to 10^{-6} cm/s) to preclude future groundwater and LNAPL flow and minimise leaching, and enough strength (50 to 200 lbs/in^2 [approx. 893 to 3,572 kg/m^2]) to form a solid mass, but not so stiff as to prevent future excavation with construction equipment (GEI, 2015c).

By using ISS, contaminated soils were treated in place and left on-site – they did not have to be disposed off-site at significant cost. Other soils throughout the site were also incorporated into the construction project. Approximately 5,000 yd^3 (approx. 3,823 m^3) of soil that had been treated and stockpiled in biopiles on-site for more than 6 years, and more than 10,000 yd^3 (approx. 7,646 m^3) of other excess soils were consolidated within the roadway right of way.

Portions of the contaminated property, including the Memorial Park, were capped with a minimum 3 ft (approx. 0.91 m) of clean soil or roadway pavement. The Memorial Park is primarily passive in nature and includes a sidewalk at the back of the kerb, parking, an end-to-end multi-use path, varied and extensive plantings, a security fence enclosure, and seating, lighting, a water fountain, bike racks, and other amenities (GEI, 2015c).

Since the clean-up was conducted under the MCP, a risk-based clean-up system, different clean-up goals were adopted for the Freight Corridor and Memorial Park. Although oil-contaminated soil, including

Figure 24.7 In situ solidification (ISS) being performed on a portion of the Former Coastal Oil property during Dedicated Freight Corridor construction.

soil solidified by ISS, remains beneath the Memorial Park, the 3 ft (approx 0.91 m) soil cap or pavement are suitable barriers to allow for recreational use.

Allowable uses, prohibitions, and obligations have been incorporated into a deed restriction that has been recorded on the CONE property deed. The noise wall constructed between the Freight Corridor and the Memorial Park is a distinct demarcation for allowable uses. The deed restriction divides the CONE property between the areas south and north of the noise wall. The area south of the noise wall is the Memorial Park. Its deed restriction allows for use of the area for recreational purposes provided that disturbance or relocation of the soils located beneath the cap is prevented. The area north of the noise wall is the Freight Corridor. Its deed restriction allows for use of the area for commercial, maritime, and industrial purposes. The deed restriction prohibits both the areas south and north of the noise wall to be used for a residence (either single or multi-family), school, daycare, or children's nursery. The deed restriction requires that specific soil management activities be performed by an appropriate professional (GEI, 2015c).

Construction of the Thomas J Butler Freight Corridor and Memorial Park was a critical enabling project for the upcoming expansion of Conley Terminal. Site contamination and regulatory compliance did not adversely affect design and construction schedule because that aspect of the project could be managed by the LSP under the privatised Massachusetts programme. In addition, in-place treatment and reuse of contaminated soils, not only beneath the roadway but also safely beneath the park, was possible using risk-based clean-up goals and a deed restriction. Once completed, the project will remove container truck traffic off the local roads and provide open space connecting the South Boston harbour, beaches, and recreational spaces with the neighbourhood.

24.6 Conclusions

The unique privatised and risk-based environmental regulatory programme in Massachusetts provides a streamlined and effective avenue for clean-up and redevelopment of contaminated sites in urban areas. The use of LSPs, who determine the remedial approach on behalf of the responsible party, limits the oversight by state employees, allowing sites to achieve regulatory closure on a more redevelopment friendly schedule. The risk-based standards allow LSPs to implement appropriate clean-up remedies, depending on current and future site uses, such as leaving soil contaminated with heavy metals beneath a protective cap to allow for the development of a new park, constructing new buildings on top of chlorinated solvent plumes with appropriate engineering controls in place to mitigate vapour intrusion, and solidifying and immobilising oil-contaminated soil in place to allow for the development of a roadway. Deed restrictions are used to limit or restrict the use of properties for certain types of activities, to ensure maintenance of systems or engineering controls and to notify future landowners of residual contamination and associated obligations. Brownfields tax credits encourage developers to take on legacy contaminated sites and put them back into productive use. The privatisation of the Massachusetts site clean-up programme has changed the urban landscape; without the implementation of a risk-based programme, many urban sites would not be cleaned up.

References

GEI (2013a) Phase IV Final Inspection Report and Completion Statement, Class A-3 Response Action Outcome and Activity and Use Limitation, RTN 3-26654, 45 Walnut Street, Peabody, Massachusetts, GEI Consultants, Inc., 4 June 2013.

GEI (2013b) Class A-2 Response Action Outcome Partial, RTN 3-27870, 400 and Portion of 370 Western Avenue, Brighton, Massachusetts, GEI Consultants, Inc., November 2013.

GEI (2014) Revised Class B-1 Response Action Outcome Partial, RTN 3-27870, 400 and Portion of 370 Western Avenue, Brighton, Massachusetts, GEI Consultants, Inc., 14 January 2014 (Revised 14 February 2014).

GEI (2015a) Final Permanent Solution with Conditions Statement, RTN 3-27870, 400 and Portion of 370 Western Avenue, Brighton, Massachusetts, GEI Consultants, Inc., September 2015.

GEI (2015b) Partial Permanent Solution with Conditions Statement, RTN 3-27870, 400 and Portion of 370 Western Avenue, Brighton, Massachusetts, GEI Consultants, Inc., September 2015.

GEI (2015c) Phase IV Final Inspection Report and Completion Statement, Partial Permanent Solution with Conditions and Activity and Use Limitation, Portion of Former CONE Terminal, 900 East First Street, South Boston, Massachusetts, GEI Consultants, Inc., 22 December 2015.

MassDEP (2014) The Massachusetts Contingency Plan, 310 CMR 40.0000, Massachusetts Department of Environmental Protection, Bureau of Waste Site Clean-up, 20 June 2014.

Massachusetts General Law (1983 et. Seq.) Chapter 21E, Massachusetts Oil and Hazardous Material Release Prevention and Response Act, 1983 and as amended.

Massachusetts General Law (1998) Chapter 206 of the Acts of 1998, An Act Relative To Environmental Clean-up and Promoting the Redevelopment of Contaminated Property, 5 August 1998.

25

Urban Pollution in China

Jianmin Ma[1] *and Jianzhong Xu*[2]

[1] *Key Laboratory for Environmental Pollution Prediction and Control Gansu Province, College of Earth and Environmental Sciences Lanzhou University, China*
[2] *State Key Laboratory of Cryospheric Sciences, Cold and Arid Regions Environmental and Engineering Research Institute, China*

25.1 Introduction

Over the last three decades, China has probably achieved the fastest economic growth and increase in personal income worldwide, which in turn increased the rate of urbanisation at the same time. The national census of China in 2015 (*Statistical Year Book*) showed that the urban residential population had reached 55.88% of the total population, 2.8 times greater than in 1985 (20% of the total population; Lamia *et al.*, 2009). While urbanisation is an important stage of modernisation in human societies, it also exerts negative effects on the urban environment. The increase in the urban population and personal income substantially enhanced demands for new housing and commercial offices to accommodate urban growth, with associated demands for electricity to serve growing cities with higher-income urbanites, substantial need for household and food products, and so on. Millions of new vehicles are being registered in cities across China, resulting in serious pollution challenges for Chinese cities. Urban pollution is represented primarily by local air pollution, greenhouse gas production (Zheng and Kahn, 2013), groundwater pollution, and soil pollution. The latter two issues are typically largely from municipal and solid wastes, toxic chemicals released from reconstruction, and the conversion of urban industrial and agricultural land to residential and commercial properties to meet the increasing demands of the expanding urban population and business.

This chapter will first address broad major problems in terms of air and soil pollution found in large cities and megacities in China. Second, using Beijing, the national capital and megacity, as an example, an insight will be given into typical urban environmental problems in China, concentrating on atmospheric and terrestrial issues. While it is understood that groundwater quality has deteriorated for the last several decades, in that 90% of it is polluted by organic and inorganic chemicals (Zhang *et al.*, 2013), it does not appear to have impacted potable supplies, and therefore this chapter will not discuss groundwater pollution.

25.2 Urban Air Pollution in China

While large cities and megacities in China are characterised by various economic modes due to geographical, historical, and political factors, the common characteristics of urban pollution in these cities are linked

Urban Pollution: Science and Management, First Edition. Edited by Susanne M. Charlesworth and Colin A. Booth.

closely with China's energy consumption, industrialisation, and growth in the urban population and their use of motor vehicles. Coal accounts for more than 70% of the total energy consumption in China, which is associated with much of the atmospheric pollutants. The emission of SO_2, a primary air pollutant released from coal combustion and a precursor of acid rain and fine particles, has become one of the major contributors to smog and acid rain production in the 1980s and 1990s (Chan and Yao, 2008). To control acid rain, the Chinese government implemented a series of regulations and measures to reduce SO_2 emissions (Li *et al.*, 2015), including the application of flue-gas desulphurisation and replacement of coal-burning facilities in industrial and domestic heating by natural gas. As a result, SO_2 emissions have declined throughout China, particularly since the 2000s. Figure 25.1a shows trends in satellite-retrieved SO_2 column density in the planetary boundary layer (unit: DU) from 2005 to 2015 by the Ozone Monitoring Instrument (OMI) on the NASA EOS Aura platform.

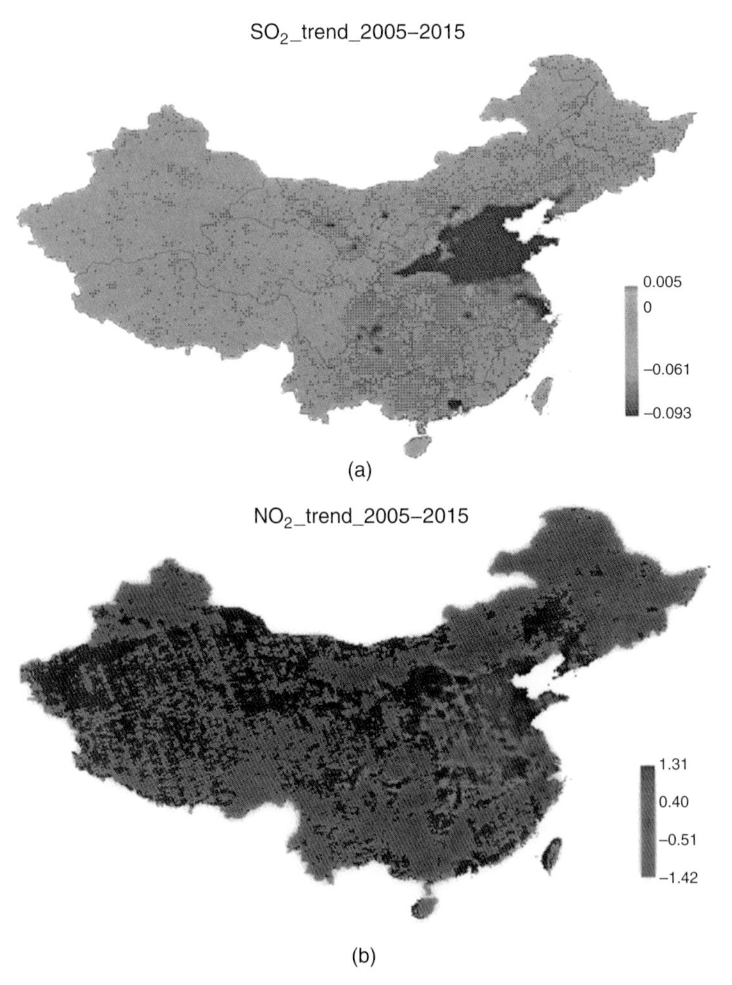

Figure 25.1 Slopes (trends) of linear regression relationships between 2005 and 2015 in OMI (10^{15} molec cm^{-2}), retrieved monthly. (a) Mean SO_2 planetary boundary-layer column density (DU) and (b) NO_x tropospheric column density. The black dots indicate the locations where the trend is statistically significant at the 95% confidence level.

The widespread decline in SO_2 column densities can be observed in most areas of China, particularly the traditionally industrialised and heavily polluted regions, such as the Northern China Plain and the Yangtze River Delta. This is good news regarding the Chinese atmospheric environment, particularly for those places suffering from acid rain. On the other hand, since the last decade, the number of motor vehicles has increased drastically, especially in large cities and megacities. In Beijing alone, the number of vehicles increased from 0.5 million in 1990 to 6 million in 2015, growing one order of magnitude in just 25 years. The substantial increase in motor vehicles further enhanced the demand for crude oil, driving the rapid development of the oil industry. All these influences together resulted in air quality problems in Chinese cities, characterised by the sharp increase in NO_x, CO, and volatile organic compound (VOC) emissions and their ambient air concentrations, which are the precursors of fine particles and ground-level ozone. Figure 25.1b shows the trend in the tropospheric column density of NO_x retrieved by OMI (10^{15} molec cm^{-2}). In contrast to SO_2, NO_x shows a widespread increase across China, due to the rapid growth in the numbers of motor vehicles, particularly in the last decade. Despite the decrease in "traditional pollutants" (e.g. SO_2), fine particulate matter ($PM_{2.5}$) and ground ozone, or combined air pollution, have now become important air pollutants in Chinese cities, threatening human health, morbidity, and mortality, and decreasing atmospheric visibility (He *et al.*, 2002).

Emerging air quality issues, such as the rapid increase in particulate matter levels in the recent decade, has resulted in significant changes to national regulations for air pollution protection and air-quality-monitoring programmes. In the 1980s and 1990s, the air quality in China was measured by TSP, SO_2, and NO_2; however, an air pollution index (API) was developed, which took three air pollutants into account: PM_{10} (replacing TSP), SO_2, and NO_2. This quantified air quality was based on the National Air Quality Standard of China, 1996 (GB3095–1996), which was released publicly on a daily basis (He *et al.*, 2002). Compared with the previously measured TSP, SO_2, and NO_2, the API was linked more closely with human health. From 1 January 2013 onwards, and in accordance with the release of the new National Air Quality Standard in 2012 (GB3095, 2012), the AQI, which measures ambient air quality based on air pollutants with an adverse effects on human health and the environment, has been implemented, and this is reported publicly on an hourly basis. The AQI added three more air pollutants, $PM_{2.5}$, ozone, and CO, to the three API pollutants. During this period, the number of Chinese cities issuing the AQI increased from 74 in 2013 to 364 in 2015, with about 1500 air-quality-monitoring stations launched in these cities, forming one of the largest air-quality-monitoring networks with high spatial resolution globally. To respond to increasing levels of particulate matter (i.e. $PM_{2.5}$ and PM_{10}) and public concern, the State Council of China issued the Air Pollution Prevention and Control Action Plan in 2013, which put forward detailed requirements and quantifiable measures to reduce PM concentrations within a five-year period (2013–2017). The action plan requested that, by 2017, urban concentrations of PM_{10} should decrease by 10% compared to those in 2012; concentration of fine particulate matter ($PM_{2.5}$) in Beijing–Tianjin–Hebei, Yangtze River Delta,and Pearl River Delta regions, which are the economy and industrial centres and the most heavily contaminated regions in China, should fall by approximately 25%, 20%, and 15% from 2012 to 2017. In mid-2016, the Chinese Academy of Engineering reported on the results of implementing the action plan since it came into effect in 2013; it assessed PM levels in 74 key cities and megacities across China from 2013 to 2015 (Ministry of Environmental Protection, 2015). The results showed that mean average $PM_{2.5}$ concentrations from the 74 cities declined from 72μg m^{-3} in 2013 to 55 μg m^{-3} in 2015, decreasing by 23.6%. The mean average PM_{10} concentration from 384 Chinese cities reduced from 97 μg m^{-3} in 2013 to 87 μg m^{-3} in 2015. These achievements are consistent with significant reductions in atmospheric SO_2, as shown in Figure 25.1a, which contributes to the composition of $PM_{2.5}$. Evidence has already revealed the change from SO_4^{2-} to NO_3^- as the dominant water soluble ion in $PM_{2.5}$ in recent years, which is contrary to earlier measurements in China and the world (He *et al.*, 2002), due to continuously increasing NO_x emissions (Figure 25.1b) and reducing SO_2 emissions (Figure 25.1a), which can alter the ratio of NO_3^- to SO_4^{2-} in $PM_{2.5}$ (Xue *et al.*, 2014; Zhao *et al.*, 2015; Wang *et al.*, 2016). The rapid increase in atmospheric NO_x and VOCs brings another air quality issue, that of ground-level ozone, which has recently become a dominant air pollutant in the summertime. Added to UV radiation, high levels of VOCs and NO_x enhance a series of photochemical chain reactions which increase oxidation in the atmosphere, causing ground ozone formation, leading to interest in this pollutant over the last two years.

To control and remediate $PM_{2.5}$ levels, extensive investigations in its source apportionment have been carried out by the Chinese Ministry of Environmental Protection (MEP). By the end of 2015, all provincial capital and major cities in China completed their respective source apportionment assessment. Figure 25.2 illustrates the contributions of local major emission sources to $PM_{2.5}$ in six key megacities, in which it was found that motor vehicles made the largest contribution in Beijing and Shanghai, the two largest megacities with populations in excess of 20 million.

Higher contributions from coal combustion and dust emission were also identified in cities located in the north (Beijing, Tianjin, and Shijiazhuang) due to winter domestic heating, lower vegetation coverage, and low rainfall. Dust and other sources released locally (emissions from biomass burning, area sources, natural sources, etc.) contributed 9%–36% to $PM_{2.5}$ mass across these cities. Except for these local sources, external sources contributed approximately 15%–36% to $PM_{2.5}$ levels via atmospheric transport. These complex sources make it difficult to predict $PM_{2.5}$ formation and evolution. Although bulk atmospheric PM levels are declining, episodically increasing concentrations are often observed and reported (Zhang and Cao, 2015). There are still large knowledge gaps regarding the formation of $PM_{2.5}$ and O_3 as primary and secondary pollutants which are stimulating intensive studies in the associations between these pollutants, atmospheric chemistry, climate, meteorological conditions, and emission sources.

A recent study revealed that residential emissions from rural areas also made significant contribution to urban pollution via atmospheric transport and diffusion, particularly during the winter domestic heating season in northern China (Liu *et al.*, 2016). Since rural residents in northern China often used faulty coal-burning equipment in cooking and heating, the use of such non-clean energy causes higher emissions of air pollutants. The application of clean energy has been strictly implemented by urban authorities; thus, migration from rural to urban areas may exert a positive influence on the urban environment by reducing the use of non-clean energy (Tao Shu, personal communication, 2016).

Compared with developed countries, considerably higher levels of aerosols, atmospheric oxidants, and organic chemicals are observed in the urban atmosphere in China. It has been reported by Zhang *et al.* (2015a) that atmospheric levels of sulphate, nitrate, organic carbon, elemental carbon, and ammonia in the Chinese urban atmosphere are factors of 2–10 times higher than those in European cities, particularly attributable to SO_2, NO_x, and VOCs, but also including other pollutants. Another distinction between China and Europe is new particle formation (NPF), which is responsible for a major fraction of particle numbers (Guo

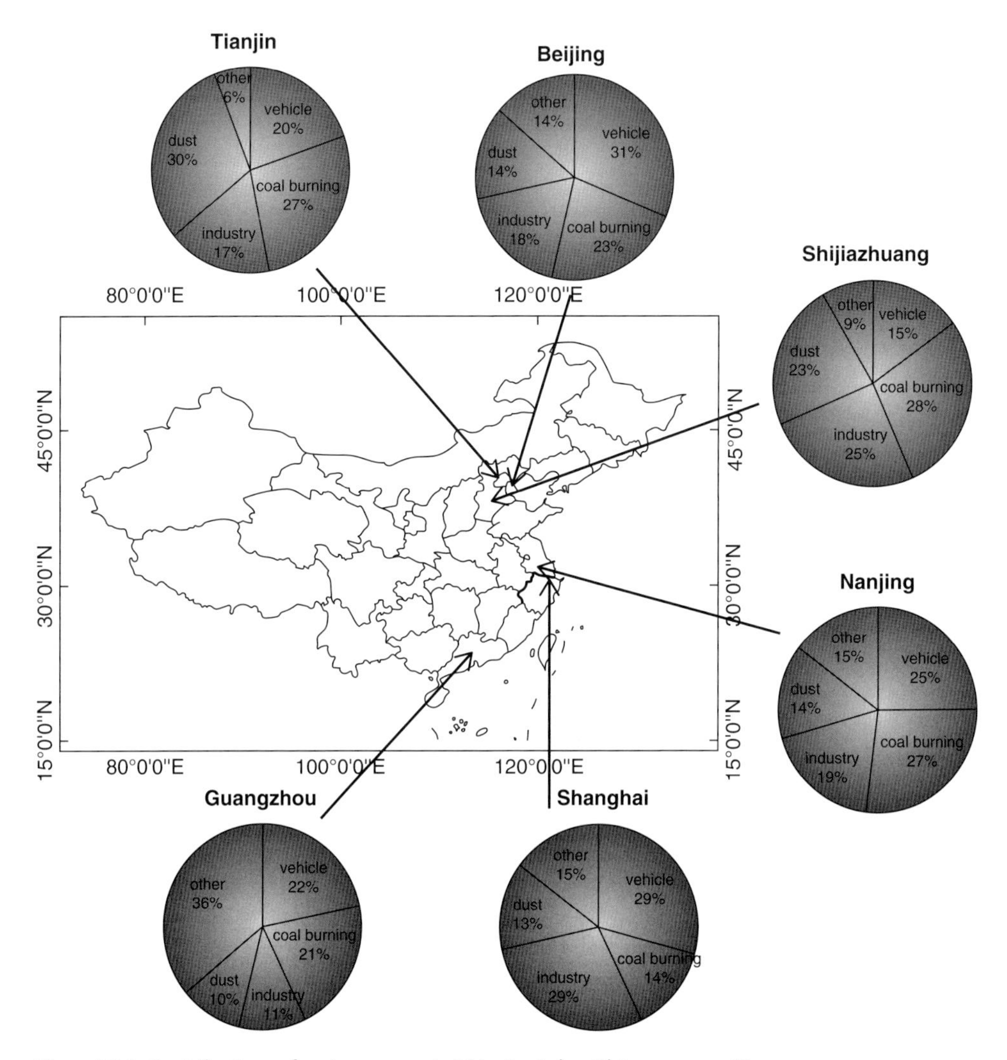

Figure 25.2 Contributions of major sources to $PM_{2.5}$ in six key Chinese megacities.

et al., 2014). Although NPF was thought to only occur rarely (Aalto *et al.*, 2001), NPF events have been frequently observed in many urban areas of China (Zhang *et al.*, 2015b). The occurrence of NPF in urban regions may be explained by the presence of high levels of aerosol nucleation precursors, such as sulphur dioxide, ammonia, amines, and anthropogenic VOCs. After undergoing atmospheric chemical reactions, these gases are converted partly to secondary aerosols and to form surface O_3 (Zhang *et al.*, 2015b). These processes lead to the complex characteristics of urban pollution in China. In recent years, the terminology 'combined air pollution' has often been used to define this, describing coal-burning-induced sulphurous-type pollution, photochemical smog/haze, and ozone production. On the one hand, the complexities of air pollution patterns, emissions, urbanisation, and interactions between economic growth and pollution control have raised challenges to improve air quality in Chinese cities. On the other, these challenges have enabled scientists to gain new insights into the mechanisms of atmospheric pollution in those countries undergoing rapid modernisation to help pave new ways of tackling air pollution in heavily contaminated cities worldwide.

25.3 Urban Land/Soil Pollution

Urban soils are subject to continuous accumulation of contaminants from both localised and diffuse sources; typically, these include trace metals (Ajmone-Marsan and Biasioli, 2010) and persistent organic pollutants (POPs) (Fabietti *et al.*, 2010), whose main sources are traditionally from industrial discharges, traffic emissions, and wastes produced from municipal activities (Wong *et al.*, 2006). Many pollutants can remain in urban soils for long periods of time, which may act as a source of further pollution in urban environments if eroded, and pose a potential threat to human health and urban ecological systems. Since the 1950s, many industrial plants and factories were built in Chinese cities under the national strategy to increase such activities. However, after many years of operation, the land surrounding these premises were poisoned by the toxic chemicals released. It was widely reported in the media (e.g. see: https://www.rt.com/news/china-benzene-water-scare-996/) during April 2014 that over 3 million residents of Lanzhou city, the capital of Gansu province in northwestern China, were ordered not to drink its tap water after benzene, a carcinogen, was found in the supply at 200 μg/L – 20 times the acceptable 'national limit'. The incident was caused by oil leaking and infiltrating the soil from a large-scale petrochemical plant which entered the mains water pipeline of the city. From the 1980s onwards, rapid population expansion and rising standards of living increased the demand for land to construct housing. At the same time, many industries closed altogether or relocated; the reconstruction and redevelopment of these heavily polluted industrial wastelands and parks located in cities led to the release of toxic chemicals to air and groundwater, with the conversion of previously industrialised land to residential and commercial use posing potential health risks to local residents. A recent national survey of soil pollution by the Chinese EPA and Ministry of Land and Resources (2014) reported that the levels of several heavy metals (zinc, mercury, lead, chromium, and arsenic), and organic chemicals such as polycyclic aromatic hydrocarbons exceeded national standards (GB15618–2009) in 20%–36% of these industrial areas. Cities located in the more developed eastern regions and old industrial zones tended to be more contaminated (Luo *et al.*, 2012). As an example of the problems encountered, Changzhou Foreign Language School was relocated in autumn 2015 to the site of a former toxic chemical plant that was home to three chemical factories producing pesticides in Changzhou city, Jiangsu Province. In spring 2016, 641 students of the approximately 2500 students were diagnosed with serious diseases including bronchitis, dermatitis, lymphoma, and leukaemia (*New York Times*, 2016). An investigation in 2012 reported that extremely high amounts of toxic substances were present in the area before the school campus was built. Dangerous levels of chlorobenzene, a colourless, flammable liquid known to cause neurotoxicity and muscle spasms in humans, exceeded the permitted amount by 78,899 times in the soil and 94,799 in the local groundwater.

25.4 Municipal Waste Contamination in Urban China

Another significant source of urban land/soil pollution due to rapid urban development and population increase is attributed to the accumulation of municipal solid waste. The World Bank (2014) estimated that the amount of solid waste collected and transported in China has increased more than fivefold nationwide from about 85,000 tons per day in 1980 to about 430,000 tons per day in 2009 and is projected to reach 1.6 million tons per day in 2030. A report by Li (2015) indicated that China produces around 820,000 tons of waste per day, the large majority of which comes from cities. These data were based on waste collected rather than waste generated. The tremendous quantity of municipal solid waste has caused significant adverse effects on the urban environment. First, given the very limited space for landfill, particularly sanitary landfill across the country, a large amount of municipal solid waste is not processed and disposed of in a timely fashion, but accumulates in suburban areas. Even worse, although there are regulations for solid waste separation, recycling is not implemented effectively, particularly for household waste, due to the lack of capacity to operate such a system. Third, due to a shortage of urban landfill, incineration is increasingly becoming the disposal method of choice in urban areas and can also generate energy waste. The number of waste incinerators in China is projected to increase from 93 in 2009 to 200 in 2015, with a corresponding increase in daily disposal capacity from 55,400 tons to 140,000 tons (World Bank, 2014). However, the expansion of waste incinerators and poorly regulated incineration have raised new concerns since it has been identified that waste incineration has become the second largest emission source of dioxin in China since 2009 (Huang *et al.*, 2014). This resulted in strong resistance from local residents to the constructions of waste incinerators, although the new generation of technologies is claimed to be much safer than before. Finally, a shortage of efficient and sufficient waste management services and disposal facilities has led to contamination of water, air, and soil, and human health. Waste disposal is a major source of land pollution with statistical data of municipal solid waste showing that, by 2011, more than 6 billion tons of solid waste produced from Chinese cities in past years had accumulated around cities, occupying about 5×10^5 acres of surrounding suburbia. Two-thirds of more than 600 Chinese cities are thus besieged by solid waste, and in addition, 1636 counties and small towns also generated about 50 million tons of solid waste; out of 650 cities, 325 do not have any waste disposal facilities. Huge amounts of garbage are simply piled up, releasing noxious odours and carbon monoxide, causing serious contamination to water, air, and soil.

25.5 A Case Study of Urban Pollution in Beijing

25.5.1 Air Pollution in Beijing

Air pollution in Beijing has attracted increasing attention from the 1980s due to high concentrations of particulate matter and sulphur dioxide (SO_2), particularly during winter and spring (Zhou and Xiang, 1984; Winchester and Bi, 1984). Coal was used as a major fuel for industry, domestic heating, and cooking, which emitted large amounts of particles and SO_2 into the atmosphere. Since Beijing is located in northern China, dust storms were also an important contributor to ambient particle concentrations during the springtime. Along with the rapid expansion of the capital city from the 1980s, air pollution in the urban area of Beijing has gradually been deteriorating. Data from the Beijing Environmental Protection Agency showed that the number of days in which ground-level O_3 concentrations exceeded 200 $\mu g/m^3$ (approximately 100 ppbv) – the Chinese standard for urban residential areas (Grade II) – reached 101 with a maximum hourly value of 384 $\mu g/m^3$ in 1998 (Hao and Wang, 2005). Fine particle ($PM_{2.5}$) levels in Beijing remain the highest among all Chinese megacities. Particular attention to air quality was paid from 2001 onward due to China's bid for the 2008 Olympic Games, which led to considerable efforts being made to elucidate the levels and sources of major air pollutants. Wang *et al.* (2005) measured fine particulate matter in Beijing and found an average mass concentration at 154.3 ± 145.7 $\mu g\ m^{-3}$ during 2001 and 2003, higher than the levels observed previously. The winter mean SO_2 air concentrations, mostly emitted from coal combustion, was as high as 163.65 μg m^{-3} during the same period, while the level of nitrogen dioxide (NO_2) was less than 100 μg m^{-3}. On the basis of these findings, the city council implemented rigorous control measures to reduce SO_2 emissions, such as the desulphurisation of all major industries and encouraged the use of anthracite coal. However, after the 2008 Olympic Games, the air pollution in Beijing became serious once more due to continued city expansion and rapid increase in the number of motor vehicles. This was evidenced by a severe and persistent haze which occurred in Beijing in January 2013 (Huang *et al.*, 2014). This acute event was accompanied by extremely poor atmospheric visibility and a sharp increase in respiratory diseases.

The formation and temporal variation of severe air pollution in Beijing is a combination of emissions, atmospheric chemistry and transport, and prevailing weather conditions (Zheng *et al.*, 2015). Figure 25.3 shows a MODIS remote sensing satellite image of northeastern China on 14 January 2013. The image shows extensive haze, low cloud, and fog in Beijing, Tianjin, and surrounding regions, implying the atmospheric transport of air pollutants from those regions in the south and west of Beijing. Several statistical models have been used to apportion sources of air pollution in Beijing, such as principal component analysis (PCA) (Sun *et al.*, 2004), chemical mass balance (CMB) (Zheng *et al.*, 2005), and positive matrix factorisation (PMF) (Sun *et al.*, 2012). The Aerodyne aerosol mass spectrometer (AMS) has been employed to gain a better understanding of the chemical processes and sources of the submicron particulate aerosol (Huang *et al.*, 2010; Sun *et al.*, 2016). Sources of the organic aerosol (OA) determined using PMF usually included 3–6 factors including hydrocarbon-like OA accounting for 10%–15% dependent on season, cooking-emitted OA (10%–20%), biomass burning OA (~10%), coal combustion OA (15%–25%), and oxygenated OA (20%–40%) (Elser *et al.*, 2016; Hu *et al.*, 2016; Sun *et al.*, 2016). These OA sources partially originated from regional emissions around Beijing, mostly in the provinces of Hebei, Shandong, Shanxi, and Henan, accounting for up to ~50% of total

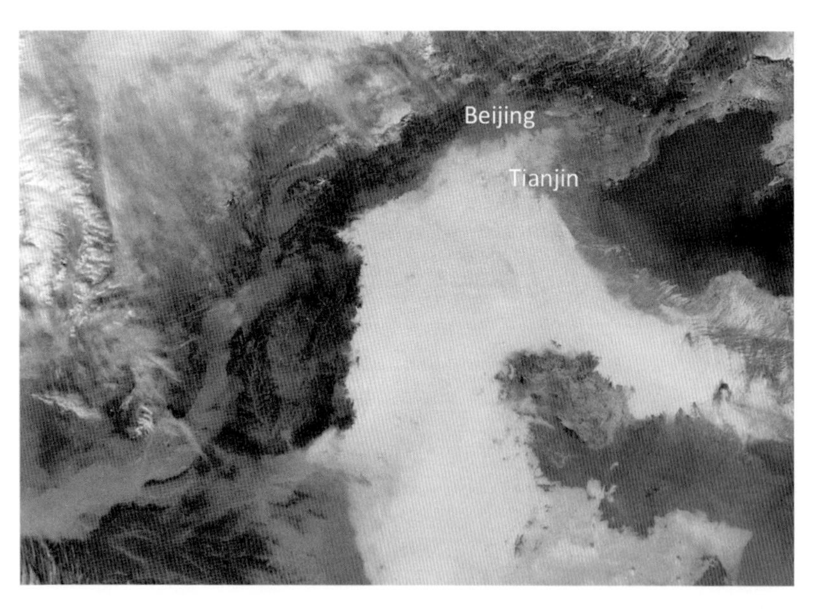

Figure 25.3 Moderate Resolution Imaging Spectroradiometer (MODIS) on NASA's Terra satellite acquired natural-colour images of northeastern China on January 14, 2013. (http://earthobservatory.nasa.gov/IOTD/view.php?id=80152)

emissions contributing to the OA levels found (Zheng *et al.*, 2015). Recent results confirmed that about 34% of air pollutants in Beijing could be traced back to sources from regions surrounding the city, such as Hebei (18%) and Shandong (4%) (http://www.caep.org.cn/index.asp). In addition to local and regional sources of particulate matter, volatile organic compounds (VOCs) and intermediate-volatility organic compounds (IVOCs) have also been identified as playing an important role in the formation of the particulate matter, contributing up to 50% of the mass concentration of the OA (Zhao *et al.*, 2016). Extensive source apportionment studies carried out around Beijing provide strong supporting evidence that air quality requires remediation. During the Asia-Pacific Economic Cooperation (APEC) summit in Beijing in 2014, the Chinese government shut down many industrial activities in Hebei and Shandong, resulting in improved air quality shown by an 'APEC blue' sky (Chen *et al.*, 2015; Li *et al.*, 2016).

In addition to the common causes of the formation of gaseous and particle pollutants worldwide, the frequent occurrence of severe haze events in Beijing is also attributed to several unique local environmental conditions:

1) Local meteorological conditions have a significant influence on the air quality; evidence has revealed that good air quality is always associated with northerly winds, whereas southerly winds and stagnation (stable) weather conditions are often accompanied by bad air quality. This is because densely populated and industrialised regions are located to the south of Beijing (Sun *et al.*, 2012).
2) The mountain-valley terrain in Beijing is also an important factor since it produces strong inversions and weak mixing boundary layers, which are usually observed during cold periods. The high levels of pollutants associated with urban and industrial emission sources at these times lead to the pollutant haze. The haze will itself prevent sunlight reaching the ground, which in turn reduces surface heating and hinders the formation of the urban boundary layer. This could reduce the dispersion and mixing of air pollutants, trapping them near the ground surface.
3) In contrast to aerosol formation typically observed in other regions of the world, efficient nucleation and growth have been observed over an extended period in Beijing (Guo *et al.*, 2014). The process of nucleation consistently occurs throughout a period of pollution, producing high concentrations of nano-sized particles. If this continues over many days, it can yield numerous larger particles with the particle mass concentration exceeding several hundred micrograms per cubic metre, causing severe and persistent pollution episodes.

25.5.2 Soil/Land Pollution in Beijing

Analogous to other urban area in China, land/soil contamination in Beijing is associated primarily with industrial activities, *solid* waste disposal, and the redevelopment of urban agricultural and industrial land. Industrialisation in Beijing began in the 1950s with electronics, manufacturing, iron and steel, thermal power plants, and chemical production. In order to lobby for and host the 2008 Olympic Games, improve the urban environment, and meet increasing demands for residential and commercial areas in the city, many industrial plants, especially heavy industry, have already closed, are being closed down, or have been moved out of the city environs. Industrial sites are often point sources and reservoirs of toxic chemicals, particularly for those with strong resistance to environmental degradation, such as heavy metals and POPs, which tend to remain in the environment for many years.

Another source of contaminants that potentially poses risks to human health is agricultural land polluted by pesticides due to their past applications, heavy metals, and other organic chemicals due to wastewater irrigation. Many of this contaminated industrial and agricultural land has been converted to residential use over the past decades due to rapid urban growth and the associated demand for housing. It has been found that the urban soils in Beijing exposed to industrial sewage and sludge were contaminated with Hg and heavy metals (Zhu, 2001). During the course of redevelopment of historically contaminated industrial and agricultural sites to commercial and residential land use, these toxic chemicals could have been released into the environment, with potential health risks to local residents. Extensive investigations have been carried out to examine and assess land/soil contaminations by toxic chemicals, primarily heavy metals, organochlorine pesticides (OCPs), and mercury (Hg) in traditional industrial, agricultural, and urban areas in Beijing (Chen *et al.*, 2005; Chen *et al.*, 2010a, b; Luo *et al.*, 2012). However, no standards for remediation and assessments for potential health risk were required in the redevelopment of this contaminated land to residential use at that time.

Heavy metal contamination in soils has been a major concern because of its impact on the health of urban residents and its cycling in the urban ecosystem. Soils were tested by Shi *et al.* (2008) for metals from different land-use types across Beijing, including business, industrial, residential, parks, cultural, and education. Results showed that Cd, Cu, Pb, and Zn had accumulated at much higher levels than their respective background values (CNEMC, 1990; Jiang *et al.*, 2011). There was particularly high Cr, Cd, Pb, and Zn contamination found in industrial sites and in agricultural land which had been irrigated with wastewater (Wang *et al.*, 2005). High metal levels were also found in parks due to the historical use of Cd, Cu, Pb, and Zn, typical urban heavy metals found in pigments, wood preservation, and brassware (Xia *et al.*, 2011). Industry and traffic were other major sources of heavy metal contamination not only in urban soil, but also in farmland soils in suburban areas around Beijing (Zou *et al.*, 2015). Jiang *et al.* (2011) carried out environmental and ecological risk assessments of seven heavy metals (Cr, Ni, Cu, Zn, As, Cd, and Pb) in these suburban areas, and found that Cd exhibited the highest environmental risk, while the remainder had low pollution levels. Overall, heavy metal contamination in urban and agricultural soils in Beijing is moderate compared with heavily polluted urban areas in other Chinese cities and other developed countries.

In addition to diffuse sources, concerning levels of OCPs were found which could have adverse effects on health for Beijing residents. These were attributed to an OCP plant which closed down in 1983, and the land subsequently became residential; even by the late 2000s, OCP levels were still posing unacceptable health risks at this site.

A soil sampling field study to assess Hg pollution in historical industrial sites in Beijing showed that about 50% of Hg samples from 57 surface soil and 108 deep soil sampling sites exceeded the "critical" concentration of 1.0 mg Hg/kg, dry weight based on the national standard (Luo *et al.*, 2012). In contrast, soils tested at other non-industrial sites found no contamination compared with Chinese government guidelines (Chen *et al.*, 2010b). Concern was also raised regarding heavy metal and POP contaminations in agricultural land, and, in a study by Zou *et al.* (2015), mean Hg was found to be almost twice as background (0.08 mg kg^{-1}). Industrial and urban domestic waste in landfill were regarded as the primary sources of Hg. A systematic field assessment of the ecological risk of agricultural soils polluted by heavy metals around Beijing has revealed that 22% of the agricultural soils reached heavy and severe risk level, and 35% of them reached moderate risk level (Jiang *et al.*, 2011). Of these heavy metals, Hg and Cd posed the highest ecological and human health risk, reflected in their highest levels in agricultural soils among the assessed heavy metals (Wei and Yang, 2010).

25.6 Conclusions

A fast-growing economy, industrial sector, and population have led to problems in urban China with pollution of the atmosphere and soils. Legislation has yet to catch up with the impacts of this contamination, in particular, with respect to the need for a more efficient waste disposal process, monitoring of the operation of landfill sites, and introduction of more widespread recycling activities to reduce waste to landfill. The rapidly increasing population of China has led to the demand for more housing, and land previously used for industry has been given over to residential use. However, the previous industry was not controlled sufficiently, and often this land is polluted, is not tested, and as a result is not remediated to provide a safe living environment for residents. An example was given in this chapter of the Changzhou Foreign Language School built on a contaminated industrial site which led to pupils being ill.

However, during the 2008 Olympic Games, certain industries were made to halt their activities, and there was an immediate improvement in environmental quality, which, however, worsened fairly rapidly as the Games finished, culminating in the 2013 haze event in Beijing, caused by a combination of polluting activities, dust storms, and fog, which caused a reduction in visibility and an increase in respiratory disease among the inhabitants.

China has acknowledged that this pollution in urban areas is prevalent, and has instituted a green agenda, including the 'Sponge City' initiative launched in 2013 (Shao *et al.*, 2016). While this is mainly targeted at the sustainable management of stormwater and flood risk, nevertheless, the fact that it includes green infrastructure means that the multiple benefits will include better air and soil quality, improving China's urban environment overall.

References

Aalto, P., Hameri, K., Becker, E., Weber, R., Salm, J., Makela, J. M., Hoell, C., O'Dowd, C. D., Karlsson, H., Hansson, H. C., Vakeva, M., Koponen, I. K., Buzorius, G., and Kulmala, M. (2001) Physical characterization of aerosol particles during nucleation events. *Tellus Series B: Chemical and Physical Meteorology*, 53, 344–358.

Ajmone-Marsan, F. and Biasioli, M. (2010) Trace elements in soils of urban areas. *Water, Air and Soil Pollution*, 213, 121–143.

Chan, C.K. and Yao, X. (2008) Air pollution in mega cities in China. *Atmospheric Environment*, 42, 1–42.

Chen, R.J., Zhao, Z.H., and Kan, H.D. (2013) Heavy smog and hospital visits in Beijing, China. *American Journal of Respiratory and Critical Care Medicine*, 188, 1170–1171.

Chen, C., Sun, Y.L., Xu, W.Q., Du, W., Zhou, L.B., Han, T.T., Wang, Q.Q., Fu, P.Q., Wang, Z.F., Gao, Z.Q., Zhang, Q., and Worsnop, D.R. (2015) Characteristics and sources of submicron Aerosols above the urban canopy (260 M) in Beijing, China, during the 2014 APEC Summit. Atmospheric Chemistry and Physics, 15, 12879–12895, 10.5194/acp-15-12879-2015.

Chen, T., Zheng, Y., Lei, M., Huang, Z., Wu, H. *et al.* (2005) Assessment of heavy metal pollution in surface soils of urban parks in Beijing, China. *Chemosphere*, 60, 542–551.

Chen, X., Xia, X., Zhao, Y., and Zhang, P. (2010a) Heavy metal concentrations in roadside soils and correlation. *Journal of Hazardous Materials*, 181, 640–646.

Chen, X., Xia, X., Wu, S., Wang, F., and Guo, X. (2010b) Mercury in urban soils with various types of land use in Beijing, China. *Environmental Pollution*, 158, 48–54.

China National Environmental Monitoring Center (CNEMC) (1990) *The Background Concentrations of Soil Elements in China*. Chinese Environment Science Press, Beijing, p. 346 (in Chinese).

Elser, M., Huang, R.J., Wolf, R., Slowik, J.G., Wang, Q., Canonaco, F., Li, G., Bozzetti, C., Daellenbach, K.R., Huang, Y., Zhang, R., Li, Z., Cao, J., Baltensperger, U., El-Haddad, I., and Prévôt, A.S.H. (2016) New insights into $PM_{2.5}$ chemical composition and sources in two major cities in china during extreme haze events using aerosol mass spectrometry. *Atmospheric Chemistry and Physics*, 16, 3207–3225, 10.5194/acp–16–3207–2016.

Fabietti, G., Biasioli, M., Barberis, R., and Ajmone-Marsan, F. (2010) Soil contamination by organic and inorganic pollutants at the regional scale: The case of Piedmont, Italy. *Journal of Soils and Sediments*, 10, 290–300.

Guo, S., Hu, M., Zamora, M.L., Peng, J., Shang, D., Zheng, J., Du, Z., Wu, Z., Shao, M., Zeng, L., Molina, M.J., and Zhang, R. (2014) Elucidating severe urban haze formation in China. *Proceedings of the National Academy of Sciences of the United States of America*, 111, 17373–17378.

He, K., Huo, H., and Zhang, Q. (2002) Urban air pollution in China: Current status, characteristics, and progress. *Annual Review of Energy and the Environment*, 27, 397–431.

He, K., Yang, F., Duan, F., and Ma, Y. (2011) *Atmospheric Particular Matters and Combined Regional Air Pollution*. China Sci Publisher, ISBN: 9787030303707, pp. 452 (in Chinese).

Hao, J.M. and Wang, T.L. (2005) Improving urban air quality in China: Beijing case study, *Journal of the Air & Waste Management Association*, 55, 1298–1305.

Huang, X.F., He, L.Y., Hu, M., Canagaratna, M.R., Sun, Y., Zhang, Q., Zhu, T., Xue, L., Zeng, L.W., Liu, X.G., Zhang, Y.H., Jayne, J.T., Ng, N.L., and Worsnop, D.R. (2010) Highly time-resolved chemical characterization of atmospheric submicron particles during 2008 Beijing Olympic Games using an aerodyne high-resolution aerosol mass spectrometer. Atmospheric Chemistry and Physics, 10, 8933–8945, 10.5194/acp–10–8933–2010.

Huang, R.-J., Zhang, Y., Bozzetti, C., Ho, K.-F., Cao, J.-J., Han, Y., Daellenbach, K.R., Slowik, J.G., Platt, S.M., Canonaco, F., Zotter, P., Wolf, R., Pieber, S.M., Bruns, E.A., Crippa, M., Ciarelli, G., Piazzalunga, A., Schwikowski, M., Abbaszade, G., Schnelle-Kreis, J., Zimmermann, R., An, Z., Szidat, S., Baltensperger, U., Haddad, I.E., and Prevot, A.S.H. (2014) High secondary aerosol contribution to particulate pollution during haze events in China. Nature, 514, 218–222, 10.1038/nature13774.

Huang, T., Jiangwan, Y., Ling, Z., Zhao, Y., Gao, H., and Ma, J. (2016). Trends in the cancer risk of Chinese inhabitants from dioxins due to changes in dietary structure: 1980–2009. *Scientific Reports*, 6, 21997; doi: 10.1038/srep21997.

Hu, W., Hu, M., Hu, W., Jimenez, J.L., Yuan, B., Chen, W., Wang, M., Wu, Y., Chen, C., Wang, Z., Peng, J., Zeng, L., and Shao, M. (2016) Chemical composition, sources, and aging process of submicron aerosols in Beijing: Contrast between summer and winter. *Journal of Geophysical Research: Atmospheres*, 121, 2015JD024020, 10.1002/2015JD024020.

Jiang, F., Sun, D., Li, H., and Zhou, L. (2011) Risk grade assessment for farmland pollution of heavy metals in Beijing. *Transactions of the CSAE*, 27 (8), 330–337 (in Chinese with English abstract).

Lamia, K.-C., Edward, L., and Zhang, R. (2009) Urban Trends and Policy in China. OECD Regional Development Working Papers, 2009/1, OECD publishing. doi:10.1787/225205036417 file://C:/lenovo%20Gdrive/Urban%20pollution%20in%20China/42607972–urban%20trend.pdf.

Li, H., Zhang, Q., Duan, F., Zheng, B., and He, K. (2016) The 'Parade Blue': Effects of short–term emission control on aerosol chemistry, *Faraday Discuss*, 189, 317–335, 10.1039/C6FD00004E.

Li, J. (2015) Ways forward from China's urban waste problem. Available and accessed on August 2018, https://www.thenatureofcities.com/2015/02/01/ways-forward-from-chinas-urban-waste-problem/.

Li, M., Zhang, Q., Kurokawa, J., Woo, J.H., He, K.B., Lu, U.,Ohara, T., Song, Y., Streets, D. G., Carmichael, Y.R., *et al.* (2015) MIX: a smosaic Asian anthropogenic emission inventory for the MICS–Asia and the HTAP projects. *Atmospheric Chemistry and Physics Discussions*, 15, 34813–34869.

Liu, J.. Mauzerall, D., Chen, Q., Zhang, Q., Song, Y., Peng, W., Klimont, Z., Qiu, X., Zhang, S., Hu, M., Lin, W., Smith, K., and Zhu, T. (2016) Air pollutant emissions from Chinese households: A major and underappreciated ambient pollution source. *Proceedings of the National Academy of Sciences of the United States of America*, 113, 7756–7761.

Luo, X., Yu, S., Zhu, Y., and Li, X. (2012) Trace metal contamination in urban soils of China. *Science of the Total Environment*, 421–422, 17–30.

New York Times. http://www.nytimes.com/2016/04/19/world/asia/china–pollution–cancer–changzhou.html

Ministry of Environmental Protection 2015 State of the Environment Report, 2015. In Chinese. http://english.mep.gov.cn/Resources/Reports/soe/

Shao, W., Zhang, H., Liu, J., Yang, G., Chen, X., Yang, Z., and Huang, H. (2016) Data integration and its application in the sponge city construction of China. *Procedia Engineering*, 154, 779–786.

Shi, G., Chen, Z., Xu, S., Zhang, J., Wang, L., Bi, C., and Teng, J. (2008) Potentially toxic metal contamination of urban soils and roadside dust in Shanghai, China. *Environmental Pollution*, 156, 251–260.

Sun, Y.L., Zhuang, G., Wang, Y., Han, L., Guo, J., Dan, M., Zhang, W., Wang, Z., and Hao, Z. (2004) The airborne particulate pollution in Beijing – Concentration, composition, distribution and sources. *Atmospheric Environment*, 38, 5991–6004.

Sun, J.Y., Zhang, Q., Canagaratna, M.R., Zhang, Y.M., Ng, N.L., Sun, Y.L., Jayne, J.T., Zhang, X.C., Zhang, X.Y., and Worsnop, D.R. (2010) Highly time- and size-resolved characterization of submicron aerosol particles in Beijing using an aerodyne aerosol mass spectrometer, Atmospheric Environment, 44, 131–140, 10.1016/j.atmosenv.2009.03.020.

Sun, Y. L., Zhang, Q., Schwab, J. J., Yang, T., Ng, N. L., and Demerjian, K. L. (2012) Factor analysis of combined organic and inorganic aerosol mass spectra from high resolution aerosol mass spectrometer measurements. *Atmospheric Chemistry and Physics*, 12 (18), 8537–8551.

Sun, Y., Du, W., Fu, P., Wang, Q., Li, J., Ge, X., Zhang, Q., Zhu, C., Ren, L., Xu, W., Zhao, J., Han, T., Worsnop, D.R., and Wang, Z. (2016) Primary and secondary aerosols in Beijing in winter: sources, variations and processes. *Atmospheric Chemistry and Physics*, 16, 8309–8329, doi:10.5194/acp–16–8309–2016, 2016.

Wang, H., Fang, Y., Wang, D., and Wu, J. (2005) Soil heavy metal pollution in Beijing City. *Urban Environment & Urban Ecology*, 18, 34–36.

Wang, Y., Jia, C., Tao, J., Zhang, L., Liang, X., Ma, J., Gao, H., Huang, T., and Zhang, K. (2016) Chemical characterization and source apportionment of PM2.5 in a semi-arid and petrochemical-industrialized city, Northwest China. *Science of the Total Environment*, doi: STOTEN_STOTEN–D–16–03331.

Wang, Y., Zhuang, G., Tang, A., Yuan, H., Sun, Y., Chen, S., and Zheng, A. (2005) The ion chemistry and the source of PM2.5 aerosol in Beijing. Atmospheric Environment, 39, 3771–3784.

Wei, B. and Yang, L. (2010) A review of heavy metal contaminations in urban soils, urban road dusts and agricultural soils from China. *Microchemical Journal*, 94, 99–107.

Winchester, J.W. and Bi, M. (1984) Fine and coarse aerosol composition in an urban setting: a case study in Beijing. *Atmospheric Environment*, 18 (7), 1399–1409.

Wong, C., Li, X., and Thornton, I. (2006) Urban environmental geochemistry of trace metals. *Environmental Pollution*, 142, 1–16.

World Bank (2014) Available and accessed on 31 July 2016. http://www.worldbank.org/en/news/press–release/2014/11/14/gef–grant–to–enhance–the–environmental–

performance–of–municipal–solid–waste–incinerators–in–chinese–cities.

Xia, X., Chen, X., Liu, R., and Liu, H. (2011) Heavy metals in urban soils with various types of land use in Beijing, China. *Journal of Hazardous Materials*, 186, 2043–2050.

Xue, J., Griffith, S.M., Yu, X., Lau, A.K.H., and Yu, J.Z. (2014) Effect of nitrate and sulfate relative abundance in $PM_{2.5}$ on liquid water content explored through half-hourly observations of inorganic soluble aerosols at a polluted receptor site. *Atmospheric Environment*, 99, 24–31.

Zhang, W., Dong, S., Wu, D., and Zhang, L. (2013) The Summary of Groundwater Contamination. *Advanced Materials Research*, 726–731, 2355–2362.

Zhang, Y.-L. and Cao, F. (2015) Fine particulate matter (PM2.5) in China at a city level. *Scientific Reports*, 5, 14884. doi: 10.1038/srep14884.

Zhang, X.Y., Wang, J.Z., Wang, Y.Q., Liu, H.L., Sun, J.Y., and Zhang, Y.M. (2015a) Changes in chemical components of aerosol particles in different haze regions in China from 2006 to 2013 and contribution of meteorological factors. *Atmospheric Chemistry and Physics*, 15, 12935–12952.

Zhang, Y., Wang, G., Guo, S., Zamora, M., Ying, Y., Lin, Y., Wang, W., Hu, M., and Wang, Y. (2015b) Formation of urban fine particulate matter. *Chemical Reviews*. doi: 10.1021/acs.chemrev.5b00067

Zhao, M., Huang, Z., Qiao, T., Zhang, Y., Xiu, G., and Yu, J. (2015) Chemical characterization, the transport pathways and potential sources of PM2.5 in Shanghai: seasonal variations. *Atmospheric Research*, 158–159, 66–78.

Zhao, B., Wang, S., Donahue, N.M., Jathar, S.H., Huang, X., Wu, W., Hao, J., and Robinson, A.L. (2016) Quantifying the effect of organic aerosol aging and intermediate-volatility emissions on regional-scale aerosol pollution in China. *Scientific Reports*, 6, 28815, 10.1038/srep28815.

Zheng, M., Salmon, L.G., Schauer, J.J., Zeng, L.M., Kiang, C.S., Zhang, Y.H., and Cass, G.R. (2005) Seasonal trends in PM2.5 source contributions in Beijing, China. *Atmospheric Environment*, 39 (22), 3967–3976.

Zheng, G., Duan, F., Su, H., Ma, Y., Cheng, Y., Zheng, B., Zhang, Q., Huang, T., Kimoto, T., Chang, D., Pöschl, U., Cheng, Y., and He, K. (2015) Exploring the severe winter haze in Beijing: the impact of synoptic weather, regional transport and heterogeneous reactions. *Atmospheric Chemistry and Physics*, 15, 2969–2983.

Zheng, S. and Kahn, M. (2013) Understanding China's urban pollution dynamics. *Journal of Economic Literature*, 51 (3), 731–772.

Zhou, Y. and Xiang, Y. (1984) Atmospheric turbidity measurement and a preliminary study of dust pollution in the Beijing area. *Environmental Science*, 5, 50–54 (in Chinese).

Zhu, G. (2001) Pollution of heavy metals on soils in east-south area of Beijing and remediation. *Agro-Environmental Protection*, 20, 164–166.

Zou, J., Dai, W., Gong, S., and Ma, Z. (2015) Analysis of spatial variations and sources of heavy metals in farmland soils of Beijing suburbs. *PLoS One*, doi:10.1371/journal.pone.0118082.

26

Urban Pollution in India

Manoj Shrivastava, Avijit Ghosh, Ranjan Bhattacharyya, and S.D. Singh

Centre for Environment Science and Climate Resilient Agriculture (CESCRA), Indian Agricultural Research Institute, India

26.1 Introduction

Indian firms/industries have expanded very rapidly in the last 30years, especially in urban areas. Rapid growth in industrialisation is leading to many environmental issues, including emission of uncontrolled pollutants (CPCB, 2010). The emissions of vehicles, municipal solid waste dumping sites, real estate development, e-waste processing sites, destruction of forests, and land degradation due to urbanisation are other sources of pollution. Several industries established near or in urban areas are polluting the environment heavily. These are aluminium (Al) and zinc (Zn) smelter industries, cement, chlorine (Cl), copper (Cu) smelters, fertiliser, iron and steel industries, distillery industries, oil refineries, pharmaceuticals and petrochemicals, and pulp and paper industries.

In India, urbanisation has witnessed unprecedented growth over the past 40 years. During the last 50 years, the urban population of India has grown nearly five times (around 400 million people live in cities, in sharp contrast to 60 million in 1947). About 140 million people will move to the cities by 2020 in India, and another 700 million by 2050. The number of Indian megacities will increase from the current three (Mumbai, Delhi, and Kolkata) to six (including Bangalore, Chennai, and Hyderabad) by 2021. That increasing population results in rapid consumption of energy and other resources, which is contributing to urban pollution. Indian urban areas contain high levels of criteria pollutants (e.g. particulate matter, SO_2, and NO_x), greenhouse gases, ozone precursors, and aerosols. The State of World Cities Report (2012) shows that Mumbai and New Delhi perform weakly on economic and environmental dimensions. Thus, Indian cities are growing in an unsustainable manner compared to other global cities, like Vienna and Tokyo (Table 26.1).

26.2 Issues Related to Urban Pollution in India

The environmental issues are unplanned settlements, waste management, natural disaster preparedness, traffic management, and degradation and pollution of water and land resources and air quality. Vegetation clearance, drainage channel modifications, and inappropriate agricultural practices cause increased water erosion, which often creates increased pollutant transport (Arora and Reddy, 2013). Alarming levels of particulate matter are reported in urban areas of India due to ever-increasing traffic, growing energy consumption, unplanned urban and industrial development, and the high influx

Urban Pollution: Science and Management, First Edition. Edited by Susanne M. Charlesworth and Colin A. Booth.

Table 26.1 Sustainability of Indian cities in comparison to global cities.

Cities	City Prosperity Index	Productivity Index	Quality of Life Index	Infrastructure Index	Equity Index	Environment Index
Vienna	0.925	0.939	0.882	0.996	0.883	0.932
Tokyo	0.906	0.925	0.931	0.989	0.828	0.936
Mexico	0.709	0.743	0.764	0.900	0.405	0.866
Moscow	0.793	0.806	0.813	0.960	0.550	0.908
Mumbai	0.694	0.645	0.739	0.745	0.715	0.632
New Delhi	0.636	0.596	0.690	0.786	0.712	0.448

Source: State of World's Cities 2012–2013, UN-Habitat.

of people into urban areas (Agarwal and Narain, 1999). An increase in vehicles and provision of roads in urban areas contributes considerably to ozone (O_3) pollution. Due to rapid urbanisation, huge quantities of vegetation is replaced with concrete buildings and low-albedo surfaces (Singh and Grover, 2015). This is a big issue in densely populated urban areas of India (Wong *et al.*, 2010). Electronic waste or e-waste and environmental contamination through persistent organic pollutants (POPs) are other increasing problems in India. Even after banning of crop residue burning, air, water and soil samples of Indian urban and sub-urban areas continue to show considerable contamination. Relatively high levels of POPs have been detected in drinking water, food products, and even human breast milk (IPEP, 2006).

Due to the application of poisonous pesticides and mineral fertilisers and due to poor garbage disposal services in both the rural and urban areas of India, land is polluted. The recent pollution status of various Indian cities is presented in Tables 26.2 and 26.3.

Table 26.2 Some of the important urban pollution case studies in India.

Study	Study area	Results	References
Water pollution	Varanasi (2001)	Mercury in the Ganga river ecosystem at Varanasi followed a trend of water<sediments (0.067 ppm)<benthic macro-invertebrates (0.118 ppm)<fish (2.638 ppm).Though Hg contamination has not reached an alarming extent, its presence in the river system is of concern.	Sinha *et al.* (2007)
	Delhi	The microbial quality of urban run-off from sub-catchments of the Delhi watershed is very poor with *faecal coliform* levels varying between 6 and 7 log orders.	Jamwal *et al.*, 2008
Air and water pollution	Pune City (2007–2008)	The average annual emission of suspended particulate matter (SPM) found at the Pune waste disposal site is 1708.3 $\mu g/m^3$. The average annual emission of SO_2 in landfill is 285.33 $\mu g/m^3$ with NO at 234.07 $\mu g/m^3$. The SPM, SO_2, and NO, concentrations are higher than those stipulated by the Indian standard limit. Leachate samples are acidic and corrosive in nature. The chemical oxygen demand (COD) and biological oxygen demand (BOD) of tested well-water samples are in the range 412 to 834 mg L^{-1}. The concentration of sodium in well-water samples varies from 2437 to 2612 mg L^{-1}, respectively. The COD, BOD, and sodium concentration in well water are higher than the limits of IS: 2291.	Dhere *et al.*, 2008

Table 26.2 (Continued)

Study	Study area	Results	References
	Puducherry	Puducherry coastal area is being polluted due to the discharge of industrial, domestic, and agricultural wastes through small tributaries and channels into the Bay of Bengal. In deeper sediments dissolved oxygen concentration ranged from 3.71 to 5.33 mg/L, and the sulphide level was high (40.43 mg/L). High sulphide content is responsible for formation of H_2S, and an offensive 'rotten egg' odour spreads out in polluted coastal areas.	Satheeshkumar *et al.* (2012)
	Patiala	There is a strong correlation between land use and development activities and resulting storm water quality. TSS, COD, and oil and grease were found to be major pollutants in surface run-off generated from commercial and urbanised catchments (all exceeded the surface water quality standards developed by the Central Pollution Control Board, India).	Arora and Reddy (2013)
Urban Heat Island (UHI)	Delhi	Being a tropical city, Delhi's UHI behaves differently in comparison to developed cold country cities. Maximum and minimum seasonal UHI intensities were observed in summer (intensity = 16.7°C) and winter (intensity = 7.4°C), respectively. Major commercial and industrial sites across the city and the airport area in the south experienced higher UHI effects and were found to be the most vulnerable locations.	Sharma and Joshi (2014)
	Chandigarh	A significant surface urban heat island has been observed in Chandigarh that varies with seasons. The average annual UHI intensity from 2009 to 2013 varied from 4.98 K to 5.43 K, and the overall average UHI intensity has been observed to be 5.2 K. The maximum value of UHI_{index} has been found to be 0.93. Pixels with an average UHI_{index} of higher than 0.90 have been considered as hot spots. The relationship between land surface temperature (LST) and percent impervious surface area (%ISA) which represents the extent of urbanisation has been used for UHI analysis, finding its relationship with LST to be season independent. A positive relationship has been found between LST and %ISA with a consistent rising trend. In the mean LST and %ISA relationship, the coefficient of correlation in winter is 0.81, which is slightly higher compared to monsoon and summer seasons. Besides other factors, elevation also plays a significant role in LST dynamics and spatial distribution.	Khodakaramiand Ghobadi (2016)
Air pollution	Coimbatore (1999–2001)	The mean quantity of heavy metals in respirable fractions of suspended particulate matter (RSPM) was of the order Zn> Cu>Pb>Ni> Cr> Cd. Concentrations of these heavy metals were BDL (below detectable level) to 2147 ng/m^3 in RSPM.	Mohanraj *et al.* (2004)
	Aizawl city (2006–07)	SPM (46–132 μg/m^3), RSPM (38–58 μg/m^3), NO_2 (3–13 μg/m^3), and SO_2 (0.8 to 1.6 μg/m^3). Vehicles and associated problems of traffic congestion are a major contributor among the various sources of air pollution.	Lalrinpuii and Lalramnghinglova (2008)
	Kolkata	Increased surface pollution can increase lightning flash rates during thunderstorms. Enhanced lightning activity intensifies the production of tropospheric NO_2.	Chaudhuri and Middey (2013)

(Continued)

Table 26.2 (Continued)

Study	Study area	Results	References
	Delhi	The concentration of PM_{10} and its chemical components including organic carbon (OC), elemental carbon (EC), water-soluble inorganic ionic components, and major and trace elements showed a strong seasonal cycle with maxima during winter (PM_{10}: 241.4 ± 50.5 μg m^{-3}; OC: 34.7 ± 10.2 μg m^{-3}; EC: 10.9 ± 3.0 μg m^{-3}) and minima during monsoon (PM_{10}: 140.1 ± 43.9 μg m^{-3}; OC: 15.5 ± 7.5 μg m^{-3}; EC: 4.9 ± 2.3 μg m^{-3}). The highest contribution to the estimated average values of PM_{10} comes from particulate organic matter (24%) with soil/crustal matter (16%), ammonium sulphate (7%), ammonium nitrate (6%), aged sea salt (5%), and light absorbing carbon (4%). The sector-wise contribution to PM_{10} mass at the observational site in Delhi was mainly from secondary aerosols (21.7%), soil dust (20.7%), fossil fuel combustion (17.4%), vehicle emissions (16.8%), and biomass burning (13.4%).	Sharma *et al.* (2014)
	Lucknow (2012)	Average levels of PM_{10} and $PM_{2.5}$ were above the permissible limits laid by WHO at densely populated and roadside sites with 189 μg/m^3 ($PM_{2.5}$ 76 μg/m^3) and 226 μg/m^3($PM_{2.5}$ 91 μg/m^3), respectively. Survey results also showed that 46% of urban people suffered from acute respiratory infections like bronchial asthma, headache, depression, and dizziness, and these people were mostly from roadside colonies.	Lawrence and Fatima (2014)
	Udaipur (2011–12)	The daily means of O_3, CO, and NO_x were in the ranges of 5–53 ppbv, 121–842 ppbv, and 3–29 ppbv, respectively. Exceptionally high levels of trace gases during the Diwali (festival) period were due to extensive use of firecrackers from evening until morning hours. The enhancements of O_3, CO, and NOx compared to normal days were about 61%, 62%, and 23%, respectively.	Yadav *et al.* (2016)
	Bangalore	Emissions from smouldering and flaming roadside trash piles were analysed for organic and elemental carbon (OC/EC), brown carbon, and redox activity. Results show high variability of chemical composition and toxicity between trash-burning emissions and characteristic differences from ambient samples. OC/EC ratios for trash-burning emissions range from 0.8 to 1500, while ambient OC/EC ratios were observed at 5.4 ± 1.8. Emissions from trash-burning piles were composed of aromatic di-acids (likely from burning plastics) and levoglucosan (an indicator of biomass burning), while the ambient sample showed high response from alkanes, indicating notable representation from vehicular exhaust. DTT content was extremely high in all trash-burning samples. Fresh trash-burning emissions are less redox-active than ambient air (13.4 ± 14.8 pmol/min/μg OC for trash burning; 107 ± 25 pmol/min/μg OC for ambient). However, overall results indicate that near trash-burning sources, exposure to redox-active particulates can be extremely high.	Vreeland *et al.*, 2016

Table 26.2 (Continued)

Study	Study area	Results	References
Land pollution	New Delhi	Both surface and sub-surface soils are affected by high metal pollution with a contamination factor (CF) > 2 and the average pollution load index (PLI) of 2.77. The geo-accumulation index values of 6.6–8.2 (pre- and post-monsoon samples) for Zn, 6.7–8.2 for Cr, and 5.6–5.1 for Cd in surface soils indicate extreme levels of pollution in the region. A total of four metal sources, industrial emissions for Fe, Mn, Cu, Ni, Co, and Cd, electroplating industry for Zn and Cr, geogenic for Al and V, and vehicular and biomass burning for Ba and Pb in the surface soils were identified. Industrial emissions are the major source of metals. Surface soils around small-scale industries are more polluted with Zn, Cr, and Cd (CFs = 25–31), and Cu and Pb (CFs = 7–11) and have high PLI (range: 3.28–8.77) compared to other sampling sites. Higher geo-accumulation indices and pollution load of metals in the urban soils are expected to have long-term impacts on human health, plants, and crop productivity in this area.	Pathak *et al.* (2015)

Table 26.3 Elemental carbon, organic carbon, and total carbon emissions in some Indian cities.

Sr. No.	City	Elemental carbon kg day^{-1}	Organic carbon kg day^{-1}	Total carbon kg day^{-1}
1	Bangalore	7,577	12,923	20,500
2	Chennai	371	2,029	2,400
3	Delhi	6,436	22,364	28,800
4	Kanpur	600	2,700	3,300
5	Mumbai	2,737	13,963	16,700
6	Pune	1,662	5,038	6,700

Source: Gargava and Rajagopalan, 2015.

26.3 Pollution from Solid Waste and Wastewater in Indian Urban Areas

Urbanisation, inadequate treatment capacity, and disposal of untreated wastes cause severe pollution in urban and peri-urban areas. During 2015, the estimated sewage generation in the country was 61,754 million litres per day (MLD) as against the developed sewage treatment capacity of 22,963 MLD. Because of the deficit in sewage treatment capacity, about 38,791 MLD of untreated sewage (62% of total sewage) is discharged directly into nearby waterbodies (CPCB, 2016). Use of such wastewater-loaded surface water as irrigation has resulted in the significant build-up of heavy metals in soils of agricultural land near several cities and towns of India (Saha and Panwar, 2013). Different polluted materials are also discharged into sewage due to higher industrialisation in residential areas, leading to environmental pollution (CPCB, 2009). It is projected that national wastewater generation will reach 122,000 MLD by 2050 (Bhardwaj, 2005). Cities around the Ganga basin are generating 2637.7 MLD of sewage, but India is in a position to treat 1174.4 MLD, i.e. 44.2% only. Apart from domestic sewage, about 13,468 MLD of wastewater is generated by industries (mostly large-scale) of which only 60% is treated (Kaur *et al.*, 2012). Although untreated sewage is being used by farmers to grow crops on urban peripheral lands due to its high nutrient contents, its use for longer periods is a matter of great concern (Saha *et al.*, 2010). Urban sewage also carries high amounts of heavy metals (Ni, Cr, Pb, Cd, and Zn) and salts, causing salinity and alkalinity hazards. The country generates ~50 million tonnes (Mt) of municipal solid wastes/year from its urban areas (CPCB, 2000). About 9%–10% of these wastes find its way into agricultural land in the form of compost, contaminating soil with heavy metals.

The types and sources of heavy metals (CPCB, 2010) are as follows:

1) Chromium (Cr)–Mining, industrial coolants, chromium salts manufacturing, leather tanning
2) Lead (Pb) – Lead acid batteries, paints, e-waste, smelting operations, coal-based thermal power plants, ceramics, bangle industry
3) Mercury (Hg) – Chlor-alkali plants, thermal power plants, fluorescent lamps, hospital waste (damaged thermometers, barometers, sphygmomanometers), electrical appliances, and so on
4) Arsenic (As) – Geogenic/natural processes, smelting operations, thermal power plants, and fuel
5) Copper (Cu) – Mining, electroplating, and smelting operations
6) Vanadium (Va) – Spent catalyst and sulphuric acid plant
7) Nickel (Ni) – Smelting operations, thermal power plants, and battery industry
8) Cadmium (Cd) – Zinc smelting, waste batteries, e-waste, paint sludge, incinerations, and fuel combustion
9) Molybdenum (Mo) – Spent catalyst Zinc (Zn) smelting and electroplating. The major heavy-metal-contaminated sites are given in Table 26.4.

Minh *et al.* (2003) reported that open-dumping sites are potential sources of organo-halogen pollutants in urban areas. E-waste burning and acid leaching areas of Indian urban areas are contaminated by PCDDs/Fs and dioxins like PCBs (Tue *et al.*, 2013). Soils from e-waste recycling sites of Chennai and Bangalore had higher levels of Cu, Zn, Ag, Cd, In, Sn, Sb, Hg, Pb, and Bi (Ha *et al.*, 2009).

26.4 Air Pollution in Urban Areas of India

Vehicle emissions produce over 90% of air pollution in urban areas in developing countries (Contreras and Ferri, 2016). The air quality index of million plus cities of India showed that >50% of cities have moderate to poor air quality (Figure 26.1).

Increased aerosol loading over the Indo-Gangetic Plains (IGP) increases the frequency of fog occurrences over the national

Table 26.4 Major heavy-metal-contaminated sites in India (CPCB, 2009). Where TN = Tamil Nadu; UP = Uttar Pradesh; MP = Madhya Pradesh; AP = Andhra Pradesh; WB = West Bengal.

Cr	Pb	Hg	As	Cu
Ranipet (TN) Vadodara(Gujarat) Talcher (Orissa) Kanpur (UP)	Ratlam (MP) Bandalamottu Mines (AP) Vadodara (Gujarat) Korba (Chattisgarh)	Kodaikanal (TN) Ganjam (Orissa) Singrauli (MP)	Tuticorin (TN) Gangetic plain (WB) Balia and other districts (UP)	Tuticorin (TN) Singbhum mines (Jharkhand) Malanjkhand (MP)

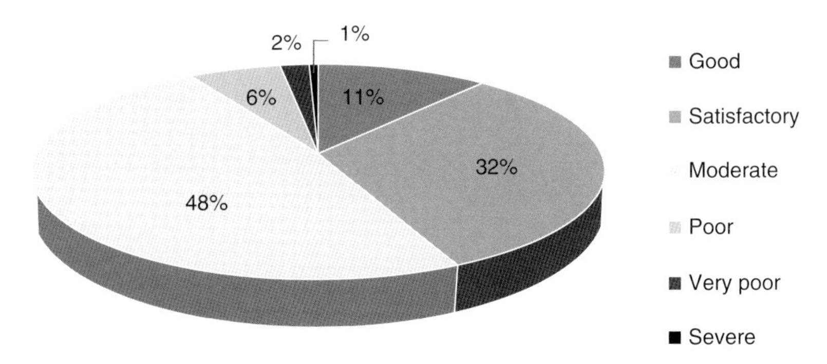

Figure 26.1 Air quality index for million plus cities during 2015. (*Source*: CPCB, 2016.)

capital region of Delhi (NCR Delhi). This further reduces the maximum temperature of Delhi during the winter months (Jenamani, 2007). Tall buildings in urban areas form 'street canyon like conditions' and hinder dispersal of air pollutants in these areas (Wong *et al.*, 2010). Roadside trash burning is also a factor that influences air quality throughout India.

Vehicles, road dust, and cooking using solid fuel are the key urban sources of air pollution. Motor vehicles are increasingly vital contributors of anthropogenic CO_2 and other greenhouse gases (GHGs). The transport sector contributes ~90% of total emissions in India. The number of motor vehicles ranged from 72.7 million in 2004 to ~141.8 million in 2011 (Khandar and Kosankar, 2014). There is a direct relationship between the road transport system and air pollution in urban areas (Shrivastav *et al.*, 2013). Table 26.5 shows the national acceptable air quality limits.

Within the vehicle group, heavy-duty diesel vehicles are the largest contributors. EC, mostly in the fine size mode, can have larger health implications. Diesel particulates are known to be carcinogenic and also have a high EC to OC ratio that signifies their global warming potential. TC emissions in the cities of Bangalore, Chennai, Delhi, Kanpur, Mumbai, and Pune are ~20.5, 2.4, 28.8, 3.3, 16.7, and 6.7 TPD, respectively. EC emissions in Bangalore, Delhi, Mumbai, and Pune are 7577, 6436, 2737, and 1662 kg day^{-1}, respectively. In Chennai (371 kg day^{-1}) and Kanpur (600 kg day^{-1}), these are comparatively low. Particulate matter emission control from the identified urban sources will offer the twin benefits of reducing health risks and global warming (Gargava and Rajagopalan, 2015).

The transport sector is responsible for generating most air pollutants (Table 26.6). The pollution load from petrol-driven vehicles is more than from diesel vehicles. Pollutants enter the cities in the form of gases, particles, or as aerosols, by evaporation of liquids or by co-evaporation of dissolved solvents from water and by wind erosion (Magnus, 1994). Major air pollutants in Indian urban areas are sulphur dioxide (SO_2), particulate matter, and nitrogen oxides (NOx). High O_3 concentra-

Table 26.5 National standards for different air pollutants.

Concentration (μg/m^3)				
	Sulphur dioxide (SO_2)	Nitrogen dioxide (NO_2)	Respirable suspended particulate matter	Carbon monoxide (CO)
National standard	80	80	100	2000

Table 26.6 Concentration of pollutants through gasoline-powered vehicles in metropolitan cities in India (in μg/m^3).

City	SO_2	NO_x	RSTM
Delhi	4	16	189
Kanpur	4	15	169
Vadodara	4	42	34
Chennai	4	13	189
Kolkata	4	34	55

Source: Sarkar and Tagore, 2011.

tion has been recorded in Delhi, with a maximum of over 600 μg per cubic meter (Misra *et al.*, 1993). Chromium, Mn, Co, Cu, In, Sn, Sb, Tl, Pb, and Bi in the air from the e-waste recycling facility were higher than the control site in Chennai city (Ha *et al.*, 2009). Sources of the different elements in street dust and soils are typically common to most urban environments (traffic, heating systems, industry, natural substrate, etc.), but their intensities and patterns of distribution vary accordingly to the peculiarities of each city.

Gurjar *et al.* (2016) studied the air pollution trends over Indian megacities, and they observed decreasing trends of SO_2 in Delhi, Mumbai, and Kolkata (all megacities), owing to decreased sulphur (S) contents in coal and diesel. However, increased NO_x was observed in all these megacities. This could be due to the increased numbers of vehicles registered. The highest emissions of suspended particulate matter (SPM) and PM_{10} have been observed at Kolkata, whereas highest ambient concentrations were found in Delhi. Fluctuating trends of SPM concentrations were noted during 1991–1998 in Mumbai and Kolkata. However, for Delhi, a variable trend was observed.

The concentration of GHGs in the atmosphere has been increasing rapidly during the last century. Major sources of GHG are deforestation, power generation (burning of fossil fuels), transportation (burning of fossil fuel), agriculture (livestock, farming, rice cultivation, and burning of crop residues), waterbodies (wetlands), industry, and urban activities (building, construction, transport, and solid and liquid waste). The GHG footprint (aggregation of CO_2 equivalent emissions of GHGs) of Delhi, Greater Mumbai, Kolkata, and Chennai are 38,633.2 Gg, 22,783.08 Gg, 14,812.10 Gg and 22,090.55 Gg, CO_2 eq., respectively. The major contributor sectors are transportation, domestic, and industry. Chennai emits 4.79 t of CO_2 equivalent emissions per capita, the highest among all megacities, followed by Kolkata (which emits 3.29 t of CO_2 equivalent emissions per capita). Also, Chennai emits the highest CO_2 equivalent emissions per GDP (2.55 t CO_2 eq./10^5rupees) (Ramachandra *et al.*, 2015).

Urban activities are significantly contributing to elevated heavy metal loads in the edible portion of vegetables. Sharma *et al.* (2008) reported that the deposition rate of Zn was the highest followed by Cu, Cd, and Pb in Varanasi city, Uttar Pradesh. Heavy metal concentrations in vegetables collected from peri-urban New Delhi were high due to atmospheric deposition and contaminated water use (Singh and Kumar, 2006).

26.5 Water Pollution in Urban Areas of India

Ever-increasing population along with swift industrialisation, urbanisation, and agricultural growth has caused water quality deterioration in India. The big Indian industries generate large-scale polluted liquid emissions that are washed out through canals into river systems. Discharge of untreated sewage in water courses, both surface and ground waters, is the most important water polluting source in India. Out of about 38,000 MLD of sewage generated, treatment capacity exists for only about 12,000 MLD. Thus, there is a large gap between generation and treatment of wastewater in India. Even the existing treatment capacity is also not effectively utilised due to operational and maintenance problems (CPCB, 2009a). Thus, of late, every freshwater body (rivers, lakes, and estuaries) is contaminated with organic and inorganic contaminants. The most harmful elements polluting river water are organic wastes, minerals, sediments, toxic chemicals, nutrients, and many more.

Pollutants are present in groundwater, rivers, and other waterbodies. Despite the ban on dichlorodiphenyltrichloroethane (DDT) and hexachlorocyclohexane (HCH) in India, residues of these POPs are extensively distributed, and their traces could be detected in waterbodies in most of areas of India (Table 26.7).

In addition, surface and groundwater in the industrial town of Patancheru, Hyderabad, had Sr, Ba, Co, Ni, and Cr, from mixed origins, with similar contributions from anthropogenic and geogenic sources. However, Fe, Mn, As, Pb, Zn, B, and Co were derived from anthropogenic activities, primarily due to uncontrolled industrial effluent discharges (Krishna *et al.*, 2009). Many studies have been conducted to quantify marine debris on the beaches of various coastal cities of India (Table 26.8). The National Oceanic and Atmospheric Administration (NOAA, 2011) defines marine debris as any persistent solid material that is manufactured or processed and directly or indirectly, intentionally or unintentionally, disposed of or abandoned into the marine environment or the Great Lakes. Marine debris is a source of serious damage to the marine environment. Data showed that most of the beaches of coastal cities are heavily polluted and need regular cleaning.

Table 26.7 Persistent organic pesticide residues (DDTs, HCHs, and Endosulfan) in water from various cities across India. NCR = National Capital Region.

Region/city	Type of waterbody	Persistent organic pesticide residues (ng/L)
A. North India		
1. Devprayag	Ganges River	Not Detected (ND)–7.24
2. Ambala	Groundwater	147.1–940.3
3. Hisar	Tapwater	13.1–83.5
4. Gurgaon	Groundwater	123.8–357.1
5. Delhi NCR	Yamuna River	310.25–387.90
6. Pilbhit	Gomti River	5.97–46.69
7. Lucknow	Rainwater	0.53–23.48
8. Unao	Dugwells	17–209
	Streams, ponds	130–1950
B. Central India		
1. Bhandara	Groundwater	60–720
	Surface water	60–80
2. Amravati	Groundwater	39–60
	Surface water	39–42
3. Yavatmal	Groundwater	Not Detected–80
C. East India		
1. Nagaon	River, ponds	4,911–6,121
	Borewell/dugwell	5,574–6,904
2. Dibrugarh	River, ponds	4,403–5,402
	Borewell/dugwell	5,168–6,549

(Continued)

Table 26.7 (Continued)

Region/city	Type of waterbody	Persistent organic pesticide residues (ng/L)
3. Kolkata	Tankwater	193–1,157
	Groundwater	10–100
D. South India		
1. Hyderabad	Groundwater	613–8,879
2. Chennai	Ennore creek	10.07–29.21
3. Thiruvallur	Open wells	4,518–10,014
E. Western India		
1. Jaipur	Tubewells	ND–7,240
2. Mumbai	Seawater	5.42–12.45

Source: Yadav *et al.*, 2015.

Table 26.8 Marine debris pollution in the seashore of different coastal cities. Where NR = not reported.

		Quantity		
Location	**Year**	**Count**	**Weight**	**References**
Caranzalem Beach, Goa	1981	50–300/m^2	NR	Nigam (1982)
Andaman and Nicobar	2003	NR	NR	Dharani *et al.* (2003)
Gulf of Cambay	2004	NR	81.43 mg/kg	Reddy *et al.* (2006)
Gulf of Mannar	2006–2008	68.5/100 m	4.34 kg/100 m	Ganesapandian *et al.* (2011)
Mumbai	2011–2012	11.67 ± 8.83/m^2	3.24±0.92 g/m^2	Jayasiri *et al.* (2013)
Chennai	2015	171.8/100 m	3.24 kg/100 m	Arun Kumar *et al.* (2016)

26.6 Soil Contamination in Urban Areas of India

Contamination of the soil environment by heavy metals is becoming rife across the globe. Rapid industrialisation and poor management of industrial effluent is increasing the risk of heavy metal pollution. Even with exposure at minute levels, heavy metals can have carcinogenic effects on humans and animals and a negative effect on soil microorganisms and crop plants. Excessive concentrations of heavy metals, that is, Cr, Cd, As, Ni, Se, and Pb, have been found in soils of agricultural land near cities, mines, and industrial areas around the world. Although a geogenic source of pollution has been observed for some trace elements in different parts of the world, including India, the secondary sources of anthropogenic pollution are more dominant, localised, and cause more soil pollution.

26.7 Noise Pollution in Urban Areas of India

Noise pollution has become a serious problem for society. In India, with expanding vehicular population, traffic noise levels have increased, which can cause serious health effects. The World Health Organization (WHO) recognised noise as one of the major pollutants affecting the health of the human population (WHO, 2011). The major sources of noise pollution are road traffic, rail, aircraft noise, construction noise, noise emitted from industrial infrastructure, honking noise from vehicles, noise emitted from household appliances, loudspeakers, community processions, and so on. (Garg and Maji, 2016). There are regulations in India not to exceed the normal range (65 decibel) of sound so that pollution could be controlled. According to studies carried out by the Central Pollution Control Board, Mumbai was found to be the noisiest city followed by Lucknow, Hyderabad, New Delhi, Chennai, Kolkata, and Bengaluru. Delhi showed an average noise level within prescribed limits (*Times of India*, 2016).

26.8 Ways to Reduce Urban Pollution in India

1) Policymakers must focus on reducing the emission intensities of production activities within cities (especially from the energy sector) if they are to avoid rapid growth in urban air pollution in future. The best option is to bring down polluted emissions and effluents to a level where nature can treat them without any negative impacts.
2) Policies encouraging higher land-use diversity, density, and transit supply have the potential to marginally decrease vehicle use in the Indian metropolis (Shirgaokar, 2016). Solar energy as a clean power source can reduce the impacts of urban pollutants. This process helps reduce exhaust air pollution in the city environment (Khodakarami and Ghobadi, 2016).
3) Untreated effluents emerging from industries must be monitored to maintain standards prescribed by the CPCB for various industries. It is suggested to check for trace metal and total dissolved solids (TDS) concentrations before letting untreated effluents into the common effluent treatment plant. Regular monitoring of the quality of groundwater should be undertaken temporally and spatially to identify the source of toxic pollutants and other inhibitory chemicals.
4) The quality of the urban environment should be improved by maintaining and enhancing forest and tree cover, rejuvenating polluted waterbodies, and upgrading urban population management skills. As urban forests both sequester CO_2 and negatively affect the emission of CO_2, these can play crucial roles in combating increasing levels of atmospheric CO_2 (Novak and Crane, 2002).
5) There should be guidelines for urban planners and foresters on green space planning, by using integrated approaches that meet the social and ecological needs of cities (Govindarajulu, 2014). Green plants play a role in air pollution attenuation (Cavanagh *et al.*, 2009). Urban forests also reduce noise pollution, control soil erosion, and enhance the aesthetic beauty of the area (Yang *et al.*, 2005). Urban forests can be developed around riverbanks, parks, gardens, roads and railways, playgrounds, cemeteries, roadside, and so on. Some plants are comparatively more tolerant to air pollutants. Pandey *et al.* (2015) reported that that *Ficus benghalensis* L., *Ficus religiosa*, *Polyalthia longifolia*, *Ficus glomerata* (Roxb.), *Anthocephalus indicus*, *Mangifera indica*, *Cassia fistula* L., *Drypetes roxburghii*, *Terminalia arjuna*, *Psidium guajava* L.,*Millingtonia hortensis*, and *Dalbergia sissoo* can be used for development of urban forests to reduce air and noise pollution.
6) The majority of marine litter originates from land-based sources and activities; a primary emphasis on controlling marine litter should focus on preventing the inflow of litter to the sea.

7) Metal recovery to prevent pollution through developing cost-effective strategies to meet environmental challenges is critical (Koelmel *et al.*, 2016).
8) Townships should be designed so that the buildings become a part of the local landscape. Solid waste should be classified, decomposed to fertilisers, and recycled to reduce landfill and avoid waste-borne hazards. Sewage should be treated before discharging into rivers.
9) Townships should live on recycled water. Efforts should be made for integrated water resource management by adopting a few towns for pilot testing of a programme on improving urban water governance capacities.
10) Community places, such as shopping malls, should be designed in such a way that people can walk or take public transportation to reach them. With a view to developing an environment-friendly urban transport system, it is necessary to develop strong national transport policies where attempts should be made to discourage the prevailing petrol-and diesel-driven vehicles to encourage bicycle traffic for short distances and to promote the uptake of battery-driven and solar-powered vehicles.
11) There is an urgent need to strengthen and facilitate environmentally sound management of POPs and other chemicals.
12) A small number of initiatives for waste treatments, for example, incineration, pyrolysis, composting, recycling, biorefining, and biogas plants, have been undertaken in the country. For sustainable solid waste management (SWM), an inclusive improvement policy and paradigm shift is necessary.
13) Most of the research around the world has been on outdoor air pollution, but in India, we have a more severe problem of indoor air pollution (IAP). As there are no specific norms for IAP in India, an urgent need has arisen for implementing the strategies to create public awareness. Moreover, improvement in ventilation and modification in the pattern of fuel use will also help mitigate this national health issue (Rohra and Taneja, 2016).

26.9 Conclusions

Most Indian cities are experiencing rapid urbanisation. Unprecedented growth of cities has brought serious challenges, including environmental degradation, loss of natural habitat and species diversity, and increased human health risks associated with heat, noise, pollution, and crowding. That means many people, and particularly children, are living and growing up in environments with increasing pollution, intense heat, and less access to diverse green spaces. Given these challenges, there is a critical need to find ways to reduce health risks and maximise opportunities for well-being in all urban communities of the country.

References

Agarwal, A. and Narain, S. (1999) *Addressing the Challenge of Climate: Equity, Sustainability and Economic Effectiveness: How Poor Nation Can Help Save the World.* Centre for Science and Environment, New Delhi.

Arora, A.S. and Reddy, A.S. (2013) Multivariate analysis for assessing the quality of storm water from different urban surfaces of the Patiala city, Punjab (India). *Urban Water Journal*, 10, 422–433.

Arun Kumar, A., Sivakumar, R., Reddy, Y.S.R., Bhagya Raja, M.V., Nishanth, T., and Revanth, V. (2016) Preliminary study on marine debris pollution along Marina beach,

Chennai, India. *Regional Studies in Marine Science*, 5, 35–40.

Bhardwaj, R.M. (2005) Status of wastewater generation and treatment in India. IWG-Env, International Work Sessionon Water Statistics, Vienna, 20–22 June 2005 pp 9. Available at http://unstats.un.org/unsd/environment/envpdf/ pap_wasess3b6india.pdf

Cavanagh, J.-A.E., Zawar-Reza, P., and Wilson, J.G. (2009) Spatial attenuation of ambient particulate matter air pollution within an urbanized native forest patch. *Urban Forestry and Urban Greening*, 8, 21–30.

Chaudhuri, S. and Middey, A. (2013) Effect of meteorological parameters and environmental pollution on thunderstorm and lightning activity over an urban metropolis of India. *Urban Climate*, 3, 67–75.

Contreras, L. and Ferri, C. (2016) Wind-sensitive interpolation of urban air pollution forecasts. *Procedia Computer Science*, 80, 313–323.

CPCB (2000) *Status of Municipal Solid Waste Generation, Collection Treatment, and Disposable in Class 1 Cities.* Central Pollution Control Board, Ministry of Environmental and Forests, Governments of India, New Delhi.

CPCB (2009) *Comprehensive Environmental Assessment of Industrial Clusters.* Ecological Impact Assessment Series: EIAS/5/2009-2010. Central Pollution Control Board, Ministry of Environment and Forests, Govt. of India, New Delhi.

CPCB (2009a) Status of water supply, wastewater generation and treatment in class–I cities and class–II towns of India. *Control of Urban Pollution Series: CUPS*, 70, 2009–10.

CPCB (2010) *Status of the Vehicular Pollution Control Program in India Program Objective Series.* PROBES/136/2010.

CPCB (2016) Air Quality Index for million plus cities during 2015. *CPCB Bulletin*, 1, 1–24.

Dharani, G., Nazar, A.K.A., Ventakesan, R., and Ravindran, M. (2003) Marine debris in Great Nicobar. *Current Science*, 85, 574–575.

Dhere, A.M., Pawar, B.B., Pardeshi, P.B., and Patil, D.A. (2008) Municipal solid waste disposal in Pune city – An analysis of air and groundwater pollution. *Current Science*, 95, 773–777.

Ganesapandian, S., Manikandan, S., and Kumaraguru, A.K. (2011) Marine litter in the northern part of Gulf of Mannar, southeast coast of India. *Research Journal of Environmental Sciences*, 5, 471–478.

Garg, N. and Maji, S. (2016) A retrospective view of noise pollution control policy in India: Status, proposed revisions and control measures. *Current Science*, 111, 29–38.

Gargava, P. and Rajagopalan, V. (2015) Source prioritization for urban particulate emission control in India based on an inventory of $PM_{10}$and its carbonaceous fraction in six cities. *Environmental Development*, 16, 44–53.

Govindarajulu, D. (2014) Urban green space planning for climate adaptation in Indian cities. *Urban Climate*, 10, 35–41.

Gurjar, B.R., Khaiwal, R., and Nagpure, A.S. (2016) Air pollution trends over Indian megacities and their local-to-global implications. *Atmospheric Environment*, 142, 475–495.

Ha, N.H., Agusa, T., Ramu, K., Tu, N.P.C., Murata, S., Bulbule, K.A., Parthasarathy, P., Takahashi, S., Subramanian, A., and Tanabe, S. (2009) Contamination by trace elements at e-waste recycling sites in Bangalore, India. *Chemosphere*, 76, 9–15.

International POPs Elimination Project (IPEP) (2006) Country Situation on Persistent Organic Pollutants (POPs) in India. PP. 57. http://ipen.org/sites/default/files/documents/4ind_india_country_situation_report–en. Pdf

Jamwal, P., Mittal, A.K., and Mouchel, J.-M. (2008) Effects of urbanization on the quality of the urban runoff for Delhi watershed. *Urban Water Journal*, 5, 247–257.

Jayasiri, H.B., Purushothaman, C.S., and Vennila, A. (2013) Quantitative analysis of plastic debris on recreational beaches in Mumbai, India. *Marine Pollution Bulletin*, 77, 107–112.

Jenamani, R.K. (2007) Alarming rise in fog and pollution causing a fall in maximum temperature over Delhi. *Current Science*, 93, 314–322.

Khandar, C. and Kosankar, S. (2014) A review of vehicular pollution in urban India and its effects on human health. *Journal of Advanced Laboratory Research in Biology*, V, 54–61.

Khodakarami, J. and Ghobadi, P. (2016) Spatial and temporal variations of urban heat island effect and the effect of percentage impervious surface area and elevation on land surface temperature: Study of Chandigarh city, India. *Sustainable Cities and Society*, 26, 264–277.

Koelmel, J., Prasad, M.N.V., Velvizhi, G., Butti, S.K., and Venkata Mohan, S. (2016) Metalliferous waste in India and knowledge explosion in metal recovery techniques and processes for the prevention of pollution. In: M.N.V. Prasad and Kaimin Shih (eds.), *Environmental Materials and Waste: Resource Recovery and Pollution Prevention*. Elsevier Inc., pp. 339–390.

Krishna, A.K., Satyanarayanan, M., and Govil, P.K. (2009) Assessment of heavy metal pollution in water using multivariate statistical techniques in an industrial area: A case study from Patancheru, Medak District, Andhra Pradesh, India. *Journal of Hazardous Materials*, 167, 366–373.

Lalrinpuii, H. and Lalramnghinglova, H. (2008) Assessment of air pollution in Aizawl city. *Current Science*, 94, 852–853.

Lawrence, A. and Fatima, N. (2014) Urban air pollution and its assessment in Lucknow City – the second largest city of North India. *Science of the Total Environment*, 488–489, 447–455.

Magnus, F.B. (1994) *Toxic Substances in the Environment*. Wiley, New York

Minh, N.H., Minh, T.B., Watanabe, M., Kunisue, T., Monirith, I., Tanabe, S., Sakai, S., Subramanian, A., Sasikumar, K., Viet, P.H., Tuyen, B.C., Tana, T.S., and Prudente, M. (2003) Open dumping site in Asian developing countries: A potential source of polychlorinated dibenzo–p–dioxins and polychlorinated dibenzofurans. *Environmental Science & Technology*, 37, 1493–1502.

Misra, D., Sreedharan, B., and Bhuyan, S. (1993) *Accelerated Degradation of Fenitrothion in Fenitrothion–or 3–methyl–4–nitrophenolacclimatized Soil Suspension*. CRRI, 1993.

Mohanraj, R., Azeez, P.A., and Priscilla, T. (2004) Heavy metals in airborne particulate matter of urban Coimbatore. *Archives of Environmental Contamination and Toxicology*, 47, 162–167.

Nigam, R.A.I.I.V. (1982) Plastic pellets on the Caranzalem beach sands, Goa, India. *Mahasagar* 15, 125–127.

NOAA (2011) Marine Debris Program. Information on Marine Debris. Office of Response and Restoration: NOAA's National Ocean Service. (http://marinedebris.noaa.gov)

Novak, D.J. and Crane, D.E. (2002) Carbon storage and sequestration by urban trees in the USA. *Environmental Protection*, 116, 381–389.

Pandey, A.K., Pandey, M., Mishra, A., Tiwary, S.M., and Tripathi, B.D. (2015) Air pollution tolerance index and anticipated performance index of some plant species for development of urban forest. *Urban Forestry and Urban Greening*, 14, 866–871.

Pathak, A.K., Kumar, R., Kumar, P., and Yadav, S. (2015) Sources apportionment and spatio–temporal changes in metal pollution in surface and sub-surface soils of a mixed type industrial area in India. *Journal of Geochemical Exploration*, 159, 169–177.

Ramachandra, T.V., Aithal, B.H., and Sreejith, K. (2015) GHG footprint of major cities in India. *Renewable and Sustainable Energy Reviews*, 44, 473–495.

Reddy, M.S., Basha, S., Adimurthy, A., and Ramachandraiah, G. (2006) Description of the small plastics fragments in marine sediments along the Alang-Sosiya ship

breaking yard, India. *Estuarine, Coastal and Shelf Science*, 68, 656–660.
Rohra, H. and Taneja, A. (2016) Indoor air quality scenario in India – An outline of household fuel combustion. *Atmospheric Environment*, 129, 243–255.
Saha, J.K., Panwar, N., Srivastava, A., Biswas, A.K., Kundu, S., and Subba Rao, A. (2010) Chemical, biochemical, and biological impact of untreated domestic sewage water use on Vertisol and its consequences on wheat (*Triticum aestivum*) productivity. *Environmental Monitoring and Assessment*, 161, 403–412.
Sarkar, P.K. and Tagore, P. (2011) An approach to the development of sustainable urban transport system in Kolkata. *Current Science*, 100, 1349-1361
Satheeshkumar, P., Manjusha, U., Pillai, N.G.K., and Senthil Kumar, D. (2012) Puducherry mangroves under sewage pollution threat need conservation. *Current Science*, 102, 13–14.
Sharma, R.K., Agrawal, M., and Marshall, F.M. (2008) Heavy metal (Cu, Zn, Cd and Pb) contamination of vegetables in urban India: A case study in Varanasi. *Environmental Pollution*, 254–263.
Sharma, R. and Joshi, P.K. (2014) Identifying seasonal heat islands in urban settings of Delhi (India) using remotely sensed data – An anomaly based approach. *Urban Climate*, 9, 19–34.
Sharma, S.K., Mandal, T.K., Saxena, M., Rashmi, Rohtash, Sharma, A., and Gautam, R. (2014) Source apportionment of PM_{10} by using positive matrix factorization at an urban site of Delhi, India. *Urban Climate*, 10, 656–670.
Shirgaokar, M. (2016) Expanding cities and vehicle use in India: Differing impacts of built environment factors on scooter and car use in Mumbai. *Urban Studies*, 53, 3296–3316.
Shrivastav, R.K., Saxena, N., and Gautam, G. (2013) Air pollution due to road transportation in India: A review on assessment and reduction strategies. *Journal of Environmental Research and Development*, 8, 69–77.
Singh, S. and Kumar, M. (2006) Heavy metal load of soil, water and vegetables in peri-urban Delhi. *Environmental Monitoring and Assessment*, 120, 79–91.
Singh, R.B. and Grover, A. (2015) Spatial correlations of changing land use, surface temperature (UHI) and NDVI in Delhi using Landsat Satellite Images. In: Singh, R.B. (ed.), *Urban Development Challenges, Risks and Resilience in Asian Mega Cities.* Advances in Geographical and Environmental Sciences; Springer, Japan, pp. 83–97.
Sinha, R.K., Sinha, S.K., Kedia, D.K., Kumari, A., Rani, N., Sharma, G., and Prasad, K. (2007) A holistic study on mercury pollution in the Ganga River System at Varanasi, India. *Current Science*, 92, 1223–1228.
Times of India, 26 April 2016: http://timesofindia.indiatimes.com/india/Mumbai–noisiest–city–Delhi–at–number–4–Central–Pollution–Control–Board/articleshow/51985961.cms
Tue, N.M., Takahashi, S., Subramanian, A., Sakai, S., and Tanabe, S. (2013) Environmental contamination and human exposure to dioxin-related compounds in e-waste recycling sites of developing countries. *Environmental Science: Process & Impacts*, 15, 1326–1331.
UN-Habitat, 'State of the World's Cities Report 2012/2013: Prosperity of 854 Cities', Website: www.unhabitat.org
Vreeland, H., Schauer, J.J., Russell, A.G., Marshall, J.D., Fushimi, A., Jain, G., Sethuraman, K., Verma, V., Tripathi, S.N., and Bergin, M.H. (2016) Chemical characterization and toxicity of particulate matter emissions from roadside trash combustion in urban India. *Atmospheric Environment*, 147, 22–30.
World Health Organization (WHO) (2011) *Burden of Disease from Environmental Noise, Quantification by Healthy Life Years Lost in Europe*. WHO Regional Office for Europe, Denmark.
Wong, M.S., Nichol, J.E., To, P.H., and Wang, J. (2010) A simple method for designation of urban ventilation corridors and its

application to urban heat island analysis. *Building and Environment*, 45, 1880–1889.

Yadav, I.C., Devi, N.L., Syed, J.H., Cheng, Z., Li, J., Zhang, G., and Jones, K.C. (2015) Current status of persistent organic pesticides residues in air, water, and soil, and their possible effect on neighboring countries: A comprehensive review of India. *Science of the Total Environment*, 511, 123–137.

Yadav, R., Sahu, L.K., Beig, G., and Jaaffrey, S.N.A. (2016) Role of long-range transport and local meteorology in seasonal variation of surface ozone and its precursors at an urban site in India. *Atmospheric Research*, 176–177, 96–107.

Yang, J., McBride, J., Zhou, J., and Sun, Z. (2005) The urban forest in Beijing and its role in air pollution reduction. *Urban Forest & Urban Greening*, 3, 65–78.

27

Urban Aquatic Pollution in Brazil

Felippe Fernandes[1], Paulo Roberto Bairros Da Silva[2], Cristiano Poleto[3], and Susanne M. Charlesworth[4]

[1] *University of Sao Paulo, Brazil*
[2] *Federal University of Santa Maria, Brazil*
[3] *Universidade Federal do Rio Grande do Sul, Brazil*
[4] *Centre for Agroecology, Water and Resilience, Coventry University, United Kingdom*

27.1 Introduction

With a population of more than 180 million and covering almost half of South America, Brazil is the world's fifth largest country in terms of both area and population (IBGE, 2010). Brazil currently has a high urbanisation rate, with around 80% of Brazilians living in urban areas. This creates significant social and environmental issues in and around these cities. São Paulo is one of the world's largest urban areas and is known for pollution, overcrowding, and poverty.

Outside of the urban areas in Brazil, there is limited infrastructure and there are low-income inhabitants. The impacts on water resources are severe, mainly because of the absence of sewage treatment systems, trash collection, and non-paved streets, which are important sources for the production of sediments in urban watersheds. On the other hand, there are also the socioeconomic and cultural components, which are closely related to the absence of environmental knowledge.

One of the greatest problems observed in Brazilian urban watersheds is the amount of solid residue, domestic sewerage, and sediments that is disposed of in rivers and streams that drain those areas (Poleto *et al.*, 2009). While sediments are essential to the functioning of aquatic ecosystems, they have long been recognised as the ultimate repository of most of the contaminants discharged into waterbodies.

It is widely accepted that sources of contaminants in Brazilian watersheds, such as organic (polycyclic aromatic hydrocarbons (PAHs) and aliphatic hydrocarbons) and inorganic pollutants (metals and metalloids) are the result of numerous human activities. In this regard, therefore, Brazil is no different from any other urban area worldwide, and has a clear need for continued scientific dialogue around the ecological risk that these contaminants might pose to aquatic biota. However, sediment pollution is not the only serious environmental problem in Brazil. The failure of a Samarco company iron mine tailings dam in Minas Gerais state in November 2015 resulted in what was likely Brazil's biggest environmental disaster ever – pouring 50 million tons of iron ore and toxic waste into the Rio Doce, killing 19 people, and impacting more than a million people. It poisoned the river along its entire 853 km (530 mile) length, immediately increased turbidity in the Rio Doce to 12,000

Urban Pollution: Science and Management, First Edition. Edited by Susanne M. Charlesworth and Colin A. Booth.

times higher than that allowed for potable uses, and reduced oxygen levels suddenly to 1 mg L^{-1}, which caused a fish kill of several tons as well as the death of many other biota (de Oliveira Neves *et al.*, 2016). Attempts to contain the disaster were unsuccessful, and 16 days after the intial failure, the mud flow had reached the Atlantic Ocean, where it severely impacted all marine life (Miranda and Marques, 2016).

This chapter offers a review of the current environmental assessment approaches in use around three Brazilian states (Figure 27.1) and some key issues in the field of sediment contamination, identifying areas requiring additional future attention.

27.2 Current Brazilian Environmental Regulations

In Brazil, since 1986, the protection of freshwater, estuarine, and marine waters against pollution has been based on Resolution No. 20 from the National Council for the Environment – CONAMA. On 13 May 2011, CONAMA issued Resolution No. 430, on the conditions and standards of effluent discharges addressing wastewater treatment systems and industrial discharges. Resolution No. 430 amended the existing effluent standards of Resolution No. 357/2005, which also extended to the classification and ecological management of water bodies.

Resolution No. 430 established standards for the discharge of effluents from sanitary sewers, consisting of residential, commercial, and publicly collected liquid wastes which can include some industrial discharges. Wastewater treatment systems that discharged directly into the ocean through submarine pipes were subject to a distinct set of standards. For industrial pollution sources, this resolution imposed a new regime of obligatory self-monitoring and testing. The requirements included collection of samples by trained professionals and their testing by laboratories specially accredited by the National Institute of Metrology, Standardization and Industrial Quality (INMETRO).

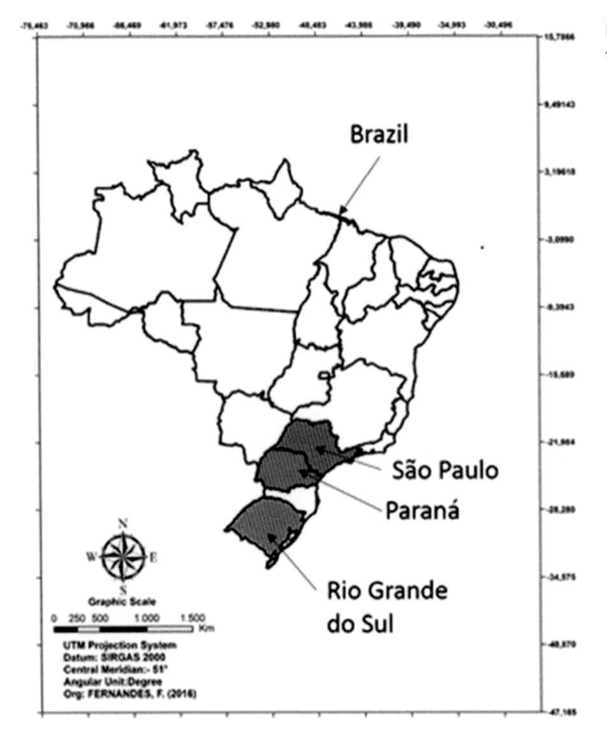

Figure 27.1 A map of Brazil, with the states focused on in this chapter shaded in grey.

On 28 December 2009, following three years of debate, the Brazilian National Environmental Council (CONAMA) issued Resolution No. 420, establishing federal standards for the environmental management of contaminated sites. The resolution provided state and municipal environmental agencies with a framework of guidelines for the management of remediation programmes. It also contained monitoring and reporting requirements that applied to Brazilian facilities. Subject to implementation by state agencies, all facilities with the potential to pollute may be required to establish soil monitoring programmes and submit technical reports on the results with each renewal of their environmental licences.

The core of the new federal standards is a multi-stage process under which potentially contaminated sites were to be identified, investigated, classified, remediated, and monitored. Responsible parties must submit to the appropriate environmental agency a plan that addresses:

- The control and elimination of the sources of contamination
- The current and future use of the area
- An evaluation of risks to human health
- Intervention alternatives considered technically and economically viable
- A monitoring programme
- Costs and time frames for implementing the intervention alternatives

The resolution also created technical criteria for use by environmental agencies, setting reference values for contaminants and procedures for determining the analytical methods to be employed by state environmental agencies. The Federal Environmental Agency (IBAMA) was also directed to create a National Database of Contaminated Sites using information obtained by state agencies.

Environmental agencies of each Brazilian state had to list the different soils in their territory and establish reference values by 2013, providing crucial information to identify contaminated areas and carry out interventions. Until now, states such as São Paulo (CETESB, 2005), Pernambuco (Biondi, 2010), and Minas Gerais (COPAM, 2010) have already carried out studies for soil reference values.

Juchen *et al.* (2014) compared the local background concentrations for trace elements in two different sets of soils from the states of Paraná and Rio Grande do Sul, in the southern region of Brazil. The authors concluded that trace element levels varied from location to location, particularly due to different classes of soils and/or parent materials. Poleto and Gonçalves (2006) reported that the specificity of each reference value is also clear when comparing the thresholds established by different guidelines.

In 2005, the São Paulo Environmental Agency (CETESB) published *Guideline Values for Soils and Groundwater*, including quality reference values (QRVs) obtained from background concentrations of trace elements in soils from the state. As well as QRVs, CETESB proposed prevention and intervention values (PRVs), above which heavy metal levels indicate potentially polluted soil and a potential risk to human health. QRVs for soils in Brazil and individual states are given in Table 27.1.

In regard to sediment quality assessment, Brazil still does not have regulations based on quality criteria for sediments. However, given the contamination of reservoirs, rivers, estuaries, and coastal areas, sediment quality evaluation started to receive more attention from scientists over the last two decades, as a means to promote conservation and remediation criteria. According to Poleto *et al.* (2009), new studies of urban sediments should provide a means of formulating management strategies focused on the way in which polluted sediment is transported in the urban environment, particularly from the perspective of Brazilian cities.

Brazilian sediment quality criteria applied to dredged material management are given by Resolution No. 344/2004 from CONAMA. However, such values were established on the basis of American and Canadian sediment

Table 27.1 Quality reference values (QRVs) for trace elements of Brazil and regional background values for Pernambuco, São Paulo, and Minas Gerais states.

	As	Cd	Ba	Cr	Cu	Ni	Pb	Sb	Se	Zn
Background (Reference value)	**mg kg^{-1} dry wt**									
Pernambuco (Biondi, 2010)	0.6	0.6	84	35	5	8.5	12	0.1	0.4	34.5
São Paulo (CETESB, 2005)	3.5	<0.5	75	40	35	13	17	<0.5	0.2	60
Minas Gerais (COPAM, 2010)	8	<0.4	93	75	49	21.5	19.5	0.5	0.5	46.5
Brazil (CONAMA, 2009)	15	1.3	150	75	60	30	72	2	5	300

quality guidelines (SQGs) and do not consider toxicity tests or contaminant bioaccumulation. Some examples of QRVs for metals in dredged materials are given in Table 27.2.

27.3 Ecological Risk Assessment Approaches in Brazil

Despite the problems associated with effluent discharges, contaminated sites, and water quality standards, ecological risk assessment approaches have only recently been introduced in South American countries, and detailed guidance on how to interpret and apply these frameworks is still generally inadequate. Usually, Brazilian studies are carried out based on the USEPA (2000) framework. A literature search of ecological risk assessment in Brazil revealed an increase in the amount of research carried out in the last five years, especially in São Paulo state. From 2005 to 2010, a total of 3443 papers were published; from 2010 to the present, the total was 6735.

Table 27.2 Quality reference values (QRVs) for dredged materials (μg g^{-1}) established by Resolution No. 344/2004, CONAMA.

		Classification levels of dredged material (in dry weight unit)			
		Freshwater		**Saline/brackish water**	
Pollutants		**Level 1**	**Level 2**	**Level 1**	**Level 2**
Metals and arsenic (mg/kg)	Arsenic (As)	5.9[a]	17[a]	8.2[b]	70[b]
	Cadmium (Cd)	0.6[a]	3.5[a]	1.2[b]	9.6[b]
	Lead (Pb)	35[a]	91.3[a]	46.7[b]	218[b]
	Copper (Cu)	35.7[a]	197[a]	34[b]	270[b]
	Chromium (Cr)	37.3[a]	90[a]	81[b]	370[b]
	Mercury (Hg)	0.17[a]	0.486[a]	0.15[b]	0.71[b]
	Nickel (Ni)	18[a]	35.9[c]	20.9[b]	51.6[b]
	Zinc (Zn)	123[a]	315[a]	150[b]	410[b]

a) Environmental Canada (1995)
b) Long *et al.* (1995)
c) FDEP (1994)

27.4 Environmental Studies in Brazil

Many of the studies carried out on urban aquatic systems in Brazil have identified the role of wastewaters and sewage in their contamination. This is reflected in high concentrations of metals, organic matter, and, at times, faecal matter mainly due to the inadequate means of treating such material at wastewater treatment plants. The following section therefore gives some examples of studies carried out in various states in Brazil which quantified the concentrations and identified the sources of some of this contamination.

27.4.1 The State of Paraná

Of macronutrients in urban water systems, generally two of them stand out: ammonia and phosphorus, as these are usually present in high concentrations and are associated with the intake of effluents (Froehner and Martins, 2008; Bem *et al.*, 2015). Ammonia is the final product of nitrogen mineralisation and can be generated by the activity of heterotrophic organisms through various biochemical reactions known as ammonification (Esteves and Amado, 2011; Santos *et al.*, 2012). In aqueous solution, it can be found as ammonium ion (NH_4^+) and ammonia gas (NH_3). Such chemical species are converted into each other, and the sum of their concentrations is the total ammonia or ammoniacal nitrogen (Santos *et al.*, 2012; Campanha *et al.*, 2014). In small concentrations, their presence is considered normal in urban aquatic environments due to natural degradation processes. However, large concentrations can be associated with contributions from domestic and industrial untreated sewage (Froehner and Martins, 2008; Bem *et al.*, 2015).

Phosphorus is a nutrient that acts as a regulator component of primary productivity in aquatic ecosystems, and although phosphates do exist in different forms in water systems, ortho-phosphate (P–ortho) assumes greater importance because it is the main form assimilated by aquatic plants, microalgae, and bacteria (Esteves and Panosso, 2011; Pantano *et al.*, 2016). Nutrients can cause deleterious effects on urban aquatic ecosystems when in excessive concentrations (Pompêo and Moschini-Carlos, 2012). Excess nutrients cause eutrophication of urban waters whereby algae proliferate, and then after death, a significantly increased number of microorganisms break down the dead algae, dramatically reducing the amount of dissolved oxygen, leading to deteriorating water quality (Pompêo *et al.*, 2015).

Using florescence and molecular analysis of dissolved organic carbon, Kramer *et al.* (2015) confirmed the presence of high concentrations of labile organic matter from domestic wastewater in several urban rivers (Iraí, Barigüi, Iguaçu, Bethlehem, and Atuba) in the Alto Iguaçu basin in the metropolitan region of Curitiba in Paraná. Once again, inflow of domestic effluents were thought to be complicit in reducing water quality due to the presence of ortho-phosphate, total phosphorus, nitrite, and ammoniacal nitrogen (Kramer *et al.*, 2015). In a similar study, of the Barigüi River, Curitiba, Frehner, and Martins (2008) evaluated total organic carbon, total nitrogen concentrations, ortho-phosphate, and total phosphorus in sediment samples. They found that faecal material from the presence of sewage contributed significantly to the organic carbon content, with organic nutrients thought to be input from domestic and industrial effluents. Bem *et al.* (2015) found similar conditions prevalent in the Iguaçu River, state of Paraná, Southern Brazil. The amounts of dissolved organic carbon in all its oxidation states were associated with the presence of domestic sewage.

The presence of metallic species in the sediments of the urban Barigüi River, which crosses municipalities in the metropolitan region of Curitiba, an urbanised area with high population and industrial densities located east of the state of Paraná, southern Brazil, was evaluated by Froehner and Martins (2008). They identified the presence of high concentrations of Zn, Pb, Ni, and Cr in the study area, associated with contributions from domestic and industrial effluents. Da Silva *et al.* (2016) sampled superficial sediment samples from the same river, finding high concentrations of Zn and Pb in comparison to regional background. However, Ni was not detected.

Juchen *et al.* (2016) studied the potential enrichment of urban sediments by metal species in the city of Toledo located west of the state of Paraná, southern Brazil. They focused on various urban elements in the Jacutinga, Panambi, Drake, Barro Preto, and Toledo urban rivers, finding them adversely impacted due to human activities and containing high levels of metals as a result. In comparison with background values, they reported percentage enrichments of individual elements as follows: Mg (18.4%), Ni (18.6%), Ba (39.3%), Cu (44.3%), Cr (56.3%), Zn (140.2%), and Na (295%). According to the calculated enrichment factor and pollution index (Chapter 7), some elements were classified as significantly enriched. This included Zn, which was probably associated with the pig production industry present in the area. Overall, it was found that the sediments were enriched in the typically urban heavy metals Zn, Cr, and Cu. The main element of concern in all of these studies would appear to be Zn, closely followed by Pb and Cu. Scheffer *et al.* (2007) evaluated the factors that govern the speciation of Cu in urban aquatic environments from the Curitiba region, specifically in the Iraí, Belem, Iguaçu, and Barigüi rivers. In common with many other studies of heavy metal speciation, it was found that Cu was preferentially attached to particulates, and that its transfer between dissolved and particulate phases was governed by pH the presence of organic matter in the environment.

Mining has also had negative impacts on the quality of urban rivers, with Melo *et al.* (2012) stating that the mining and metallurgy industries are the most environmentally concerning in the region. This has led to levels of Pb of up to 795.3 $\mu g\ L^{-1}$ in river water and 24,300 $mg\ kg^{-1}$ in sediments of the Alto Vale do Ribeira, in the municipality of Adrianópolis, Paraná. However, as reported by Abessa *et al.* (2014), there is some evidence that this contamination may be reducing since mining ceased in the 1990s. Historical tailings dumped along the riverbanks, related to the original 89,000 m^3 of waste material, still has the potential to release Pb after storms, which will contribute to contamination of the surrounding area and the downstream environment.

In recent years, the presence of new and emerging pollutants (NEPs) in urban water systems has gained importance (e.g. Mozeto and Zagatto, 2008). NEPs include medicines, antibiotics, endocrine disrupters (hormones), toiletries, beauty products, illegal drugs, nanomaterials, insect repellents, and sunscreens (Cunha *et al.*, 2016). They are not monitored by traditional water quality evaluation programmes, and are generally believed to not be removed by traditional processes in water and sewage treatment plants. The main source of these compounds in municipal water systems is linked to sewage, treated and untreated, contributions which eventually return for human consumption due to the low effectiveness of NEP water treatment processes (Kramer *et al.*, 2015). In research conducted on water and sediment samples from five urban rivers located in the Alto Iguaçu basin, Kramer *et al.* (2015) confirmed the presence of the anti-inflammatory drugs diclofenac, ibuprofen, and paracetamol. Measured using gas chromatography, diclofenac and ibuprofen were found in concentrations of up to 285 $ng\ L^{-1}$ and 729 $ng\ L^{-1}$ in water samples, respectively. Paracetamol was found in relatively high concentrations associated with sediments, reaching 6,896 $g\ g^{-1}$ in the Barigüi River. Unplanned urbanisation was thought to be part of the problem associated with the presence of these NEPs. Osawa *et al.* (2015) conducted a study in the Upper Iguaçu basin, known to be particularly affected by industrial and domestic effluents. They focused on quantifying the presence of the antihypertensive drugs metoprolol, propranolol, and nadolol in the surface waters of three tributaries of Alto Iguaçu using GC-MS. The Atuba river had the highest concentrations of these drugs, with 4658.2, 3877.91, and 123.8 ng L^{-1} of metoprolol, propranolol, and nadolol, respectively, found downstream of the Atuba-Sul wastewater treatment plant. This illustrates the lack of treatment of these compounds at the treatment plant, but also the interaction of these drugs with the larger environment: whether they are transported to groundwater, contaminate potable water supplies, or impact the biosphere is largely unknown, and merits further study.

27.4.2 The State of Rio Grande do Sul

The water resources of the Rio Grande do Sul State have been the subject of numerous studies over the past decades, with regard to morphological sedimentological, physical, chemical, and biological aspects. Currently, many studies have been conducted in this state and point to the production and enhancement of pollutants in urban watersheds, reflecting the health of the lacustrine ecosystems that are part of them. Usually, those pollutants are carried by run-off and/or urban drainage systems, having as their ultimate destination the waterbodies in which they cause degradation.

The degradation processes in urban waterbodies by pollutants are caused mainly through the three main producing sources: industrial, residential, and rural pollution. Additionally, it occurs through the transport, deposition, accumulation, and release of those pollutants and aggregated particles.

The distribution and classification of particle sizes is an important parameter for the characterisation of the constituent elements of the samples. Granulometric techniques are employed for characterisation of materials with various sources, such as industrial, pharmaceutical, chemical, and food, as well as in soils and sediments (Bortoluzzi *et al.*, 2006). For sediments, the analysis of particle size assists studies on particle aggregation, sediment transport in rivers, dissolution of fine particles (≥63μm), and analysis of sources of urban sediments.

Most studies in urban watersheds of Rio Grande do Sul show that sediments accumulated in urban impermeable areas, such as drainage systems, streets, and avenues, are the main source of sediments during rainy periods (Charlesworth *et al.*, 2003; Poleto and Merten, 2008). Poleto *et al.* (2009) studied the sub-urban watershed in the municipality of Viamão/RS, using fingerprinting methodology, finding that, on average, 46% of river sediments in suspension originate from paved areas, 23% from unpaved roads (dirt roads or just unpaved roads), and 31% originate from the river channel itself due to the action of erosion because of the hydrological changes caused by urban anthropogenic interventions, totalling 100% of sediments found in a river and its origins.

Martínez (2010) and Fernandes & Poleto (2016) conducted a study on the enrichment of urban sediments by heavy metals in the city of Porto Alegre/RS. The author noted that the concentrations of heavy metals found in the sediment exceeded those adopted as local background, thus indicating enrichment of natural concentrations resulting from human activities.

Lucheta *et al.* (2010) reviewed the quality of sediments (organic matter, xenobiotics, and metals) in the Gravataí River, which covers the municipalities of Santo Antonio da Patrulha, Glorinha, Gravataí, Cachoeirinha, Alvorada, Viamão, Canoas, Porto Alegre, and part of Taquara municipal areas, applying toxicological testing protocols using *Daphnia magna*. They identified inadequate quality conditions for the local lake ecosystem, with toxicology tests showing that the survival and reproduction of some organisms were affected. It is likely, therefore, that the Gravataí River is contaminated, with previous studies of physical and chemical parameters between 1992 and 1994 including BOD, COD, turbidity, total solids, nitrogen, and phosphate indicating decreasing water quality downstream due to contributions from household, industrial, and rural untreated effluents. According to Santos *et al.* (2008), discharges of sewage into waterbodies have become one of the greatest environmental problems in Brazil. They state that only about 14% of the cities have any sort of sewage treatment, and that is mostly based on activated sludge processes. In a study of Vieira creek, which discharges into Mangueira Bay, they found that treated effluent from the wastewater treatment plant located along the creek was an important source of nutrients; in particular, ammonium was of prime concern.

Savage *et al.* (2004) showed that domestic wastewater is high in nitrogen and can trigger eutrophication. According to Routh *et al.* (2004), organic matter is an important component of sediment due to its association with the biota, nutrient cycles, and geochemical processes. Van Metre and Mahler (2003) showed that there is a wide range of pollutants arising in urban areas, from organic compounds to highly toxic metals. Many of these contaminants are particulate-bound and are fairly widespread in urbanised areas, leading to reduced water quality in waterbodies. However, three contaminants groups stand out and are often identified in urban aquatic sediments, showing toxicity to aquatic biota: heavy metals, hydrocarbons, and organochlorine compounds.

According to Porto *et al.* (2004), the study and understanding of the dynamics of the process of urbanisation and the production and quality of the sediments are of great importance, since they serve as a starting point to develop measures aimed at maximising the efficient use of available water resources. In terms of designing integrated water resource management, the risk from soil, riverbed, river ecosystem, and estuary degradation, or sediment contamination, greater attention needs to be given to problems arising from the changes that result from continued urbanisation in Brazil.

27.4.3 The State of São Paulo

Urban watercourses in São Paulo state suffer from the same environmental problems as those discussed for both Rio Grande Do Sul and Paraná. This is reflected in a study by Campanha *et al.* (2014), in which superficial samples and sediment cores were taken from the urban Preto River during both wet and dry seasons. The results demonstrated anthropogenic impacts with the discovery of high Zn concentrations, similar to other sites in Brazil. These results are important for the nearly half a million people living in São José do Rio Preto since the river provides almost 30% of the public water supply to the city.

The Piracicaba River basin is the most contaminated river basin in the state because of the amount of agricultural deposits such as herbicides and fertilisers draining from the extensive sugarcane plantations into the river. The area is also highly urbanised and industrialised with sewage treatment at an early stage of development, resulting in relatively high levels of organic and inorganic contaminants being discharged into the river system, among which heavy metals, in both water and sediment, have been found with concerning levels of bioavailability (Rodgher *et al.*, 2005). In an evaluation of the potential for this contamination to affect biota, Piña *et al.* (2009) tested the blood of 37 Geoffroy's side-necked turtles, *Phrynops geoffroanus*, for 13 heavy metals. Blood As, Co, Cr, Se, and Pb levels varied between sites, whereas Sn appeared to vary according to the gender of the turtles. While the serum level of Cu and Pb (2194 and 1150 ng g^{-1}, respectively) were the highest ever found in any reptile, there were no obvious signs of resultant pathological effects presenting clinical symptoms.

Da Silva *et al.* (2002) undertook a study of the speciation of heavy metals in the sediments of the Tietê–Pinheiros river system – a large system that passes through the metropolitan area of São Paulo. They used a three-step sequential extraction protocol to determine the availability of the metals in the environment and also their potential release if environmental conditions were to change. The study focused on four reservoirs: Billings, Pirapora, Rasgão, and Barra Bonita, the latter located 270 km downstream of the metropolitan area of São Paulo, but all connected by the Tietê–Pinheiros river system. It was found that the Billings, Pirapora, and Rasgão reservoirs were substantially contaminated with heavy metals, with Cu, Pb, Cd, and Ni being of particular concern due to their potential availability should environmental conditions change. These elements were associated with acid soluble and reducible phases, which would suggest that they could be released with changes of pH or Eh. Barra Bonita, on the other hand, while still contaminated with high levels of metals in terms of their total concentrations, did not exhibit associations that would be readily released, since they were mostly bound to the residual fraction. This indicates they may have been preferentially transported to the reservoir in association with fine particulate material.

A more recent study of the same river system (Rocha *et al.*, 2010), but this time examining the distribution of PAHs and of AhR-mediated toxicity, confirmed that the source of most of the toxicity was from São Paulo and that pollutants accumulated downstream such that each reservoir is degraded in turn. Sediment PAH concentrations were high, comparable to those found in studies of contaminated European rivers, such as the upper Danube in Germany.

Concentrations of NEPs, such as the pharmaceuticals and endocrine disruptors acetaminophen, acetylsalicylic acid, diclofenac, ibuprofen, caffeine, 17β-estradiol, estrone, progesterone, 17α-ethynylestradiol, levonorgestrel, diethylphthalate, dibutylphthalate, 4-octylphenol, 4-nonylphenol, and bisphenol A, were investigated in the Atibaia River, São Paulo State, by Montagner and Jardim (2011). The river is the main source for potable water for Campinas City, and thus it is of concern that 10 of the 15 NEPs were found at least once, with caffeine recorded at a high value of 127 μg L^{-1}, although ibuprofen was not detected in any of the samples. Of the endocrine disruptors, dibutylphthalate was most common, as well as bisphenol A; the hormones 17b-estradiol and 17a-ethynylestradiol were also present in many of the samples. The trend in contaminants along the Atibaia River was for an increase in their presence, mainly associated with the City of Campinas, with a gradual decrease downstream. Once again, the presence of these NEPs indicate the lack of suitable sewage treatment from centres of population, with caffeine being shown as a reliable marker able to indicate the potential presence of other contaminants of interest.

27.5 A Case Study of Curitiba, Paraná

The Global Sustainable City Award was given to Curitiba, Paraná, in 2010, a city of some 2 million people (Figure 27.2). The award was given because those cities excelled in sustainable urban development. Curitiba is a city with highly ambitious plans to improve urban life, with boulevards extending from central areas with public amenities and industrial districts including pedestrianisation, strict controls on urban development, and an affordable and efficient public transport system (Rabinovitch and Leitmann, 1993). The latter was key to its sustainable development as were the provision of green spaces of 52 m^2 per person and addressing aspects of waste and water pollution, with parks located along rivers including artificial lakes, which are able to absorb and retain water, minimising floods, and also providing treatment of the stormwater, thus potentially improving water quality.

The metropolitan region of Curitiba lies in the Upper Iguaçu basin comprising five main rivers of which the Iguaçu is the most well known. In crossing the city, it becomes polluted by industrial activity, which releases up to 30 tons/month of toxicants, including domestic sewage (Martínez *et al.*, 2016). The Caximba Landfill also released about 240 million litres/year of slurry into the river, leading to noxious smells and a darkening of the river water, which breached environmental legislation standards 60 times.

Figure 27.2 Curitiba street (public domain).

In terms of sewage (Chapter 9), this has been of prime concern in the contamination of Brazil's urban environment, with about 60 tons of raw domestic sewage from the upper Iguaçu discharged into the river. In fact, pollution of the river became so serious that the Iguaçu was recognised as Brazil's second most contaminated river by the Sustainable Development Indices in 2008, 2010, and 2012. In the metropolitan region of Curitiba, while 91% of households have connection to the public water supply, only 58% are connected to the sewer network, and only 35% of sewage that is collected has any form of treatment before it is disposed of into rivers (Macedo, 2004). Markers can be used to track this contamination, and this technique was used by Froeher *et al.* (2010) in the Barigüi River, a tributary of the Iguaçu. Human habitation markers are generally those that are eliminated by the body and include substances such as caffeine and coprostanol. The latter is a breakdown product of cholesterol, and both are excreted in urine; thus, they can be used as markers, or tracers, of untreated sewage in the environment. Results of the study found that there was a strong correlation between caffeine and coprostanol, and also between BOD and nitrate. The limit for coprostanol concentration in sediments for it to be considered polluted is >0.1 $\mu g\ g^{-1}$; thus, with a range of 0.08 to 196 $\mu g\ g^{-1}$ the sediments of the Barigüi River were considered to be contaminated. The use of caffeine as a marker for hormones in water with oestrogen effects was reported in Orsi (2013), where it was found in 20 Brazilian state capitals, including Curitiba, in high concentrations. The highest levels of caffeine in water supplies were from Porto Alegre (2,257 ng L^{-1}), Campo Grande (900 ng L^{-1}), and Cuiabá (222 ng L^{-1}). While 25 parks were designed in the urban area of Curitiba to include lakes in order to retain excess surface water, they were not sufficient to prevent flooding, and with the municipal Drainage Master Plan still being revised in spite of being in existence for several years, water management has not been implemented.

27.6 Conclusions

Sediment ecological risk assessment is widely used and will continue to be used to protect the environment and prioritise remedial actions around the world. As sediment pollution continues to grow at a phenomenal pace, Brazilian environmental authorities should establish a standard framework for risk assessment in sites posing some risk, taking into account the great variability of biomes and Brazil's enormous territory. Experience can be acquired with the system by testing the USEPA (2000) basic approach in practical situations at a number of characteristic sites, aiming to provide important information to help regular utilisation of risk assessment processes to support site restoration and reclamation decisions in Brazil.

Brazil is a country that has yet to solve issues around the provision of basic sanitation. Cunha *et al.* (2016) quote a figure of 82.7% in terms of provision of safe running water for the population. Probably of greater concern is the 51.7% who are not connected to a sewage system, and of the amount of sewage produced, only 38.7% receives any treatment before it is dumped into receiving waterbodies. It is therefore unsurprising that urban rivers are contaminated due to the impacts of human activity, and it is rather unrealistic to expect any improvement before basic human needs are met.

While the Global Sustainable City Award was given to Curitiba, Paraná, in 2010, the highly ambitious plans to improve urban life remain unfinished, with drainage, in particular, not achieved, and disposal of sewage into local rivers still remaining an issue that needs urgent attention.

References

Abessa, D.M.S., Morais, L.G., Perina, F.C., Davanso, M.B., Rodrigues, V.G.S., Martins, L.M.P., and Sigolo, J.B. (2014) Sediment geochemistry and climatic influences in a river influenced by former mining activities: The case of Ribeira de Iguape River, SP–PR, Brazil. *Open Journal of Water Pollution and Treatment*, 1, 43–53.

Bem, C.C., Higuti, J., and Azevedo, J.C.R. (2015) Qualidade da água de um ambiente lótico sob impacto antropogênico e sua comunidade bentônica. *Revista Brasileira de Recursos Hídricos*, 20 (2), 418–429.

Biondi, C.M. (2010) *Teores naturais de metais pesados nos solos de referência do Estado de Pernambuco*. 70 p. Tese de Doutorado. Universidade Federal Rural de Pernambuco, Recife.

Bortoluzzi, E.C., Poleto, C., and Merten, G. (2006) Metodologias para estudos de sedimentos: ênfase na proporção e na natureza mineralógica das partículas. *Qualidade de Sedimentos. Porto Alegre, ABRH*, 80–140.

Campanha, M.B., Romera, J.P., Coelho, J., Pereira-Filho, E.R., Moreira, A.B., and Bisinoti, A.C. (2014) Use of Chemometric tools to determine the source of metals in sediments of the rivers of the Turvo/Grande hydrographical basin, São Paulo State, Brazil. *Brazilian Chemical Society*, 25, 665–674.

CETESB – São Paulo Environmental Agency. (2005) Board Decision No. 195–2005–E, November 23, 2005. Guiding Values for Soils and Groundwater in the State of São Paulo,. http://www.cetesb.sp.gov.br/Solo/relatorios/tabela_valores_2005.pdf. In Portuguese.

Charlesworth, S.M., Everett, M., McCarthy, R., Ordoñez, A., de Miguel, E. (2003) A comparative study of heavy metal concentration and distribution in deposited street dusts in a large and a small urban area: Birmingham and Coventry, West Midlands, UK. *Environment International*, 29, 563–573.

CONAMA. (2004) Brazilian National Environmental Council. Resolution N° 344 of May 7, 2004. http://www.mma.gov.br/port/conama/legiabre.cfm?codlegi=445. In Portuguese.

CONAMA. (2005) Brazilian National Environmental Council. Resolution N° 357 of March 17, 2005. http://www.mma.gov.br/port/conama/res/res05/res35705.pdf. In Portuguese.

CONAMA. (2009) Brazilian National Environmental Council. Resolution N° 420 of December 28, 2009. http://www.mma.

gov.br/port/conama/legiabre.cfm?codlegi=620. In Portuguese.

CONAMA. (2011) Brazilian National Environmental Council. Resolution N° 430 of May 13, 2011. http://www.mma.gov.br/port/conama/legiabre.cfm?codlegi=646. In Portuguese.

COPAM. (2010) State Council of Environmental Politics. Normative Resolution COPAM/CERH n° 02, from September 08, 2010, Republication–Daily Executive Report. http://www.siam.mg.gov.br/sla/download.pdf?idNorma=14670. In Portuguese.

Cunha, D.L., Silva, S.M.C., Bila, D.M., Oliveira, J.L.M., Sarcineli, P.N., and Larentis, A.L. (2016) Regulation of the synthetic estrogen 17α-ethinylestradiol in water bodies in Europe, the United States, and Brazil. *Cadernos de Saúde Pública*, 32 (3), 01–06.

da Silva, I.S., Abate, G., Lichtig, J., and Masini, J.C. (2002) Heavy metal distribution in recent sediments of the Tietê–Pinheiros river system in São Paulo state, Brazil. *Applied Geochemistry*, 17, 105–116.

da Silva, P. R. B., Makara, C.N., Munaro, A.P., Schnitzler, D.C., Wastowski A.D., and Poleto, C. (2016) Comparison of the analytical performance of EDXRF and FAAS techniques in the determination of metal species concentrations using protocol 3050B (USEPA), *International Journal of River Basin Management*, 14 (4), 401–406. doi: 10.1080/15715124.2016.1203792

de Oliveira Neves, A.C., Nune, F.P., de Carvalho, F.A., and Fernandes, G.W. (2016) Neglect of ecosystems services by mining, and the worst environmental disaster in Brazil. *Natureza and conservação*, 1 (4), 24–27.

Esteves, F.A. and Amado, A.M. (2011) Nitrogênio. Fundamentos de Limnologia. Rio de Janeiro: Interciência.

Esteves, F.A. and Panosso, R. (2011) Fósforo. Fundamentos de Limnologia. Rio de Janeiro: Interciência.

Fernandes, F. and Poleto, C. (2016) "Geochemical enrichment of metals in sediments and their relation with the organic carbon." International Journal of River Basin Management, 15 (1), 69–77.

Froehner, S. and Martins, R.F. (2008) Avaliação da Composição Química de Sedimentos do Rio Barigui na Região Metropolitana de Curitiba. *Química Nova*, 31 (8), 2020–2016.

IBGE. Censo Demográfico 2010. http://www.censo2010.ibge.gov.br.

Juchen, C.R., CervI, E.C., Vilas Boas, M.A., Charlesworth, S., and Poleto, C. (2014) Comparative of local background values for trace elements in different Brazilian tropical soils. *International Journal of Environmental Engineering and Natural Resources*, 1, 255–261.

Juchen, C.R., Vilas Boas, M.A., Poleto, C., and Juchen, P.T. (2016) Enrichment of sediments in urban rivers by heavy metals. *Global NEST Journal*, 18 (3), 643–651.

Kramer, R.D., Mizukawa, A., Ide, A.H., Marcante, L.O., Santos, M.M., and Azevedo, J.C.R. (2015) Determinação de anti-inflamatórios na água e sedimento e suas relações com a qualidade da água na bacia do Alto Iguaçu, Curitiba–PR. *Revista Brasileira de Recursos Hídricos*, 20 (3), 657–667. doi: 10.21168 / rbrh.v20n3.

Lucheta, F., Feiden, I. R., Gonçalves, S.P., Gularte, J.S., and Terra, N.R. (2010) Evaluation of the Gravataí River sediment quality (Rio Grande do Sul–Brazil) using *Daphnia magna* (Straus, 1820) as the test-organism for toxicity assays. *Acta Limnologica Brasiliensia*, 22.4, 367–377.

Macedo, J. (2004) City profile: Curitiba. *Cities*, 21 (6), 537–549.

Martínez, J.G., Boas, I., Lenhart, J., and Mol, A.P.J. (2016) Revealing Curitiba's flawed sustainability: How discourse can prevent institutional change. *Habitat International*, 53, 350–359.

Melo, V.F., Andrade, M., Batista, A.H., Favaretto, N., Grassi, M.T., and Campos, M.S. (2012) Chumbo e zinco em Águas e Sedimentos de Área de Mineração e Metalurgia de Metais. *Química Nova*, 35 (1), 22–29.

Miranda, L.S. and Marques, A.C. (2016) Hidden impacts of the Samarco mining waste dam collapse to Brazilian marine fauna – An example from the staurozoans

(Cnidaria). *Biota Neotropica*, 16 (2), e20160169.

Montagner, C.C. and Jardim, W.F. (2011) Spatial and seasonal variations of pharmaceuticals and endocrine disruptors in the Atibaia River, São Paulo State (Brazil). *Journal of the Brazilian Chemical Society*, 22 (8), 1452–1462.

Mozeto, A.A. and Zagatto, P.A. (2008) *Introdução de Agentes Químicos no Ambiente.* Ecotoxicologia Aquática: Princípios e Aplicações. São Carlos: Editora RiMa.

Orsi, C. (2013) Água de vinte capitais tem contaminantes emergentes. *Jornal da Unicamp*, p.03.

Osawa, R.A., Ide, A.H., Sampaio, N.M.F.M., and Azevedo, J.C.R. (2015) Determinação de fármacos anti-hipertensivos em águas superficiais na região metropolitana de Curitiba. *Revista Brasileira de Recursos Hídricos*, 20 (4), 1039–1050. doi: 10.21168 / rbrh.v20n4

Pantano, G., Grosseli, G.M., Mozeo, A.A., and Fadini, P.S. (2016) Sustentabilidade no uso do fósforo: uma questão de segurança hídrica e alimentar. *Química Nova*, 39 (6), 732–740.

Piña, C.I., Lance, V.A., Ferronato, B.O., Guardia, I., Marques, T.S., and Verdade, L.M. (2009) Heavy metal contamination in *Phrynops geoffroanus* (Schweigger, 1812) (Testudines: Chelidae) in a River Basin, São Paulo, Brazil. *Bull Environ Contam Toxicol.*, 83, 771–775.

Poleto, C. and Gonçalves, G.R. (2006) Qualidade das amostras e valores de referência. In: Poleto, C. and Merten, G.H. (Orgs.), *Qualidade dos sedimentos.* Porto Alegre, ABRH.

Poleto, C. and Merten, G.H. (2008) Estudos de Zn e Ni em sedimentos fluviais em suspensão e o risco potencial aos recursos hídricos. *Revista Brasileira de Recursos Hídricos*, 13 (3), 147–154.

Poleto, C., Bortoluzzi, E.C., Charlesworth, S.M., and Merten, G.H. (2009) Urban sediment particle size and pollutants in Southern Brazil. *Journal of Soils and Sediments*, 9, 317–327.

Poleto, C., Merten, G.H., and Minella, J.P. (2009) The identification of sediment sources in a small urban watershed in Southern Brazil: An application of sediment fingerprinting. *Environmental Technology*, 30 (11), 1145–1153.

Pompêo, M.L.M., Nishimura, P.Y., Cardoso, S.C., and Moschini-Carlos, V. (2015) Kit Clorofila: Uma proposta de método de baixo custo na estimativa do índice de estado trófico com base nos teores de clorofila. *Ecologia de reservatórios e interfaces.* São Paulo: Instituto de Biociências da USP. 1, 411–420.

Porto, R., Zahed, K., Tucci, C., and Bidone, F. (2004). Drenagem Urbana. In: Tucci, C.E.M. (Org.), *Hidrologia: Ciência e Aplicação.* Porto Alegre: Ed. Universidade/UFRGS, 805–847.

Rabinovitch, J. and Leitmann, J. (1993) Environmental innovation and management in Curitiba, Brazil. UNDP/UNCHS (HabitatWorld Bank). Urban management and the environment, Working Paper Series 1. https://www.ircwash.org/sites/default/files/827–BRCU93–13268.pdf

Rocha, P.S., Azab, E., Schmidt, B., Storch, V., Hollert, H., and Braunbeck, T. (2010) Changes in toxicity and dioxin-like activity of sediments from the Tietê River (São Paulo, Brazil). *Ecotoxicology and Environmental Safety*, 73, 550–558.

Rodgher, S., Espíndola, E.L.G., Rocha, O., Fracácio, R., Pereira, R.H.G., and Rodrigues, M. H.S. (2005) Limnological and ecotoxicological studies in the cascade of reservoirs in the Tietê River (São Paulo, Brazil). *Brazilian Journal of Biology*, 65 (4), 697–710.

Routh, J., Meyer, P.A., Gustafsson, Ö., Baskaran, M., Hallbeg, R., and Schöldström, A. (2004) Sedimentary geochemical records of human-induced environmental changes in the Lake Brunnsviken watershed, Sweden. *Limnology and Oceanography*, 49 (5), 1560–1569. doi:10.4319/lo.2004.49.5.1560

Santos, I.R., Costa, R.C., Freitas, U., and Fillmann, G. (2008) Influence of effluents from a wastewater treatment plant on nutrient distribution in a coastal creek from southern Brazil. *Brazilian Archives of Biology and Technology*, 51 (1), 153–162.

Santos, S., Oliveira, L.C., Santos, A., Rocha, J.C., and Rosa, A.H.R. (2012) Poluição Aquática. *Meio Ambiente and Sustentabilidade*. Porto Alegre: Editora Bookman. 1, 17–46.

Savage, C., Leavitt, PR., and Elmgren, R. (2004) Distribution and retention of effluent nitrogen in surface sediments of a coastal bay. *Limnology and Oceanography*, 49 (5), 1503–1511. doi:10.4319/lo.2004.49.5.1503

Scheffer, E.W., Sodré, F.F., and Grassi, M.T. (2007) Fatores que governam a especiação de cobre em ambientes quáticos urbanos: Evidências da contribuição de sulfetos solúveis. *Química Nova*, 30 (2), 332–338.

USEPA. United States Environmental Protection Agency (2000) Stressor Identification Guidance Document. Washington DC: USEPA, Office of Water. EPA/822/B–00/025.

Van Metre, P.C. and Mahler, B.J. (2003) The contribution of particles washed from rooftops to contaminant loading to urban streams. *Chemosphere*, 52. 1727–1741.

28

Potentially Toxic Metal-Bearing Phases in Urban Dust and Suspended Particulate Matter: The Case of Budapest, Hungary

Péter Sipos[1], Tibor Németh[1], Viktória Kovács Kis[2], Norbert Zajzon[3], Chung Choi[4], and Zoltán May[5]

[1] *Institute for Geological and Geochemical Research, Research Centre for Astronomy and Earth Sciences, Hungarian Academy of Sciences, Hungary*
[2] *Institute of Technical Physics and Materials Science, Centre for Energy Research, Hungary*
[3] *Institute of Mineralogy and Geology, Faculty of Earth Science and Engineering, University of Miskolc, Hungary*
[4] *Department of Earth Sciences, University of Bristol, United Kingdom*
[5] *Institute of Materials and Environmental Chemistry, Research Centre for Natural Sciences, Hungarian Academy of Sciences, Hungary*

28.1 Introduction

Airborne particulate matter (PM) has been widely associated with health disorders primarily due to its fine particles but also due to its toxic components (Kim *et al.*, 2015). Recent attention has been focused on the characterisation of its very fine-size fractions (below 10 μm) due to their easy penetration to the innermost regions of the lung (Samet *et al.* 2000). However, particles with a diameter up to 100 μm can be inhaled or ingested, and those below 32 μm may reach the bronchial tubes (UNEP and WHO, 1992). Airborne particulate matter can be divided into two types: the urban dust sediment and the suspended particles (Remeteiova *et al.*, 2007). Urban dust is created by particles with great sedimentation power, and their delay time in the atmosphere is very short, causing generally near-source pollution. Suspended particulates, however, may travel great distances due to their small particle size, resulting in contamination far away from their sources. Both of these materials generally show significant enrichment in several potentially toxic elements in the urban environment. Thus, after sedimentation, these particles can also contaminate soils, groundwater, and even the food chain (Seiler *et al.*, 1988).

Studies on the sources, compositions, and distribution of airborne PM components are necessary for their risk assessment of atmospheric quality, ecology, and human health. This is especially true for the urban environment, where population and traffic density are relatively high, and the harmful effect of airborne PM is expected to be significantly increased (Vardoulakis *et al.*, 2003). Environmental risk assessment of metals associated with PM has usually been based on the analysis of their total concentrations. However, it is a poor indicator of metal bioavailability, mobility, and toxicity, because these properties depend on the geochemical association of the trace elements with the different components of the solid matrix (Dabek-Zlotorzynska *et al.*, 2003). Unfortunately, there is no known universal analytical technique capable of identifying as well as quantifying all metal species present in airborne PM. Furthermore, owing to the chemical complexity, extremely small particle sizes, and typically small total sample sizes, such

Urban Pollution: Science and Management, First Edition. Edited by Susanne M. Charlesworth and Colin A. Booth.

materials can pose significant problems for analysis (Huggins *et al.*, 2000).

A combination of several direct mineralogical and indirect geochemical analyses, however, was found to be an effective tool to study the potential host phases of potentially toxic elements in the airborne PM (e.g. Sipos *et al.*, 2012). The aim of this chapter is to support this finding through summarising the results of our study on the host phases of potentially toxic elements in urban dust and total suspended particulate matter from Budapest, Hungary, based on published data.

28.2 Materials and Methods

28.2.1 Sampling

Urban dust (UD) samples were collected according to the Hungarian standard using glass pots of 2 L containing 500 mL distilled water seasonally for two years. Altogether eight sampling pots were placed at the front and the back sides of a high building at 2, 9, 21, and 33 m heights next to a very busy road. Dust and water samples were separated using vacuum filtering by a 2 μm Millipore filter. The dust samples were separated from the filters in an ultrasonic bath. More details on sampling and the characterisation of the sampling site and the samples can be found in Sipos *et al.* (2014).

Total suspended particulate matter (TSP) samples were collected from the air filters placed in the air supply channels of methane-heated turbines in four thermal power stations. Altogether 11 samples were collected from four sampling sites close to industrialised areas. The filters were in use for between 3 and 15 months. Samples were removed from the filters mechanically, and the material was passed through a 0.5 mm sieve. More details on the sampling and the characterisation of sampling sites and samples can be found in Sipos *et al.* (2013a). TSP material collected from such filters was also found to be useful for studying metals' speciation because the mineralogical and geochemical characteristics of metals are similar to those of samples collected by conventional air samplers (Sipos *et al.*, 2016).

28.2.2 Geochemical and Mineralogical Analyses

The total concentrations of major and potentially toxic elements (PTEs) in UD (Ba, Ca, Cr, Cu, Fe, Mn, Pb, Sr, and Zn) and TSP (Ba, Ca, Cr, Cu, Fe, K, Mn, Ni, Pb, Rb, S, Sr, Ti, V, Zn, and Zr) samples were analysed by X-ray fluorescence spectrometry from pressed powder samples (Philips PW2404 and Thermo Niton XL3 type spectrometers). Enrichment of metals in the samples was calculated by geoaccumulation indexes after Ji *et al.* (2008) using the data of the geochemical map of Hungary as geochemical background values (Ódor *et al.*, 1997). Details of the chemical analyses can be found in Sipos *et al.* (2013b and 2013c).

Single (water extraction and aqua regia dissolution) and sequential (BCR method after Rauret *et al.*, 2001) chemical extractions were also carried out on the TSP samples. The residual materials after the BCR extraction were subjected to a four-acid digestion, as well. Solutions of the single-step extractions were analysed by ICP-MS (Perkin-Elmer Elan 9000), whereas those from sequential extraction were analysed by the ICP-OES technique (Agilent 710) for the concentrations of Al, Ca, Fe, K, Mg, Mn, Na, P, S, Ti, Cd, Cr, Cu, Ni, Pb, and Zn. Details of the analyses can be found in Sipos *et al.* (2016).

The bulk mineralogical composition of the samples was characterised by X-ray powder diffractometry (XRD) (Philips PW 1710). Loss on ignition of the TSP materials was studied by thermogravimetry (MOM derivatograph). High-resolution transmission electron microscopy (TEM) and selected area electron diffraction analyses were carried out to analyse the mineralogy and composition of single particles both in UD and TSP samples (Philips CM20 equipped with a Noran energy dispersive spectrometer (EDS)). The chemistry and morphology of particles in the TSP samples were also studied by scanning electron microscopy (SEM) equipped with EDS (Jeol 8600). Details of the analyses can be found in Sipos *et al.* (2013c and 2014).

28.3 Results and Discussion

28.3.1 Urban Dust (UD)

28.3.1.1 Metals' Enrichment

The average concentrations of Ba, Ca, Cr, Cu, Fe, Mn, Pb, Sr, and Zn and their ranges (see detailed data in Sipos and May, 2013b) in the UD samples were found to be similar to those as found in other Central European towns (e.g. Krolak, 2000). On the basis of the geoaccumulation indexes of the studied metals, two main groups were distinguished, which could be also subdivided further (Figure 28.1).

The first group was that of Ca, Fe, Mn, and Sr, showing no significant enrichment in the dust. Within this group, Ca and Sr showed moderate enrichment in some cases. These latter two metals are considered to be the marker of construction dust in urban areas on the one hand (Ji *et al.*, 2008), while on the other hand, their slight enrichment can be also due to the dolomitic geological environment. Although these elements are generally attributed to natural sources, the potential of their contamination from anthropogenic sources cannot be excluded (Tahri *et al.*, 2012)

A characteristic example of which is iron. Despite the fact that this metal showed similar concentrations in the dust to those found in the background geological formations, magnetic analyses by Márton *et al.* (2011) showed the same characteristics for magnetite particles in the dust and in traffic-originated PM, suggesting that the major source of magnetite (and so the magnetite-derived Fe) is vehicular traffic in this case.

The second group consisted of the other studied metals, which generally showed moderate enrichment in the dust. However, Pb and Zn showed heavy contamination in several cases. Ji *et al.* (2008) found that the total fraction of the UD (<100 μm) exhibits practically similar metal concentrations to natural geological formations, while the fine fractions (<10 μm) were mostly heavily enriched with Cr, Co, Cu, Pb, and Zn as shown by a study conducted on 15 Chinese cities. Literature data show that most of these metals may originate from traffic sources, but they are also the common components of the built environment (Sutherland, 2000).

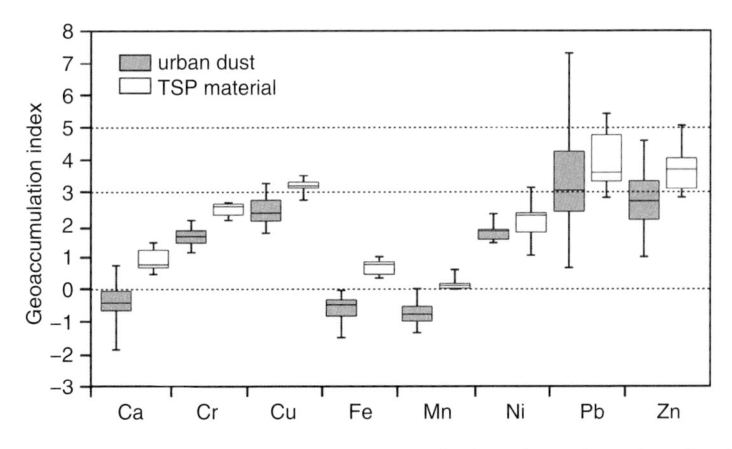

Figure 28.1 Comparison of the enrichment of selected metals in urban dust (n = 64) and TSP (n = 11) materials from Budapest, Hungary. Geoaccumulation index values were calculated after Ji *et al.* (2008). Values below zero refer to uncontaminated, those between zero and three to moderately contaminated, between three and five to heavily contaminated, and values over five extremely contaminated materials.

Accumulation of metals in the dust exhibited both spatial and temporal variance (Sipos and May, 2013b). For example, Cu and Zn showed the highest enrichment at the high-traffic side and lower heights of the building as expected, whereas Pb could be characterised by high enrichment at the low-traffic side, too. Moreover, Pb and Zn may have shown strong enrichment even at 33 m. On the other hand, the highest metal enrichments were found generally in the summer and spring samples. These characteristics, however, can be evaluated more effectively when metal deposition rates are considered, as the diluting effect of non-metal-bearing natural particles can be excluded (Sipos *et al.*, 2014).

Both the seasonal and the vertical deposition of Fe, Cu, Pb, and Zn showed large similarities to that of the dust. As this phenomenon was observed for each metal, it could not be explained by the changes in metal sources. The weather conditions favouring dust deposition and the increased rate of resuspended material in the dust in summer and spring may have affected the metal deposition together and resulted in its increase. Other studies, however, did not find any significant seasonal variation in metal deposition (Odabasi *et al.*, 2002), or they found slightly higher values in the heating season (Krolak, 2000). In our case, however, the contribution of soil/dust resuspension was found to be the dominant source of the dust (primarily in the summer season) and may even overcome the effect of recent anthropogenic activities. Moreover, the resuspended material contains the contribution of past anthropogenic activities, which could have resulted in much higher metal concentrations than recent activities.

28.3.1.2 Phase Composition of the Dust Material

The major mineral components of the dust material reflected that of the geological background (Sipos *et al.* 2013d). Their most frequent components were found as follows: quartz (60%–90 wt%), dolomite (0%–20 wt%), calcite (0%–15 wt%), feldspar (5%–10 wt%), illite (1%–5 wt%), chlorite (1%–5 wt%), and occasionally smectite (1%–5 wt%). These phases are the most common natural components of airborne PM (Farkas and Weiszburg, 2006) although both silicates and carbonates may also originate from anthropogenic sources (Zajzon *et al.*, 2013). In the fine particle size fraction (<20 μm), the ratio of clay minerals and amorphous materials significantly increased, and gypsum also appeared. This latter originated probably from the construction materials and/or the reaction of sulphuric acid and calcic material in the air (Panigrahy *et al.*, 2003). The major mineral constituents of the UD samples identified by XRD were also observed by TEM analyses (Sipos *et al.*, 2014). Using the latter technique, magnetite was also identified as a frequent component of the dust. This mineral often formed aggregates consisting of nanosized (up to 100 nm) magnetite attached to large silicate particles, and was also present as relatively large (up to a few micrometres) spherular or xenomorphic polycrystalline particles. Magnetite was also identified as the only magnetic phase in the dust by magnetic analyses (Márton *et al.*, 2011). These results showed that magnetite particles were present primarily as superparamagnetic (<30 nm) particles. Both its size and morphology suggested that this mineral originated primarily from anthropogenic sources (Zajzon *et al.*, 2013). Additional iron minerals, such as hematite and ilmenite, were also found by TEM analyses but in much smaller amounts than magnetite. The most spectacular spatiotemporal variation in mineral ratios was found for the carbonate minerals (Sipos *et al.*, 2014). Much higher dolomite ratios were found at the front of the building than at the back despite the presence of the dolomite hill behind the building. Due to the vicinity and morphology of the dolomite hill and the building, formation of a wake interface flow could be expected between them, resulting in the transport and deposition of the hill's material on the leeward (front) side of the building in an isolated separation bubble (Oke, 1988). The amount of carbonate minerals was generally higher in the periods of large dust deposition, which was also observed for clay minerals. This suggests that a significant local source is resuspension of street dust and local soil. However, clay minerals were the only phases showing characteristic vertical differences in their deposition: their amount decreases with increasing sampling height at the back, and much more uniform pattern was found at the front of the building. The spatiotemporal variation in the clay mineral's deposition suggested the presence of remote sources in addition to local ones. These sources may have antagonistic effects at the different sides and levels of the building.

28.3.1.3 Metal-Bearing Phases

The most important potentially toxic metal-bearing phases were found to be magnetite and clay minerals in the dust materials as shown by the TEM-EDS analyses by Sipos *et al.* (2012) and Sipos *et al.* (2014). Additionally, Zn was found to be associated with a calcite particle in one–single case. Zinc could be associated both with clay and Fe oxide particles (Figure 28.2a, c), while lead primarily to the latter particles (Figure 28.2b). The silicate and oxide particles often formed aggregates with each other. The Zn content of clay minerals could be as high as 5 wt%, while Fe oxides were characterised by a slightly lower Zn content (up to 2.5 wt%). The Pb content of the latter phases was generally between 2 and 3 wt%, and they also contained a small amount of Mn (around 0.5 wt%). Among Fe oxide particles, both magnetite and hematite were identified (Figure 28.2b). Additionally, ilmenite and titanite were also found in the samples, but they do not contain a significant amount of PTEs (except one ilmenite particle, which contained 0.5 wt% Mn). Urban anthropogenic particles are often enriched in toxic trace metals (Maher, 2009). Magnetite particles in the dust may have originated from the anthropogenic emissions, while clay particles may be derived rather from the resuspension of urban soils. Magnetite particles are resistant to weathering, releasing their toxic components slowly to the environment. However, its close association with hematite suggests its oxidation, which may proceed already in the anthropogenic combustion process as showed by the results of Zajzon *et al.* (2013), who found a close association between magnetite and hematite in vehicle exhaust materials. This latter phase is much less resistant than magnetite (Silva *et al.*, 2007), and together with layer silicates (and carbonates), they could be the potential sources of mobile toxic metals in the dust material (Sipos *et al.*, 2014).

Hierarchical cluster analysis based on the linear correlation between the metal deposition rates at the different sampling sites showed that Cu, Pb, and Zn showed slightly different spatiotemporal deposition characteristics (Sipos *et al.*, 2014). Such differences in deposition patterns may have even suggested differences in host phases for the metals. The highest similarity was found for Fe and Zn, and these two metals showed more similar deposition characteristics to Pb than to Cu. This suggested that Zn and Pb may be associated with Fe at a higher proportion than Cu.

28.3.2 Suspended Particulate Matter (PM)

28.3.2.1 Metals' Enrichment

Concentrations and ranges of 16 chemical elements (with Cd, Cu, Cr, Ni, Pb, and Zn among them being PTEs; see details in Sipos *et al.*, 2013c) were found to be similar to those reported for street dusts (Sysalova and Szakova, 2006) and even for fine PM (Feng *et al.*, 2009), although the general observation is that such elements tend to concentrate in the fine PM. This is, however, largely dependent on the prevailing local metal source as some of them may also emit large-sized particles (Zajzon *et al.*, 2013).

On the basis of the geoaccumulation index values of the studied PTEs, Pb, Zn, and Cu showed heavy contamination, whereas Cr, Cd, and Ni showed moderate contamination in the TSP material (Figure 28.1), suggesting

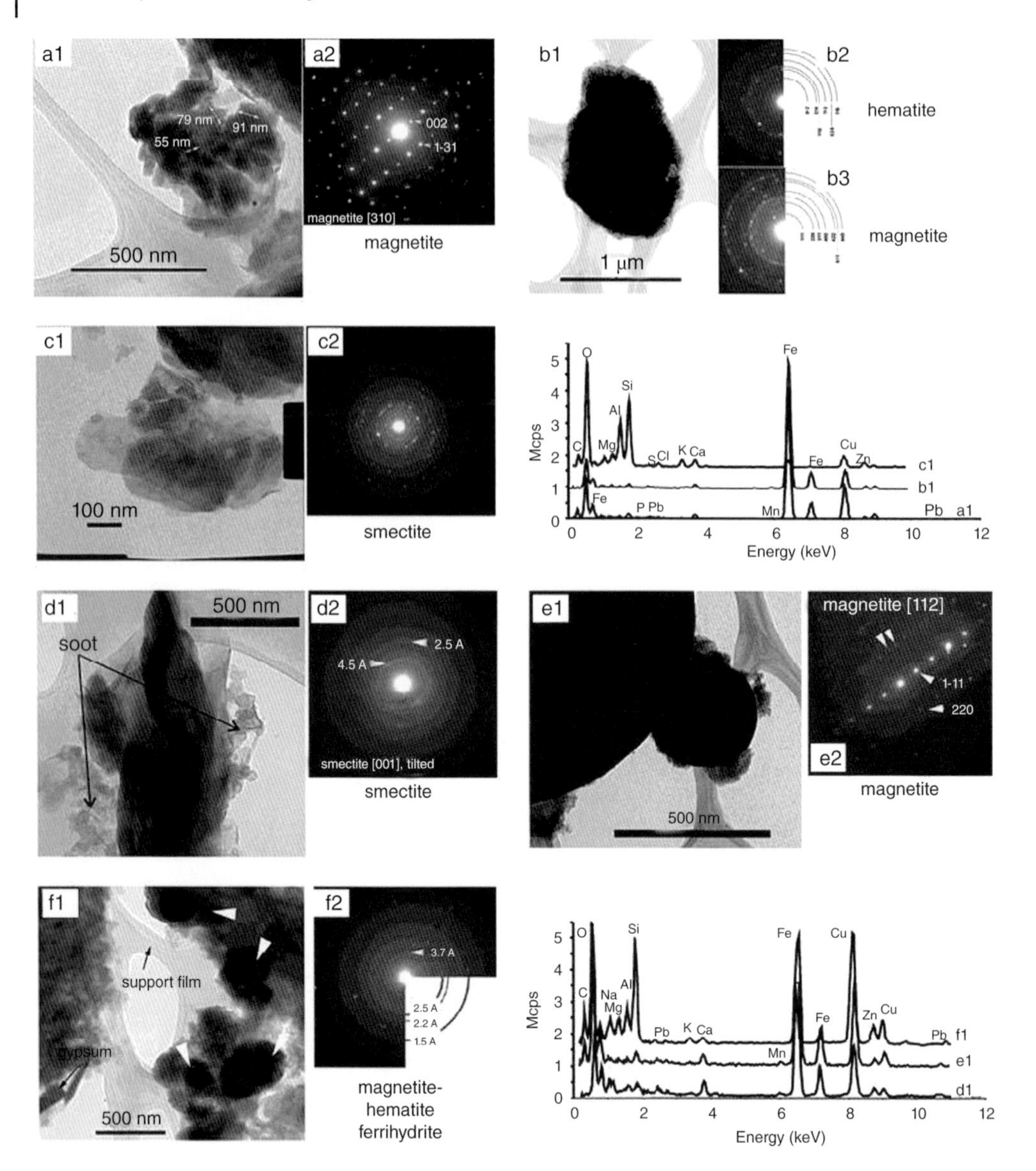

Figure 28.2 Transmission electron microscopy micrograph (1) and electron diffraction patterns (2) of mineral particles observed in the urban dust (a, b, and c) and suspended particulate matter (d, e, and f) materials by TEM-EDS analyses: (a) an aggregate consisting of nano-sized magnetite single crystals with 2.49 wt% Zn content; (b) a polycrystalline association of hematite and magnetite making up a large Fe oxide particle with 2.88 wt% Pb; (c) relatively large smectite plates containing 4.89 wt% Zn; (d) smectite particle with 4.04 wt% Zn and 0.41 wt% Pb associated with soot aggregates composed of nano-spheres; (e) magnetite spheres with 1.84 wt% Zn and 2.38 wt% Pb; (d) dense aggregates of magnetite, hematite (3.7 A) and poorly crystalline ferrihydrite with 2.55 wt% Zn, 2.5 wt% Pb, and 0.47 wt% Mn. Cu on the EDS spectra comes from the instrument.

the dominating presence of anthropogenic source(s) primarily for the former elements (Richter *et al.*, 2007). The earliest studies on the metal concentrations in the TSP material from Budapest also found similar enrichment characteristics for the studied metals (Hlavay *et al.*, 1996). Differences in metal concentrations and enrichments among the sampling sites suggest differences in the local metal sources there and/or in their distances as well, as local sources generally prevail for TSP material (Mateus *et al.*, 2013). Salma *et al.* (2001) found that Cd, Cu, Pb, and Zn exhibit very high enrichment both in coarse and fine PM in Budapest, Hungary, whereas Cr and Ni show moderate enrichment when compared to their crustal abundances. However, they showed also high variability among various sampling sites, which could be related to the variation of local sources. Kukkonen *et al.* (2003) estimated the relative contribution of source categories to ambient trace elements concentrations in the PM from Budapest. They found that substantial amounts of Cd, Pb, and Zn are originated from waste incineration (65%–70%), whereas the contribution of traffic ranges from 11% to 17%. Other sources (like coal or oil burning) account for 5%–7% additional contribution. Sipos *et al.* (2016) found that the highest Zn content in the TSP material may be related to the vicinity of a waste incinerator, as such objects are the major emission sources for several metals, especially for Zn (Liu *et al.*, 2005). On the other hand, a high amount of Pb in some locations was related to vehicular traffic as busy road environment are huge reservoirs of former (but also of recent) Pb emission of traffic origin (Johansson *et al.*, 2009). On the other hand, sampling points with low metal content could be related to the fact that they are placed next to the Danube River, which provides the major wind channel in Budapest.

28.3.2.2 Phase Composition of PM

The bulk mineralogical composition of the samples was dominated by the presence of minerals characteristic of the geological environment of the sampling sites (Sipos *et al.*, 2013c). The samples consisted mostly of 15–20 wt% quartz, 5–20 wt% carbonates, 5–10 wt% clay minerals, and 5–10 wt% (plagioclase) feldspar. Among the clay minerals, illite dominated, but chlorite and smectite also appeared. Such phases are characteristic natural components of UD, representing primarily the fraction depositing fast (Grobéty *et al.*, 2010).

The relatively high loss on ignition values showed the presence of a large amount of volatile components. This could be primarily related to the presence of organic matter and soot, which was calculated to be between 14 and 25 wt% there. This also corresponded to the results of XRD analyses, with which large amounts of amorphous phases were identified. Transmission electron microscopic analyses also showed that soot aggregates consisting of nano-sized soot particles are common in this material. It is widely known (Grobéty *et al.*, 2010) that soot is a common anthropogenic component of urban airborne PM as a result of different combustion (vehicle, heating, industrial, etc.) processes. Besides soot, the dominant anthropogenic phases in the samples were found to be the iron oxides, and gypsum and halite also appeared in large amounts occasionally (Sipos *et al.*, 2013c). The XRD analyses showed the presence of large amounts of magnetite (5–10 wt%) at most of the sampling sites. This phase was also shown by magnetic and TEM analyses together with ferrihydrite and hematite, but the frequency of the latter two phases was much lower than that of magnetite. They could be the oxidation/weathering products of magnetite. The oxidation may proceed already in the anthropogenic combustion process, as suggested by the results of Zajzon *et al.* (2013), who found a close association between magnetite and hematite in vehicle exhaust samples.

28.3.3.3 Metal-Bearing Phases

Both indirect and direct methods were used to study the metals' speciation in the TSP material. Hierarchical cluster analyses based on the linear correlation of the total concentrations of 16 elements showed that Ca, Sr, Pb, and S can be probably related to carbonates and sulphates; Fe, Mn, Cu, Ni, and V to iron oxides; Ba to both of the former groups; and K, Rb, Cr, Ti, Zn, and Zr to silicates and oxides (Sipos *et al.*, 2013c). These results suggested that elements showing significant enrichment can be associated with different components with varying origin.

Chemical extractions of PTEs and major chemical elements were also found to be useful in studying the potential host phases for metals (Sipos *et al.*, 2016). Significant ratios (10%–15% of their totals) of Zn (and Cd) was found in water-soluble forms together with Na, K, Mg, Ca, and S, forming mostly sulphates (and probable nitrates and chlorides, as well). Such phases may arise, for example, from the fly ash of fossil fuel combustion or waste incineration (Pinzani *et al.*, 2002). The weak acid-soluble fraction mobilising carbonate bound and sorbed elements was also characteristic of Zn and Cd (40%–50% of the totals), suggesting that high mobility could be related to a large variety of potential phases. The strongly similar behaviour of Zn and Cd could be due to the common Cd impurities in Zn materials which seem to survive anthropogenic processes as well (Monaci *et al.*, 2000). Although Pb and Cu showed lower fractionation in the easily soluble fractions (below 10% of their totals), our SEM-EDS analyses supported the presence of their water-soluble (Cu–sulphate; Figure 28.3e) and weakly acid-soluble (Pb sulphate; Figure 28.3d) forms in the TSP material. Additionally, TEM-EDS analyses showed that clay minerals (primarily smectites) also contained small amounts of Pb and Zn (up to a few wt%) (Figure 28.2e). These metals are probably sorbed on the surface of clay minerals in the soil, whose resuspension may have contributed to the airborne PM, which is a well-documented phenomenon in the urban environment (Laidlaw and Filippelli, 2008).

The reducible fraction is expected to mobilise not only metal oxides dissolving at redox conditions but also metal carbonates susceptible to stronger pH drop (Fernandez *et al.*, 2000). This fraction was mostly characteristic of Pb, Zn, and Cd (30%–50% of their totals) and partly for Ca, Cu, Fe, Mn, Mg, and P. Mobilisation of Fe (and Mn) suggested the presence of their hydroxides, whereas that of carbonates could be related to the appearance of Ca and Mg (but also of Fe and Mn). Metal compounds, like sulphate, carbonate, and (hydr)oxide, are often found in emission sources including waste incineration and smelters in the urban environment (Sobonska *et al.*, 1999). Our TEM analyses showed the occasionally close association of magnetite, hematite, and ferrihydrite (Figure 28.2f). This can be due to the weathering of resistant Fe oxides, and the presence of ferrihydrite can be related to the reducible fraction of metals and the release of the detected Zn, Pb, and Mn through the reducible extraction step. Also, a single particle of calcite containing Pb (4.88 wt%) and that of a smithsonite were also found by TEM, suggesting the presence of Pb and Zn in the form of (acid-soluble) carbonates. Such phases were found as a common component of urban airborne PM, and their presence can be due to the contribution of fly ashes from fossil fuel combustion, waste incineration, or industrial (smelter) dust (Wichmann *et al.*, 2000) to the urban airborne material.

Among the studied PTEs, Cu could be characterised by the highest association to organic matter in TSP on the basis of its large ratio (around 30% of the total) in the oxidisable fraction together with P. We observed the frequent presence of relatively large (several tens of micrometres) aggregates composing of silicates, Fe oxides, and organic matter as suggested by their high P and S content by SEM-EDS. Such aggregates often contained detectable amount Zn besides Cu (Figure 28.3c).

The residual fraction was dominated by elements composing of silicates and resistant oxides (Al, K, Mg, Na, Fe, and Ti). The very high ratios of Cr and Ni (around 80% of the totals) in this, as well as in the aqua regia non-soluble fraction together with Fe suggest their association to magnetite. This phase is a widespread phase of anthropogenic origin in the airborne PM and exhibits high compatibility with these metals under various genetic conditions (Dare *et al.*, 2014). Nevertheless, Cu and Pb exhibited a moderate fraction (40% of their totals), whereas Zn and Cd exhibited a small ratio (10%–20% of their totals) in this fraction. Our SEM-EDS analyses showed the presence

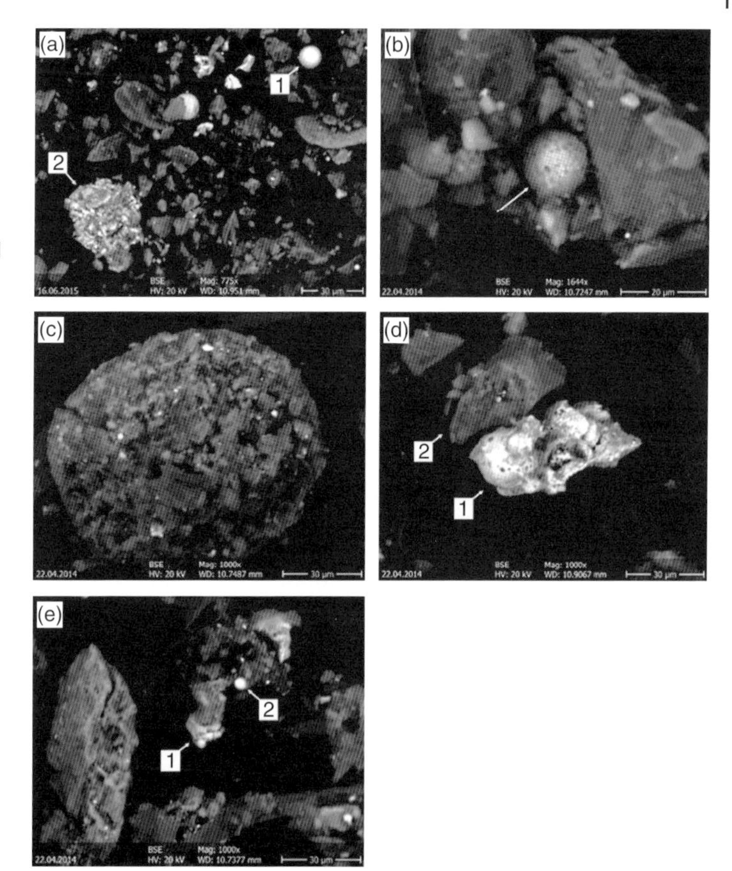

Figure 28.3 Backscattered electron images of PTE-bearing particles observed in the TSP material as studied by SEM-EDS: (a) an Fe oxide spherule with detectable Cu and Zn content (1) and a silicate aggregate containing Cr (2); (b) a silicate spherule with detectable Pb and Zn content; (c) an aggregate composing of silicates (grey particles), Fe oxides (bright particles), and probably of organic matter (as suggested by the large amount of P and S) containing Zn and Cu; (d) a particle composing of Pb and S (probably lead sulphate) (1) and a silicate particle containing Zn; (e) a particle composing of Cu and S (probably copper sulphate) (1) surrounded by silicates (grey particles) and Fe oxide spherules (2).

of both Fe oxide and silicate spherules, whose shape suggests their anthropogenic origin (Figure 28.3a, b). These particles often contained Cu, Zn and Pb, Zn, respectively. Both spherular and xenomorphic magnetite particles were often found by TEM analyses as well, which were found to be the most important Pb-bearing phases in the TSP material (Sipos *et al.*, 2013a). They sometimes contain 2–3 wt% of Pb and Zn (and occasionally less Mn) (Figure 28.2e, f). Several authors (e.g. Gautam *et al.*, 2005) found a significant linear relationship between magnetic susceptibility (primarily due to magnetite) and the Pb content of urban PM. The magnetite particles often form aggregates and are closely associated with soot and/or clay minerals (Figure 28.2e). In samples with high magnetite content, metal-free magnetite spherules up to a few micrometres in size also appeared. Moreover, a single ZnO particle was also found in the sample with the highest Zn content, which could also contribute to the residual Zn fraction.

28.4 Conclusions

Several potentially toxic metals (like Cu, Pb, and Zn) show moderate or even strong enrichment in the UD material when compared to their geochemical background. They are mostly of anthropogenic origin; however, this cannot be excluded for elements (like Ca, Fe) with no or slight enrichment either. Similar characteristics were found for the suspended PM, where metals with strong enrichment exhibit much higher concentrations. This phenomenon can be related to the small particle size of air pollutants of anthropogenic origin.

On the basis of the metal deposition characteristics in the UD, resuspension of dust and soil, seasonal variation of weather, and the morphology of the natural and built environment affect the metal loading simultaneously, where all of these factors together may overcome the metal contribution from recent anthropogenic activities. In the suspended PM, however, the local anthropogenic sources showed a much higher effect on the type and concentration of metals, and the other factors mentioned for metals in UD dominate only in special cases.

The mineralogical composition is dominated by those from the geological environment for both types of material. As long as the anthropogenic phases are present at higher ratios among the major components of the suspended PM, they can be detected primarily in the fine fraction of the UD. Magnetite is their most frequent anthropogenic phase, but gypsum and soot are also often detectable. Not only the composition but the morphology of particles can be related to their origin, as spherules generally originate from anthropogenic sources. The ratio of natural phases in UD may supply additional information on the variation in the contribution of local and remote dust sources, supporting the evaluation of variation in metal deposition, as well.

The most important potentially toxic metal-bearing phases in the UD material are magnetite, clay minerals, and carbonates. Magnetite can be related to the contribution of (recent) anthropogenic activities; the presence of clay particles, however, can be due to the resuspension of urban soil after sorption of metals by them, resulting in metal contribution from former (inactive) anthropogenic sources. Such metal-bearing phases could also be observed in the suspended PM. Due to its higher metal concentrations and the combination of direct and indirect speciation analyses, however, more detailed characterisation of host phases could be provided. Mobile fractions (like water-soluble phases, metals sorbed on clay surfaces, metal carbonates) could be observed for each metal, showing strong enrichment in the suspended PM with primary significance for Zn and Cd. Besides its carbonates, Pb could be associated with iron oxides, showing transformation or weathering from magnetite probably. Copper showed significant relation to the organic fraction of the suspended PM, although its detection with direct methods has failed. Very resistant phases, like magnetite and other metal oxides, contained the dominant proportion of Cr and Ni, but significant amounts of Pb and Zn were also detected in such particles.

Iron and other metal oxide particles are relatively resistant to weathering, releasing their potentially toxic components to the environment slowly. On the other hand, water-soluble metal phases, metals sorbed on the surface of clay particles, and metal (containing) carbonates represent the potential sources of mobile components in the airborne PM. The transformation of magnetite to ferrihydrite observed in the suspended PM suggests that potentially toxic components associated with iron oxides cannot be considered exclusively immobile either.

References

Dabek-Zlotorzynska, E., Kelly, M., Chen, H., and Chakrabarti, C.L. (2003) Evaluation of capillary electrophoresis combined with BCR sequential extraction for determining distribution of Fe, Zn, Cu, Mn and Cd in airborne particulate matter. *Analitica Chimica Acta*, 498, 175–187.

Dare, S.A.S., Barnes, S.J., Beaudoin, G., Meric, J., Boutroy, E., and Potvin-Doucet, C. (2014) Trace elements in magnetite as petrogenetic indicators. *Mineralium Deposita*, 49, 785–796.

Farkas, I. and Weiszburg, T. (2006.) Ülepedő és szálló por ásványtani vizsgálata a romániai

Kolozs megyéből. *Földtani Közlöny*, 136, 547–572.
Feng, X.D., Dang, Z., Huang, W.L., and Yang, C. (2009) Chemical speciation of fine particle bound trace metals. *International Journal of Environmental Science and Technology*, 6, 337–346.
Fernandez, A.J., Ternero, M., Barragan, F.J., and Jimenez, J.C. (2000) A approach to characterization of sources of urban airborne particles through heavy metals speciation. *Chemosphere*, 2, 123–136.
Gautam, P., Blaha, U., and Appel, E. (2005) Magnetic susceptibility of dust-loaded leaves as a proxy of traffic-related heavy metal pollution in Kathmandu city, Nepal. *Atmospheric Environment*, 39, 2201–2211.
Grobéty, B., Gieré, R., Dietze, V., and Stille, P. (2010) Airborne particles in the urban environment. *Elements*, 6, 229–234.
Hlavay, J., Polyák, K., Bódog, I., Molnár, Á., and Mészáros, E. (1996) Distribution of trace elements in filter-collected aerosol samples. *Fresenius Journal of Analytical Chemistry*, 354, 227–232.
Huggins, F.E., Huffman, G.P., and Robertson, J.D. (2000) Speciation of elements in NIST particulate matter SRMs 1648 and 1650. *Journal of Hazardous Materials*, 74, 1–23.
Ji, Y., Feng, Y., Wu, J., Zhu, T., Bai, Z., and Duan, C. (2008) Using geoaccumulation index to study source profiles of soil dust in China. *Journal of Environmental Sciences*, 20, 571–578.
Johansson, C., Norman, M., and Burman, L. (2009) Road traffic emission factors for heavy metals. *Atmospheric Environment*, 43, 4681–4688.
Kim, K.H., Kabir, E., and Kabir, S. (2015) A review on the human health impact of airborne particulate matter. *Environmental International*, 74, 136–143.
Krolak, E. (2000) Heavy metals in falling dust in Eastern Mazowieckie province. *Polish Journal of Environmental Studies*, 9, 517–522.
Kukkonen, J., Bozó, L., Palmgren, F., and Sokhi, R.S. (2003) Particulate matter in urban air. In: Moussiopoulos, N. (ed.), *Air Quality in Cities*. Springer, Berlin, pp. 91–120.
Laidlaw, M.A.S. and Filippelli, G.M. (2008) Resuspension of urban soils as a persistent source of lead poisoning in children: A review and new directions. *Applied Geochemistry*, 23, 2021–2039.
Liu, F., Liu, J., Yu, Q., Jin, Y., and Nie, Y. (2005) Leaching characteristics of heavy metals in municipal solid waste incinerator fly ash. *Journal of Environmental Science and Health A*, 40, 1975–1985.
Maher, B.A. (2009) Rain and dust: Magnetic record of climate and pollution. *Elements*, 5, 229–234.
Mateus, V.L., Monteiro, I.L.G., Rocha, R.C., Saint'Pierre, T.D., and Gioda, A. (2013) Study of the chemical composition of particulate matter from the Rio de Janeiro metropolitan region, Brazil, by inductively coupled plasma-mass spectrometry and optical emission spectrometry. *Spectrochimica Acta Part B*, 86, 131–136.
Márton, E., Sipos, P., Németh, T., and May, Z. (2011) Transport of pollutants around a high building: integrated magnetic, mineralogical and geochemical study. In: *Conference Proceedings and Exhibitor's Catalogue, 6th Congress of Balkan Geophysical Society* (3–6 October 2011, Budapest, Hungary), European Association of Geoscientists and Engineers, B9 1–6.
Monaci, F., Moni, F., Lanciotti, E., Grechi, D., and Bargagli, R. (2000) Biomonitoring of airborne metals in urban environments, new tracers of vehicle emission, in place of lead. *Environmental Pollution*, 107, 321–327.
Odabasi, M., Muezzinoglu, A., and Bozlaker, A. (2002) Ambient concentrations and dry deposition fluxes of trace elements in Izmir, Turkey. *Atmospheric Environment*, 36, 5841–5851
Ódor, L., Horváth, I., and Fügedi, U. (1997) Low-density geochemical mapping in Hungary. *Journal of Geochemical Exploration*, 60, 55–66.
Oke, T.R. (1988) Street design and urban canopy layer climate. *Energy and Buildings*, 11, 103–113.
Panigrahy, P.K., Goswami, G., Panda, J.D., and Panda, R.K. (2003) Differential

comminution of gypsum in cements ground in different mills. *Cement and Concrete Research*, 33, 945–947.

Pinzani, M.C.C., Somogyi, A., Siminovici, A.S., Ansell, S., Steenari, B.M., and Lindovist, O. (2002) Direct determination of Cadmium speciation in municipal solid waste fly ash by synchrotron 22 radiation induced μ-X-ray fluorescence and μ-X-ray absorption spectroscopy. *Environmental Science and Technology*, 36, 3165–3169.

Rauret, G., Lopez-Sanchez, J.F., Lück, D., Yli-Halla, M., Muntau, H., and Quevauviller, P. (2001) The certification of extractable contents (mass fractions) of Cd, Cr, Cu, Ni, Pb and Zn in freshwater sediment following a sequential extraction procedure. BCR information, Reference Materials BCR–701, European Commission, EUR 19775 EN.

Remeteiova, D., Smincakova, E., and Florian, K. (2007) Study of the chemical properties of gravitation dust sediments. *Microchimica Acta*, 156, 109–113.

Richter, P., Grino, P., Ahumada, I., and Giordano, A. (2007) Total element concentration and chemical fractionation in airborne particulate matter from Santiago, Chile. *Atmospheric Environment*, 41, 6729–6738.

Salma, I., Maenhaut, W., Zemplén-Papp, É., and Záray, G. (2001) Comprehensive characterisation of atmospheric aerosols in Budapest, Hungary: Physicochemical properties of inorganic species. *Atmospheric Environment*, 35, 4367–4378.

Seiler, H., Sigel, H., and Sigel, A. (1988) *Handbook on Toxicity of Inorganic Compounds*. Marcel Dekker Inc., New York.

Silva, M., Kyser, K., and Beauchemin, D. (2007) Enhanced flow injection leaching of rocks by focused microwave heating with in-line monitoring of released elements by inductively coupled plasma mass spectrometry. *Analitica Chimica Acta*, 584, 447–454.

Sipos, P., Kovács Kis, V., Márton, E., Németh, T., May, Z., and Szalai, Z. (2012) Lead and zinc in the suspended particulate matter and settled dust in Budapest, Hungary. *European Chemical Bulletin*, 1, 449–454.

Sipos, P., Németh, T., and Kovács Kis, V. (2013a) Lead isotope composition and host phases in airborne particulate matter from Budapest, Hungary. Central European Geology 56, 39–57.

Sipos, P. and May, Z. (2013b) Vertical distribution of metal deposition rates next to a major urban road in Budapest, Hungary. *Carpathian Journal of Earth and Environmental Sciences*, 8, 5–12.

Sipos, P., Márton, E., Németh, T., Kovács Kis, V., May, Z., and Szalai, Z. (2013c) Mineral phases containing heavy metals in the suspended dust from Budapest, Hungary. *E3S Web of Conferences*, 1: 20010.

Sipos, P., Németh, T., Kovács Kis, V., and Szalai, Z. (2013d) Seasonal and vertical variation of mineralogy and particle size distribution of settling dust along a high building. In: Zákányi, B. and Faur, K.B. (eds.), *IX. Kárpát-medencei Környezettudományi Konferencia* (Miskolc, 2013. június 13–15), Miskolci Egyetem, Műszaki Földtudományi Kar, 499–504.

Sipos, P., Márton, E., May, Z., Németh, T., and Kovács Kis, V. (2014) Geochemical, mineralogical and magnetic characteristics of vertical dust deposition in urban environment. *Environmental Earth Sciences*, 72, 905–914.

Sipos, P., Choi, C., and May, Z. (2016) Combination of single and sequential chemical extractions to study the mobility and host phases of potentially toxic elements in airborne particulate matter. *Chemie der Erde – Geochemistry*.

Sobonska, S., Riqo, N., Laboudigue, A., Bremard, C., Laureyns, J., Merlin, J.C., and Wignacourt, P. (1999) Microchemical investigation of dust emitted by a lead smelter. *Environmental Science and Technology*, 33, 1334–1339.

Sutherland, R.A. (2000) Bed-sediment-associated trace metals in an urban stream, Oahu, Hawaii. *Environmental Geology*, 39, 611–627.

Sysalova, J. and Szakova, J. (2006) Mobility assessment and validation of toxic elements in tunnel dust samples – Subway and rod using sequential chemical extraction and

ICP–OES/GF AAS measurements. *Environmental Research*, 101, 287–293.
Tahri, M., Bounakhala, M., Bout, A., Banyaich, F., Noack, Y., and Essaid, B. (2012) Application of nuclear analytical techniques (XRF and NAA) to the evaluation of air quality in Moroccan cities – Case of Meknes city. *Carpathian Journal of Earth and Environmental Sciences*, 7, 231–238.
Vardoulakis, S., Fisher, B.E.A., Pericleous, K., and Gonzalez-Fresca, N. (2003) Modelling air quality in street canyons: A review. *Atmospheric Environment*, 37, 155–182.
Wichmann, H., Spenger, R., Wobst, M., and Bahadir, M. (2000) Combustion induced transport of heavy metals in the gas phase: A review. *Fresenius Environmental Bulletin*, 9, 72–125.
Zajzon, N., Márton, E., Sipos, P., Kristály, F., Németh, T., Kovács Kis, V., and Wieszburg, T. (2013) Integrated mineralogical and magnetic study of magnetic airborne particles from potential pollution sources in industrial–urban environment. *Carpathian Journal of Earth and Environmental Sciences*, 8 (1), 179–186.

29

The Role of Urban Planning in Sub-Saharan Africa Urban Pollution Management

Kwasi Gyau Baffour Awuah

School of the Built Environment, The University of Salford, United Kingdom

29.1 Introduction

Urban pollution remains one of the global environmental concerns in the twentieth-first century (Corburn, 2015; Ngo *et al.*, 2017) because of its adverse public health implications. For example, air pollution has a strong correlation with diseases such as pneumonia, respiratory tract infections, asthma, and other cardiovascular diseases. Indeed, the World Health Organization (WHO) estimated in 2014 that air pollution alone accounts for 7 million premature deaths annually (Ngo *et al.*, 2017).

Traditionally, urban pollution such as air and noise pollution and their outcomes have been usually associated with cities in the global north. For example, WHO notes that air pollution ranks second among a series of environmental stressors for public health impact in European countries (Gozalo *et al.*, 2016). Air pollution is also identified as a major environmental cause of premature death in Europe. However, recent evidence shows that urban pollution, in particular, both indoor and outdoor air pollution, in Sub-Saharan African (SSA) cities has been increasing, and air quality continues to deteriorate (Amegah and Agyei-Mensah, 2017; Ngo *et al.*, 2017). Brauer *et al.* (2012) report that some of the highest fine particle levels in the world have been recorded in SSA cities and other developing regions, and that $PM_{2.5}$ concentrations in SSA cities has been estimated at around 100 mg/m^3 compared to <20 mg/m^3 in most European and North American cities. Further, the cost of air pollution in African cities is estimated to be in the region of about 2.7% of the GDP (Akumu, 2016).

Several factors account for the growing urban pollution in SSA cities, such as rising levels of urbanisation and urban growth, expansion of industrial and construction activities, and growth in vehicular ownership (Amegah and Agyei-Mensah, 2017; Ngo *et al.*, 2017). Vehicle emissions, for example, account for 90% of urban air pollution in developing economies, including those of SSA (United Nations Environmental Programme (UNEP), 2005). As a consequence, UNEP has been promoting initiatives such as better regulation and capacity building. Also, improvement of air quality constituted one of the 4 out of the 17 Sustainable Development Goals (SDGs) which was recently adopted for implementation. Nevertheless, urban planning is perceived as a critical tool which could be used to help address the growing urban pollution menace in the region's cities (UN-Habitat, 2009; UN-Habitat, 2014a). A fundamental function of urban planning is to allocate land resources among the various desired land uses in an efficient manner. Urban planning

Urban Pollution: Science and Management, First Edition. Edited by Susanne M. Charlesworth and Colin A. Booth.

also promotes certain allocation of financial and human resources (Baffour Awuah, 2013). This implies that urban planning could be used to influence activities in SSA cities to help manage urban pollution. Yet, how planning could be leveraged to help address urban pollution in the region has not been fully demonstrated.

This chapter seeks to contribute to an understanding as to how planning could be used to help manage urban pollution in SSA on the basis of a critique of the literature. After this introduction, the chapter provides an overview of urban pollution in SSA in the next section. This is followed by ways in which planning could be leveraged to help address urban pollution and then concludes with some lessons for SSA urban planning in the context of the growing urban pollution in the region.

29.2 Overview of Urban Pollution in Sub-Saharan Africa (SSA)

Urban pollution can simply be defined as the creation and collection of pollutants in cities and urban areas. Several types of urban pollution can thus be identified. These include air, water, noise, visual, soil, radioactive, and thermal pollutions. However, this work focuses on the first four types of urban pollution, especially given the recent concern over these types of pollutions in the developing world.

Air pollution refers to the release of pollutants into the air that are detrimental to human health and the planet as a whole (Mackenzie, 2016). It can thus be regarded as a gas or a liquid or solid released through air in quantities that damages the health of or kills human beings, animals, and plants, or damages or disrupts aspects of the environment, such as causing buildings or civil works to crumble, reducing visibility, or producing an unpleasant odour. Victoria (2016) states that the Environmental Protection Authority (EPA) defines clean air as predominantly comprising nitrogen and oxygen, and some other gases such as argon and carbon dioxide (CO_2) in minute quantities. Therefore, it is the concentration of or quantities of chemicals in the air that determine whether or not it is harmful – polluted or harmless. For example, CO_2 is noted to be usually at a concentration of 0.05% in air. Consequently, a CO_2 concentration of say 5% in air around a neighbourhood will mean the air is polluted.

Although natural air pollution is possible, urban air pollution has traditionally stemmed from gases released through burning fossil fuels and exhaust fumes generated by vehicles, factories, and power generators, as well as harmful fumes from chemicals including paint, plastic, and toxic spills, among others. According to Lindén *et al.* (2012), air pollutants that are usually of grave concern to human health include particulate matter (PM), carbon monoxide (CO), nitrogen oxides (NO_x, comprising NO and NO_2), volatile organic compounds (VOCs including benzene and toluene) and ozone (O_3).

Like air pollution, water pollution arises when there is a concentration of chemicals or particles – gas, liquid, or solid in waterbodies that harm human beings and the environment. For example, waterbodies such as groundwater, lakes, and rivers may be polluted by run-off from factories, farmlands, chemical spills, vehicle exhausts, raw sewage, industrial waste, and household garbage washed away by rain and mixed with water resources. Noise pollution refers to excessive noise in urban environments, which often result in adverse effects such as headaches, insomnia, hearing loss, and disruption in standard of living. Visual pollution often results from unsightly scenes in urban environments often caused by temporary structures such as advertisement banners and giant billboards.

Although the literature on air pollution is prolific in the developed world, such as the United States and some European countries, the same cannot be said of the developing world like the economies of SSA. Indeed, long-term air pollution studies and air quality monitoring are scarce in SSA countries (Lindén *et al.*, 2012; Ngo *et al.*, 2017). However, some recent studies and WHO monitoring show that both indoor and outdoor air pollution in SSA cities continue to soar. For example, Gatari *et al.* (2009) found $PM_{2.5}$ levels of (22.9 μg/m^3) and (30 μg/m^3) in Accra, Ghana, and Nairobi, Kenya, respectively during vehicle traverses. During the same period, Boman *et al.* (2009) also found $PM_{2.5}$ level of (77.9 μg/m^3) in Ouagadougou, Burkina Faso. These increasing levels of air pollution in SSA cities are said to be partly responsible for several health problems including asthma, acute respiratory infections, heart disease, and lung cancer (Lindén *et al.*, 2012; Ngo *et al.*, 2017). The estimated number of premature deaths due to outdoor air pollution in the developing world is 0.8 million per annum (Cohen *et al.*, 2005). Further, Akumu (2016) notes that the cost of air pollution in African cities is about 2.7% of the GDP.

Varied causal factors are implicated in the rising levels of air pollution in SSA cities. However, prominent among the causal factors in the literature include dust (Arku *et al.* 2008; Eliasson *et al.*, 2009; Boman *et al.*, 2009; Dionisio *et al.*, 2010; Lindén *et al.*, 2012; Ngo *et al.*, 2017), vehicular emissions (Adjei-Mensah, 2017; Ngo *et al.*, 2017), meteorological factors (Lindén *et al.*, 2012), industrial emissions (Adjei-Mensah, 2017; Ngo *et al.*, 2017), bush burning (including burning of rubbish/trash), smoke from cooking, and sewerage/dirty water (Ngo *et al.*, 2017). Dust as a causal factor of air pollution in SSA cities comes from two main sources, namely, unpaved roads and transport from the Sahara Desert. Studies, such as Arku *et al.* (2008; on Accra), Dionisio *et al.* (2010; on Ouagadougou and Dar es Salaam), and Eliasson *et al.* (2009; on Gaborone), give ample evidence of how dust from unpaved roads contributes to air pollution in the cities mentioned in parentheses. Transported dust from the Sahara Desert mostly affects regions close to the Sahara, such as Ouagadougou and Nouakchott. Vehicular emission remains one of the major contributors to air pollution in SSA. According to UNEP (2005), it accounts for 90% of urban air pollution in developing economies. Meteorological factors relate to how concentration of pollutants during changes in atmospheric conditions contributes to air pollution (Lindén *et al.*, 2012). The literature also highlights that increasing levels of urbanisation and urban growth, and poverty, as well as rise in vehicle ownership and industrial activities, have intensified the manifestation of most of these causal factors and by extension air pollution in the region (Ngo *et al.*, 2017).

Africa, and SSA for that matter, has scarce water resources, and there are efforts to improve the continent's water situation (UNEP, 2016; Emenike *et al.*, 2017; Lapworth *et al.*, 2017). For example, upon the expiration of the date (2015) for the achievement of the UN's Millennium Development Goals (MDG), which in many respects were not realised in the case of Africa's water situation, the UN introduced the SDGs. Target six of the SDGs aims to achieve unhindered access to safe and economical drinking water and sanitation for all. A major challenge to this goal is the continuous pollution of SSA water resources, both ground and surface water, particularly in urban areas (Lapworth *et al.*, 2017). Safe water resources are mostly achieved through good and adequate sanitation. The MDG's sanitation target was to reduce by half the proportion of the population without sustainable access to basic sanitation by 2015. However, only about 10 out of the 53 countries in Africa were on schedule to achieve this target, and most of these 10 countries were in North Africa (UNEP, 2016). This signifies a substantial water resources contamination or pollution in SSA. Indeed, available studies suggest that most of the waterborne diseases, such as cholera epidemics in SSA, result from contamination or pollution of ground and surface water. For example, Dagdeviren and Robertson (2009) report the widespread outbreak of cholera in informal settlements in Abidjan, Côte d'Ivoire, and in Dar es Salaam, Tanzania, due to groundwater contamination from inadequate sanitation facilities. The United Nations Development Programme (UNDP) (2006) also reported that one of the worst epidemics to sweep SSA in recent years was the claim of more than 400 lives in a month in Angola due to water pollution.

Water pollution in SSA cities and urban areas result from several sources. Lapworth *et al.* (2017) identify several of these sources, particularly for groundwater pollution (Table 29.1). According to the authors, water pollution is often determined by the concentration of nitrates, but the sources outlined in the table lead to the concentration of the substance. The authors noted that the major contributors to groundwater pollution in SSA cities and urban areas include:

- Municipal/domestic waste: for example, pit latrines, septic tanks, sewer leakage, sewage effluent, sewage sludge, urban road run-off, landfill/waste dumps, and health-care facilities
- Industrial sources and waste: for example, process waters, plant effluent, stored hydrocarbons, and tank and pipeline leakage
- Urban agriculture: for example, leached salts, fertilisers, pesticides, and animal and human waste
- Mining activities: including both current and historical solid and liquid waste

Table 29.1 Sources and pathways for urban groundwater contamination in Sub-Saharan Africa (SSA).

Component	Category	Risk factors
Regional considerations		Population density Land use and land cover Physical relief/slope Rainfall amount and intensity
Sources	Municipal and household level sources, including domestic livestock and urban agriculture	Surface sources: Open defecation from humans and animals Surface waste sites and incineration sites Fertilisers and pesticides and waste use (solid/liquid) Atmospheric deposition of combustion products Subsurface sources: Pit latrines Septic tanks Soak-aways Waste pits Cemetery or other burial sites Open sewers/drains – most common type in SSA Reticulated sewers – very limit coverage Other potential sources: Market places, abattoir waste, both liquid and solid
	Hospital or treatment centre	Surface and subsurface sources: Liquid waste discharge to soak- aways/surface channels Solid medical waste disposal Latrines/septic tanks on site
	Industry, e.g., mining	Surface and subsurface sources: Process plant effluent Solid waste disposal sites Storage tanks including petroleum products Site run-off and leaching from mine spoil

(Continued)

Table 29.1 (Continued)

Component	Category	Risk factors
Pathways	Horizontal and vertical pathways in unsaturated and saturated zone	Shallow sub-horizontal pathways in tropical soil: Tropical soils, e.g., Plithosol/Ferrasol horizons present Shallow depth of water table Thin soils and low organic matter content Natural rapid bypass from tree roots and burrows Vertical and horizontal pathways in saturated zone: Thin low-permeability zone above weathered basement Thickness and maturity of weathered basement zone Fracture size, length, and density in the more competent bedrock below weathered basement
	Local/headwork pathways	Lack of dugwell headwall and/or lining Lack of well cover Use of bucket and rope – soil/animal/human contact Gap between apron and well lining Damaged well apron Propensity for surface flooding Gap between borehole riser/apron Damaged borehole apron Eroded or de-vegetated spring backfill

Of course, the situation is exacerbated by current and future climate extremes, and increased urbanisation and urban growth, which are already causing increased flood risks and related disease outbreaks (Nsiah-Gyabaah, 2003).

Like air and water pollutions, noise and visual pollutions are rising in SSA cities. These are also being compounded by increased urbanisation (Nsiah-Gyabaah, 2003; Mpofu, 2013). The noise pollution predominantly stems from traffic jams and vehicular activities, such as horns; commercial activities including music from recording studios, restaurants, and discotheques and night clubs; construction activities, such as road works; worship centre activities; and industrial activities, particularly light industrial activities such as those of mechanic and joinery shops. Conversely, visual pollution emanates from irregular developments and unsightly scenes like those from the rising levels of informal and unauthorised developments including indiscriminate mounting of temporal structures, such as billboards (Oosterbaan *et al.*, 2012) and mountains of solid refuse often found in various locations in SSA cities. These pollutions often cause discomfort and sometimes lead to accidents in the case of visual pollution.

29.3 Urban Planning as a Panacea

Urban planning is widely recognised as a tool that could be effectively deployed to help redress not only the pollution taking place in SSA urban environments, but all the environmental challenges that confront the world and promote sustainable development (Godschalk, 2004; UN-Habitat, 2009; Baffour Awuah *et al.*, 2014; Corburn, 2015). Indeed, central to the achievement of the SDG target and proposed actions in relation to air quality as summarised in Table 29.2 is the role of urban planning.

Urban planning thus has a critical role to play to redress the ongoing urban pollution in SSA. In fact, the causes of urban pollution discussed in the preceding section and relevant studies demonstrate that in spite of variation in cities, urban morphology or design and human action or activities are

Table 29.2 Air quality targets of the Sustainable Development Goals (SDGs) and proposed actions for achieving the targets.

SDG	Air Quality Target	Proposed Actions
Ensure healthy lives and promote well-being for all at all ages	By 2030, substantially reduce the number of deaths and illnesses from hazardous chemicals and air, water and soil pollution and contamination	1) Strengthening countries' capacities in air quality monitoring 2) Raising awareness of air pollution and surveillance of air-pollution-related illnesses 3) Seeking technical assistance for developing air quality management systems 4) Promoting and disseminating clean cooking solutions
Build resilient infrastructure, promote inclusive and sustainable industrialisation, and foster innovation	By 2030, upgrade infrastructure and retrofit industries to make them sustainable, with increased resource-use efficiency and greater adoption of clean and environmentally sound technologies and industrial processes, with all countries taking action in accordance with their respective capabilities	1) Curbing industrial expansion in cities 2) Setting higher emission standards for factories such as installation of scrubbers 3) Prohibiting the importation of obsolete technologies
Make cities and human settlements inclusive, safe, resilient, and sustainable	By 2030, reduce the adverse per capita environmental impact of cities, including by paying special attention to air quality and municipal and other waste management	1) Seeking financial assistance for easing vehicular traffic congestion 2) Outlawing importation of over-aged vehicles and promotion of non- motorised transport 3) Ensuring sound waste management practices 4) Mobilising the community for development of urban neighbourhood roads
Ensure sustainable consumption and production patterns	By 2020, achieve the environmentally sound management of chemicals and all wastes throughout their life cycle, in accordance with agreed international frameworks, and significantly reduce their release to air, water, and soil in order to minimise their adverse impacts on human health and the environment	1) Ensuring sound waste management practices

fundamental to the activities that lead to pollution (Roy, 2009; Panagopoulos *et al.*, 2016; Perera and Emmanuel, 2016; Rosalia *et al.*, 2017). For example, Rosalia *et al.* (2017) in their study modelled six distinct types of urban designs to determine their impact on waste management and carbon emissions. The study found that urban morphology has a substantial impact on the use of public services and appliances, and represents up to 2.6% of the total carbon emission per capita. In particular, the study noted that the most contaminating urban designs are those based on single-family buildings (detached and semi-detached), as during their use stage they emit up to 155% more carbon emissions per inhabitant. This was because designs based on single-family houses comparatively call for longer waste collection service routes, as well as more collection points because dwellings are more dispersed. Kozlovtseva *et al.* (2016) in their study on indoor air quality of multi-storey educational buildings noted that people spend about 90% of the day in confined areas and found that the dust particle diameter decreases when building height increases. The authors therefore concluded among others that building use designs can be choreographed to reduce dust pollution. Also, Lindén *et al.* (2012) in their air quality study in Ouagadougou found that, in broad terms, rapidly growing spontaneous settlements are highly associated with air pollution and that unpaved and unirrigated residential areas are likely to be exposed to much higher levels of PM_{10} compared to paved and irrigated central residential areas. Further, the preceding section noted that climate change could cause both air (through concentration of particles in the air) and water (through flooding) pollution. However, adaptation strategies through urban design using knowledge of significant land use–land cover elements, and estimated urban heat islands, as well as their effects on local climate could be used to mitigate these pollutions (Perera and Emmanuel, 2016).

Urban planning is the main tool for the design of cities and urban areas. Also, as stated in the introduction section of the chapter, planning allocates land resources among the various desired land uses and also promotes certain allocation of financial and human resources (Baffour Awuah, 2013). This presupposes that, on the basis of information such as those presented in the preceding paragraph, planning could be used to design or redesign cities and promote the use of land resources and properties in certain ways to prevent or minimise urban pollution. This could be done through several plans, such as strategic, development, and local plans. Thus, the above discussed knowledge could be incorporated into the decision relating to the formulation and implementation of these plans. For example, a city plan could incorporate a ring road to reduce traffic jams in the inner city. This will also reduce noise pollution and CO_2 emissions.

Urban pollution management could be further carried out as part of development control using regulatory frameworks under a country's planning system. This may be done through the use of a zoning ordinance, which specifies land for various uses like residential, commercial, educational, industrial, and recreational, and instruments like the floor area ratio (FAR) and building setbacks to control the intensity of development (Mathur, 2015; Baffour Awuah, 2016). For example, urban development regulations could be used to set noise limits for music from recording studios, restaurants, and discotheques. It could also be used to promote the plantation of trees and vegetation cover within residential neighbourhoods and industrial areas, along unpaved roads and waterbodies to mitigate dust and other greenhouse gases (in particular CO_2) emissions, and water pollution through land degradation.

A planning system could also provide education through, for example, capacity development, to enlighten urban sector stakeholders on urban pollution, its sources and effects or impact, as well as interventions – land use decisions and practices, and development materials that can prevent air, water, and noise and visual pollution. These may include development based on eco-effective principles; the use of green and air cleansing development materials, such as green doors and roofs; and renewable energy. Indeed, stakeholders should be made part of the solution to the problem and should feel involved in the solution process (Ngo *et al.*, 2017). Connected to this is the need for the provision of incentives, contrived or instinctive, by planning systems to impel urban sector stakeholders to take up eco-effective development principles and the use of environmentally enhancing materials. The foregoing implies that planning as a tool to address urban pollution should be based on sustainability principles and be informed by scientific, environmental, economic, and sociocultural principles, and acceptable to relevant communities (Rosalia *et al.*, 2017). This further implies that a planning system devoted to addressing urban pollution, among other things, should have the mandate, capacity (both human and material resources), the requisite data, and be based on collective action involving all stakeholders, as well as encourage compliance with its regulations.

29.4 Lessons for SSA Urban Planning

The preceding section has demonstrated how urban planning could be used to help address urban pollution and some requirements needed for doing so. However, SSA planning systems in their current form largely appear incapable of addressing the rising levels of urban pollution taking place in the region's cities in the face of rapid urbanisation, as they are inadequate. The planning systems are largely relics of colonialism. They are not based on broad-based participation and rely on complex bureaucratic processes and procedures, and often lack the capacity, human and material resources, as well as the requisite data to formulate and implement plans (UN-Habitat, 2009; Baffour Awuah, 2013). For example, to address urban pollution in the region, it may be useful to integrate climate-sensitive design with local planning processes to manage warming trends. However, as Perera and Emmanuel (2016) noted, although an understanding of the interaction between the physical form and climate context is now emerging, there are challenges with data needs and methods of analysis to translate them into practical applications. Further, findings from urban climatology are not communicated in ways that are meaningful or understandable to urban planners. It is also common knowledge that due to inadequate planning schemes and resources, planning in the region continues to chase developments, resulting in sprawls and informal developments. These imperatives undoubtedly do not positively correspond with the essentials required for planning to contribute to addressing pollution in the region.

Nevertheless, there are ongoing initiatives in SSA mostly between governments and development agencies to improve planning practice in the region. These, for example, have resulted in the passage of new generation urban policies across the region (UN-Habitat, 2014b) and the adoption of the Sustainable Cities Programmes (SCP), which seek to promote sustainable urban development and management based on collective action (UN-Habitat, 2008). On the basis of the preceding discussions, however, this chapter proposes that in the context of mitigation of rising SSA urban pollution, ongoing initiatives to improve planning in the region should place more emphasis on:

- Strengthening the capacities of planning authorities and institutions, as well as providing them with adequate human and material resources
- Ensuring that planning institutions and urban planners have access to requisite volumes of data in a form understandable to them
- Ensuring effective coordination between planning authorities, institutions, and other relevant organisations to design programmes to mitigate urban pollution
- Constant engagement of planning authorities and institutions with communities and urban sector stakeholders, and educating them on pollution prevention and management activities and practices
- Providing incentives to stakeholders to take up pollution prevention and management practices
- The need for planning systems to encourage compliance with proposals and regulations to mitigate urban pollution

References

Akumu, J. (2016) Improving Air Quality in African Cities. Nairobi: UNEP. Available online at: http://www.unep.org/Transport/astf/pdf/Overview_AirPollution_Africa.pdf. (Accessed: 15 March 2017).

Amegah, A.K. and Agyei-Mensah, S. (2017) Urban air pollution in Sub-Saharan Africa: Time for action. *Environmental Pollution*, 220, 738–744.

Arku, R.E., Vallarino, J., Dionisio, K.L., Willis, R., Choi, H., Wilson, J.G., Hemphill, C., Agyei-Mensah, S., Spengler, J.M., and Ezzati, M. (2008) Characterizing air pollution in two low-income neighbourhoods in Accra, Ghana. *Total Environment*, 402, 217–231.

Baffour Awuah, K.G. (2013) A Quantitative Analysis of the Economic Incentives of Sub-Saharan Africa Urban Land Use Planning Systems: Case study of Accra, Ghana. PhD Thesis. University of Wolverhampton, UK.

Baffour Awuah, K.G. (2016) Leveraging Rising Land Values to Finance Urban Infrastructure Development in Ghana: A Case Study of Accra. Final Report: UWE/GLGS.

Baffour Awuah, K.G., Hammond, F.N., Booth, C.A., and Lamond, J.E. (2014) Evolution and development of urban land use planning: Analysis from human action theory perspective. *Theoretical and Empirical Researches in Urban Management*, 9 (2), 35–67.

Boman, J., Lindén, J., Thorsson, S., Holmer, B., and Eliasson, I. (2009) A tentative study of urban and suburban fine particles (PM2.5) collected in Ouagadougou, Burkina Faso. *Xray Spectrom*, 38, 354–362.

Brauer, M., Amann, M., Burnett, R.T., Cohen, A., Dentener, F., Ezzati, M., Henderson, S.B., Krzyzanowski, M., Martin, R.V. Van Dingenen, R., van Donkelaar, A., and Thurston, G.D. (2012) Exposure assessment for estimation of the global burden of disease attributable to outdoor air pollution. *Environmental Science & Technology*, 46 (2), 652–660.

Cohen, A.J., Anderson, H.R., Ostro, B., Pandey, K.D., Krzyzanowski, M., Kunzli, N., Gutschmidt, K., Pope, A., Romieu, I., Samet, J.M., and Smith, K. (2005) The global burden of disease due to outdoor air pollution. *Journal of Toxicology and Environmental Health*, 68, 1301–1307.

Corburn, J. (2015) City planning as preventive medicine. *Preventive Medicine*, 77, 48–51.

Dagdeviren, H. and Robertson, S. (2009) Access to Water in the Slums of the Developing World. Policy Centre for Inclusive Growth Working Paper 57. Available online at: UNDP.http://www.ipc–undp.org/pub/IPCWorkingPaper57.pdf (Accessed: 15 March 2017).

Dionisio, K.L., Rooney, M.S., Arku, R.E., Friedman, A.B., Hughes, A.F., Vallarino, J., Agyei-Mensah, S., Spengler, J.M., and Ezzati, M. (2010). Within neighborhood patterns and sources of particle pollution, mobile monitoring and geographic information system analysis in four communities in Accra, Ghana. *Environmental Health Perspectives*, 118, 607–613.

Eliasson, I., Jonsson, P., and Holmer, B. (2009). Diurnal and intra-urban particle concentrations in relation to windspeed and stability during the dry season in three African cities. *Environmental Monitoring and Assessment*, 154, 309–324.

Emenike, C.P., Tenebe, I.T., Omole, D.O., Ngene, B.U., Oniemayin, B.I., Maxwell, O., and Onoka, B.I. (2017). Accessing safe drinking water in sub-Saharan Africa: Issues and challenges in South–West Nigeria. *Sustainable Cities and Society*, 30, 263–272.

Environmental Protection Authority (EPA) Victoria (2016) Your Environment. Available online at: http://www.epa.vic.gov.au/your–environment/air (Accessed: 15 March 2017).

Gatari, M.J., Boman, J., and Wagner, A. (2009) Characterization of aerosol particles at an industrial background site in Nairobi, Kenya. *Xray Spectrom*, 38, 37–44.

Godschalk, D.R. (2004) Land use planning challenges: Coping with conflicts in visions of sustainable development and liveable communities. *Journal of the American Planning Association*, 70 (1), 5–13.

Gozalo, G.R., Morillas, R.M.B., Carmona, J.T., González, D.M., Moraga, P.A., Escobar, P.G., Vílchez-Gómez, R., Sierra, J.A.M., and Prieto-Gajardo, C. (2016) Study on the relation between urban planning and noise level. *Applied Acoustics*, 111, 143–147.

Kozlovtseva, E.Y., Loboyko, V.F., and Nikolenko, D.A. (2016). Monitoring of fine dust pollution of multistory buildings air environment as an adoption factor of town-planning decisions. *Procedia Engineering*, 150, 1954–1959.

Lapworth, D.J., Nkhuwa, D.C.W., Okotto-Okotto, J., Pedley, S., Stuart, M.E., Tijani, M.N., and Wright, J. (2017). Urban groundwater quality in sub-Saharan Africa: Current status and implications for water security and public health. *Hydrogeology Journal*, 25(4), 1093-1116.

Lindén, J., Boman, J., Holmer, B., Thorsson, S., and Eliasson, I. (2012). Intra-urban air pollution in a rapidly growing Sahelian city. *Environment International*, 40, 51–62.

Mackenzie, J. (2016). Air Pollution: Everything You Need to Know (NRDC). Available online at: https://www.nrdc.org/authors/jillian–mackenzie (Accessed: 15 March 2017).

Mathur, S. (2015). Sale of development rights to fund public transportation projects: Insights from Rakjot, India, BRTS project. *Habitat International*, 50, 234–239.

Mpofu, T.P.Z. (2013) Urbanization and urban environmental challenges in Sub-Saharan Africa. *Research Journal of Agricultural and Environmental Management*, 2 (6), 127–134.

Ngo, N.S., Kokoyo, S., and Klopp, J. (2017) Why participation matters for air quality studies: Risk perceptions, understandings of air pollution and mobilization in a poor neighbourhood in Nairobi, Kenya. *Public Health*, 142, 177–185.

Nsiah-Gyabaah, K. (2003) Urbanization, Environmental Degradation and Food Security in Africa. Paper presented at the meeting of the Global Environmental Change Research Community, Montreal, Canada, 16–18 October.

Oosterbaan, C., Arku, G., and Asiedu, A.B. (2012) Conversion of residential units to commercial spaces in Accra, Ghana: A policy dilemma. *International Planning Studies*, 17 (1), 45–66.

Perera, N.G.R. and Emmanuel, R. (2016) A 'Local Climate Zone' based approach to urban planning in Colombo, Sri Lanka. *Urban Climate*, 1–16.

Panagopoulos, T., Duque, J.A.G., and Dan, M.B. (2016) Urban planning with respect to environmental quality and human well-being. *Environmental Pollution*, 208, 137–144.

Rosalía, P., Julio Roldán, J., Gago, E.J., and Ordó˜nez, J. (2017) Assessing the relationship between urban planning options and carbon emissions at the use stage of new urbanized areas: A case study in a warm climate location. *Energy and Buildings*, 136, 73–85.

Roy, M. (2009) Planning for sustainable urbanisation in fast growing cities: mitigation and adaptation issues addressed in Dhaka, Bangladesh. *Habitat International*, 33, 276–286.

United Nations (UN) (2015) Transforming Our World: The 2030 Agenda for Sustainable Development. In: Seventieth General Assembly, New York, 15 Septembere2 October. Resolution A/RES/70/1. New York: United Nations. Available online at: https://sustainabledevelopment.un.org/post2015/transformingourworld/publication (Accessed: 15 March 2017).

UNDP (2006). Human Development Report 2006. United Nations Development Programme. Available online at: http://hdr.undp.org/en/reports/global/hdr2006/(Accessed: 15 March 2017).

UNEP (2005). Urban Air Pollution. Available online at: http://www.unep.org/urban_environment/Issues/urban_air.asp (Accessed: 10 October 2014).

UNEP (2016). Emerging Issues of Environmental Concern. UNEP Frontiers 2016 Report. Nairobi: UNEP.

UN-Habitat (2008). *The Sustainable Cities Nigeria Programme (1994–1996): Building Platforms for Environmentally Sustainable Urbanisation*. Nairobi: UN-Habitat.

UN-Habitat (2009). *Planning Sustainable Cities: Policy Directions, Global Report on Human Settlements* (Abridged Edition). London: Earthscan.

UN-Habitat (2014a). *The State of African Cities 2014:Re-imagining Sustainable Urban Transitions*. London: Earthscan.

UN-Habitat (2014b). *New Generation of National Urban Policies.* Nairobi: UN-Habitat.

30

Water Pollution and Urbanisation Trends in Lebanon: Litani River Basin Case Study

Jamal M. Khatib[1,3], Safaa Baydoun[2], and A. A. ElKordi[1]

[1] *Faculty of Engineering, Beirut Arab University, Lebanon*
[2] *Research Centre for Environment and Development, Beirut Arab University, Lebanon*
[3] *Faculty of Science and Engineering, School of Architecture and Built Environment, University of Wolverhampton, United Kingdom*

30.1 Introduction

Urban growth in Lebanon has been rapid over the last few decades. The recent expansion of major cities constituting more than 85% of the Lebanese population has become a national concern due to the environmental impacts and increased needs for services and facilities necessary for human welfare (MoE/UNDP/ECODIT, 2011). Although the country is blessed with reasonable water availability, pressures on this important resource have increased in recent decades; the country is now experiencing considerable water stress and serious quality deterioration (MoE/UNDP/ECODIT, 2011; gef/WB/Plan Bleu, 2015). While the country's population has more than doubled, from 2,703,019 in 1990 to 5,850,743 in 2015 (World Bank Group, 2016), urbanisation has largely outpaced the institutional capacity of the Lebanese government to efficiently manage wastewater, creating severely polluted water bodies. It is estimated that up to 90% of all wastewater is discharged untreated directly into rivers, lakes, or the sea, causing major environmental degradation and health risks. Numerous recent studies have identified a broad spectrum of pollutants, high biochemical oxygen demand (BOD) levels, and microbiological counts affecting both surface and groundwater (gef/WB/Plan Bleu, 2015; MoE/UNDP/ECODIT, 2011).

Lebanon is situated on the eastern shore of the Mediterranean Sea at the meeting point of the Fertile Crescent and the Mediterranean Basin, with Syria to the north and east (Figure 30.1). It is a mountainous country with an area of 10,452 km^2 featuring sharp changes between lowland and highland topography. The country is divided into five geo-morphological regions:

1) The coastal zone, including the shoreline and continental shelf, the narrow strip of coastal plain, and the foothills of Mount Lebanon. This zone represents 13% of the total territory and is where most of the population lives.
2) Mount Lebanon range, representing 47% of the territory, which rises steeply to reach the highest altitudes in the country at 3,088 m above sea level.
3) Bekaa Valley, fertile land representing 14% of the territory located between the Eastern and Western mountain range at an average elevation of 762 m.
4) The Eastern mountain range along the Syrian borders in the eastern part of the country representing 19% of the total territory. It gradually rises in the south to reach

Urban Pollution: Science and Management, First Edition. Edited by Susanne M. Charlesworth and Colin A. Booth.

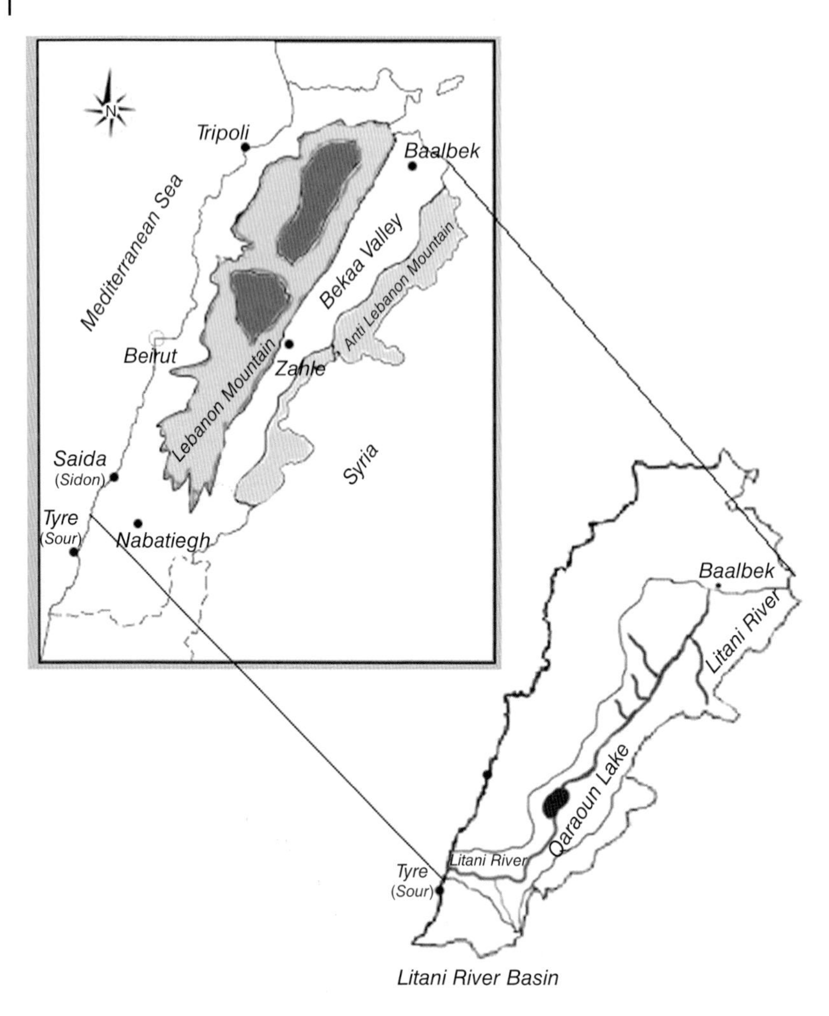

Figure 30.1 Lebanon map.

its highest altitude at Mount Hermon (2,814 m).

5) South Lebanon, an elevated plateau that represents 7% of the territory and extends between the coastal zone in the south and the foothills of Mount Hermon in the southeast.

Lebanon's climate is typically Mediterranean, characterised by four distinct seasons: a long, semi-hot, and dry summer, a rainy and snowy winter, and two transitional seasons of fall and spring which have moderate temperatures and little rain. Most of the rainfall occurs between November and January, with the annual precipitation varying between 600 and 900 mm along the coast and 1400 mm over the mountains an average of 800 mm. While precipitation decreases to 400 mm in eastern regions and less than 200 mm in the northeastern parts of the country, snowmelt at high altitudes (>2000 m) constitute a major contribution to the total run-off (Assaf, 2010).

With water at the very core of the UN 2030 Agenda for Sustainable Development, Goal 6 sets out to 'ensure availability and sustainable management of water and sanitation for all', and it urges developing countries like Lebanon to devote more attention to improving water management for future sustainable development and economic growth (UN-Water, 2014). In recent decades, Lebanon has made noteworthy strides in the water sector. Major reforms to create autonomous and accountable utilities and considerable investment to expand and improve services have been achieved. However, the reforms are not yet fully in place, and the efficiency of investment has been relatively low (gef/EU/Plan Bleu, 2015). As a result, wastewater management is still very poor, and the severe urban pollution is seriously affecting water quality. This chapter will provide an overall review of the current critical status of water quality of the Litani River in view of concurrent urbanisation. It will explain recent efforts of wastewater management and the future perspectives necessary for the integrated management of the basin.

30.2 Water Resources in Lebanon

To date, estimations of the hydrological budget in Lebanon vary according to the source of information and assumptions made, and are mostly based on the UNDP study of groundwater (UNDP, 1970). These estimates do not take much regard of recent trends in water demand, overexploitation, and the impact of climate change on the water cycle. Nevertheless, estimates of precipitation range between 8,200 and 9,881 Mm^3, and net renewable and exploitable amounts range between 2,400 and 2,767 Mm^3 (MoE/UNDP/ECODIT, 2011; MoEW and UNDP, 2014). Lebanon has 17 perennial and 23 seasonal rivers with a total annual flow of 3,900 Mm^3 on average. Of them all, the perennial Litani, Ibrahim, and El Assi Rivers have the highest flows (MoE/UNDP/ECODIT, 2011). The groundwater balance, the difference between water recharge and discharge, is estimated at 2,140 Mm^3 in dry years, and up to 4,675 Mm^3 in wet years (MoEW and UNDP, 2014). Annual water demand in 2010 (MoE/UNDP/ECODIT, 2011) was estimated to range between 1,473 to 1,530 Mm^3, with agriculture being the major consumer (60%). These values will have changed in recent years due to the sudden increase in population with the influx of Syrian refugees and as more water is being diverted for domestic and industrial consumption. In its *State and Trends of the Lebanese Environment 2010*, the Ministry of Environment (MoE) importantly stressed that these estimates do not reflect the water deficit that usually occurs during the dry seasons of summer and fall. Such a deficit was estimated to range between 220 to 388 Mm^3 in 2010 (MoE/UNDP/ECODIT, 2011). Even before the Syrian conflict, it was evident that Lebanon had already entered a critical period of water stress whereby demands had been estimated to reach 2,055 Mm3 in 2020 and 2,818 Mm^3 in 2030 to exceed the exploitable renewable resources (MoE/UNDP/ECODIT, 2011). Meanwhile, agriculture has been facing major challenges, and agricultural lands have become fragmented to be ultimately abandoned (Darwish, 2012; Haydamous and El Hajj, 2016).

With the Syrian conflict now entering its seventh year (2011–2017), the unexpected increase in the Lebanese population by as much as 37% has resulted in extra pressure on the already fragile water resources. According to the United Nations High Commissioner for Refugees (UNHCR), the number of registered Syrian refugees only reached 1,011,366 at the end of 2016, with tens of thousands more awaiting registration or residing in Lebanon illegally (UNHCR, 2016). At the end of 2014, an increase of 43–70 Mm^3 (8%–12%) in domestic water demand was reported (MoE/EU/UNDP, 2014).

As well as all these major uncertainties, the impact of climate change is likely to have substantial impacts, increasing water demand while shrinking water supplies, due to projected temperature increases of around 1°C–2°C in 2040 and 3.5°C–5°C in 2090 accompanied by 10%–20% decreases in rainfall by 2040 and 25%–45% by 2090. Significantly less wet and substantially warmer conditions will result in intensified temperature and precipitation extremes and more frequent and extended drought periods, jeopardising food, water, and energy security

(MoE/UNDP/GEF, 2011; 2015). With this grim and alarming picture, treated wastewater presents the potential to contribute to the water balance on the supply side and partially meet industrial and agricultural water needs.

30.3 Urbanisation Trends

Urbanisation rates in Lebanon have exceeded the service capacities of large cities; notably, in the capital city of Beirut, also known as the Greater Beirut Area, water and wastewater infrastructures are increasingly challenged, and water pollution continues to be one of the most serious threats to water resources (gef/WB/Plan Bleu, 2015; MoE/UNDP/ECODIT, 2011). In spite of data limitations and significant discrepancies in urbanisation estimates, it is apparent that urban planning in Lebanon is exclusively geared towards maximising land use coefficients and commercial profit, and is not adequately addressing sustainability, access to services, and environmental standards (MoE/UNDP/ECODIT, 2011). Accurate recent data on total zoning is lacking in the country. In 2004, the total area approved by the Higher Council of Urban Planning (HCUP) was 1,693 km^2, or 16.2 % of the total territory (MoE/UNDP/ECODIT, 2011). Since then, no data on the number and extent of urban master plans has been reported. Nonetheless, the Center for Remote Sensing has made attempts to understand urbanisation trends and land use on the basis of topographic maps and multi-date high spatial resolution satellite images (Darwish *et al.*, 2004; Faour, 2015; Faour and Mhawej, 2015). Findings show drastic increases in the urbanisation rate, from 221 to 465 km^2 between 1963 and 1994 to reach 741 km^2 in 2005, and the rate is projected to hit 884 km^2 in 2030 (Table 30.1). Beirut shows the highest rate, with 98% of its total area already urbanised. The city, which started experiencing high rates of urban development in 1963, rapidly spread to bordering towns and villages. Other major cities such as Tripoli, Sour, Saida, Zahle, Baalbek, and Nabatiye also exhibit substantial urban expansion. Uncontrolled urban expansion has not only created large agglomerations around major cities, but has also caused the emergence of large suburbs around Beirut and Tripoli associated with the coastal zone that have gradually became slums (UN habitat, 2003, 2016). The formation of these slums largely took place in waves associated with rural-urban migration and mass displacement during the civil war of 1975–1990, Israel's 18-year occupation of south Lebanon between

Table 30.1 Urban area in (km^2) of major urban cities and the whole country, 1963–2005.

Agglomeration		1963	1994	2005	Growth Factor
Beirut		63.139	113.191	121.421	1.92
Tripoli		4.337	8.376	14.113	3.22
Zahle		1.849	4.699	5.287	2.85
Baalbek		1.54	4.366	5.967	3.87
Saida		0.764	3.278	4.178	5.46
Sour		0.85	2.3	3.34	3.92
Nabatiye		0.966	1.298	3.829	3.96
Jounieh		5.366	23.929	38.101	7.1
Lebanon	Urban Area km^2	221.47	464.98	741.2	
	Urban Area (%)	2.16	4.53	7.22	

Source: Faour, 2015.

1982 and 2000, and the establishment of camps or low-cost housing for international refugees. Nowadays, these slums house a large population and lack proper wastewater services, while raw sewage is commonly discharged to nearby waterbodies or abandoned groundwater wells (MoE/UNDP/ECODIT, 2011; UN habitat, 2003, 2016). With the lack of an evidence-based approach to urban dynamics combined with inadequate standards in urban planning, it is very likely that Lebanon will further witness more uncontrolled migration to urban centres, to challenge inadequate wastewater services and infrastructure.

30.4 Wastewater Management

Since networks are inadequate and currently suffer from major leakages and maintenance problems, the data for wastewater flow remains inaccurate. The Ministry of Energy and Water estimates that in 2012 Lebanon generated about 310 Mm^3 of wastewater, of which 250 Mm^3 was from municipal and domestic establishments and 60 Mm^3 was from industrial enterprises (MoEW, 2012). Since early 1990, more than $1.4 billion has been invested in wastewater facilities, as a result of which considerable treatment capacity has been installed, and about 60% of the population, before the Syrian conflict, was connected to wastewater collection networks (gef/WB/Plan Bleu, 2015). The 2011 Country Environmental Assessment by the World Bank estimated that the cost of environmental damage from untreated wastewater was 1.08% of the GDP annually (WB, 2011). The World Bank (WB, 2011) reported that wastewater generated from households was estimated at 249 Mm^3/year with approximately 119,348 tons of BOD. On the basis of the Central Administration of Statistics, only 52% of buildings were provided with access to improved sewage networks in 2004, while the remaining 48% depended on septic tanks and cesspits, which are very common in rural areas (MoE/UNDP/ECODIT, 2011). Network coverage is generally adequate in large cities and urban centres such as Beirut and Mount Lebanon. Incomplete treatment plants or lack of operational and maintenance capacity, however, complicate wastewater management and induce environmental and water pollution (MoE/UNDP/ECODIT, 2011). The management of specifically industrial wastewater is almost totally absent in Lebanon, causing the addition of an estimated 60 Mm^3 of untreated industrial wastewater to municipal sources, thus increasing pollution loads to receiving waterways. In spite of the lack of recent data on both the quantity and the quality of effluents, estimates from 1998 show that about 5000 tons of BOD and an array of inorganic toxic pollutants is annually discharged through industrial wastewater (METAP/Tebodin, 1998; Mouneimne and Gemayel, 2012).

In response to the Barcelona Convention and its amendments, including the Protocol for the Protection of the Mediterranean Sea against Pollution from Land-Based Sources, in 2005, Lebanon developed a Sectoral Plan for the Reduction of Pollution of the Mediterranean Sea. The increase in urbanisation in the coastal area, however, has considerably jeopardised the ecological and socio-economic integrity of the coast and the subsequent successful implementation of the plan. Out of 53 major sewage outfalls, only two are connected to operational treatment plants, at Ghadir and Saida, to receive preliminary treatment before discharge to the Mediterranean Sea (MoE/UNDP/ECODIT, 2011). Large solid waste dumpsites are also located in the heavily populated and urbanised coastal strip, increasing pollution risk into the sea (Nader *et al.*, 2012).

Due to the destruction caused by the civil war in 1975–1990, wastewater management was generally absent in Lebanon for many years. Soon afterwards, a National Emergency Reconstruction Program (NERP) for wastewater management was launched, setting the goal to connect nearly 80% of the population to 113 major sewage treatment plants by 2020 (WB, 1993). To date, the construction of sewer networks and treatment stations, and the rehabilitation achieved by NERP as well as efforts of the Council for Development and Reconstruction (CDR) have mainly focused on metropolitan areas. Villages and mountainous regions therefore still lack an integrated water and wastewater management scheme. In 2012, the Council of Ministers endorsed the New National Water Sector Strategy (2010–2020) (NWSS) set by MoEW which addresses infrastructural concerns relating to transmission, distribution, and wastewater treatment, as well as management issues related to institutional, financial, legal, and environmental concerns (MoEW, 2012). In 2012, MoEW, in an effort specifically dedicated to wastewater management, launched a National Strategy for the Wastewater Sector, setting goals to increase wastewater collection from 60% to 80% by 2015, treatment from 8% to 95% by 2020, increase reuse of treated effluent from 0% to 50% by 2020, and recover operation and maintenance costs fully by 2020 (WB, 2012). To define a plan for its involvement in the implementation of these strategies during 2012–2016, the World Bank developed a Country Water Sector Assistance Strategy (WB, 2012) that indicated the need to assign clear responsibilities for the various institutions involved and the development of local capacities to complete installed services. Unfortunately, there is no formal mechanism in place yet to monitor the implementation status of these strategies or to track overall progress towards achieving the objectives set. In fact, the Strategic Environmental Assessment of the NWSS showed that the majority of wastewater management and implementation of treatment plants remains incomplete (gef/WB/Plan Blue, 2015). Sluggish economic performance, hostile terrorist attacks, and the surge of Syrian refugees has stretched the infrastructure of the country to its limits, drastically worsening the vulnerability of wastewater management.

30.5 Water Quality

Water quality deterioration of rivers and groundwater resources is one of the main threats to water sustainability in urban developed areas. Although technical solutions have improved significantly, wastewater treatment and sanitation constitute a major problem in cities in Lebanon. Decreasing water quality is also caused by degradation and land use changes in the catchment area. Unchecked wastewater discharges into the Mediterranean Sea from agglomerations and industry are affecting coastal water quality. Lebanon's coastline receives around 162 Mm^3/ year (equivalent to 276,000 m^3/ day) of untreated or partially treated sewage from at least 53 outfalls spread along Lebanon's 240 km coastline, of which 16 lie within the Beirut area (gef/WB/Plan Blue, 2015). On the basis of its seawater monitoring programmes, the National Centre for Marine Sciences profiled five public beaches during 20082010. Not unexpectedly, results showed very high bacteriological contamination at Beirut and Saida beaches (gef/WB/ Plan Blue, 2015).

Most springs and river systems in Lebanon also continue to be adversely impacted by raw sewage generated from residential and industrial areas, resulting in dangerously high biological loads and unacceptable high levels of total coliform and *E. coli* (BAMAS, 2005; Hamze *et al.*, 2005; Houri and El Jeblawi, 2007; LRBMS, 2010; LMTA, 2013; LRBMS, 2013; Haydar *et al.*, 2014b) (Table 30.2). Similarly, groundwater wells in highly urbanised zones also show faecal contamination and increased salinity in inland and coastal zone aquifers (El Moujabber *et al.*, 2006; Halwani *et al.*, 2000; LRBMS, 2010; LRBMS, 2012a). Generally, coastal wells are subject to severe salt water intrusion. This situation is particularly acute in the Beirut area (Korfali and Jurdi, 2010).

With the Syrian conflict, the dramatic increase in wastewater and solid waste dumping in rivers and along the coastline have

Table 30.2 Quality parameters for selected rivers (dry season).

River	BOD_5 (mg/L)	Total coliform (CFU/100 mL)	Faecal coliform (CFU/100 mL)	Faecal Streptococcus (CFU/250 mL)	*E. coli* (CFU/100 mL)
Kabir [1]	14.4	900	—	—	20
Bared [1]	28.2	610	—	—	17
Abou Ali [1]	39.3	26,500	—	—	3,000
Ibrahim [1]	62.8	3,500	—	—	200
Antelias [1]	53.2	28,000	—	—	6,000
Damour [1]	21.3	490	—	—	15
Awali [1]	33.4	710	—	—	1
Qasmieh[1]	22.5	80	—	—	0
Zahrani [2]	—	500	350[a)]	80	—
Kabir [3]	—	540,091	78,438	—	—
Litani [4a,b]	48.46	—	2,234,877	—	—
	547.65	—	71.61	—	—
Hasbani[5]	5.9	835	658	—	—

a) CFU/250 mL.
Source: [1]Houri and El Jeblawi, 2007; [2]ELRAD, 2006; [3]Hamze *et al.* 2005; [4a]BAMAS, 2005; [4b]LRBMS, 2010; [5]Badr *et al.*, 2014.

caused drastic increases in pollution levels. The incremental pollution load of wastewater generated by refugees was estimated at an additional 40,000 tons BOD per year, representing an increase of around 34% at the national level (FAO, 2016; MoE/EU/UNDP, 2014). This was accompanied by sharp rises in communicable diseases among refugees and the hosting Lebanese communities (APIS, 2016; MoE/EU/UNDP, 2014; MoPH, 2017).

30.6 The Case of the Litani River Basin

The Litani River presents an example of the influence of urbanisation on water quality in Lebanon. Improperly managed and highly contaminated domestic and industrial wastewaters are discharged into the river along with agricultural run-off, rendering a valuable water resource unusable. The Litani River is one of the most important waterbodies in Lebanon. Running for about 170 km in length, it provides an average annual flow estimated at 770 Mm^3, accounting for about 30% of the total water flow in all Lebanese rivers. The river originates near the ancient city of Baalbek, and runs in a southwesterly direction through Bekaa Valley to the Mediterranean Sea to the north of Tyre. The Litani River Basin covers about 2000 km^2 and supports the livelihoods of 20% of the Lebanese territory (Figure 30.1). In the 1950s, the Qaraoun Dam was constructed with a reservoir of approximately 220 Mm^3 storage capacity for hydropower generation and irrigation during the dry season. The Qaraoun reservoir divides the river basin into the Upper Litani sub-basin (ULB), which covers about 1500 km^2 of the Bekaa Valley (1000 m elevation), and the lower sub-basin (LLB), which occupies the remaining 500 km^2. The Litani River has 16 tributaries, the majority of which are situated in Upper Litani Basin (LRA, 2017).

On the basis of the most recent estimates of water balance in Upper Litani Basin by LRBMS (2011), the total run-off in the river is estimated at some 300 Mm^3/year. Competing demands of 210 Mm^3/year in 2010, 60 Mm^3/year from surface withdrawal, and 150 Mm^3/year from groundwater pumping have led to difficult allocation decisions which have the potential to limit both economic growth and sustainable development, not only just regionally, but nationally. Excessive groundwater is now facing a significant annual drawdown estimated at 70 Mm^3, and this is projected to increase in the short term, especially in the dry summer seasons as the rivers flow decrease to 70 Mm^3, mainly made up of wastewater/sewage (LRBMS, 2011). In spite of this fact, the plan is for the Qaraoun reservoir to provide water for domestic use in Beirut (Bisri-Awali project) and to irrigate large areas in the south (Canal 800 project).

The geological history and climate reflects the variation in surface cover, soil type, and hydrology and has resulted in a variety of human activities in the river basin, including agriculture, industries, and trade and services. As many as 250 villages and small towns are located there, most of which are concentrated in the ULB, which had an estimated population of 1.06 million in 2012, with 77% rural and 23% urban (SWIM, 2013). A significant proportion of the Syrian refugees reside in the Upper Basin, and the current population may make up one third of the total living there (MoE/EU/UNDP, 2014). Apart from a few villages, the Lower Litani River Basin does not have any urban centres, and land use is mostly natural and some traditional agricultural activities; the main cities of Baalbeck, Zahleh, Anjar, Bar Elias, Joub Jenine, and Qabb Elias are situated in the ULB. The Litani River Basin has a high poverty ranking as 25% of the population is characterised as poor or very poor; employment is below 40% (65% for males, 10% for females), with agriculture contributing 20% to the income (SWIM, 2013). According to the USAID programme 'Litani River Management Support System' (LRBMS) in 2012, land use in ULB consist of 50% natural land, notably on the hillsides of Mount Lebanon and Anti-Lebanon; 40% agricultural land, mostly in the Bekaa region; and 10% urban and peri-urban areas, industry, quarries, roads/highways, and other artificial made structures (LRBMS, 2012a). These values are likely to have significantly changed since the large increase in population resulting from the Syrian conflict, which has significantly changed the socio-economic state of affairs. According to the UN, nearly half a million refugees are living in Bekaa, bringing the issue of poverty in ULB to a critical level (UNHCR, 2016) and placing tremendous pressure on water and agricultural productivity, making it difficult to fulfil the needs of both the Lebanese and the Syrian refugees (MoE/EU/UNDP, 2014).

30.7 Urbanisation and Water Pollution Trends

The current deterioration in water quality in the LRB is a threat to livelihoods and a growing source of conflict between competing users in different sectors. Random urbanisation and poor wastewater management combined with the uncontrolled use of fertilisers have combined to accelerate eutrophication and increase levels of various contaminants in the Litani River and Qaraoun reservoir, particularly ULB (Baydoun *et al.*, 2016; Haydar *et al.*, 2014a; LRBMS, 2010; SWIM, 2013). These problems are exacerbated by climate change, as extended dry periods and higher temperatures are likely to have increased bacterial activity and pollution levels. Faour (2015) evaluated urban expansion in Lebanon between 1963 and 2005 using topographic maps and satellite images, and showed that the main cities of Zahle and Baalbek in ULB exhibited among the most rapid urbanisation rates (2.85 and 3.87, respectively) (Table 30.1). An estimated 45.4 Mm^3 of untreated wastewater was discharged into the ULB with a BOD load of 16,600 tons/year (SWIM, 2013); this was expected to reach 72.9 Mm^3/year. About 71% of the households are connected to a sewer, while the remaining 29% (30,674) discharge into un-maintained septic tanks and directly into waterbodies (SWIM, 2013). The Syrian conflict will have substantial impacts on these figures with ULB being affected the most, with surface and groundwater pollution estimated to increase to around 13,840 tons BOD per year (MoE/EU/UNDP, 2014).

Industrial wastewater generated from about 294 medium- and small-scale industrial establishments including agribusiness, plastics manufacturing, detergents, batteries, and paper manufacturing as well as dyeing and tanning was estimated at 4 Mm^3 in 2011. This waste contains a wide spectrum of biodegradable organic pollutants and non-biodegradable inorganic pollutants such as hydrocarbons, phenols, and detergents as well as various nitrogen and phosphate chemicals (MoE/UNDP/ECODIT, 2011). With the lack of incentives to treat their wastewater, industries discharge it mostly untreated into the sewer network or directly into the Litani River. The situation becomes aggravated during the summer months, when the concentration of untreated wastewater increases, transforming the river into 'an open sky sewer'. As indicated in Table 30.3, seven wastewater treatment plants (WWTPs) with a total installed capacity of 39,500 m^3/day have been completed in ULB. As these plants face major constraints of incomplete networks, lack of technical capacity, and poor management, only 3.88% of ULB's total municipal waste effluents was treated in 2012 (SWIM, 2013). As for the LLB, three WWTPs are currently under construction (collectively 35,000 m^3/day) with a few others being planned on its tributaries (gef/WB/Plan Blue, 2015; SWIM, 2013).

Table 30.3 State of municipal wastewater treatment and discharge at the Upper Litani Basin.

Wastewater treatment plants (WWTPs) completed, ongoing, and planned	Installed capacity (m^3/day)	Operating capacity (m^3/day)	% of installed discharge capacity
	Completed		
Baalbeck (Iaat)	12,500	1,250	10%
Aitanit, Forzol and Ablah	8,000	4,320	54%
Yammouneh	1,000	500	50%
Saghbine	7,500	750	10%
Jib Janine	10,500	1,050	10%
Subtotal	***39,500***	***7,870***	***19.92%***
	Under construction		
Zahle	18,000		
Subtotal	***18,000***		
	Planned		
Tamnine	24,534		
El Marj	120,000		
Maaraboun	383		
Subtotal	***144,917***		
Total completed, ongoing, and planned capacity	**202,417**		**3.88%**

Source: Adapted from SWIM, 2013; CDR, 2015.

A few small-scale WWTPs were established on some of the tributaries in the late 1990s, but none are currently operational (CDR, 2015; SWIM, 2013).

Municipal and industrial solid waste dumped or directly thrown into Litani River and Qaraoun Lake or deposited along their banks and across the whole basin is also a major source of water pollution. A total of 680 t/day of solid waste is generated and dumped on 60 sites in ULB (SWIM, 2013). Leachate from dumped waste containing heavy metals and organic chemicals can eventually reach the river or percolate through the soil to groundwater. A large number of quarries and stone processing industries generate large quantities of stone sediments, which form another pollution source affecting water quality in ULB. Zahle is the only city that has one operational sanitary landfill, which was financed by both the World Bank and USAID, and could dispose of 200 t/day from its 12 neighbouring municipalities. In Baalbeck, another sanitary landfill is now under construction and is expected to be in operation very soon.

Assessment of water quality of ULB has been the subject of extensive research studies and monitoring programmes. The findings of four major water quality programmes are as follows:

- *Water Quality Assessment of the Upper Litani River Basin and Lake Qaroun* (2003); *Basin Management Advisory Services Project* (BAMAS, 2005–2007)
- *Business Plan for Combating Pollution of the Qaraoun Lake* (MoE and UNDP, 2011)
- *Litani River Basin Management Support Program* (LRBMS, 2009–2012)

The above studies provide a scientific basis for understanding the impact of wastewater and solid waste on water quality to support decision-making related to the management and restoration of the river basin. Tenfold increases in BOD values were recorded in 2010 as compared to values in 2005, confirming that poor wastewater and solid waste management are among the major contributors to the degradation of the water quality of the river (LRBMS, 2010; LRBMS, 2013) (Table 30.4).

Table 30.4 Upper Litani Basin water quality.

	BAMAS 2005		LRBMS 2010		Haydar *et al.* 2014b		Baydoun *et al.* 2016	
Indicator	**Min.**	**Max.**	**Min.**	**Max.**	**Min.**	**Max.**	**Min.**	**Max.**
Litani River								
BOD (mg/L)	2	624	2.5	2530	5	12	0	26
COD (mg/L)	—	—	—	—	20	97	3.88	1,027
DO (mg/L O_2)	0	8	0.38	9.4	—	—	0.18	3.77
Nitrates (mg/L)	3	62	0.1	4.9	0.08	12	2.67	17.67
Ammonia (mg/L)	0	133.33	0.1	83.32	0.08	11.33	3.14	69.2
Total coliform (CFU/100 mL)	—	—	—	—	0	300,000	—	—
Faecal coliform (CFU/100 mL)	0	1,500,000	1	400	0	300,000	—	—
Staph (CFU/100 mL)	—	—	—	—	1,000	5,000	—	—
Cadmium (mg/L)	—	—	0.005	0.079	ND	ND	—	—
Copper (mg/L)	—	—	—	—	0.05	0.1	—	—
Chromium (mg/L)	—	—	—	—	0.01	0.09	—	—

Table 30.4 (Continued)

	BAMAS 2005		LRBMS 2010		Haydar *et al.* 2014b		Baydoun *et al.* 2016	
Indicator	**Min.**	**Max.**	**Min.**	**Max.**	**Min.**	**Max.**	**Min.**	**Max.**
Iron (mg/L)	—	—	—	—	2	7.2	—	—
Qaraoun Lake								
TDS (mg/L)	120	196	221	256	—	—	—	—
pH	6.5	7.5	8.2	8.3	—	—	—	—
BOD (mg/L)	2	4	2	3.3	—	—	—	—
DO (mg/L O_2)	1.3	7.7	7.22	9.41				
Nitrates (mg/L)	16.1	31.2	0.8	1.2	—	—	—	—
Ammonia (mg/L)	0.01	0.95	0	0.33	—	—	—	—
Phosphates (mg/L)	0.01	0.35	0	0.24	—	—	—	—
Faecal coliform (CFU/100 mL)	0	450	0	400	—	—	—	—
Cadmium (mg/L)	—	—	0.0007	0.021	—	—	—	—
Groundwater								
TDS (mg/L)	—	—	170	863	—	—	—	—
pH	6.5	7.2	7	8.7	—	—	—	—
DO (mg/L O_2)	—	—	4.1	7.77	—			
Nitrates (mg/L)	3	171	0.2	29	—	—	—	—
Ammonia (mg/l)	—	—	0	39.75				
Phosphates (mg/L)	0	12	0.1	6.4	—	—	—	—
Faecal coliform (CFU/100 mL)	0	400	0	400	—	—	—	—
Manganese (mg/L)	—	—	0.03	0.54	—	—	—	—

ND: Not detected.

The drop noted in faecal contamination in 2010 may represent an underestimate of the actual contamination level due to the low levels of dissolved oxygen (DO) and high levels of toxic ammonia on the survival of microbial organisms. In spite of the considerable uncertainty over the magnitude and rates of nitrogen cycling in the river, the high levels of toxic ammonia exceed many times the healthy waterbody criteria. Adding moderate levels of nitrates – that is, active denitrification – may be the process controlling water quality in the ULB under low DO conditions (Baydoun *et al.*, 2016).

Elevated levels of toxic heavy metals such as nickel, arsenic, mercury, and chromium as well as certain organic pollutants such as total petroleum hydrocarbons and p-chlorophenol in the water and sediments of the river and reservoir further illustrate the environmental impacts of gas run-off and industrial effluents (Haydar *et al.*, 2014a; MoE and UNDP, 2011).

In recent years, there has been an increasing concern over public health associated with water contamination by heavy metals. In addition, high levels of nutrients (nitrogen and phosphorus) due to excessive use of fertilisers have resulted in severe eutrophication of the Litani River and Qaraoun reservoir, necessitating the use of copper sulphate on a continual basis by the Litani River Authority to keep algae and weeds in check (BAMAS,

2005). Cyanobacterial blooms have also been reported in the reservoir despite the high levels of ammonia (Fadel *et al.*, 2014; Slim *et al.*, 2014). Not only surface water resources but also groundwater in the basin has also been shown to be impacted by urban pollution and contaminants from intensive agricultural activities in the Bekaa Valley. These pollutants have infiltrated into the groundwater encouraged by the karst limestone structures and high levels of fissures and fractures characterising water aquifers in Lebanon. High levels of faecal and total coliforms as well as nitrates, chloride, and sulphates have been recorded in different studies (Baydoun *et al.*, 2015; LRBMS, 2010; LRBMS, 2013; Saadeh *et al.*, 2012).

In spite of the limitation in the number of studies, water quality in LLB is generally better when compared to that of ULB. Nonetheless, moderate to high levels of faecal coliforms exceeding drinking water standards and nitrogen due to domestic wastewater pollution and excessive fertiliser application have recently been recorded (Baydoun *et al.*, 2016; Nehme *et al.*, 2013; Nehme *et al.*, 2014a; OPTIMA/ELARD/CNRS, 2006). Moreover, elevated levels of heavy metal concentrations in the river sediments were recently recorded at different points along its course (Nehme *et al.*, 2014b).

30.8 Pollution Impact

Over the last decade, there has been increasing national concern over the public health impacts attributed to high levels of contaminants in Litani River and the Qaraoun reservoir. It is no longer a secret that most of the Bekaa Valley's agricultural fields are irrigated with water that is highly contaminated with dangerous pollutants (Halablab *et al.*, 2011). Faced with growing concern, the Ministry of Agriculture and local municipalities intervened on several occasions to ban the use of Litani water for irrigation. Since 2005, the national epidemiological surveillance programme at the Ministry of Public Health has registered significant increasing trends in notifiable water and food-borne diseases (MoPH, 2017) with 243 recorded cases in 2005, 28 in 2010, 2,068 in 2014, and 909 cases in 2016 in Bekaa, in contrast with 45, 112, 122, and 74 cases, respectively, in Beirut. In the specific case of viral Hepatitis A, which is a common infection in Lebanon, the increase in Bekaa was particularly notable, recording 284 in 2016, 54.7% of all cases in the whole country, while Beirut recorded 11. These values give an indication of the influence of the Syrian conflict on deteriorating water quality and the spread of water and food-borne diseases through the already vulnerable Lebanese host communities. More specifically, the recent assessment of the prevalence of diarrhoea and subsequent mortality in the Litani Basin due to poor water quality degradation by EU estimates 0.3 cases of deaths on 13 newborns per 1,000 inhabitants in 2012 (SWIM, 2013). The prevalence of diarrhoea was 2.3 cases in all children less than 5 years of age and 0.45 cases in those aged 5 years or more.

Information about wildlife along the Litani River and Qaraoun reservoir is very limited, with some groups such as macrophytes and fish fairly well studied and others such as amphibians, reptiles, and macroinvertebrates not being subject to any studies. In spite of data limitations, there are strong indications that not many organisms still live in and around the river. Reduced oxygen supply due to severe eutrophication in ULR and the Qaraoun reservoir has resulted in extensive degradation in vegetation biodiversity (Baydoun *et al.*, 2016), making these water resources almost totally dead, and fish can now be seen floating on the surface and shores of the reservoir (El Zein and Hanna, 2010). The high levels of toxins produced by cyanobacteria therefore pose major environmental risk.

There have been some moves to mitigate these problems, for example, the establishment of a pilot wetland project established by USAID (LRBMS, 2012b) for phytoremediation and to mitigate functions lost from the river ecosystem. Being the first of its kind in the country, the project has been shown to efficiently improve water quality, restore the integrity of the river's ecosystem, and provide valuable habitat for wildlife (Amacha *et al.*, 2017).

Increasing recognition of the economic cost of water quality degradation of Litani River Basin has recently become an important socio-economic issue. Estimations from the EU show that the cost of such environmental degradation in the ULB reached US$ 227 million in 2012, equivalent to 2.2% of the GDP in the ULB, and 0.5% of the national GDP of Lebanon (SWIM, 2013). Specifically, the cost associated with water quality was 77% of the total (US$ 176 million).

30.9 Current Management Efforts and Strategies

In spite of the efforts of the Lebanese government aided by international development programmes to control pollution in and around the Litani River Basin, the problem has become more severe as urbanisation planning and effective wastewater and solid waste management are still lacking.

Most recently, the MoE has launched a business plan to assist the government in identifying the major sources of pollution in the Qaraoun reservoir and has recommended appropriate solutions to mitigate them (MoE/UNDP/ELARD, 2011; WB, 2016). Out of the US$255 million programme identified in the plan, the first tranche approved for financing by a loan from the World Bank (US$ 60 million) focuses on reducing municipal wastewater pollution by extending the sewage network and connecting it to wastewater treatment plants that are either functional or will soon become functional (WB, 2016). In 2014, in an attempt to tackle industrial pollution and encourage green investments through a combination of regulations and incentives, the Central Bank of Lebanon and MoE supported by the World Bank and the Italian Cooperation Agency initiated the Lebanon Environmental Pollution Abatement Program (LEPAP) (MoE and UNDP, 2017). Finally, a seven-year plan of around $733 million to combat the catastrophic pollution of the river was very recently approved by the Lebanese Parliament. In spite of all the promising projects and plans, Lebanon's track record of implementation, enforcement, and compliance has always been poor. Implementation has been sporadic, held back by shared and overlapping responsibilities between different institutions and lack of operational and maintenance capacity.

30.10 Conclusions and Recommendations

As urbanisation and population growth are increasing in Lebanon, extensive water pollution will continue to impact water resources in the country. With the lack of urban planning and current sluggish performance in the implementation of wastewater management strategies and enforcement of compliance systems, water pollution is envisaged to become worse. Consequently, the health of many thousands of people will be affected, and further losses in biodiversity and ecosystem resilience will undermine prosperity and efforts towards a more sustainable future. Lebanon needs a more comprehensive approach to wastewater management that takes into account urbanisation plans and integrated management of municipal and industrial wastewater. There is an urgent need for alignment between urban development and water management. Governmental institutions must be more proactive in implementing urban water planning that considers sustainable wastewater management as an integral part of urban planning and considers all steps from water source to re-entry into the environment, including reuse of treated water and sludge. Also, more innovative methods in wastewater treatment and utilisation of various applications such as developing construction material are required to provide promising low-cost and eco-friendly sustainable technology. Recent research suggests that partial treatment of raw sewage sludge can be used as water replacement in concrete applications (Hamood and Khatib, 2012; Hamood *et al.*, 2013; Hamood and Khatib, 2016; Hamood *et al.*, 2017). The use of waste, local or low energy intensive materials as partial cement replacement or aggregate replacement in concrete together with raw sewage sludge is possible (Dvorkin *et al.*, 2018; Egenti *et al.*, 2013; Herki and Khatib, 2016; Khatib and Mangat, 2003; Khatib *et al.*, 1996; Khatib, 2009; Khatib *et al.*, 2012; Khatib *et al.*, 2018; Khatib, 2014; Mangat and El-Khatib, 1992; Mangat *et al.*, 2006; Negim *et al.*, 2014; Siddique and Khatib, 2010; Sonebi *et al.*, 2015a; Sonebi *et al.*, 2015b; Wild *et al.*, 1996). This would save the much-needed fresh water for essential applications such as irrigation and drinking. Policies, measures, and programmes oriented towards mitigating the impact of Syrian refugees on water resources while meeting water needs should be also at the top of national priorities.

References

Amacha, N., Karam, F., Jerdi, M., Frank, P., Houssein, D., Kheireldeen, S., Viala, E., and Baydoun, S. (2017) Assessment of the efficiency of a pilot constructed wetland on the remediation of water quality of Litani River, Lebanon. *Environment Pollution and Climate Change*, 1, 119.

APIS (2016) Health Consulting Group Report. Syrian Refugees Crisis. Impact on Lebanese Public Hospitals. Financial Impact Analysis: Generated Problems and Possible Solutions. file:///C:/Documents%20and%20Settings/BAU/My%20Documents/Downloads/SyrianRefugeesImpactonPublicHospitals-APISReport%20(1).pdf.

Assaf, H. (2010) Water resources and climate change. In: *Arab Environment: Water, Sustainable Management of a Scarce Resource*, 2010 report (AFED) of the Arab Forum for Environment and Development, pp. 25–38.

Badr, R., Holail, H., and Olama, Z. (2014) Water quality assessment of Hasbani River in South Lebanon: Microbiological and chemical characteristics and their impact on the ecosystem. *Journal of Global Biosciences*, 3, 536–551.

BAMAS. (2005) Canal 900 Algae Control: Testing and validation.

Baydoun, S., Darwich, T., Nassif, M.H., Attalah, T., Puig, J., Kamar, M., Amacha, N., Molle, F., and Closas, A. (2015) Spatial and Temporal Variations in Groundwater Quality in Upper Litani Basin. Les Troisièmes Journées Franco-Libanaises JFL3: La Recherche au Service de la Communauté, Lebanese University, Lebanon, 29–31 October. p. 7.

Baydoun, S., Ismail, H., Amacha, N., Apostolides Arnold, N., Kamar, M., and Abou-Hamdan, H. (2016) Distribution pattern of aquatic macrophytic community and water quality indicators in upper and lower Litani river basins. *Lebanon Journal of Applied Life Sciences International*, 6 (2), 1–12.

Council for Development and Reconstruction (CDR) (2015) Basic Services. Waste Water.

Darwish, T. (2012) Country Study on Status of Land Tenure, Planning and Management in Oriental Near East Countries, Case of Lebanon. F A O – S N O, Cairo, Egypt.

Darwish, T., Faour, G., and Khawlie, M. (2004) Assessing soil degradation by landuse-cover change in coastal Lebanon. *Lebanese Science Journal*, 5 (1), 45–59.

Dvorkin, L., Lushnikova, D., Bezusyak, O., Sonebi, O., and Khatib, J. (2018) Hydration characteristics and structure formation of cement pastes containing metakaolin, *MATEC Web of Conferences 149*, 01013 https://doi.org/10.1051/

matecconf/201814901013 CMSS-, 1-8, https://www.matec-conferences.org/articles/matecconf/pdf/2018/08/matecconf_cmss2018_01013.pdf

Egenti, C., Khatib, J.M. and Oloke, D. (2013) Appropriate Design and Construction of Earth Buildings: Contesting Issues of Protection Against Cost, *African Journal of Basic & Applied Sciences*, 5(2), 102–106, ISSN 2079-2034 DOI: 10. 5829/idosi.ajbas.2013.5.2.1130

El Moujabber, M., Bou Samra, B., Darwish, T., and Atallah, T. (2006) Comparison of different indicators for groundwater contamination by seawater intrusion on the Lebanese coast. *Water Resources Management*, 20, 161–180.

El Zein, G. and Hanna, D. (2010) Premières données sur l'inventaire et la distribution de l'ichtyofaune du bassin du Litani au Liban. INOC-Tischreen University, International Conference on Biodiversity of the Aquatic Environment.

Fadel, A., Atoui, A., Lemaire, B., Vinçon-Leite, B., and Slim, K. (2014) Dynamics of the toxin cylindrospermopsin and the cyanobacterium Chrysosporum (Aphanizomenon) ovalisporum in a Mediterranean eutrophic reservoir. *Toxins*, 6, 3041–3057.

FAO (2016) Assessment of Treated Wastewater for Agriculture in Lebanon, Final report. Prepared by: Salman, M., Abu Khalaf, M., Del Lungo, A. With contribution from: Maacaroun, A. and Roukoz, S.

Faour, G. (2015) Evaluating urban expansion using remotely-sensed data in Lebanon. *Lebanese Science Journal*, 16 (1), 23–32.

Faour, G. and Mhawej, M. (2014) Mapping urban transitions in the Greater Beirut Area using different space platforms. *Land*, 3, 941–956. doi:10.3390/land3030941.

gef/World Bank/Plan Bleu (2015) Regional Governance and Knowledge Generation Project: Strategic Environmental Assessment for the New Water Sector Strategy for Lebanon.

Halablab, M., Sheet, I., and Holail, H. (2011) Microbiological quality of raw vegetables grown in Bekaa Valley, Lebanon. *American Journal of Food Technology*, 6, 129–139.

Halwani, J., Ouddane, B., Crampon, N., and Wartel, M. (2000) Saline contamination in groundwater of Akkar plain in Lebanon. *Journal Européen d'Hydrologie*, FAO (France).

Hamood, A. and Khatib, J.M. (2012) The Use of Raw Sewage Sludge (RSS) As a Water Replacement in Cement-Based Mixes. *2nd International Conference on Sustainable Design, Engineering and Construction (ICSDEC)*. ICSDEC 2012: Developing the Frontier of Sustainable Design, and Construction. Section: Water Resources & Efficiency, (Editor: WaiKiong Oswald Cjong, Jie Gong, Jae Chang and Mohsin Khalid). 7–9 November 2012, pp. 1001–1008, Fort Worth, Texas, USA. ASCE, ISBN: 9780784412688. http://ascelibrary.org/doi/abs/10.1061/9780784412688.118

Hamood, A., Khatib, J.M., and Williams, C. (2013) Engineering Properties of Cement-Based Materials Incorporating Raw Sewage Sludge (RSS) and High Loss on Ignition (LOI) Fly Ash (365 Days Curing). *1st International Conference & Exhibition, Application of Efficient & Renewable Energy Technologies in Low Cost Buildings and Construction*, 16–18 September 2013, Gazi University, Ankara, Turkey, pp. 322–330. http://renewbuild.org/?page=homepage, ISBN: 978-975-507-268-5

Hamood, A. and Khatib, J.M. (2016) Sustainability of sewage sludge in construction. Chapter in Khatib, J.M. (ed.), *Sustainability of Construction Materials*, 2nd edition, Elsevier Publishing Limited, Cambridge, England, August 2016, pp. 625–641 [ISBN 9780081009956].

Hamood, A., Khatib, J.M. and Williams, C. (2017) The effectiveness of using raw sewage sludge (RSS) as a water replacement in cement mortar mixes containing uprocessed fly ash (u-FA), Construction and Building Materials Journal, 147, 30 Aug, 27-34, http://dx.doi.org/10.1016/j.conbuildmat.2017.04.159

Hamze, M., Hassan, S., Thomas, R.L., Khawlie, M., and Kawass, I. (2005) Bacterial indicators of faecal pollution in the waters of the El-Kabir River and Akkar watershed in Syria and Lebanon. *Lakes & Reservoirs: Research and Management*, 10, 117–126.

Haydamous, P. and El Hajj, R. (2016). Lebanon's Agricultural Sector Policies: Considering Inter-Regional Approaches to Adaptation to Climate Change. *Policy Brief #2/2016*, AUB policy Institute.

Haydar, C., Nehme, N., Awad, S., Koubaissy, B., Fakih, M., Yaacoub, A., Toufaily, J., Villeras, F., and Tayssir Hamieh, T. (2014a) Water Quality of the Upper Litani River Basin, Lebanon. *Physics Procedia*, 55, 279–284. Elsevier.

Haydar, C., Nehme, N., Awad, S., Koubayssi, B., Fakih, M., Yaacoub, A., Toufaily, J., Villieras, F., and Hamieh, T. (2014b) Physiochemical and Microbial Assessment of Water Quality in the Upper Litani River Basin, Lebanon. *Journal of Environment and Earth Science*, 4 (9), 87–97.

Herki B. A. and Khatib, J. M. (2016) Valorisation of waste expanded polystyrene in concrete using a novel recycling technique, *European Journal of Environmental and Civil Engineering*, Taylor & Francis pub, 1384–1402, http://dx.doi.org/10.1080/19648189.2016.1170729 ISSN: 19648189

Houri, A.W.S. and El Jeblawi, J. (2007) Water quality assessment of Lebanese coastal rivers during dry season and pollution load into the Mediterranean Sea. *Water and Health*, 5 (4), 615–623.

Khatib, J. M., Sabir, B.B. and Wild, S. (1996) Some properties of Metakaolin Paste and Mortar, International Conference Concrete in the Service of Mankind: Concrete for Environment Enhancement and Protection, (Eds. R. K. Dhir and T. D. Dyer), Dundee, 24-28 June 1996, 637-643. ISBN 0-419-21450-x

Khatib, J. M. (2009) Low curing Temperature of MK Concrete, American Society of Civil Engineers (ASCE) - Materials in Civil Engineering Journal, 21 (8), August, pp 362-367, ISSN 0899-1561/2009/8-362-367 DOI: 10.1061/(ASCE)0899-1561(2009)21:8(362)

Khatib, J. M. and Mangat, P. S. (2003) Porosity of Cement Paste Cured at 45°C as a Function of Location Relative to Casting Position." *Journal of Cement and Concrete Composites*, 25 (1), 97–108.

Khatib J. M., Wright, L., Mangat, P.S. and Negim, E. M. (2012) Porosity and Pore Size Distribution of Well Hydrated Cement-Fly Ash- Gypsum Pastes." *American-Eurasian Journal of Scientific Research*, 7 (4), 142–145. ISSN 1818-6785, DOI: 10.5829/idosi.aejsr.2012.7.4.65131.

Khatib, J.M., Baalbaki, O. and Elkordi, A.A. (2018) Metakaolin, Chapter 15 in "Waste and Supplementary Cementitious Materials in Concrete", Siddique R. & Cachim P. (Editor), Elsevier Publishing Limited, Woodhead, Cambridge, England, 15 June 2018, pp 493-511. [ISBN: 978-0-08-102156-9] https://doi.org/10.1016/B978-0-08-102156-9.00015-8.

Khatib, J. M. (2014) Effect of Initial Curing on Absorption and Pore Size Distribution of Paste and Concrete Containing Slag, *Korean Society of Civil Engineers (KSCE) Journal*, 18 (1), 264–272, January. ISSN 1226-7988, eISSN 1976-3808 DOI 10.1007/s12205-014-0449-7

Korfali, F.I. and Jurdi, M. (2010) Deterioration of coastal water aquifers: Causes and impacts. *European Water*, 29, 3–10.

LRA (2017) The Litani River Authority, Lebanese Republic. http://www.litani.gov.lb/en/.

LMTA (2013) Lebanon Mountain Trail Association.

LRBMS (2010) Litani River Basin Management Support Program. Water Quality Survey – Dry Season. Volume I – Main report.

LRBMS (2011) Litani River Basin Management Support Program Water Balance Report.

LRBMS (2012a) Litani River Basin Management Plan Volume I: Current Situation.

LRBMS (2012b) Litani River Basin Management Support Program. Litani River constructed treatment wetland design report contract no.: epp-i-00-04-00024-00 order no 7.

LRBMS (2013) Litani River Basin Management Support Program Water Quality Index.

Mangat, P.S., and El-Khatib, J.M. (1992) Influence of Initial Curing on Pore Structure and Porosity of Blended Cement Concrete, American Concrete Institute SP-132, Proceedings 4th International Conference on Fly ash, Slag and Natural Pozzolans in Concrete (Ed: V. M. Malhotra), 1, May, 813–833.

Mangat, P.S., Khatib, J.M., and Wright, L. (2006) Optimum Utilisation of Flue Gas Desulphurisation (FGD) Waste in Blended Binder for Concrete, *Construction Materials Journal - Proceedings of the Institution of Civil Engineers.* 1 (2), 60–68.

METAP and Tebodin (1998) Ministry of Environment, Industrial Pollution Control Lebanon.

MoE and UNDP (2011) Business plan for combating pollution of the Qaraoun Lake. http://test.moe.gov.lb/Documents/Final%20Business%20Plan%20Document%2013062011%20Main%20Report%20-%20Part1.pdf.

MoE/UNDP/ECODIT (2011) State and Trends of the Lebanese Environment.

MoE/UNDP/ELARD (2011) Business Plan for Combating Pollution of the Qaraoun Lake, Progress Report II: Draft Business Plan. Prepared by ELARD for MOE and UNDP, April 2011.

MoE/UNDP/GEF (2011) Lebanon's Second National Communication to the UNFCCC. http://www.lb.undp.org/content/dam/lebanon/docs/Energy%20and%20Environment/Publications/Lb_SNC_final.pdf.

MoE / UNDP / GEF (2015) Economic Costs to Lebanon from Climate Change: A First Look. Beirut, Lebanon.

MoE/EU/UNDP (2014) Lebanon Environmental Assessment of the Syrian Conflict & Priority Interventions. file:///C:/Documents%20and%20Settings/BAU/Desktop/Dr.Lamis/EASC-ExecutiveSummaryEnglish.pdf.

MoE and UNDP (2017) Lebanon Environmental Pollution Abatement Project (LEPAP). Report No. PAD629-LB. Middle East and North Africa Region.http://www.lb.undp.org/content/lebanon/en/home/operations/projects/environment_and_energy/support-to-the-lebanon-environmental-pollution-abatement-project.html

MoEW/UNDP (2014) Assessment of Groundwater Resources of Lebanon.

MoEW (2010) National Water Sector Strategy, 2010-2020. http://www.databank.com.lb/docs/National%20Water%20Sector%20Strategy%202010-2020.pdf.

MoEW (2012) National Strategy for Wastewater Sector. Beirut.

MoPH (2017) Lebanese Ministry of Public Health. Epidemiologic Surveillance Department. http://www.moph.gov.lb/Prevention/Surveillance/Pages/PastYears.aspx

Mouneimne, A.H. and Gemayel, E. (2012) Waste water management in Lebanon. In: Kouyoumjian, H. and Hamze, M. (eds.), *Review and Perspectives of Environmental Studies in Lebanon*, INCAM–EU/CNRS, pp. 95–116.

Nader, M., Jazi, M., Abou Dagher, M., and Indary, S. (2012) Environmental Resources Monitoring in Lebanon – ERML; Component A (i): Improved Understanding, Management and Monitoring in the Coastal Zone. www.cbd.int/doc/meetings/mar/ebsaws-2014-03/other/ebsaws-2014-03-submission-lebanon-02-en.pdf.

Negim, E. S. M., Rakhmetullayeva, R. K., Yeligbayeva, G. Zh., Urkimbaeva, P. I., Primzharova, S. T., Kaldybekov, D. B., Khatib, J. M., Mun, G. A. and Williams, C. (2014) Improving biodegradability of polyvinyl alcohol/starch blend films for packaging applications, *International Journal of Basic and Applied Sciences*, 3(3), 263–273, Science Publishing Corporation, www.sciencepubco.com/index.php/IJBAS, doi: 10.14419/ijbas.v3i3.2842.

Nehme, N., Haydar, C., Koubaissy, B., Fakih, M., Awad, S., Toufaily, J., Villieras, F., and Hamieh, T. (2013) Evaluation of the physicochemical characteristics of water in the Lower Litani Basin, Lebanon. *Elixir*

International Journal, Aquaculture, 62, 17478–17484.

Nehme, N., Haydar, C., Koubaissy, B., Fakih, M., Awad, S., Toufaily, J., Villieras, F., and Hamieh, T. (2014a) The distribution of heavy metals in the Lower River Basin, Lebanon. *Eighth International Conference on Material Sciences*, CSM8-ISM5.

Nehme, N., Haydar, C., Koubaissy, B., Fakih, M., Awad, S., Toufaily, J., Villieras, F., and Hamieh, T. (2014b) Study of the correlation of the physicochemical characteristics of the Litani Lower River Basin. *Physics Procedia*, 55, 451–455.

OPTIMA/ELARD/NCRS (2006) Optimisation for Sustainable Water Resources Management. Case Study: Lower Litani Basin – Lebanon. http://www.ess.co.at/OPTIMA/FTP/D09.1.pdf.

Saadeh, M., Semerjian, L., and Amacha, N. (2012) Physicochemical evaluation of the Upper Litani river watershed, Lebanon. *The Scientific World Journal, Vol.* 2012, 8 pages.

Siddique, R., and Khatib, J. M. (2010) Mechanical properties and Abrasion Resistance of HVFA Concrete, *Materials and Structures - RILEM*, 43(5), June, 709-718. print ISSN: 1359–5997 online ISSN 1871-6873 doi: 10.1617/s11527-009-9523-x.

Slim, K., Fadel, A., Atoui, A., Lemaire, B.J., Leite, B.V., and Tassin, B. (2014) Global warming as a driving factor for cyanobacterial blooms in Lake Karaoun, Lebanon. *Desalination and Water Treatment*, 52, 2094–2101.

Sonebi, M., Garcia-Taengua, E., Hossain, K.M.A., Khatib, J.M., and Lachemi, M. (2015a) Effect of nanosilica addition on the fresh properties and shrinkage of mortars with fly ash and superplasticizer, *Construction and Building Materials Journal*, 84, 1 June, 269–276; doi:10.1016/j.conbuildmat.2015.02.064.

Sonebi, M., Wana, S., Amziane, S., and Khatib, J. M. (2015b) Investigation of the mechanical performance and weathering of hemp concrete, Proceedings of the 1st International *Conference on Bio-based Building Materials (ICBBM 2015)*, Eds. Amziane & Sonebi, 21–24 June, Claremont-Ferrand, France. RILEM publication, 416-421, ISBN PRO 99: 978-2-35158-154-4 https://sites.google.com/site/icbbm2015/home

Sustainable Water Integrated Management (SWIM). (2013) Support Mechanism Project Funded by the European Union. Lebanon Cost Assessment of Water Resources Degradation of the Litani Basin. http://www.swim-sm.eu/files/Assessment_Litani_EN.pdf.

UN Habitat (2003) Summary of City Case Studies, 193–228.

UN Habitat (2016) *Tripoli City Profile* 2016. https://unhabitat.org/tripoli-city-profile-2016/

UNDP (1970) *Liban – Etude des eaux souterraines*. New York: United Nations Development Programme.

UNHCR (2016) Syria Regional Refugee Response. https://data.unhcr.org/syrianrefugees/region.php?id=90&country=122

UN-Water (2014) A Dedicated Water Goal. http://www.unwater.org/sdgs/a-dedicated-water-goal/en/

WB (1993) World Bank. Lebanon – Emergency Reconstruction and Rehabilitation Project. http://documents.worldbank.org/curated/en/617201468055442496/Lebanon-Emergency-Reconstruction-and-Rehabilitation-Project

World Bank Group (2016) Lebanon. http://data.worldbank.org/country/lebanon

WB (2011) World Bank. Republic of Lebanon Country Environmental Analysis. Report No. 62266-LB. http://test.moe.gov.lb/Documents/WB%20-%20Final%20CEA.pdf

WB (2012) World Bank. Lebanon – Country Water Sector Assistance Strategy (2012–2016). Sustainable Development Department Middle East and North Africa Region. Report No. 68313-LB. http://documents.worldbank.org/curated/en/401211468088175955/Lebanon-Country-water-sector-assistance-strategy-2012-2016

WB (2016) World Bank. Lake Qaraoun Pollution Prevention Project. Report No. PAD860-LB. Middle East and North Africa Region. http://documents.worldbank.org/curated/en/937711478610994157/Lebanon-Lake-Qaraoun-Pollution-Prevention-Project-P147854-Implementation-Status-Results-Report-Sequence-01.

Wild, S., Khatib, J.M., Sabir, B.B., and Addis, S.D. (1996) The Potential of Fired Brick Clay as a Partial Cement Replacement Material, *International Conference - Concrete in the Service of Mankind: Concrete for Environment Enhancement and Protection,* (Eds. R. K. Dhir and T. D. Dyer), Dundee, 24–28 June 1996, pp 685-696. ISBN 0-419-21450-x.

31

Closing Comments on Urban Pollution

Susanne M. Charlesworth[1] *and Colin A. Booth*[2]

[1] *Centre for Agroecology, Water and Resilience, Coventry University, United Kingdom*
[2] *Architecture and the Built Environment, University of the West of England, United Kingdom*

31.1 Introduction

The different chapters and international expertise of the authors in this book illustrate clearly the complexity of processes and systems in urban environments. A statistic that is mentioned in Chapter 1, the introduction to this volume, and is repeated many times in other chapters, probably due to the alarming nature of the statement, is that by 2050, 66% of all human beings on planet Earth will live in cities (UN, 2014); that is, for every three persons, two will live in an urban area. In fact, since 2008, a majority of the world's population has lived in cities; Table 31.1 shows the statistics behind these projected increases in global urban population, and also gives the population of today's megacities. These, too, are projected to grow, with Tokyo increasing to 37 million, but Delhi rapidly increasing by almost one third to 36 million people by 2030. The global urban population is expected to grow approximately 1.84% per year between 2015 and 2020, 1.63% per year between 2020 and 2025, and 1.44% per year between 2025 and 2030 (WHO, 2016).

The impact of this growth has been substantial, and the chapters in this book give an indication of the challenges faced by cities worldwide in terms of sources of contamination, their transport, deposition, and impacts. This includes not only pollutants generally associated with processes supporting urban life such as industry and transport, but also light pollution and its impacts on human health (Chapter 11), pathogens in wastewater, their treatment, and impacts worldwide (Chapter 9) and also the management and disposal of greywater (Chapter 16). To a certain extent, these impacts can be addressed, and Chapters 20 (groundwater bioremediation) and 21 (soil bioremediation) show this in the use of bioremediation techniques to address pollution of groundwater, which may be used for potable water supplies, and also the remediation of soils, important should the expanding city wish to make use of contaminated land in future construction efforts. Chapter 24 (the United States) promotes the use of licensed site professionals who have limited state oversight, a strategy trialled in Massachusetts as a means to streamline efforts in remediating contaminated sites, making them available for future building. Importantly, this strategy is underpinned by payments through Brownfields Tax Credits, which attracts developers who would previously not have considered redeveloping such sites. However, very often, it is the construction activities themselves which are polluting, particularly if virgin materials are sourced, which can spread the impacts of urbanisation beyond the city environs.

Urban Pollution: Science and Management, First Edition. Edited by Susanne M. Charlesworth and Colin A. Booth.

Table 31.1 Global urbanisation statistics, 2014 (UN, 2014).

% global population living in urban areas	1950	2014	2050		
	30	54	66		
% urbanised by region	North America	Latin America, Caribbean	Europe	Africa	Asia
	82	80	73	40	48
% of world's urban population	Asia	Europe	Latin America, Caribbean		
	53	14	13		
Megacities	Tokyo	Delhi	Shanghai	Mexico City, Mumbai, Sao Paulo	
Population (millions)	38	25	23	21	

Chapter 18 (construction waste) proposes the use of a large array of waste materials to replace those used in construction from less sustainable sources. A current focus for research has been the 'nexus', a means of establishing links between two or more things (Oxford dictionaries: https://en.oxforddictionaries.com/definition/nexus); this approach may be an opportunity to establish the nexus between industry, waste, and construction. The chapter from Lebanon (Chapter 30) in fact tries this to a limited extent by proposing the extraction of water from sewage sludge and its use in construction activities.

The likelihood is that the original contamination was due to historical pollution-emitting activities (Chapter 2), and while a particular success was the introduction of unleaded petrol and its subsequent reduction in solid material such as urban dusts, sediments, and soil (Chapter 7), a historical legacy still remains. In Chapter 28, a specific case study from Budapest, Hungary, concentrates on fine particulate matter carried in urban air, which has the potential to have significant impacts on human health. The conclusions of this chapter ask the question of what pollutants carried in association with the dusts are mobile, and as such available once breathed into the human lung. Chapter 8 (on bioaccessibility) seeks to address this by undertaking risk and bioaccessibility assessments of dusts in children's playgrounds. Children are the most vulnerable in society to such fine material not least, since, though pica, or hand–mouth activities, they are likely to ingest, inhale, and/ or absorb pollutants. However, the authors query the validity of some of these assessments due to the plethora of protocols available; the potency of potential toxins, whereby it is difficult to estimate in the mixtures found in urban environments; and the site-specific nature of these mixtures found which will reflect the type of industry, patterns of development, and traffic movements characteristic of the individual city. This chapter started with a comment on the complexities of urban environments, which Chapter 8 illustrates well. As soon as one pollutant is tackled, another becomes problematic, sometimes the impact of which is largely unknown, such as many new and emerging pollutants (Chapter 27 on Brazil) including hormones, medication, and personal care products. Many urban-related pollutants are common to cities worldwide, regardless of continent, stage of development, or size, but Chapter 29 (on Ghana) shows that as developing countries continue to expand and industrialise, the undesirable effects are reflected in poorer environmental quality, particularly in cities. As wealth increases, so do construction, car ownership, and increased migration to the cities. In India, for example

(Chapter 26; Table 26.1), the urban population increase is staggering, bringing with it related and unsustainable consumption of resources, producing pollutants which the authors call 'alarming'. In China, too (Chapter 25), urban population increase has outstripped their huge cities' abilities to specifically deal with issues of waste disposal, leading to problems at landfill sites which legislation is too weak to address, and a lack of the infrastructure to encourage recycling, which could go some way to not only reduce the amount of waste as a whole, but also reduce the amount landfilled. The Sponge City concept initiated by the Chinese Government could well be copied by other cities across the world in a strategy that utilises green infrastructure (GI) and associated ecosystem services (ES; Chapter 15) to improve environmental quality as a whole.

Conflicts impact cities bringing particular impacts, which are also centred on population increase, but in the case of Lebanon (Chapter 30) the impact is due to refugees or displaced persons swelling local populations to such an extent that they can no longer bear the pressures of coping with a sudden influx of people. This chapter calls for better wastewater management strategies and more innovative use of alternative water sources to reduce potable use.

Chapters 3, 4, and 5 covered policy as it is applied to soil, air, and water. Concentrating mostly on the United Kingdom and Europe, they show, for example, how complicated legislating to improve water quality is when the final destination of water, or its use, is not known. Legislating for air quality is more complicated to a certain extent with the effects of transboundary pollution having been recognised for decades; this is exemplified by the emission of greenhouse gases worldwide and their implications for global climate change (Chapter 15). Cities do have the ability to adapt to change, but the rapid, unplanned, and increasingly dense growth of urban areas does not support sustainable development, and tends to block the use of less technological solutions in the rush to deal with the associated urban sprawl, where many people live in substandard conditions in environmentally degraded areas.

Tighter restraints on pollution-emitting processes could potentially reduce certain contaminants, and WHO (2016) suggests that inclusive, participatory processes in decision-making, as well as the co-design of local interventions, will encourage citizens to take up the responsibility of managing their own environment, leading to the sustainability of such strategies, development of communities, and their empowerment. In Chapter 23, citizen science is explored in the context of the monitoring of air quality. This chapter shows the power of this approach and the array of case studies that have developed alongside technological advancements, enabling mobile and interactive monitoring unheard of a few years ago. Fly-tipping, waste dumping, and the discharge of effluents from a variety of industries is the scourge of cities everywhere, with Interpol being involved in producing strategy documents to address the problem. Forensics (Chapter 12) now can fingerprint and trace the source of much of this illegal activity and bring the perpetrators before the courts. However, once again the local community can be involved in protecting their local environment by identifying and reporting incidences of waste dumping.

While not specifically addressing urban pollution, the 8 Millennium Development Goals (MDGs) included as Goal 7: 'To ensure environmental sustainability', which did aim to reduce CO_2 emissions and the use of ozone-depleting substances. The MDGs were superseded in 2015 by the Sustainable Development Goals (SDGs): *Transforming Our World: The 2030 Agenda for Sustainable Development*. Based on the Rio+20 Conference held in Rio de Janeiro, 2012, the document itself is actually non-binding, but it is the first time that cities have been identified as a priority when considering the development strategy globally (WHO, 2016). The SDGs include 17 'Global Goals' with 169 targets between them, which became known as 'The Future We Want'.

Chapter 9 ('Wastewater and Sewage Disposal') quotes Goal 6, which tackles Clean Water and Sanitation; it aims to reduce wastewater, thus improving water quality. Another Goal (number 11) is 'Sustainable Cities and Communities', which contains 10 targets; urban pollution is not specified, but clean air is identified under the following:

> By 2030, reduce the adverse per capita environmental impact of cities, including by paying special attention to air quality. (http://www.undp.org/content/undp/en/home/sustainable–development–goals/goal–11–sustainable–cities–and–communities/targets/)

Many of the chapters in this book address how air pollution can be tackled using GI, such as Chapter 14 ('Urban Meadows on Brownfield Land') and Chapter 10 ('Living Green Roofs'), and the use of ES in Chapter 15. Chapter 3 outlines policy related to the quality of air, but legislation may target current issues in reducing urban contamination. However, there still remains the necessity of addressing historical pollution associated with construction, industrial processes, and transport. Chapters 19–21 detail the means of tackling urban contamination in a variety of ways, including the use of biological and chemical processes, and transforming land which cannot be used to potentially have a useful role. Chapter 14 shows how what was previously relatively unpleasant areas can be turned into urban meadows, with many benefits, and this would appear to have become popular in many cities recently, with roadside banks and central reservations (medians) becoming a riot of colour with wildflowers and flowering shrubs. This shows the potential for urban areas to address SDG 11:

> By 2030, provide universal access to safe, inclusive and accessible, green and public spaces, in particular for women and children, older persons and persons with disabilities. (http://www.undp.org/content/undp/en/home/sustainable–development–goals/goal–11–sustainable–cities–and–communities/targets/)

Green spaces and GI provide ES (Chapter 15), among which is their ability to reduce the storm peak, improve water quality, provide opportunities for biodiversity, and also provide areas for city dwellers to have access to amenity. The latter provides an opportunity for those living in cities to become more physically active, and the green space improves perceptions of human health and also promotes mental health (WHO, 2016). The multiple benefits of this approach will serve to enrich city living, making cities healthier places to live and work, improving sustainability and resilience to change. Chapter 13 ('River Ecology') shows the application of the ES approach in terms of river ecology, and its ability to 'self-purify' when impacted by organic pollutants. It also provides a means of monitoring the extent of pollution, by providing biological indicators in the form of invertebrate species. Of particular interest is Chapter 10 ('Living Green Roofs'), where the capability of green roofs to improve air quality and soil and water quality is explained. Up to 50% of available, unused space in cities comprises the roofs of buildings, and depending on their structural strength, these spaces could help tackle air pollution, and also, with other strategies such as wet pavements (Charlesworth, 2010), the urban heat island effect, because of which many cities such as Tokyo have currently become 'unliveable' during the summer months.

Finally, Chapter 22 details an overarching Environmental Management Strategy to minimise the impacts that industry can have on its surroundings, but this is dependent on those industries taking responsibility for, engaging with, and implementing the strategy. As stated in Chapter 22, organisations have to acknowledge the potential benefits of a system, not only in terms of economics, but also in terms of the perceptions of that company by its customers, suppliers, and workers.

31.2 The Future for Towns and Cities

The future outlined above sounds grim, particularly for the poor, and certainly the threats of global climate change, new and emerging pollutants, and uncontrolled urban growth appear to represent the perfect storm. However, evolving in parallel with pollutant production are strategies for pollution remediation, and ultimately human society has to ensure that its activities do not damage the environment on which its life and livelihoods depend to such an extent that it can no longer be supported. A strategy which singles out impacts, focussing on just one impact and attempting to address it in isolation, will be inefficient and will ultimately fail. Management of contamination in urban areas needs to be multi-functional and flexible; it must engage local communities and encourage them to become responsible for their environment, with support from the government and suitable legislation. The changes that have occurred in the urban environment as it has developed and spread indicate the means of its amelioration: increase GI with its multiple ES provision, and replace hard, impermeable surfaces with those that can infiltrate water and support the GI. This can to some extent address the polluting activities associated with industry and traffic, and will encourage people to interact with, value, and protect their environment.

Reference

Charlesworth, S. (2010) A review of the adaptation and mitigation of Global Climate Change using Sustainable Drainage in cities. *Journal of Water and Climate Change*, 1, 165–180.

Index

Urban Pollution: Science and Management, First Edition. Edited by Susanne M. Charlesworth and Colin A. Booth.

C